Electromagnetic Environments
and Health in Buildings

Also available from Spon Press

Creating the Productive Workplace
Edited by Derek Clements-Croome

Spon Press   Pbk: 0-419-23690-2

Clay's Handbook of Environmental Health – 19th edition
Edited by W.H. Bassett

Spon Press   Hbk: 0-415-31808-4

Air Pollution – 2nd edition
Jeremy Colls

Spon Press   Pbk: 0-415-25565-1

Air Quality Assessment and Management – a practical guide
D. Owen Harrop

Spon Press   Hbk: 0-415-23410-7
                          Pbk: 0-415-23411-5

Information and ordering details

For price availability and ordering visit our website **www.ergonomicsarena.com**
Alternatively our books are available from all good bookshops.

Electromagnetic Environments and Health in Buildings

Edited by

Derek Clements-Croome

Spon Press
Taylor & Francis Group

LONDON AND NEW YORK

First published 2004
by Spon Press
11 New Fetter Lane, London EC4P 4EE

Simultaneously published in the USA and Canada
by Spon Press
29 West 35th Street, New York, NY 10001

Spon Press is an imprint of the Taylor & Francis Group

© 2004 Spon Press

except 'Biological Effects of Electromagnetic Fields' and 'Exposure
Guidelines for Electromagnetic Fields and Radiation' © National
Radiological Protection Board; 'Electromagnetic Environments and Health
in Buildings' © Nicola Hughes and Mike Dolan; 'Moving Beyond EMF
Public Policy Paralysis' © Blake Levitt

Publisher's Note
This book has been prepared from camera-ready copy supplied
by the editor.

Printed and bound in Great Britain by
TJ International Ltd, Padstow, Cornwall

British Library Cataloguing in Publication Data
A catalogue record for this book is available
from the British Library

Library of Congress Cataloging in Publication Data
A catalog record for this book has been requested

ISBN 0-415-31656-1

Contents

Contributors   viii
Preface   xxii

Part I   Setting the Scene

1   Healthy Buildings   3
Derek Clements-Croome

2   Evidence for Nonthermal Electromagnetic Bioeffects: Potential
Health Risks in Evolving Low-Frequency & Microwave Environments   35
W. Ross Adey

3   Effects of Electromagnetic Fields in the Living Environment   53
Cyril W. Smith

4   An Introduction to the Naturally Occurring and the Man Made
Electromagnetic Environment   119
A.C. Marvin

Part II   Electromagnetic Fields, the Environment & the Law

5   Mast Action UK – Legal Services   137
Alan Meyer

6   Environmental Impact of Electrosmog   149
G.J. Hyland

7   Product Liability, Product Safety and the Precautionary Principle   167
Simon Pearl

   Questions & Discussion   179

Part III   Emissions & Standards

8   What are We Exposed to? The Most Significant Sources of
Exposure in the Built Environment   187
Philip Chadwick

9   Biological Effects of Electromagnetic Fields Leading to Standards   195
Zenon Sienkiewicz

10   Biological Effects of Low Frequency Electromagnetic Fields   207
Magda Havas

11 Reduction and Shielding of RF and Microwaves   233
    Dietrich Moldan and Peter Pauli

12 Are We Measuring the Right Things? Windows, Viewpoints & Sensitivity 241
    Alasdair Philips

    Questions & Discussion   257

Part IV   Health Effects of Electromagnetic Environments

13 Biological Effects of Neutralising Vaccines: the Effects of Weak
    Electromagnetic Fields and the Concordance between the Two   267
    Jean A. Monro

14 Electroclinical Syndromes – Live Wires in Your Office?   281
    Anne C. Silk

15 Health Effects of High Voltage Powerlines   295
    Denis L. Henshaw and A. Peter Fews

16 Effects of 50 Hz Magnetic Field Exposure on Mammalian Cells
    in Culture   307
    *Barry D. Michael, Kevin M. Prise, Melvyn Folkard, Stephen Mitchell
    and Stuart Gilchrist*

17 Electromagnetic Fields - Interactions with the Human Body   313
    Jeffrey W. Hand

18 Evidence to Support the Hypothesis that Electromagnetic Fields
    and Radiation are a Ubiquitous Universal Genotoxic Carcinogen   331
    Neil Cherry

19 Static Electricity in the Modern Human Environment   365
    Jeremy Smallwood

20 Screen Dermatitis and Electrosensitivity: Preliminary Observations
    in the Human Skin   377
    Olle Johansson

21 Theoretical and Experimental Evidences where Present Safety
    Standards Conflict with Reality   391
    V.N. Binhi and M. Fillion-Robin

22 Mobile Phones and Cognitive Function   405
    Alan Preece

    Questions & Discussion   413

Part V   Awareness

23 Electromagnetism and the Insurance Industry   431
    Alastair Speare-Cole

24 PTSD or PTSD^2 Assessing & Responding to Pre and Post
    Telecoms Stress Disorder   435
    Ray Kemp

25 Electromagnetic Environments and Health in Buildings   437
    Nicole Hughes and Michael Dolan

    Questions & Discussion   441

Part VI   The Future

26 Exposure Guidelines for Electromagnetic Fields and Radiation:
    Past, Present and Future   447
    Zenon Sienkiewicz

27 Research on Mobile Phones and Health   455
    Sakari Lang

28 New Site Sharing Technology for 3G and the City of Tomorrow   467
    Mike Smith

29 Mobile Communications and Health   487
    Peter Grainger

30 Moving Beyond EMF Public Policy Paralysis   501
    B. Blake Levitt

    Questions & Discussion   519

Appendix: Catania Resolution   525
Index   527

Contributors

Professor Derek Clements-Croome

Professor of Construction Engineering and director of research in the School of Construction Management & Engineering at the University of Reading. With industry set up an undergraduate course in Building Services Engineering Design and Management in 1989. Set up new MSc in Intelligent Buildings in 1996 funded by EPSRC. Past Chairman and founder of the South East Centre of the Built Environment in Construction. Chairman of European Intelligent Building Group. Visiting Professor at five universities in China. Member of EPSRC Innovative Construction Research Centre at Reading University.

Author of several books: including *Airconditioning and Ventilation of Buildings,* Pergamon Press, 1981(Second Edition), also published in Chinese and Russian; *Noise, Buildings and People*, Pergamon Press, 1977; Editor and part author of *Naturally Ventilated Buildings* Spon (1997) and *Creating the Productive Workplace (2000)* Spon; published over 200 papers on various aspects of his research and education projects.

The University of Reading
School of Construction Management and Engineering
Whiteknights, PO Box 219
Reading, RG6 6AW
UK
E-mail: d.j.clements-croome@reading.ac.uk

Professor Ross Adey

William Ross Adey received degrees of M.B & B.S. (1943) and M.D. (1949) University of Adelaide, Australia. Royal Australian Navy, 1944-1946. Reader in Anatomy, University of Adelaide 1947-49. University of California, Los Angeles and UCLA Brain Research Institute, 1954-1977. Chief of Research Svc., Veterans Affairs Med. Centre, Loma Linda CA 1977-1997. University Calif. Riverside 1997-2000.

Since 1965, he has studied cell and molecular, physiological and behavioural effects of nonthermal environmental electromagnetic field exposures with emphasis on RF/microwave low-frequency modulation. These studies support a key role for field transduction in cooperative processes at cell membranes that may involve free radicals and calcium-dependent enzymatic mechanisms, organised hierarchically. Nuffield Foundation Fellow, University of Oxford, 1950-51. D'Arsonval Medallist, Bioelectromagnetics Soc., 1989. Sechenov Medallist, Russian Acad. Medical Sciences, 1996. Wellcome Trust Visiting Professor, Royal

Society of Medicine, 1996. Hans Selye Medallist, American Institute of Stress, 1999.

Distinguished Professor of Physiology
School of Medicine
Loma Linda University
31866 Third Avenue
Redlands, California 92373
USA
E-mail: Radey43450@aol.com

Dr Cyril Smith

Cyril was born in London in 1930. His career started in Radar at T.R.E. Malvern; his Ph.D was on X-Ray image intensification at Imperial College, London; He taught physics before moving to Salford in 1964. With over 100 publications and the book Electromagnetic Man, there have been four mainstreams of his research: instrument technology, dielectric liquids, bio-medical electronics; electromagnetic hypersensitivity and electromagnetic effects in living systems, bio-molecules and water. Since retiring in 1990, he has continued writing, research and lecturing.

Honorary Senior Lecturer
University of Salford
School of Acoustics and Electronics
Manchester, M30 9EA
UK
E-mail: cyril.smith@which.net

Professor Andrew Marvin

A Lecturer and Senior Lecturer at the University of York from 1979 – 1995. Appointed Professor of Applied Electromagnetics 1995. Founder and Leader of the Applied Electromagnetics Research Group, Department of Electronics, University of York. Technical Director of York EMC Services Limited. Research interests are Electromagnetic Compatibility Measurements and Metrology, Associate Editor of IEEE Trans EMC. Chairman of Cost 261, EMC in Complex and Distributed to Systems.

Applied Electromagnetics
Department of Electronics
University of York
Heslington, York, YO10 5DD
UK
E-mail: acm@ohm.york.ac.uk

Alan Meyer

Alan Meyer has been a practising lawyer in Westminster since the early 1960s and from 1976 to 2001 was the Senior Partner of a medium sized Private Client firm. He is now a full time Consultant with the large Private Client firm, Bircham Dyson Bell.

In 1997 he was instructed to oppose a mast application on behalf of Mr Mohammed Al Fayed. This led to a judicial review which found that the planning process was flawed. However, it was necessary to appeal the decision to the Court of Appeal with regard to the High Court Judge's discretion. At that Court of Appeal hearing, it was stated by the Court of Appeal that it was common ground that genuine public fear and concern was a material planning consideration.

Alan Meyer acted initially as Legal Adviser for NIFATT (Northern Ireland Families Against Television Towers) and in that capacity gave evidence in both Belfast and London to the Stewart Independent Expert Group on mobile telecommunications. Since then, he has been involved in the government/industries link further research programme. From December 2000 he became Legal Director of Mast Action UK and in legalistic terms would be best described as an objector's lawyer although he has assisted more than one local planning authority with regard to the law and how it affects planning guidance.

Senior Consultant, Legal Director of Mast Action UK
Bircham Dyson Bell (Solicitors)
50 Broadway
London, SW1H OBL
UK
Web site: www.mastaction.co.uk

Dr Gerard Hyland

Honorary Associate Fellow of the University of Warwick, UK, and a member of the International Institute of Biophysics, in Neuss-Holzheim, Germany specialising in biophoton research. During the last 6 years, a particular concern has been potential health hazards associated with non-thermal influences of the low intensity, pulsed microwave radiation used in GSM telephony, and how to render this technology more electromagnetically biocompatible.

He is regularly invited to speak on these subjects at international conferences, and gave evidence to the IEGMP. Recent publications include a review article for The Lancet (November 2000) – Physics and Biology of Mobile Telephony – and a report commissioned by the Scientific and Technological Options Assessment (STOA) Programme of the Directorate General for Research – The Physiological and Environmental Effects of Non-ionising Electromagnetic Radiation – which was published by the European Parliament in March 2001.

Physicst Department
University of Warwick
Coventry, CV4 7AL
UK
E-mail: gjhyland@hyland.demon.co.uk

Simon Pearl

A Partner at Davies Arnold Cooper with 25 years practice experience in product
liability litigation including the leading cases of the Whooping Cough Vaccine test
case and HIV/AIDS and Hepatitis C virus in blood. He is currently working on the
MMR Vaccine litigation.

Partner
Davies Arnold Cooper (Solicitors)
6-8 Bouverie Street
London, EC4Y 8DD
UK
Web site: www.dac.co.uk

Dr Philip Chadwick

Philip Chadwick has worked on the interaction of EMF with the human body since
1985, within the University system, at NRPB and at the Department of Health. He
is currently Technical Director of MCL, an independent research and consultancy
organisation. He is Chairman of CENELEC TC106X and four of its sub-
committees, a member of IEEE SCC28 and is a Consulting Member of ICNIRP.

Technical Director
Microwave Consultants Limited
17B Woodforsd Road
London E12 2EL
UK
E-mail: phil.chadwick@mcluk.org
Web site: www.mcluk.org

Dr Zenon Sienkiewicz

Dr Sienkiewicz is from the Radiation Effects Department of the NRPB and has an
interest in behavioural effects of power frequency magnetic fields and radio
frequency fields. He has written many papers, reviews and articles on the subject
of electromagnetic fields. He is also a consultant to AGNIR and a member of the
UK MTHR programme management committee.

National Radiological Protection Board (NRPB)
Chilton, Didcot, Oxon, OX11 ORQ
UK
E-mail: Zenon.Sienkiewicz@nrpb.org.uk
Web site: www.nrpb.org

Dr Magda Havas

Magda Havas is an Associate Professor in Environmental and Resource Studies, Trent University, Peterborough, Canada. She received her B.Sc. and Ph.D. at the University of Toronto and then did PostDoctoral Research at Cornell University.

Her research, on the biological effects of environmental pollutants, includes acid rain, metal pollution (aluminum, copper, nickel), drinking water quality, and electromagnetic fields both low frequency and radio frequency. She also has a keen interest and has published in the area of environmental education.

Environmental & Resource Studies
Trent University
Peterborough
Ontario, K9J 7B8
Canada
E-mail: mhavas@trentu.ca

Dr Ing Dietrich Moldan

Graduated in 1987 at the University of Aachen.worked from 1986-2001 in different construction material producing companies. Until end of 2001, he had been Head of the Department of Environment and Health at Gebr. Knauf Westdeutsche Gipswerke, Germany. Since 1996 he has a consulting office for environmental analysis (technical fields and waves; indoor air pollution). His paper entitled Reduction and Shielding of RF and Microwaves in the Building Industry was written with co-author Prof. Dipl.-Ing Peter Pauli from the German Armed Forces University Munich.

Consulting Office for Environmental Analysis
Baubiologe IBN, Am Henkelsee 13
D-97346 Iphofen
Germany
E-mail: dietrich.moldan@dr-moldan.de

Alasdair Philips

Alasdair is a professional EMC engineer and scientist and is widely recognised as one of the few truly "independent" UK experts on the possible adverse health effects of non-ionising radiation. He started working in the telecommunications industry in 1969.

He has been studying and writing on the subject of the possible biological effects of electromagnetic fields since 1986. He has given professional advice to many local authorities and presented Expert Evidence on these matters to the High Court, local Courts, Public Inquiries and Hearings and has been on Regulator committees. In 1987 he was a founder member of Powerwatch, a consumer information service on the possible adverse health effects of non-ionising radiation.

UK Powerwatch
2 Tower Road, Sutton, Ely
Cambridgeshire, CB6 2QA
UK
E-mail: aphilips@gn.apc.org
Web Site: www.powerwatch.org.uk

Dr Jean Monro

Dr Monro has a background in hospital general medicine and worked at the National Hospital for Nervous Diseases, London, researching migraine and multiple sclerosis. She entered full-time practice in Environmental Medicine in 1982 and in 1988 established the Breakspear Hospital for Allergy and Environmental Medicine. She has many publications to her name and regularly speaks at conferences worldwide. Her primary areas of interest are nutrition, medicine and immunology, metabolic function and environmental medicine.

Breakspear Hospital
Lord Alexander House
Waterhouse Street
Hemel Hampstead, Herts, HP1 1DL
UK
E-mail: jmonro@breakspear.org

Anne Silk

Anne Silk is an environmental health researcher with special interest in Electromagnetic Fields. She is a retired contact lens practitioner with 30 years experience with referred patients, many with allergies and atypical ocular conditions.

The Bostons
Optician Health Researcher
Vice Chairman of Electromagnetic Biocompatibility Association
Thorn Close
Whiteleaf
Princess Risborough, HP27 0LU
UK
E-mail: annesilk@waitrose.com

Professor Denis Henshaw

Professor Henshaw holds a Medical Research Council Programme Grant leading a team specialising in the effects of low level radiation on the human body. Although much of this work has concerned with low level ionising radiation, more recent work has concentrated on the ability of high voltage powerline electric fields to mediate increased exposure to air pollution. The findings suggest that people living near high voltage powerlines are at increased risk of illnesses known to be associated with air pollution.

HH Wills Physics Laborotary
University of Bristol,
Tyndall Avenue, Bristol, BS8 1TL
UK
E-mail: d.l.henshaw@bris.ac.uk

Professor Barry Michael

PhD in Radiation Physics. Professor Michael has been mainly interested in radiation biophysics (ionising and non-ionising), radiation and cancer treatment, radiation risk, DNA damage, micro-irradiation techniques. He is a visiting Professor at Queen Mary and Westfield College, London and a Member International Commission on Radiation Units and Measurement.

Joint Executive, Head of Cell and Molecular Biophysics Group,
Gray Cancer Institute,
PO Box 100,
Mount Vernon Hospital,
Northwood, Middlesex, HA6 2JR
UK
E-mail: michael@gci.ac.uk
Web site: www.graylab.ac.uk

Professor Jeffrey Hand

Jeff Hand is Head of the Radiological Sciences Unit within the Imaging Directorate at Hammersmith Hospitals NHS Trust and Professor in Imaging Physics within the Faculty of Medicine at Imperial College, London. He has worked on medical applications of RF and microwave fields for 25 years and authored more than 170 scientific papers. He has PhD and DSc degrees from the University of Newcastle upon Tyne, and has worked at Stanford University and the University of Arizona in the USA, and at Shimane Medical University at Izumo in Japan. He is a Fellow of the Institute of Physics, a Fellow of the Institution of Electrical Engineers and a Fellow of the Institute of Physics and Engineering in Medicine. He is also a Chartered Engineer, a Chartered Physicist and a registered Clinical Scientist.

Head of Radiological Sciences Unit
Hammersmith Hospital
Due Cane Road,
London, W12 OHS
UK
E-mail: j.hand@ic.ac.uk

Dr Neil Cherry

Dr Neil Cherry was the Associate Professor of Environmental Health at Lincoln University, New Zealand. This academic position involves using his experience of research and teaching or climate change, wind and solar energy resources, energy efficiency, human biometeorology, biophysics, environmental epidemiology in health effects weather, air pollution, viruses and electromagnetic radiation health effects. For his advanced and innovative work in this area of integrative environmental health science, Dr Cherry was awarded a Royal Honour in 2002, Officer of the New Zealand Order of Merit. He was a well-known independent international academic presenter of over 30 papers to international conferences on the health effects of EMR/EMF. Dr Neil Cherry died on 24 May 2003, aged 56, from motor neurone disease.

Dr Jeremy Smallwood

After seven years in electronics R&D, Jeremy Smallwood worked on contract research in the Electrostatics Research Group at the University of Southampton from 1987 to 1994, studying electrostatic discharge measurements and ignition risks, which became the subject of his PhD. He subsequently worked for ERA Technology Limited as Electrostatics Section Leader before forming his specialist R&D and consultancy company Electrostatic Solutions Ltd.

Dr Smallwood is Chairman of the Institute of Physics Statics Electrification Group, and the International Electrotechnical Committee TC101 (Electrostatics).

Electrostatic Solutions Limited
13 Redhill Crescent,
Bassett, Southampton,
Hampshire, SO16 7BQ
UK
E-mail: jeremys@static-sol.com
Web site: www.static-sol.com

Professor Olle Johansson

Olle Johansson works in the field of basal and clinical neuroscience at the Karolinska Institute in Stockholm. He is the head of the Experimental Dermatology Unit in the Department of Neuroscience and he specialises in radiobiology, ionising and non-ionising radiation. He is particularly interested in the effects of microwaves on the human body as well as the clinical issue of electrosenstivity. Author of more than 450 articles within the field of neuroscience and dermatoscience.

Experimental Dermatology Unit,
Karolinska Institute,
17177 Stockholm, Sweden
E-mail: olle.johansson@neuro.ki.se

Dr Vladimir Binhi

Dr V.N. Binhi graduated form the Moscow Physical Technical Institute in 1977; he was awarded his PhD (maths & physics) degree at the P.N. Lebedev Physical Institute, USSR Academy of Sciences in 1982. From 1982 to 1994 he worked in the General Physics Institute USSR Academy of Sciences and other physics institutes. From 1994–2000 he was Head of Biophysics department in the International Institute of Theoretical and Applied Physics, Moscow, and then from 2000 – Head of GPI RAS Laboratory. Research expertise: magnetic processes in molecular systems, theoretical modelling of biological effects of electromagnetic fields, proton dynamics and structure defects in liquid water, magnetic measurements.

Head of Radiobiology Laboratory
General Physics Institute,
Russian Academy of Sciences,
38 Vavilova Street, Moscow, 119991
GSP-1 Russian Federation
E-mail: infor@biomag.info

Professor Alan Preece

Alan Preece is a professor in Oncology and Medical Physics at the University of Bristol. He specialises in radiobiology, non-ionising radiation and physiological measurement and is particularly interested in the interactions of microwaves and with the human body.

Bristol Haematology and Oncology Centre
University of Bristol,
Level C, Horfield Road
Bristol, BS2 8ED
UK
E-mail: a.w.preece@bristol.ac.uk

Alastair Speare-Cole

Alastair Speare-Cole graduated from Oxford in 1981 having read geology. His expertise lies in: liability reinsurance, motor reinsurance, alternative risk transfer, E.M.F., subsidence, pollution and property.
History: FCII Chartered Insurance Institute, Fellow of Geological Society, M.A. Oxon: Geology and Mineralogy.

Managing Director
AON Limited
8 Devonshire Square
London, EC2M 4PL
UK
E-mail: alastair.speare-cole@aon.co.uk
Web site: http://www.aon.co.uk

Professor Ray Kemp

Prof. Ray Kemp BA, MSc, PhD, MRTPI is Head of the Risk Management Group at Galson Sciences Limited. He holds an honorary visiting professorship in Risk Management and Communication at the University of Surrey, Guildford, UK. Ray has 20 years research and consulting experience in corporate risk assessment, risk management and risk communication for controversial projects. He was President of the Society for Risk Analysis in Europe from 1995 to 19996. Prof. Kemp's most recent work has been in relation to risk communication regarding the health risks of mobile phone base stations. He has provided guidance on environmental planning, consultation and professional training in risk communication both in the UK and in Europe. He has acted as chairman and independent facilitator of a series of stakeholder roundtable meetings on the mobile phone health issue for the UK mobile phone sector. Ray is also an advisor to the World Health Organisation in Geneva, chairing an international expert group on risk communication in relation to the health risks of electromagnetic fields.

Head of Risk Management Group
Galson Sciences Limited
5 Grosvenor House
Melton Road, Oakham
Rutland, LE 15 6AX
UK
E-mail: rvk@galson-sciences.co.uk
Web site: www.galson-sciences.co.uk

Michael Dolan

Michael Dolan is Director of the Mobile Telecommunications Advisory Group (MTAG) within the Federation of the Electronics Industry (FEI), which represents he UK mobile manufacturers and network operators on health and related issues including planning. Michael has held this position since 1 September 1999. He was formerly Deputy Director of the Electricity Supply Association of Australia where he was responsible over a period of nine years for electricity industry management of the power frequency health issue. A lawyer by profession, Michael has extensive experience in issues management and media interface.

Director
Mobile Telecoms Advisory Group,
Federation of the Electronics Industry,
London,
UK
E-mail: mdolan@fei.org.uk

Nicole Hughes

Nicole Hughes joined the FEI in October 2001 as Public Affairs Manager. The FEI represents the five mobile phone companies in the UK on health and planning related matters. During her time with the FEI, Ms Hughes has presented at numerous conferences on the issue of risk management and the precautionary principle relating to mobile health.

Ms Hughes has extensive experience in public affairs, journalism and issues management. She currently works on policy development and public affairs interface.

Public Affairs Manager
Mobile Telecoms Advisory Group
Federation of the Electronics Industry
London,
UK
E-mail: nhughes@fei.org.uk
Web site: http://www.fei.org.uk

Dr Sakari Lang

He was awarded an MSc in Environmental Sciences (1988) and a PhD in Radiation Biology (1994), both from the University of Kuopio, Finland. Post-doctoral fellow at the Department of Radiological Health Sciences, Colorado State University 1995-1996. Assistant professor in radiation biology at University of Kuopio 1997-1998. Principal scientist and research manager (responsible for bioelectromagnetics research and radiation and safety issues at Nokia) at Nokia Research Centre, Helsinki, Finland, from 1998 to date. Sakari Lang has studied effects of both ionising and non-ionising radiation.

Research Manager
Nokia Research Centre
PO Box 407, FIN-0045 Helsinki
Finland
E-mail: sakari.lang@nokia.com

Mike Smith

Mike Smith initially trained in the Royal Air Force as an engineering technician. His subsequent technical and management career within the RAF spanned 22 years and covered a variety of engineering, management, quality and health & safety roles. During his last 4 years within the RAF he fulfilled the role of Ministry of Defence HS&Q Manager (Cryogenics) for mainland Europe. Upon retirement from the RAF, he joined the Defence Evaluation & Research Agency's (now under

PPP renamed QinetiQ) Tactical Communications Division working in a variety of project management roles before taking up the post of Health, Safety & Environment Manager. For the past two years, Mike has been seconded to Quintel S4 Ltd with the post of Project Director for Health, Safety, Environment & Planning with specific responsibility for liaising with local and central government, regulators, the general public, pressure groups and others both in the UK and abroad. He is a corporate member of the Institute of Occupational Safety and Health and the International Institute of Risk and Safety Managers.

Project Director of Health and Safety, Environment & Planning
Quintel S4 Ltd
Building A57
Cody Technology Park
Farnborough
Hants, GU14 0LX
UK
E-mail: mike.smith@quintels4.com
Web site: www.quintels4.com

Dr Peter Grainger

Peter Grainger is a graduate of the University of Bristol and London. He has been researching the effects of non-ionising electromagnetic energy and health for seven years. Recent work has included mathematical modelling of personal exposure to magnetic fields.

Bristol Haematology and Oncology Centre
Medical Physics University Research Centre
Bristol University
Horfield Road
Bristol, BS2 8ED
UK
E-mail: drpetergrainger@yahoo.co.uk

B. Blake Levitt, Journalist

Blake Levitt is an award-winning medical and science journalist. A former New York Times writer, she is the author of Electromagnetic Fields, A Consumer's Guide to the Issues and How to Protect Ourselves (Harcourt Brace 1995) for which she won an Award of Excellence from the American Medical Writers Association. She is also the editor of Cell Towers, Wireless Convenience? or Environmental Hazard? Proceedings of the Cell Towers Forum, State of the Science/State of the

Law (New Century Publishing, 2001). She lectures widely on environmental energy issues.

355 Lake Road
P.O. Box 2014,
New Preston
CT. 06777
USA.
E-mail: Blakelevit@cs.com

Preface

Electromagnetic fields are an intrinsic part of the universe. Animals and plants conduct electricity and with this there are associated magnetic fields. Over the last hundred years there has been a rapid rise in the use of electrical equipment inside and outside buildings. Does the rapid increase in electropollution affect the health of people? We do know some people are hypersensitive to electromagnetic fields, referred to in this book as electrosensitive people. We do not however understand the mechanisms which underlie this phenomenon. Knowledge on the subject is very fragmented. Our education at school about hybrid areas such as bio-physics and electro-medicine is relatively poor.

The *Electromagnetic Environments and Health in Buildings Conference* took place on 16th and 17th May 2002 at the Royal College of Physicians in London. About 150 people attended including scientists, people from various professions, various manufacturers, electromagnetic field networkers, private individuals and representatives from the media. The main aim of the conference was to bring people from different disciplines and all walks of life together, to discuss many different points of view and hear from those that carry out research, those that design buildings, those medics that treat patients, as well those who collect knowledge and distribute it internationally via various networks.

The problem of hypersensitivity is making us think deeply about the validity of the knowledge that we have. On one hand some scientists are saying that there are no known mechanisms for understanding the interaction of the human body with electromagnetic fields. But there are others who believe that there are known mechanisms, which have been developed and replicated by others. The book *Magnetobiology* by Binhi (2002) gives detailed scientific explanations of such mechanisms. There are a number of people who suffer from electrosensitivity and we need to look at their problems sympathetically even though they describe what scientists usually term *anecdotal (or case study) evidence*, which many scientists dismiss. There is the possibility that different issues are being confused. Case study evidence clearly does not propose a mechanism for the problem and so the cause of electromagnetic hypersensitivity still eludes us. On the other hand the effect exists; this is no different from a patient going to the doctor with some illness and describing the symptoms for which the doctor may be able to give some help. The fact that a condition may be ameliorated does not mean that the doctor understands the mechanisms behind the illness in all cases. It is quite possible to have an effect demonstrated without understanding the cause but this does not mean the effect should be dismissed. One recalls that initially the health effects of smoking, asbestos or repetitive strain injury from working with computers were all dismissed. Several leading scientists have supported the September 2002 Catania Resolution which basically states electromagnetic fields do adversely affect health (see Appendix).

In general we are better at dealing with reductionist ideas and analysis but not so confident when dealing with holistic ideas requiring a synthesis of knowledge from a range of disciplines. However, scientists should acknowledge that even physics and mathematics have limitations. Confucius wrote that:

To be uncertain is uncomfortable
To be certain is ridiculous

Government bodies tend to be cautious and manufacturers tend to be profit oriented. We need to beware of being so excited by the opportunities afforded by new technologies that matters of health are forgotten. In the November issue of Physics World 2002 some research is reported that has developed miniature antennae which produce a much lower energy density than current devices. Mobile phones and computers are used in the workplace and by very young children in the home as well as in school. People want to communicate with anyone they like wherever they are. The limitations of technology are also becoming apparent. Ultimately face to face communication is preferable but many times this is not possible, and understandably the technology which makes communication possible is therefore welcomed, but then possibly overused in situations where it is not necessary.

This book is largely based on the Conference with subsequent additions and covers health effects of electromagnetic fields; emissions and standards; and offers some glimpses into the future. It concludes that there is much more valid scientific knowledge available but it is fragmented. There is clearly a need for much more research using a range of methodologies. Physicists, biologists, medics, physiologists, epidemiologists and sociologists need to work together on scientific research programmes.

There remains the problem of who pays for the research. The health of nations is partly a responsibility of governments but the computer, mobile phone and multi-media industries are developing high technology products to sell in vast volumes to young and the old alike. The convenience of communication offered by their products is seductive and attractive. Health issues however are given scant attention. There is a need for public-private sector financed research programmes to ensure health is not endangered.

I would like to thank my colleagues Anne Silk (Vice Chairman of the Electromagnetic Biocompatibility Association, EMBA), Dr Peter Grainger (formerly at the Department of Oncology at the University of Bristol), Dr Cyril Smith (formerly at the University of Salford) who helped me plan the programme. Additional thanks to other peer reviewers Professor Anthony Barker (Royal Hallamshire Hospital); Professor Lawrie Challis (University of Nottingham); Dr Zenon Sienkiewicz (National Radiological Protection Board); ABACUS for undertaking the administrative arrangements and the financial risk for the conference under the leadership of Peter Russell; all the speakers who gave their time at and after the conference to help produce this book; to the audience that attended the Conference and for the many who wrote to me afterwards about how they had found it valuable and enjoyable; to EMBA for supporting the marketing of the book. My special thanks to Gülay Ozkan for preparing the camera-ready copy for the publishers with such care, diligence and dedication; to Maureen Taylor and Göksenin Inalhan for manuscript preparation; to Anne Silk Vice-Chairman of EMBA for proof reading.

Professor Derek Clements-Croome
School of Construction Management and Engineering
University of Reading
October 2002

Part I
Setting the Scene

CHAPTER 1

Healthy Buildings

Derek Clements-Croome

1.1 BACKGROUND

What do we mean by a healthy building? There are health and safety regulations but these merely set a basic threshold for environmental safety conditions. They define nothing about the real quality of the human sensory experience.

We live through our senses which continually receive stimuli from the world around us, from which particular ones are selected to deal with the activity being undertaken at a particular time. The remaining stimuli still exist but our mind, whilst aware of them, places them in the background. Stimuli can be thought of as packets of information received by the body's receptors and processed from them via the nervous system to the brain which results in our perceived experiences or so-called qualia.

There are *moving* and *static* elements in our environment. Air, light and sound patterns vary in space and time, in other words they *flow* through the space. Structural elements such as walls and floors are usually *fixed* but the way they are arranged, the materials from which they are made, their colours, their textures affect the sound and the visual environments. Materials can emit gases or particulates to the air so they can also affect the air quality. There are moveable components such as furniture which affect our skeletal comfort.

Pollutants are formed from dust and gases in the air; these arise from people, clothing, furnishings, coverings of objects, air entering the space from outside or even the dirty air from poorly maintained ventilation systems.

The effects of electromagnetic fields are less understood hence the reason for this book (Binhi, 2002; Brune *et al.*, 2001). Visible light is a very small but important part of the electromagnetic spectrum. Non-ionising and ionising radiations are important too although the latter are usually at a negligible level in most buildings. Non-ionising radiation is emitted from computers, mobile phones and power equipment, which are an increasingly prominent part of our living and working environments. It is important to increase our understanding of the interaction between the body's internal electromagnetic fields and those around the outside of the body.

Health means good well-being and refers to the states of mind and the physical body. There must be an overall satisfaction and awareness of the environment with a degree of comfort and freshness, but also a keenness to concentrate on the task in hand. It is easy to disturb this balance by stuffy air conditions, bland surroundings, shabby decor, lack of daylight or non-ergonomic furniture. Of course the nature of the work being undertaken, the organisation and the social ambience are also key factors.

Our responses are conditioned by our past experiences and our expectations which arise from them within particular situations. They are partly physiological and partly psychological so that there is an adaptive physical objective context but also a more elusive fuzzy subjective one. Environment can affect our mood, our feelings as well as our physical reactions.

Poor quality environments – physical, social and organisational – can disturb or distract the mind and lead to negative stress. Stress can weaken the body's defence system (i.e. the immune system) and thus increase the likelihood of building sickness symptoms taking root. These symptoms include respiratory ailments; skin irritations; tired eyes; headaches; and lethargy.

1.2   ARCHITECTURE AND THE SENSES

The idea of taking into account the senses of a building occupant has led to our research into how we smell, touch and see a building, as well as our psychological interactions with it. Architecture deals not only with the materials and form but also with people, emotion, space and relationships between them (Clements-Croome, 2000). Buildings should be a multi-sensory experience. Pallasmaa (1996) elegantly describes this belief in his book *The Eyes of the Skin* and also in association with Holl and Perez-Gomez (1994) in the book *Questions of Perception.*

During the Renaissance, the five senses were understood to form a hierarchical system from the highest sense of vision down to touch. It is by vision and hearing that we acquire most of our information from the world around us. But one should not underestimate the importance of the other senses. Olfactory enjoyment of a meal or the fragrance of flowers, and responses to temperature provide a bank of sensory experience which help to mould our attitudes and expectancies about the environment. The senses not only mediate information for the judgment of the intellect, they are also channels which ignite the imagination. This aspect of thought and experience through the senses, which trigger the body and mind, is stimulated not only by the environment and people around us but when inside a building, it is the architecture of the space which sculpts the outline of our reactions. Merleau-Ponty (1964) said that the task of architecture was to make visible how the world touches us.

At the heart of architecture is the fundamental question of how buildings in their design and use can confront the questions of human existence in space and time and thus express and relate to man's being in the world. If this question is ignored the result is soulless architecture which is a disservice to humanity. There is a danger, for example that the ever-increasing pace of technology is distorting natural sociological change. Such distortion makes it difficult for modern architecture to be coherent in human terms.

Buildings must relate to the language and wisdom of the body. If they do not they become isolated in the cool and distant realm of vision. But in assessing the value of building, how much attention is made to the quality of the environment inside the building and its effects on the occupants? The qualities of the environment affect human performance inside a building and these should always be given a high priority. This can be considered as an invisible aesthetic that together with the visual impact makes up a total aesthetic.

Buildings should provide a multi-sensory experience for people and uplift the spirit. A walk through a forest is invigorating and healing due to the interaction of all sense modalities; this has been referred to as the polyphony of the senses. Architecture is an extension of nature into the man-made realm and provides the ground for perception, a basis from which one can learn to understand the world. Buildings filter the passage of light, air and sound between the inside and outdoor environments; they also mark out the passage of time by the views and shadows they offer to the occupants. Pallasmaa (1996) gives as an example to illustrate this point. He believes that the Council Chamber in Alvar Aalto's Säynätsalo Town Hall recreates a mystical and mythological sense of community where darkness strengthens the power of the spoken word. This demonstrates the very subtle interplay between the senses and how environmental design can heighten the expression of human needs within a particular context.

Although the five basic senses are often studied as individual systems covering visual, auditory, taste - smell, orientation and the haptic sensations, there is interplay between the senses. Sight, for example, collaborates with the other senses. All the senses can be regarded as extensions of the sense of touch because the senses as a whole define the interface between the skin and the world. The combination of sight and touch allows a person to get a scale of space, distance or solidity.

However, qualitative attributes in building design are often only considered at a superficial level. For example, in the case of light the level of illuminance, the glare index and the daylight factor are normally taken into account. But in great spaces of architecture there is a constant deep breathing of shadow and light; shadow inhales, whereas illumination exhales light. The light in Le Corbusier's Chapel at Ronchamps for example gives the atmosphere of sanctity and peace. How should we consider hue, saturation and chroma in lighting design for example? Buildings provide contrast between interiors and exteriors. The link between them is provided by windows. The need for windows is complex. It includes the need for an interesting view and for contact with the outside world; at a fundamental level, it provides contrast for people working in buildings. Much work today is done at computers in close quarters and requires eye muscles to be constrained to provide the appropriate focal length, whereas when one looks outside towards the horizon the eyes are focused on infinity and the muscles are relaxed. There are all kinds of other subtleties, such as the need to recreate the wavelength profile of natural light in artificial light sources, which need to be taken into account. Light affects mood. How can this be taken into account in design?

The surfaces of the building set the boundaries for sound. The shape of the interior spaces and the texture of surfaces determine the pattern of sound rays throughout the space. Every building has its characteristic sound of intimacy or monumentality, invitation or rejection, hospitality or hostility. A space is conceived and appreciated through its echo as much as through its visual shape, but the acoustic concept usually remains an unconscious background experience. It is said that buildings are composed as the architecture of space, whereas as music represents the architecture of time. The sense of sound in buildings combines the threads of these notions. Without people and machines, buildings are silent. Buildings can provide sanctuary or peace and isolate people from a noisy fast moving world.

The ever increasing pace of change can temporarily be slowed down by the atmosphere created in a building. Architecture emancipates us from the embrace of the present and allows us to experience the slow healing flow of time. Again buildings provide the contrast between the passing of history and the pace of life today.

The most persistent memory of any space is often its odour. Every building has its individual scent. Our sense of smell is acute, and strong emotional experiences are awakened by the olfactory sense. Odours can influence cognitive processes that affect creative task performance as well as personal memories. Creative task performance is influenced by moods, which odours can also affect (Warren and Warrenburg, 1993; Erlichman and Bastone, 1991; Baron, 1990).

Various parts of the human body are particularly sensitive to touch. The hands are not normally clothed and act as our touch sensors. The skin reads the texture, weight, density and temperature of our surroundings. There is a subtle transference between tactile, taste and temperature experiences. Vision can be transferred to taste or temperature senses; certain odours, for example, may evoke oral or temperature sensations. The remarkable, world-famous percussionist Evelyn Glennie is deaf but senses sound through her hands and feet and other parts of her body. Architectural experience brings the world into intimate contact with the body.

The body knows and remembers. The essential knowledge and skill of the ancient hunter, fisherman and farmer, for example, can be learnt at any particular time, but more importantly, the embodied traditions of these trades have been stored in the muscular and tactile senses. Architecture has to respond to behaviour that has been passed down through the genes. Sensations of comfort, protection and home are rooted in the primordial experiences of countless generations. The word "habit" is too casual and neglects the history embedded in us.

The interaction between humans and buildings is more complex than we imagine. As well as simple reactions that we can measure, there are many sensory and psychological reactions that are very difficult to understand and quantify.

1.3 HEALTHY BUILDINGS

Indoor environment can be defined by physical features of the environment such as lighting, colour, temperature, air quality and noise. These factors have been studied extensively with regard to their impact on task performance and satisfaction. Sundstrom (1987) reports laboratory studies that show with high consistency that ambient conditions do have real and meaningful influences on behaviour in 150 out of a total of 185 experiments. A complete analysis of indoor environmental quality would take into consideration not only indoor air quality and thermal comfort but also lighting, floor lay-out, colour scheme, building materials, noise level, disruption, weather, management styles, space, employee and customer backgrounds employee/customer satisfaction and employee motivating factors. Any field study must account for all of these factors.

Interior finishes can have a marked effect on air quality. Lisbet (1994) shows that occupants working in offices with carpets complained significantly more about the sensation of dry and stuffy air than occupants working in offices with linoleum flooring, Research by Rundnai (1994) shows the prevalence of the most frequent symptoms among people living in buildings with different types of construction in Budapest. Brick construction shows a lower incidence of sick building syndrome (SBS) than concrete; the brick buildings were warmer and drier, whereas the concrete ones had poor heating control and possibly low infiltration. People living in mixed concrete and breeze block homes suffered from higher concentrations of formaldehyde. Clearly, materials and construction are important.

Ricci-Bitti (1994) distinguishes between SBS and mass psychogenic illness (MPI). The former is a temporary failure to cope with the environment whereas the latter represents a collective stress response. Building related illness (BRI) and Neurotoxic disorders (NTD) are deemed to be caused by environmental pollution, although minor forms of the these syndromes may be related to stress. The principal difference between these phenomena is that neurotoxic disorders and building related illness (such as legionnaires disease) persist whether the person is in the workplace or away from it, but in the case of sick building syndrome and mass psychogenic illness the effects only occur in the working environment and tend to decay quickly on leaving it. Most attention has been given to the sick building syndrome.

The World Health Organisation (WHO, 1983) describes the term sick building syndrome as a collection of non-specific symptoms expressing a general malaise associated with occupancy of the workplace. Symptoms include irritations of the eyes, respiratory system, skin rashes, as well as mental discomforts due to lethargy or headaches. The ancient Chinese believed that invisible lines of energy, known as *chi* run through our bodies and the environment. A smooth flow of chi means wellbeing, but if it is blocked then ill health, expressed in various ways results. Feng Shui is the art of freeing and circulating chi. Equipment such as computers emit electromagnetic fields and these together with underground water and geological faults can disturb the earth's natural electromagnetic field. There remains the possibility that geopathic stress could be partially responsible for sick building syndrome.

There have been numerous studies on *sick building syndrome*, but a lack of coordinated research means the methodology of investigation has varied, making it difficult to compare results. Research by Jones (1995), together with the work on a standard questionnaires by Raw (1990, 1995), Burt (1997), Baldry (1997) and Berglund (2000), will allow future investigations to be carried out in a more consistent manner. Sick building syndrome conditions are prevalent in the work environment but disappear when people leave it.

It is, for example, very difficult to make any strong conclusions about the effect of ventilation with respect to sick building syndrome. According to work by Jaakkola (1991) and also Sterling (1983) ventilating a building with 25% or 100% outdoor air makes little difference. Likewise other work referred to by Hedge

(1994) quotes work where the ventilation rate has been increased from 10 l/s per person to 25 l/s per person with no beneficial effects on SBS symptoms; Sundell (1994) and also found that ventilation rates above 10 l/s per person have no discernable effect on the symptoms.

Jones (1995) classifies healthy offices as those which average 1-2 symptoms per worker, while relatively unhealthy offices have 4-5 symptoms per worker. Headaches and lethargy appear to be nearly always the most frequently reported symptoms.

Hedge (1994) carried out an investigation in six office buildings and found that the prevalence of eye, nose and throat symptoms were higher in airconditioned offices than naturally ventilated ones, but headaches did not show any consistent pattern. Other studies by Burge *et al.* (1987); Mendell and Smith (1990); Wilson and Hedge (1987) confirmed that SBS symptoms were less prevalent in naturally ventilated buildings than airconditioned ones, but some mechanically ventilated buildings did not give rise to any problems either. Robertson *et al.* (1985) compared SBS symptoms in adjacent air-conditioned and naturally ventilated offices and found that the symptoms were more prevalent among workers in the airconditioned offices, although measurement of a variety of physical environmental factors failed to show any significant differences in the environmental conditions between buildings. Other work described by Hedge (1994) confirms these findings.

However, care should be taken to ensure that the concentrations of volatile organic compounds are measured and compared. Some studies have found that symptoms maybe associated with suspended particulate matter, but there is contrary evidence about this issue also. The lack of a consistent association between symptoms and they physical environment suggests that building sickness syndrome remains elusive and so maybe there are a number of other factors like geopathic stress, individual factors, perceived control and occupational factors are more important. Again psychosocial factors are probably not directly related to the SBS symptoms, but Hedge (1994) argues that psychosocial variables may trigger off patterns of symptom reporting. Hedge (1994) describes a large study that he has undertaken in 27 air-conditioned offices. Dry eyes and headaches were found to be weakly associated with formaldehyde; mental fatigue was found to be weakly associated with formaldehyde and particulate concentrations; complaints about stale air were associated with carbon dioxide levels. The building users age, job grade and smoking status was not associated with SBS symptoms. Results showed that more SBS were reported by women; full-time computer users; building occupants with high job stress or low job satisfaction; people who perceived the indoor air quality to be poor; the 18-35 years old age group; occupants who have allergies; people who have migraines; users who wore spectacles or contact lenses, and finally smokers. Hedge (1994) concluded that reports about sick building syndrome symptoms are influenced by several individual and occupational interacting factors. Other studies by Hedge (1994) attempted to establish if a certain personality type was susceptible to sick building syndrome but did not establish any connection. Likewise the effect of circadian rhythms did not appear to be important. Sick buildings syndrome appears to arise from a set of multiple risk factors, some of which are environmental, but biological, perceptual and occupational aspects are also important. The general conclusion can be made that environmental, occupational and psychosocial factors interact to induce sick building syndrome symptoms.

Jones (1995) considers health and comfort in offices and states that the symptoms commonly referred to as sick building syndrome (SBS) show a wide range of variation in reported symptoms between buildings and also between zones in any one building. Only about 30% of the variation in symptoms has been explained by the built environment, by an individuals characteristics or by the job.

In common with other work he concludes that the symptoms of SBS arise from a combination of factors such as building environment, nature of the work being undertaken, as well as the individual characteristics of the person. One important conclusion reached by Jones (1995) is that when thermal conditions are perceived as comfortable, the office is not necessarily healthy. This is also in some ways analogous to earlier work, already discussed, which suggests that the most productive environments are slightly less than comfortable. He calls for improved standards of maintenance, spatial simplicity and flexibility, as well as adaptability of services, good management and recognition that the building, its environment and activities all interact dynamically.

The Health and Safety Executive published a booklet in 1995 entitled How to Deal with Sick Building Syndrome. This booklet helps building owners to identify and investigate buildings, besides giving advice on how to create a good work environment. As regards productivity, it is pointed out that although the SBS symptoms are often mild they do not appear to cause any lasting damage. They can affect attitudes to work and can result in reduced staff efficiency; increased absenteeism; staff turnover; extended breaks and reduced overtime; lost time due to complaining. Advice is offered in the HSE Booklet on building services and indoor environment; finance; and job factors, including management systems and work organisation.

Whitely *et al.* (1995b) suggests *equity theory* may explain the range of individual differences in symptom reporting in the same building and between buildings. Equity theory suggests that the way workers are treated may affect productivity. If this is true then a person in an organisation may intuitively assess their level of reward from the organisation and put the amount of effort in which is related to this. The physical environment in which a person works, can be seen as part of the reward system. The organisation can influence people's attitudes depending on whether any attempt is made to improve conditions in the workplace.

Whitely *et al.* (1995a) describes a well established, relatively stable, personality measure which is related to perceived control called locus of control, which measure the general tendency to attribute outcomes of behaviour due to internal or external causes and goes on to conclude that, the locus of control and job satisfaction appear to explain the perceptions of people to the environment as a whole. These factors also significantly influence the way people report sick building syndrome, environmental conditions and productivity. Whitely states that self reports of productivity change due to the physical environment must be treated with extreme caution, as job satisfaction is a major factor also. Research has already been described which shows that the perception of control over the environment is important with regard to productivity.

1.4   WELL-BEING AND PRODUCTIVITY

Myers and Diener (1997) have been carrying out systematic studies about awareness and satisfaction with life among populations. Psychologists often refer to this as *subjective well-being.* Findings from these studies indicate broadly that those who report well-being have happy social and family relationships; are less self-focused; less hostile and abusive; and less susceptible to disease. It appears also that happy people typically feel a satisfactory degree of personal control over their lives, whether in the workplace or at home. It is probably fair to assume that it is more likely that the work output of a person will be higher if their well-being is high. A study by Jamison (1997) reviews research which links manic-depressive illness and creativity. Many artists, such as the poets Blake, Byron and Tennyson, the painter Van Gogh and the composer Robert Schumann, are well-known examples of manic-depressives. The work output from such people is distinguished but lacks continuity. Mozart and Schubert were not classified as manic-depressives, and their work output was consistently high throughout their short lives. In contrast, Robert Schumann, who suffered hypo-mania throughout 1840 and 1849, was prolific during 1839 - 41 and 1845 - 53. Between these periods he suffered from severe depression; and before 1838 and after 1853, he made suicide attempts.

In the workplace one does not expect creativity at such levels of genius. Rather, it is hoped there will be a consistently high standard of work performance. Townsend (1997), in an article *How to Draw Out All the Talents*, states that 25% of us enjoy our work but the rest of us do not. Productivity suffers when the workplace is a site of conflict and dissatisfaction. Lack of productivity shows up in many ways, such as absenteeism, arriving late and leaving early, over-long lunch breaks, careless mistakes, overwork, boredom and frustration with the management and the environment. Townsend believes that we are all capable of focusing completely on the task at hand and that when we succeed in doing this the whole body feels different. Townsend also says that people in the workplace can be encouraged to use both halves of their brain. The left side is concerned with logic, whereas the right side is concerned with feeling, intuition and imagination (Ornstein, 1973). When logic and imagination work together, problem-solving becomes more enjoyable and more creative.

It is from the body that we orient ourselves in the world. Many measurements used in architecture are originally derived from measurements of parts of the body – a foot, a stride of three feet (a yard), or the size of a brick according to the hand. Many abstract structures for thinking and understanding also originate in bodily experiences of perception, movement and interaction with physical objects. We experience these structures when encountering the environment, and then we project them onto other situations, which helps us to organise shared understanding and knowledge. Our ways of inhabiting the world physically, as well as psychologically and intellectually, extend from our bodies outward. There are information exchanges between us and our surroundings, which are defined by nature, the built environment and people. These form physical and social environments. Their impact on our system is physiological, psychological, emotional and social.

Well-being reflects feelings about oneself in relation to the world. Warr (1998) proposes a view of well-being which comprises three scales: pleasure to displeasure; comfort to anxiety; enthusiasm to depression. There are work and non-work attributes which characterise one's state of well-being at any time, and these can overlap with one another. Well-being is only one aspect of mental health; other factors include personal feelings about one's competence, aspirations and degree of personal control. Good architecture extends and enhances human capacities. Buildings moderate climates, which help to keep the body healthy and enhance well-being. Some buildings demand closely controlled environments, and various equipment can be installed in order to achieve this but many buildings can take advantage of the body's ability to adapt and interact in a compensatory way with other senses.

We all have circadian rhythms, physiologically and psychologically, and these change as we carry out different activities during the course of a day. There is a large variation in these needs, and also behaviour patterns, between one individual and another. It is important that when people are within buildings they have contact with the outside world throughout their working day, and also have the means to adjust their environment according to changing needs. The built environment therefore has to be sensitive to these requirements and allow individuals to control their surroundings, as well as provide adaptability to changing needs.

Architecture is given life and spirit by all the qualities that touch the human senses *and* the human soul. If the functional nourishes our physical needs, the poetic nourishes our soul. This nourishing has been referred to by Franck and Lepori (2000) as *the animism of architecture.*

Stone, marble, brick and concrete create solidity and mass, as well as darkness and enclosure within. Glazed openings bring in light, and provide views and contact with the outside world. Materials have character. Glass is transparent and a mirror extends space by reflection. Brick suggests absorption rather than reflection because of its permeable and porous nature and its closeness to earth. Marble seems to be more aloof because of its coolness, hardness and exotic beauty. In contrast, concrete is a much more plastic building material that allows ease of construction through malleability. Steel structures provide the bones and skeleton of buildings and can be sculpted to different shapes to support or contain structures or spaces.

Lightweight structures that have been developed in the last decades remind us that the skins of buildings are like another layer of clothing. Tents of nomadic tribes or the *ger* of Mongolia are perfect examples of how lightweight materials have been used in everyday living in various parts of the world, whether cold or hot, over many centuries.

Wood is very special because it reminds us of trees and nature. It is flexible and responds to the vibrations of sound, in addition to providing a natural source of colour and pattern. String instruments like the violin, the viola, the cello and the double-bass demonstrate this perfectly.

Architecture supports our living activities and also provides an important ingredient in our perception of the world. We are surrounded by many examples of souless architecture, but there are also some buildings which exhibit great sensitivity to the human senses. The Kiasma Museum of Contemporary Art in Helsinki, designed by Steven Holl, and the lecture hall at the Institute of Technology in Otaniemi, Finland, by Alvar Aalto are two examples. Other examples are the Ronchamp Chapel by Le Corbusier; the Pyramid at the Louvre by IM Pei; the Philharmonie in Berlin by Hans Scharoun; and the house Falling Water, in Pennsylvania, by Frank Lloyd Wright. The work of Hassan Fathy also provides a rich source of human architecture. In all these examples their richness is demonstrated by an imaginative combination of function, form and aesthetics. Buildings are designed for the soul and the spirit not just the function, convenience and form.

A good working environment helps provide users with a good sense of well-being, inspiration and comfort. The main advantages of good environments are reduced investment in upgrading facilities, reduced sickness absence, an optimum level of productivity and improved comfort levels. Individuals respond very differently to their environments, and research suggests a correlation between worker productivity, well-being, environmental, social and organisational factors.

Research shows occupants who report a high level of dissatisfaction about their job are usually the people who suffer more work and office environment related illnesses which affect their well-being, but not always so. Well-being expresses overall satisfaction. There is a connection between dissatisfied staff and low productivity; and a good sense of well-being is very important as it can lead to substantial productivity gain (Clements-Croome 2000). If the environment is particularly poor, people will be dissatisfied irrespective of job satisfaction.

Health is the outcome of a complex interaction between the physiological, psychological, personal and organisational resources available to individuals and the stress placed upon them by their physical and social environments, work and home life. A deficiency in any area increases stress and decreases human performance. Weiss (1997) suggests that the mind can affect the immune system. Stress can decrease the body's defences and increase the likelihood of illness, resulting in lowering of well-being. Stress arises from a variety of sources: *the organisation, the job, the person* and *the physical environmental conditions.* It can affect the mind and body, weaken the immune system and leave the body more vulnerable to environmental conditions. In biological terms, the hypothalamus reacts to stress by releasing the hormone ACTH; then the hormone cortisol in the blood increases to a damaging level possibly affecting the brain cells involved in memory. This chain of events interferes with human performance, and productivity falls as a consequence.

1.5 INDOOR ENVIRONMENT AND PRODUCTIVITY

Typical mental health factors related to a job include: stress (which may cause physical symptoms or emotional and psychological difficulties); dissatisfaction with the job itself or with organisational design and structure; unhappiness with achievement and growth; problems with personal relationships; and overall dissatisfaction with the physical indoor environment (Clements-Croome and Li, 2000).

Traditionally thermal comfort has been emphasised as being necessary in buildings but is comfort compatible with health and well-being? The mind and body need to be in a state of health and well-being for work and concentration, which are a prime prerequisite for productivity. Good productivity brings a sense of achievement for the individual as well as increased profits for the work organisation. The holistic nature of our existence has been neglected because knowledge acquisition by the classical scientific method has dominated research and is controlled but limited in application, whereas the world of reality is uncontrolled, subjective and anecdotal but nevertheless is vitally important if we are to understand systems behaviour. It is possible to reconsider comfort in terms of the quality of indoor environment and employee productivity.

Dorgan (1994) has analysed some 50,000 offices in the USA. He found (i) 20% were *Healthy Buildings* (always met ASHRAE Standards 62-1989 and 55-1992 during occupied periods); (ii) 40% *Generally Healthy Buildings* (meet ASHRAE Standards 62-1989 and 55-1992 during most occupied periods); (iii) 20% were *Unhealthy Buildings* (fail to meet ASHRAE Standards 62-1989 and 55-1992 during most occupied period); (iv) 20% were *Buildings with Positive SBS* in which more than 20% of occupants complain of more than two SBS symptoms, and frequently 6 of the more common 18 SBS symptoms.

Health is the outcome of a complex interaction between the physiological, personal and organisational resources available to the individual and the stress placed upon them by their physical environment, work, and home life. Symptoms occur when the stress on a person exceeds the ability to cope and where resources and stress both vary with time so that it is difficult to predict outcomes from single causes. Sickness building syndrome is more likely with warmer room conditions and this can lead to decreased productivity. Higher temperatures also mean wasteful energy consumption. When the temperature reaches uncomfortable levels, output is reduced. On the other hand, output improves when high temperatures are reduced by air-conditioning. When temperatures are either too high or too low, error rates and accident rates increase. While most people maintain high productivity for a short time under adverse environmental conditions, there is a temperature threshold beyond which productivity rapidly decreases. Mackworth (1946) stated that overall the average number of errors made per subject per hour increased at higher temperatures and showed that the average number of mistakes per subject per hour under the various conditions of heat and high humidity was increased at higher temperatures especially above 32°C.

Vernon (1936) demonstrated relative accident frequencies for British munitions plant workers at different temperatures; the accident frequency was a minimum at about 20°C in three munition factories.

Pepler (1963) showed that individual productivity fell as indoor air temperature increased. His work shows that variations in productivity in a non-airconditioned mill were influenced by temperature changes, although absenteeism was apparently not related to the thermal conditioning. On average an 8% productivity increase occurred with a 5°C decrease in temperature (Pepler, 1963).

Factors, such as poor lighting, both natural and artificial, poorly maintained or designed air conditioning, and poor spatial layouts are all likely to affect performance at work. This may be evidenced by lower performance. In a survey of 480 UK offices occupiers, Richard Ellis (1994) states that 96% were convinced that the design of a building affects productivity and when asked an open ended question on how aspects of design tend to lead to this effect; 43% used words such as attractive, good visual stimulus, colours and windows; 41% mentioned good morale, 'feel good' factor and contented happy staff; 19% said more comfortable, relaxing, restful conditions to work; 16% said increases in motivation and productivity; 15% said improved communications; 3% or less said reduced stress. All these aspects help to promote a well designed building. The importance of various factors are summarised in the following table and it can be seen that natural daylight and ventilation are rated highly, but green issues and the use of atria are also significant.

Table 1.1 Importance ratings of various design factors (Ellis, 1994)

Feature	Very	Quite	Not very	Not-at-all
Best use of natural daylight	57%	31%	10%	1%
Ventilation using windows	30%	41%	25%	3%
Thermal design for building	12%	40%	36%	6%
Energy-saving green design	15%	36%	37%	8%
use of atria & glazed streets	4%	20%	52%	18%

Clearly any building that does not maximise its natural daylighting is likely to be unpopular with office occupiers. The high value attributed to the use of windows rather than air-conditioning partly reflects the generally low level of effectiveness achieved by air-condtioning in many buildings, but also more fundamentally the inherent need for natural light and good views out of the building.

Wilkins (1993) reports that good lighting design practice, particularly the use of daylight, can improve health without compromising efficiency. Concerns about the detrimental effects of uneven spectral power distribution and low-frequency magnetic fields are not as yet substantiated. Wilkins states that several aspects of lighting may affect health, including (i) low-frequency magnetic fields; (ii) ultra-violet emissions; (iii) glare; and (iv) variation in luminous intensity; and (v) flicker frequency. The effects of low-frequency magnetic fields on human health are uncertain. The epidemiological evidence of a possible contribution to certain cancers cannot now be ignored, but neither can it be regarded as conclusive but the effect of electromagnetic fields on cognitive functions may be significant.

The ultraviolet light from daylight exceeds that from most sources of artificial light. Its role in diseases of the eye is controversial, but its effects on skin have been relatively well documented.

The luminous intensity of a light source, the angle it subtends at the eye, and its position in the observer's field combine to determine the extent to which the source will induce a sensation of discomfort or impair vision. Glare can occur from the use of some lower intensity sources, such as the small, low-voltage, tungsten-halogen lamps. It is reasonable to suppose that in the long-term, glare can have secondary effects on health and that is visibly flickering can have profound effects on the human nervous system. At frequencies below about 60 Hz it can trigger epileptic seizures in those who are susceptible. In others it can cause headaches and eyestrain. Wilkins concludes that the trend towards brighter high-efficiency sources is unlikely to affect health adversely, and may indeed be advantageous. The trend could have negative consequences for health were it to be shown that the increasing levels of ambient light at night affect circadian rhythms. Improvements in brightness and the evenness of spectral power may be beneficial. In particular, the move towards a greater use of daylight is likely to be good for both health and efficiency.

In many buildings, users report most dissatisfaction with temperature and ventilation, while noise, lighting and smoking feature less strongly. The causes lie in the way temperature and ventilation can be affected by changes at all levels in the building hierarchy, and, most fundamentally, by changes to the shell and services. In comparison, noise, lighting and smoking are affected mainly by changes to internal lay out and work station arrangements which can often be partly controlled by users.

There are some indications that giving occupants greater local control over their environmental conditions improves their work performance and their work commitment and morale which all have positive implications for improving overall productivity within an organisation. Building users are demanding more control at their workstations of fresh air, natural light, noise and smoke. Lack of control can be significantly related to the prevalence of ill health symptoms in the office environment, and there is widespread agreement that providing more individual control is beneficial. Work by Burge *et al.* (1987) demonstrates the relationship between self-reports of productivity and levels of control over temperature, ventilation and lighting.

Intervention to ensure a healthy working environment should always be the first step towards improving productivity. There are very large individual differences in the tolerance of sub-optimal thermal and environmental conditions.

Even if the average level of a given environmental parameter is appropriate for the average worker, large decrements in productivity may still be taking place among the least tolerant. Environmental changes which permit more individual adjustment will reduce this problem. Productivity is probably reduced more when large numbers work at reduced efficiency than when a few hypersensitive individuals are on sick leave. Wyon (1993) states that commonly occurring thermal conditions, within the 80% thermal comfort zone, can reduce key aspects of human efficiency such as reading, thinking logically and performing arithmetic, by 5-15%.

Lorsch and Abdou (1993) summarise the results of a survey undertaken for industry on the impact of the building indoor environment on occupant productivity, particularly with respect to temperature and indoor air quality. They also describe three large studies of office worker productivity with respect to environmental measurements, and discuss the relationship between productivity and building costs.

It is felt in general that improving the work environment increases productivity. Any quantitative proof of this statement is sparse and controversial. There are a number of interacting factors which affect productivity, including privacy, communications, social relationships, office system organisation, management, as well as environmental issues. It is a much higher cost to employ people who work than it is to maintain and operate the building, hence spending money on improving the work environment may be the most cost effective way of improving productivity. In other words if more money is spent on design, construction and maintenance and even if this results in only small decreased absenteeism rates or increased concentration in the workplace, then the increase in investment is high cost effective (Clements-Croome, 2000).

In one case study reported by Lorsch and Abdou (1994b) it is not clear if the drop in productivity was due to a reduction in comfort, by the loss of individual control of frustration due to being inconvenienced. According to Pepler and Warner (1968), people work best when they are slightly cool, but perhaps not sufficiently so to be termed discomfort, and should not be cool for too long.

Lorsch and Abdou (1994b) conclude that temperatures which provide optimum comfort may not necessarily give rise to maximum efficiency in terms of work output. The difficulty here is that this may be true for relatively short periods of time, but if a person is feeling uncomfortable over a long period of time it may lead to a decrement in work performance. However, there is a need for more research in this area. It almost seems that for optimum work performance a keen sharp environment is needed which fluctuates between comfort and slight cool discomfort.

According to a report by the National Electrical Manufacturers Association in Washington (1989) increased productivity occurs when people can perform tasks more accurately and quickly over a long period of time. It also means people can learn more effectively and work more creatively, and hence sustain stress more effectively. Ability to work together harmoniously, or cope with unforeseen circumstances, all point towards people feeling healthy, having a sense of well-being, high morale and being able to accept more responsibility. In general people will respond to work situations more positively. At an ASHRAE Workshop on Indoor Quality held at Baltimore in September 1992 the following productivity measures were recommended as being significant.

- Absence from work or workstation. Health costs including sick leave, accidents and injuries.

- Interruptions to work.
- Controlled independent judgements of work quality.
- Self assessments of productivity.
- Speed and accuracy of work.
- Output from pre-existing work groups.
- Cost for the product or service.
- Exchanging output in response to graded reward.
- Volunteer overtime.
- Cycle time from initiation to completion of process.
- Multiple measures at all organisational levels.
- Visual measures of performance, health and well-being at work.
- Development of measures and patterns of change over time.

Rosenfeld (1989) describes that when airconditioning was first conceived it was expected that the initial cost of the system would be recovered by an increased volume of business. He quotes an example where the initial cost of the airconditioning system for an office building is about £100 per m^2, so that if the average salary is £3,000 per m^2 and there is an occupancy of 10 m^2 per person, then adding 10% to the cost of the system is justified if it increases productivity by as little as 0.33%. Such small differences are difficult to measure in practice. Rosenfeld shows the relationship between the savings in working hours and the incremental initial cost of the system for a range of salaries.

Rosenfeld (1989) shows that improvement in indoor air quality can be more than offset by modest increases in productivity. This leads to the conclusion that in general, high quality systems which will have higher capital costs can generate a high rate of return in terms of productivity. In addition systems will be efficient, be effective, have low energy consumptions and consequently achieve healthier working environments in buildings with a low CO_2 emission.

Holcomb and Pedelty (1994) attempt to quantify the costs of potential savings that may accrue by improving the ventilation system. The increase in cost can be offset by the gain in productivity resulting from an increase in employee work time. Higher ventilation rates generally result in improved indoor air quality. Collins reported that over half of all acute health conditions were caused by respiratory conditions due to poor air quality. Cyfracki (1990) reported that a productivity increase of 0.125% would be sufficient to offset the costs of improved ventilation. It should be mentioned again that some studies have shown a decrease in SBS symptoms with increased ventilation rates while others have not conclude that although there is some inconsistency there is still sufficient evidence to suggest that there is an association between ventilation rates, indoor air quality, sick building syndrome symptoms and employee productivity.

It is easier to assess the effects of temperature on physical performance, but much more difficult to test the effects on mental performance. For example, the lowest industrial accident rate occurs around 20°C and rises significantly above or below this temperature. The other problem is the interaction with other factors which contribute towards the productivity. Motivated workers can sustain high levels of productivity even under adverse environmental conditions for a length of time which will depend on the individual.

Lorsch and Abdou (1994c) analyse several independent surveys which show that when office workers find the work space environment comfortable, productivity tends to increase when air-conditioning is introduced by as much as 5-15%, in the opinion of some managers and researchers. These are however, only general trends and there is little hard data and some findings are contradictory. Kobrick and Fine (1983) conclude that it is difficult to predict the capabilities of groups of people, never mind individuals, in performing different tasks under given sets of climatic conditions.

A study for the Westinghouse Furniture Systems Company in Buffalo, New York, entitled The Impact of the Office Environment on Productivity and .the Quality of Working Life suggested that the physical environment for office work might count for a 5-15% variation in employee's productivity. And the general conclusion was that people would do more work on an average workday if they are physically comfortable.

Woods (1987) reported that satisfaction and productivity vary with the type of heating, ventilation or air-conditioning system. Central systems appeared to be more satisfactory than local ones, the most important factor being whether there was cooling or not. In one study on user controlled environmental systems by Drake, the ability to have local control was important in maintaining or improving job satisfaction, work performance and group productivity, while reducing distractions from work. For example, some users reported that they wasted less time taking informal breaks compared to times when environmental conditions were uncomfortable. They were also able to concentrate more intensively on their work. The gain in-group productivity from the user controlled environmental system amounted to 9%. A number of studies suggest that a small degree of discomfort is acceptable, but it has to be confined to a point where it does not become a distraction.

Work by Kamon ((1978) and others shows that heat can cause lethargy which can increase the rate of accidents, and also affect productivity. Bedford (1949) concluded that there was a close relationship between the external temperature and the output of workers. Deteriorating performance is partially contributable to insomnia due to heat. Schweisheimer (1962) carried out some surveys concerned with establishing the effect of air-conditioning on productivity at a leather factory in Massachusetts, an electrical manufacturing company in Chicago and a manufacturing company in Pennsylvania. In all cases after the installation of air-conditioning the production increased by between 3-8.5% during the Summer months. On the basis of these investigations Schwisheimer concluded that the average performance of workers dropped by 10% at an internal room temperature of 30°C, by 22% at 32°C and by 38% at 35°C. Konz and Gupta (1969) investigated the effects of local cooling of the head on mental performance in hot working environments. The subject had to create words in ten minutes from one of two sets of 8 letters, which were type printed on a blank form. Poor conditions without cooling resulted in the creation of works dropping by some 20% in the hot condition, whereas with cooling the reduction was only some 12%.

Abdou and Lorsch (1994) studied the effects of indoor air quality on productivity. It was concluded that productivity in the office environment is sensitive to conditions leading to poor indoor air quality and this is linked to sick building syndrome. It is recognised that any stress is also influenced by management and other factors in the workplace. Occupants having local control over their environment generally have an improvement in their work effort, but in a more general way there is a synergistic effect of a multitude of factors which effect the physical and mental performance of people. Abdou and Lorsch (1994) conclude that in many case studies occupants have been highly dissatisfied with their environment, even though measurements have indicated that current standards were being met. This highlights the need to review standards and the basis on which they are made. Exactly the same conclusion is made by Donnini *et al.* (1994).

Although, it is difficult to collect hard data which would give a precise relationship between the various individual environmental factors and productivity there is sufficient evidence to show that improved environment decreases peoples complaints and absenteeism, thus indirectly enhancing productivity. The assessment of problems at the work place, using complaints is unreliable, because there is little mention of issues that are working well, and also the complaints may be attributable to other entirely different factors. Abdou and Lorsch (1994) contend that the productivity of 20% of the office work force in the USA could be increased simply, by improving the air quality of offices, and this is worth approximately $60 billion per year.

Work by Vernon (1936) shows that there is a clear relationship between absenteeism and the average ventilation grading for a space, which was judged by the amount of windows on various walls, so that windows on 3 sides had the highest grading and windows on one side only had the lowest. Abdou and Lorsch (1994) give the following causes as being the principal ones contributing to sick building syndrome:

- Building occupancy higher than intended.
- Low efficiency of ventilation.
- Renovation using the wrong materials.
- Low level of facilities management.
- Condensation or water leakage.
- Low morale and lack of recognition.

In this case lower efficiency of ventilation means that the supply air is not reaching the space where the occupants are, hence the nose is breathing in recirculated stale air. It is important to realise that even if the design criteria are correct for ventilation, the complete design team are responsible for ensuring that the systems can be easily maintained; the owner and the facilities manager also need to ensure that maintenance is carried out effectively. The tenant and occupants should use the building as intended. When new pollutant sources are introduced, such as new materials or a higher occupancy density, then the ventilation will become inadequate.

Burge (1987) conducted a study of building sickness among 4373 office workers in 42 office buildings having 47 different ventilation conditions in the United Kingdom. The data was further analysed by Raw (1990). The principal conclusions were that as individuals reported more than two symptoms, the subjects reported a decrease in productivity; none of the best buildings in this survey were airconditioned and these had fewer than two symptoms per worker on average, whereas the best airconditioned buildings had between two and three symptoms; women recorded more symptoms than men, but there was no overall difference in productivity; individual control of the environment has a positive effect on productivity; the productivity is increased by perceived air quality; productivity, however, only increases with perceived humidity up to a certain point and then appears to decrease again. Evidence supporting the importance of individual control of environment is again provided by Preller (1990). It should be said that some contrary evidence exists concerning some of these factors, which points to the need for a systems approach to studying the effects of environment in buildings such as that proposed by Jones (1995).

Productivity can be related to quality and satisfaction of the service or functional performance. Studies have shown that productivity at work bears a close relationship to the work environment. Burge demonstrates that there is a strong relationship between self-reports of productivity and ill health symptoms related to buildings: productivity decreases as ill health symptoms increase. There is a slightly less marked trend relating productivity and air quality but there is a significant effect.

Dorgan (1994) defines productivity as the increased functional and organisational output including quality. This increase can be the result of direct measurable decreases in absenteeism, decreases in employees leaving work early; or reductions of extra long breaks and lunches. The increase can also be the result of an increase in the quantity and quality of production while employees were active; improved indoor air quality is an important consideration in this respect. There is general agreement that improved working conditions, and the office environment is certainly one of the more important working conditions, tend to increase productivity. However, determining a quantitative relationship between environment and productivity proves to be highly controversial. While some researchers claimed they reliably measured improvements of 10% or more, others present data showing that no such relationships exist. Since the cost of the people in an office is an order-of magnitude higher than the cost of maintaining and operating the building, spending money on improving the work-environment may be the most cost-effective way to improve worker productivity.

In 1994, the energy use in an average commercial office building in the United States costs approximately $20/m^2/year, whereas the functional cost is approximately $3,000/m^2/year. The functional cost includes the salaries of employees, the retail sales in a store, or the equivalent production value of a hotel, hospital, or school. This means that 1% gain in productivity ($30/m^2/year) has a larger economic benefit than a 100% reduction ($20/m^2/year) in energy usage. In addition, the productivity gains will increase the benefits such as repeat business in hotels, faster recovery times in hospitals, and attainment of better jobs due to a better education in schools. A small gain in worker productivity has major economic impacts and it makes sense to invest in improving the indoor environment to achieve productivity benefits. Dorgan (1994) states that productivity gains of 1.5% in generally healthy buildings, and 6% in sick buildings, can be easily achieved. As typical payback costs of improvement in the indoor air ranges from less than 9 months to 2 years, the benefits clearly offset the renewal cost resulting in a very favourable cost-to-benefit ratio. The 1.5% improvement is conservative. Some literature indicates that this may be as high as 5 to 10%. However, achieving such productivity gains may require using advanced active or passive environmental control as well as personal controls. Examples of productivity gains in the order of 1-3% are found in several studies. Informal (unpublished) and anecdotal reports on productivity gains have been performed in supermarkets, fast food outlets, retail department store, schools, and office buildings resulting in estimated gains in sales ranging from 4 to 15% in retail stores during summer months .

By focusing on the productivity benefits, projects which improve the indoor environment are moved away from an energy-saving viewpoint and more towards productivity-increase issues. Even if a proposed project improves the indoor environment but increases the energy cost by 5%, the project may still be economically feasible if the productivity increase is greater than 0.04%. Wyon states that the *leverage* of environmental improvements on productivity is such that a 50% increase in energy costs of improved ventilation would be paid for by a gain of only 0.25-0.5% in productivity, and capital investments of $50/m^2 would be paid for by a gain of only 0.5% in productivity. The pay-back time for improved ventilation is estimated to be as low as 1.6 years on average, and to be well under 1 year for buildings with ventilation that is below currently recommended standards.

An increase in productivity can be achieved with either (i) no increase in energy usage or even a decrease (ii) with an increase in funding for the given level of technology. The use of energy recovery systems, and the increased use of such technologies as advanced filtration, dehumidification, thermal storage, natural energy and personal environmental control systems are all examples of energy improving technologies the cost of which can be offset by increase in productivity. The increased building services budget can allow for the introduction of the best system, not the cheapest. Any indoor environment productivity management program should be able to include reducing energy consumption as one of the design objectives. Improving indoor environment will provide a high return on investment through productivity gains, health saving and reduced energy use. The benefits of improved indoor environment are improved productivity, increased profits, greater employee-customer-visitors health and satisfaction, and reduced health costs. The potential productivity benefits of improved indoor environment are so large that this opportunity cannot be ignored. There are indirect long-term, and social benefits.

Hawkins (2001) states that productivity in the USA, in terms of GDP per worker, is currently 1.4 times greater than the UK. Massive investment in information and communications technologies is seen as the main driver in the surge in US productivity. Indeed, recent studies have shown that the UK is only

15th in an international league table of worker productivity. For 2000 UK construction was valued at £69.5 billion; of this the wage bill for construction project personnel amounted to £17 billion. Each 1% global improvement in workforce performance would therefore generate a saving of £170 million in labour costs and enable construction programme times to be reduced.

Fisk (2000) reports that in the US respiratory illnesses cause the loss of about 176 million workdays and the equivalent of 121 million days of substantially restricted activity. Fisk (1999) and Clements-Croome (2000) state that in office buildings, the salaries of workers exceed the building energy and maintenance costs and the annual construction rental costs by a factor of at least 25. This means that small increases in productivity, of 1% or less, are sufficient to justify additional capital expenditure to improve the quality of the building's services. Ultimately, this will result in healthier working environments as well as reduced energy and maintenance costs.

People spend about 90% of their lives in buildings, so the internal environment has to be designed to limit the possibilities of infectious disease; allergies and asthma; and building related health symptoms, referred to as sick building syndrome symptoms. Buildings should provide a multi-sensory experience, and therefore anything in the environment which blocks or disturbs the sensory systems in an unsatisfactory way will affect health and work performance. Thus, lighting, sound, air quality and thermal climate are all conditions around us that affect our overall perception of the environment. Air quality is a major issue because it only takes about four seconds for air to be inhaled and for its effect to be transmitted to the bloodstream and hence the brain. Clean, fresh air is vital for clear thinking, but it is not the only issue to be considered. Fisk (1999) discusses linkages between infectious disease transmission, respiratory illnesses, allergies and asthma, sick building syndrome symptoms, thermal environment, lighting and odours. He concludes that in the USA the total annual cost of respiratory infections is about $70bn, for allergies and asthma $15bn, and reckons that a 20-50% reduction in sick building syndrome symptoms corresponds to an annual productivity increase of $15-38bn. The linkage between odour and scents and work performance is less understood, but Fisk (1999) concludes that the literature provides substantial evidence that some odours can affect some aspects of cognitive performance. He refers to work by Rotton (1983), Dember *et al.* (1995), Knasko (1993), Baron (1990) and Ludvigson (1989). The application of scents has been used by the Kajima Corporation in their Tokyo office building, as reported by Takenoya in Clements-Croome (2000). Fisk (1999) goes on to consider the direct linkage between human performance and environmental conditions and writes that for US office workers there is a potential annual productivity gain of $20-200bn. His conclusions are that there is relatively strong evidence that characteristics of buildings and indoor environments significantly influence the occurrence of respiratory disease, allergy and asthma symptoms, sick building syndrome and worker performance. Langston and Ding (2001) briefly describe a number of case studies which demonstrate how the environment can improve worker productivity.

There are only crude estimates of the magnitude of productivity gains from improvements of indoor environments. Those for the US office workforce are shown in Table 1.2.

Productivity depends on four cardinal factors: *personal*, *social*, *organisational* and *environmental*. There are preferred environmental settings which decrease dissatisfaction and absenteeism, thus indirectly enhancing productivity. The assessment of problems at the workplace based on numbers of complaints is unreliable, because there is little mention of positive aspects and because complaints may be attributable to other, entirely different factors.

In the Summer 1997, the Journal of the British Council for Offices, entitled Office, reckoned that *advanced building intelligence* can increase the productivity of occupants by 10% annually and improve efficiency to satisfy owner-occupiers. In contrast, *standard building intelligence* can improve efficiency by 8% annually and improve efficiency to result in a payback within two to four years. The argument is that in an intelligent building there is less illness and absenteeism.

Table 1.2 Estimated potential productivity gains from improvements in indoor environments (Fisk, 1999)

Source of Productivity Gain	Potential Annual Health Benefits	Potential U.S. Annual Savings or Productivity Gain (1996 $U.S.)
Reduced respiratory disease	16 to 37 million avoided cases of common cold or influenza	$6 - $14 billion
Reduced allergies and asthma	10% to 30% decrease in symptoms within 53 million allergy sufferers and 16 million asthmatics	$2 - $4 billion
Reduced sick building syndrome symptoms	20% to 50% reduction in Sick building syndrome health symptoms experienced frequently at work by approximately 15 million workers	$15 - $38 billion
Improved worker performance from changes in thermal environment and lighting	Not applicable	$20 - $200 billion

Productivity depends on good concentration, technical competence, effective organisation and management, a responsive environment and a good sense of well-being. The economic assessment of environment, both in terms of health (medical treatment, hospitalisation) and of decreases in productivity has received very little attention by researchers as yet. However, this assessment is absolutely necessary in order to assess the effectiveness of improved design and management protocols (Barbatano, 1994). Until now there have not been any standard procedures to measure productivity. Thus it has been difficult to persuade clients to accept the concept of a relationship between economic productivity benefits and indoor environment. The challenge is to investigate productivity and develop a strong methodology to assess the link between indoor environment and productivity.

Several methods of performance measurement have been reported in the published literature. For example, Ilgen (1991) classified the methods of performance measurement into three categories: *physiological, objective* and *subjective.* The rational for using physiological methods is based on the reasoning that physiological measures of activation or arousal are associated with increased activity in the nervous system, which is equated with an increase in stress on the operator. However, physiological measures of workload have received wide criticism regarding their validity, the sensitivity of measures to contamination and the intrusive nature of the measures themselves.

Objective measures (O'Donnell and Eggemeier, 1986) are frequently used to infer the size of workloads, both mental and physical. A further class of workload measures comprises subjective measures (Cyfracki, 1990). Subjective measures of workload are applied to gain access to the subjects' perceptions of the level of load they face in task performance. Rating scales, questionnaires, and interviews are used to collect opinions about the workload. While these methods may not have the empirical or quantitative appeal of physiological or objective measures, it is often argued that subjective measures are more appropriate and realistic since individuals are likely to work in accordance with their feelings, regardless of what physiological or behavioural performance measures suggest. There needs however to be a distinction between *being busy* and being *work effective* as assessed by quality indicators. Working long hours for example does not guarantee high productivity.

1.6 CONCLUSIONS

We live through our senses and the environment we provide for them to interact with is important. A building and its environment can help people produce better work, because they are happier and more satisfied when their minds are concentrated on the job at hand; building design can help achieve this. At low and high arousal or alertness levels, the capacity for performing work is low; at the optimum level the individual can concentrate on work while being aware of peripheral stimuli from the physical environment. Different work requires different environmental settings to achieve an optimum level of arousal. It is necessary to assess if a sharper or leaner indoor environment is required for the occupants' good health and high productivity and to redefine comfort in terms of *well-being.* There are three current standards providing guidance for the assessment of occupant comfort: ASHRAE standard 55-92; ASHRAE standard 62-89; and ISO standard 7730 (ISO, 1984). All emphasise thermal comfort rather than overall sensory comfort.

Kline (1999) believes it is important to design *thinking environments.* The most important aspect is to have places for work *where people feel that they matter.*

When that becomes a guideline for architectural interior design she argues a very different place emerges, than when some abstract standard of opulence and furnishing for pure functionality is adopted as a guideline. Building regulations and codes to practice are only a basic foundation for providing health and safety in the workplace; they do not guarantee producing an environment which is conducive to well-being and this includes feelings and emotions.

Froggatt (2001) enunciates eight principles for workplace design:

- the *initiative* to explore remote and mobile work strategies
- *trust* employers to work out of sight of management
- encourage *joy* in the workplace (Cabanac, 2000)
- value *individuality*
- emphasise *equality* more than *hierarchy*
- engage in open honest *dialogue*
- epitomise *cognitivity* between all the stakeholders in the business
- provide access to a wide range of *workplace options.*

Froggatt (2001) states that:-

The physical infra-structure of workplaces and the technological infra-structure of cyberspace are both critical elements of the new world of virtual work. One will not cause the demise of the other; rather, they re-enforce the need for each other. Knowledge work will continue to be a combination of solo contribution and collaborative team effort. While these activities can happen anywhere, they do have to happen somewhere. People will still need places where they can gather for face-face interaction, places where they can share resources, and places for solo work.

Offices of the future will be thought of as organisms which are developing in response to changes in technology and ways of working. It is important that office spaces allow people to work in teams, but at the same time bond to individual needs for motivation to stimulate productivity. Some examples of futuristic offices are those of Cellular Operations Ltd., in Swindon; the Pittard Sullivan and the TWA Chiat Day Company buildings both in Los Angeles. These buildings have some common properties in that they are designed for people to enjoy working in them and have a happy experience. Common descriptors used for such buildings are: *views; flood of light; the sound and sight of water; perfumed air; use of colour and abstract forms;* and *fresh air.* Good environments can enrich the work experience. Stimulating environments can help people to think creatively and buildings have a role here because spaces have an emotional content (Farshchi, 2000). It is important that the built environment is designed to respect feelings of people as well as the functional aspects.

The Health and Safety Executive list the following causes of sick building syndrome (Turner, 2001):

- Poor building and office design
- Deep plan or open plan offices of more than ten workstations

- Large areas of soft furnishing, open shelving and filing
- New furniture, carpets and painted surfaces
- Poor building services and maintenance
- Poor air quality
- Lighting (particularly the type and positioning which causes high glare and flicker) and insufficient daylight
- Low level of user control over ventilation, heating and lighting
- Poor design and maintenance of building services
- Poor standards of general repair
- Insufficient or badly organised office cleaning
- High temperature or excessive variations in temperature during the day
- Very low (<40%) or high (>60%) humidity
- Chemical pollutants - tobacco smoke, ozone, volatile organic compounds from building materials and furnishings
- Dust particles and fibres in the atmosphere
- Job factors
- Routine clerical work
- Work with display screen equipment

Prevention is better than cure. Consideration of all these factors at the design stage can prevent these problems occurring later.

Working with display screen equipment deserves particular mention, it can give rise to body postural problems, repetitive strain injuries and eyestrain. There can also be effects from low frequency electromagnetic fields and this is the subject of this book.

1.7 REFERENCES

Abdou, O.A., Lorsch, H.G., 1994c, The Impact of the Building Indoor Environment on Occupant Productivity: Part 3: *Effects of Indoor Air Quality,* ASHRAE Trans.1, 100(2), pp 902-913.

Allwright, P., 1998, Basics of Buddhism (Taplow Press)

Anderson, D., *et al.*, 1990, Statistics for Business and Economics, 4th Edition, West Publishing Company, USA.

Anon, Mongolia Land of the Five Animals, Greater London Council.

Arnold, J., Cooper, C.L., Robertson, I.T., 1998, Understanding Human Behaviour in the Workplace, 3rd Edition, *Financial Times* – Pitman Publishing.

Attenborough, D., 1979, Life on Earth, Collins.

Baldry, C *et al.*, 1997, *Work Employment Society,* 11 (3), 519-539.

Ballast Wiltshier, 1999, Landscape of Change: Built Environment of the Digital Age, *Bartlett School of Architecture at University College,* London

Barbatano, L. *et al.*, 1994, *Proceedings of Healthy Air '94,* Italy, 395-403.

Baron, R.A., 1990, Environmentally Induced Positive Affect: Its Effect on Self-Efficacy Task, Performance, and Conflict, *Journal of Applied Social Psychology,* **20, 5**, 368-384.

Baron, R.A, 1990, *J. App. Soc. Psych.*, **16**, 16-28.

Bedford, T., 1949, *Air* Conditioning and the Health of the Industrial Worker, *Journal of the Institute of Heating and Ventilating Engineers* **17** (166), pp.112-146.

Berglund, B., Gunnarsson, A.G., 2000, Relationships between Occupant Personality and the Sick Building Syndrome Explored, *Indoor Air*, **10**, 152-169.

Binhi, V.N., 2002, Magnetobiology, Academic Press.

Bochsler, K., 1994 Computerised Facilities Management, *Sulzer Technical Review* **76**, (**3**), 14-23.

Bordass, W.T., 1995, Comfort, Control and Energy Efficiency in Offices, *BRE Information Paper* IP3/95.

Bosti, 1982 The Impact of the Office Environment on Productivity and the Quality of Working Life, *Buffalo*, N.Y: Westinghouse Furniture Systems.

Briggs, J., 1992, Fractals; The Patterns of Chaos, Thames and Hudson.

Brune, D., *et al.*, 2001, Radiation at Home, Outdoors and in the Workplace, (Scandinavian Science Publisher).

Bundy, A., 1988, Artificial Intelligence: Art or Science? *Royal Society of Arts Journal*, July, p 557-569.

Burge, S.A., *et al.*, 1987, Sick Building Syndrome: A Study of 4373 Office Workers, *Annals of Occupational Hygiene*, **31**, pp. 493-504.

Burt, T.S., 1997, Thesis for Licentiate of Engineering, Institute for Uppvarmnings-Och Ventilationsteknik, ISSN 1100-8997. Also PhD Thesis, Royal Institute of Technology, Stockholm.

Busigiani, A., 1972, Gröpius, Hamlyn.

Chalmers, D.J., 1996 The Conscious Mind; *In Search of a Fundamental Theory* (Oxford University Press).

CIB Working Group W98, 1997, *Proceedings of Second International Congress on International Buildings*, Tel Aviv, 4-6 March, Edited A Lustig. (ISSN 0793 2448).

Clements-Croome, D.J., 2000, *Creating the Productive Workplace*, Editor and Author, published by E & FN Spon.

Clements-Croome, D.J., Li, B., 2000, *International Conference on Healthy Buildings 2000*, August 6-10, Helsinki.

Clements-Croome, D.J., 2000a, Future Horizons for Construction, *Proceedings of Conference Technology Watch and Innovation in the Construction Industry*, Brussells, April 5-6.

Cohen, J., Cohen, P., 1983, *Applied Multiple Regression - Correlation Analysis for Behaviourial Sciences*, 2nd Edition, Lawrence Earle Baum Associates, New Jersey/London.

Collins, J.G., 1989, Health characteristics by occupation and industry: United States, 1983-1985. *Vital Health and Statistics* **10(170)**. Hyattsville, MD: National Center for Health Statistics.

Construction Industry Board, 1999, *Construction: A 2020 Vision*.

Conti, F., 1978, *In Architecture as Environment*, (HBJ Press Inc.).

Cook, P., 1985, *Richard Rogers Plus Architects*, Academy Editions.

Cooper, C.L., 1988, *Occupational Stress Indicator Management Guide,* NFER-Nelson,

Cresti, C., 1970, *Le Corbusier,* Hamlyn. Second Ed.

Crick, F., Koch, C., 1997 *The* Problem of Consciousness, Scientific American Special Issue, January *(Mysteries of the Mind),* 19-26.

Croome, D.J., 1990 - The Invisible Architecture, *Building Services Engineering,* Building Serv. Eng. Res. Technol., **11, (1),** 27-31.

Croome, D.J., 1990 - The Invisible Architecture, *Building Services Engineering,* Building Serv. Eng. Res. Technol., **11, (1),** 27-31.

Croome, D.J., 1985, Covered Northern Township, *International Journal Ambient Energy,* **6, (4),** 171-186.

Croome, D.J., 1985, Services: Internal Environment of Buildings, *Proceedings, Symposium Building Appraisal, Maintenance and Preservation,* July 10-12, Bath University 234-257.

Croome, D.J., Roberts, B.M., 1981, *Airconditioning & Ventilation of Buildings,* Second Edition, Pergamon.

Croome, D.J., 1977, *Noise, Buildings and People,* Pergamon

Cyfracki, L., 1990, Could upscale ventilation benefit building occupants and owners alike*? Proceedings of Indoor Air '90: 5th International Conference on Indoor Air Quality and Climate,* Vol.5, pp. 135-141. Aurora, ON: Inglewood Printing Plus.

Davies, M., 1987, the Richard Rogers Partnership, Design For the Information Age, *The World Teleport Association Conference.*

Dember, W.N., Warm JS, Parasuraman R., 1995. Olfactory stimulation and sustained attention. In: *Gilber* an editor. Compendium of Olfactory Research: explorations in aroma-chology: investigating the sense of smell and human response to odours. 1982-1994. Iowa: Kendall Hunt Pub.Co.p39-46.

Donnini, G., *et al.,* 1994, Office Thermal Environments and Occupant Perception of Comfort, *La Riforma Medica,* Vol.109, Supp **1, (2),** pp.257-263.

Dorgan, C.E. *et al.,* 1994 Productivity Link to the Indoor Environment Estimated Relative to ASHRAE 62-1989, *Proceedings of Healthy Buildings'94,* Budapest, pp.461-472.

Drake, P., 1990, Summary of Findings from the Advanced Office Design Impact Assessments. *Report to Johnson Controls, Inc.* Milwaukee, WI.

Ehrlichman, H., Bastone, L., 1991, Odour Experience as an Affective State, *Report to the Fragrance Research Fund,* New York.

Fisk, W.J., 1999, Estimates of Potential Nationwide Productivity and Health Benefits from Better Indoor Environments: An update, Lawrence Berkeley National Laboratory Report, LBNL-42123 and also see *Indoor Air Quality Handbook,* Edited by Spengler et al., (McGraw-Hill).

Fisk, W.J., 2000, Review of Health and Productivity Gains from Better IEQ, Proceedings of Healthy Buildings 2000, Helsinki, Vol 4, Pages 24-33.

Fitch, J.M., 1976, *American Building: The Environmental Forces That Shape It,* Schocken Books.

Frank, K.A., and Lepori, R.D., *Architecture Inside Out,* published by Wiley-Academy, 2000. ISBN 0471 984 663.

Frisch, K. Von, 1975, *Animal Architecture,* Hutchinson.

Gardner, H., 1997, *Extraordinary Minds, Weidenfeld and Nicolson: London.*

Gleick, J., 1988, *Chaos,* Heinemann

Greenfield, S., 1997 How might the Brain Generate Consciousness? Communication Cognition, Vol.30, **3-4**, 285-300.

Guidoni, E., 1975, *Primitive Architecture,* Faber & Faber.

Habraken, N.J., *The Structure of the Ordinary,* MIT Press, 1998.

Haghighat, F., Donnini, G., 1999, Impact of Psycho-social Factors on Perception of the Indoor Air Environment Studies in 12 Office Buildings, *Building and Env.,* 34, 479-503.

Hartmann, J., 1988, Thermal and Solar Optical Properties of Silica Aerogel for Use in Insulated Windows, Solar 87, *Twelfth Passive Solar Conference,* Portland, Oregan, July 11th-16th p42-47.

Henderson, M., 1997 Mental Athletes Tone their Bodies to Keep their Minds in Shape, *The Times,* August 19, page 6.

Health & Safety Executive, 1995, *How to Deal with Sick Building Syndrome,* ISBN 0 7176 0861 1

Hedge, A., 1994, Sick Building Syndrome: Is It An Environment Or A Psychological Phenomenon? *La Riforma Medica,* Vol. 109, Supp.1, (2), 9-21.

Holcomb, L.C., Pedelty, J.F., 1994, Comparison of Employee Upper Respiratory Absenteeism Costs with Costs Associated with Improved Ventilation, *ASHRAE Trans.,* **100** (2), 914-920.

Holl, S., Pallasmaa, J., Perez-Gomez ,A., 1994, *Questions of Perception.* (Special Issue by Architecture and Urbanism).

Humphreys, M.A., 1995, Chapter 1 in *Standards for Thermal Comfort,* edited by Nicol *et al.,* E & FN Spon.

IBE, 1992, *The Intelligent Building in Europe,* DEGW (London) and Teknibank (Milan) with the European Intelligent Building Group, Report for British Council for Offices.

Ilgen, D.R., Schneider, J., 1991, *International Review of Industrial and Organisational Psychology,* Vol.6, Chapter 3, 71-108, Edited by Cooper and Robertson, John Wiley & Sons Ltd.

Intelligent Building Institution, 1994, Washington, *High Tech High Touch Buildings,* 88, 2.

ISO, 1984, *Standard 7730: Conditions for Thermal Comfort,* Geneva, Switzerland: International Standards Organisation.

Jaakkola, J.K.J., Heinonen, O.P., and Seppänen, O., 1991, Mechanical Ventilation in Office Building Syndrome. An Experimental Epidemiological Study. *Indoor Air* 2, pp. 111-121.

Jamison, K.R., 1997, The Problem of Consciousness, *Scientific American Special Issue,* January, 44-49.

Jones, P., 1995, Health and Comforting Offices, *The Architects Journal,* June 8th, 33- 36.

Kamon, E., 1978, Physiological and Behavioural Responses to the Stresses of Desert Climate. *In Urban Planning for Arid Zones,* G., Golany, ed. New York: John Wiley and Sons.

Kell, A., 1996, Intelligent Buildings Now, *Electrotechnology*, October /November, 26-27.

Knasko, S.C., 1993, Performance Mood and Health During Exposure to Intermittent Odours, *Archives of Environmental Health*, 48, 5, 305-308.

Kobrick, J.L., and Fine, B.J., 1983, Climate and Human Performance. In *the Physical Environment at Work*, D.J. Osborne and M.M. Gruneberg, eds., pp.69-197. New York: John Wiley and Sons.

Konz, S., Gupta, V.K., 1969, Water-Cooled Hood Affects Creative Productivity. *ASHRAE Journal* 40:43.

Kroner, W.M., Stark-Martin, J.A., 1992, *Environmentally Responsive Work Stations and Office Worker Productivity*, Presented at Workshop on Environment and Productivity, June.

Kroner, W.M., 1988, *A New Frontier: Environments for Innovation, Proceedings International Symposium on Advanced Comfort Systems for the Work Environment*, May, Centre for Architectural Research, Rensselaer Polytechnic Institute, Troy, New York 12180.

Laing, A., *et al.*, 1998, *New Environments for Working*, E & F.N.Spon.

Lambot I., 1986, The New Headquarters for the Hong Kong and Shanghai Banking Corporation, Drake & Scull with Ian Lambot and The South China Printing Company.

Langen, G.M., 1987, Smart Walls: A Potential Kinetic Response Architecture, Solar 87, *Twelfth Passive Solar Conference Proceedings*, Portland, Oregon, July 11th-16th p538-541.

Langston, C.A., Ding, G.K., 2001, *Sustainable Practices in the Built Environment*, 2nd Edition (Butterworth-Heinemann).

Li, B., 1998, *Assessing the Influence of Indoor Environment on Productivity in Offices*, PhD Thesis, The University of Reading.

Lisbet, B., and Gunnar, L., 1994, Indoor Climate Complaints And Symptoms of the Sick Building Syndrome in Offices, *Proceedings of Healthy Buildings' 94*, Budapest, pp.391-396.

Littlefair, P., 1988, Innovative Daylighting Systems: A Critical Review, *National Lighting Conference*, p367-391

Lorsch, H.G., and Ossama, A.A., 1993, The Impact of the Building Indoor Environment on Occupant Productivity, *ASHRAE Transactions*, Vol.99, Part 2, USA.

Lorsch, H.G.H., and Abdou, O.A., 1994b, The Impact of the Building Indoor Environment on Occupant Productivity: Part II: Effects of Temperature, *ASHRAE Trans.* **100(2)**, pp.895-901.

Lorsch, H.G., Abdou, O.A., 1994c, *The Impact of the Building Indoor Environment on Occupant Productivity: Part 3: Effects of Indoor Air Quality*, ASHRAE Trans.2, **100(2)**, pp. 902-913

Ludvigson, H.W., Rottman TR. 1989. Effects of odours of lavender and cloves on cognition, memory, affect, and mood. *Chemical Senses* **14(4)**: 525-536.

Mackworth, N.H., 1946, Effects of Heat on Wireless Operators Hearing and Recording Morse Code Messages, *Bri. J.Ind.Med.* **3**: pp. 143-158.

Markus, T.A., 1967, The Role of Building Performance, Measurement and Appraisal in Design Method, Architects Journal **146**, 1567-1573

Mendell, M.J., and Smith, A.H., 1990, *Consistent Pattern of Elevated Symptoms in Airconditioned Office Buildings*, **80**, pp.1193-1199.

Merleau-Ponty, M., 1964, *Eye and Mind in Primary of Perception* (Evanston, Northwestern University Press).

Michell, G., 1978, Architecture of the Islamic World, Thames & Hudson

Moray, N., *et al.*, 1979, *Final Report of the Experimental Psychology Group, Mental Workload: Its Theory and Measurement*, edited by Moray N., New York: Plenum, pp.101-11463.

Myers, D.G., Diener, E., 1997 The Pursuit of Happiness, *Scientific American Special Issue*, January, (Mysteries of the Mind), 40-49.

Nema, 1989, Lighting and Human Performance: A Review. A report sponsored by the Lighting Equipment Division of the National Electrical Manufacturers Association, Washington, DC, and the Lighting Research Institute, New York.

Nouvel, J., 1988, Institut Du Monde Arabe, Champ Vallon.

O'Donell, R.D., Eggemeier, F.T., 1986 *Handbook of Perception and Human Performance: Cognitive Processes and Human Performance,* edited by Boff, K.R., *et al.* New York: Wiley.

Olgyay, V, 1963, Design with Climate, Princeton University Press.

Ornstein, R.E., 1973 *The Nature of Consciousness* (Viking).

Pallasmaa, J., 1996, *The Eyes of the Skin: Architecture and the Senses* (Academy Editions).

Pashler, H.E., 1998 *The Psychology of Attention* (MIT Press).

Pepler, R.D., 1963, Temperature: Its Measurement and Control. *in Science and Industry,* 3, ed.Hardy, 319 (Rheinhold).

Preller, L., *et al.*, 1990, Indoor Air '90, *Fifth International Conference on Indoor Air Quality and Climate 1:* pp. 227-230.

Piaget, J., 1980, *Language and Learning*, Routledge, Kegan, Paul.

Piano, Renzo, 1984, Projects and Buildings 1964-1983, The Architectural Press Ltd.

Pinker, S., 1997 *How the Mind Works* (Allen Lane: The Penguin Press).

Putnam, H., 1981, *Reason, Truth and History*, Cambridge University Press.

Raw, G.J., Roys, M.S., and Leaman, A., 1990, Further Findings From the Office Environment Survey: Productivity. *Indoor Air Quality '90, Fifth International Conference on Indoor Air Quality and Climate* **1**: pp.232-236.

Raw, G.J., 1995, A Questionnaire for Studies of Sick Building Syndrome, *BRE Report* br 287, ISBN 1 86081 0195.

Ricci-Bitti, P.E., Caterina, R., 1994, La Riforma Medica, Vol. 109, Supp **1**, **(2)**, pp.215-223.

Richard; J.M., *et al.*, 1985, *Hassan Fathy*, The Architectural Press Ltd.

Robertson *et al.*, Comparison of health problems related to work and environment measurements in two office buildings with different ventilation systems. *British Medical Journal*, 291, 373-376.

Rogers, R., 1988, Belief in the Future is Rooted in the Memory of the Past, *Royal Society of Arts Journal*, November, p875-884.

Rosenfeld, S., 1989, Worker productivity: Hidden HVAC Cost. *Heating/Piping Air Conditioning*, September pp.69-70.

Rotton, J., 1983, *Affected and Cognitive Consequences of Malodorous Pollution, Basic and Applied Psychology,* **4, 2,** 171-191.

Rudnai, P., *et al.,* 1994, *Proceedings of Healthy Buildings,* 1994, Budapest, **1,** 487-497.

Rudofski, B., 1964, *Architecture Without Architects,* Academy Editions

Saaty, T.L., 1972, *Analytic Hierarchy Process,* McGraw-Hill, New York.

Samimi, B.S., and Seltzer, J.M., 1992, Sick building syndrome due to insufficient and/or nonuniform fresh air supply in a multi storey office building. *IAQ '92: Environments for People,* pp 319-322. Atlanta American Society of Heating, Refrigerating and Airconditioning Engineers inc.

Schweisheimer, W., 1962, Does Air Conditioning Increase Productivity? Heating and Ventilating Engineer **35 (419):** pp.669.

Schweisheimer, W., 1966, *Erhaehung und Leistung und Produktion,* Warme Luftungs und Gesundheitstechnik, No., 278-265.

Sterling, E.M., and Sterling, T., 1983, The Impact of different ventilation levels and fluorescent lighting types on building illness: an experimental study. Canadian Journal of Public Health, **74,** pp.385-392.

Sundell, J., 1994, *On the Association Between Building Ventilation Characteristics, Some Indoor Environmental Exposures, Some Allergic Manifestations, and subjective symptom reports.* Indoor Air, Supp.2.

Sundstrom, E., 1987, Work Environments: Offices and Factories, *Handbook of Environmental Psychology,* Vol.1, New York: pp.733-783.

Steele, J., 1988, *Hassan Fathy,* Academy Editions.

Townsend, J., 1997, How to Draw Out all the Talents, *Independent Tabloid,* Thursday July 24, 17.

Turner, T., 2001, *The Phantom Menace,* Director, September, 90-94.

Vitruvius, M.V.P., 1960, *The Ten Books of Architecture,* tr. Morgan (Dover: New York).

Vernon, H.M., *et al.,* 1926, A Physiological Study of the Ventilation and Heating in Certain Factories. *Rep. Industry, Fatigue Res. Bd.,* No.35. London.

Vernon, H.M., *et al.,* 1930, A Study of Heating and Ventilation in Schools. *Rep. Industry. Health Res.Bd.,* No.35.London.

Vernon, H.M. 1936, *Accidents and Their Presentation,* Cambridge University Press.

Worthington, J., 1998, *Reinventing the Workplace,* Architectural Press.

Warr, P., 1998 What is our Current Understanding of the Relationships between Well-Being and Work, *Economics and Social Sciences Research Council Seminar Series at Department of Organizational Psychology,* Birkbeck College, London (R. Briner), September 22nd and Journal of Occ. Psychology 1990, **63,** 193-210

Warren, C., Warrenburg, S., 1993, Mood Benefits of Fragrance, *Perfumer and Flavourist,* **18,** March/April, 9-16.

Weiss, M.L., 1997, PhD Thesis, *Division of Behaviour and Cognitive Science,* Rochester University, New York.

Whitley, T.D.R., *et al.,* 1995a, Organisation and Job Factors in Sick Building Syndrome, *Proceedings of Healthy Buildings '95,* Milan, 11-14, Sept.

Whitley, T.D.R., *et al.*, 1995b, The Environment, Comfort and Productivity, *Proceedings of Healthy Buildings '95*, Milan, 11-14 Sept.

Wilkins, A.J., 1993, *Health and Efficiency in Lighting Practice*, Energy, Vol. 18, No.2. pp.123-129.

Windsor. C.L., 1988, *Occupational Stress Indicator Management Guide*, NFER-Nelson, Windsor.

Wilson, S., and Hedge, A., 1987, *The Office Environment Survey: A Study of Building Sickness*, Building Use Studies Ltd., London, U.K.

Woods, J.E., 1989, *State of the Art Reviews* **4(4)**, eds. Cone and Hodgson, and also *Problem Buildings: Buildings Associated Illness and the Sick Building Syndrome*, pp. 753-770, Philadelphia: Hanley & Belfus.

Woods, J.E., *et al.*, 1987, Office Worker Perceptions of Indoor Air Quality Effects on Discomfort and Performance. *Fourth International Conference on Indoor Air and Climate*, pp. 464-468, Berlin, Germany.

World Health Organisation, 1983, Indoor Air Pollutants Exposure and Health Effects, EURO Reports and Studies **78**, World Health Organisation.

Wyon, D.P., 1993, The Economic Benefits of a Healthy Indoor Environment, *Proceedings of Healthy Air' 94*, Italy, pp. 405-416.

CHAPTER 2

Evidence for Nonthermal Electromagnetic Bioeffects: Potential Health Risks in Evolving Low-Frequency & Microwave Environments

W. Ross Adey

2.1 INTRODUCTION

In our solar system, the natural electromagnetic environment varies greatly from planet to planet. In the case of the planet Earth, a semiliquid ferromagnetic core generates a major and slowly migrating *static* geomagnetic field. Concurrently, there are much weaker natural *oscillating* low-frequency electromagnetic fields that arise from two major sources: in thunderstorm activity in equatorial zones of Central Africa and the Amazon basin; and in lesser degree, from solar magnetic storms in years of high activity in the 11-year solar sunspot cycle.

2.1.1 A Comparison of Natural and Man-Made Electromagnetic Environments

All life on earth has evolved in these fields. Defining them in physical terms permits direct comparison with far stronger man-made fields that have come to dominate all civilized environments in the past century. Energy in the *oscillating* natural fields is almost entirely in the extremely-low-frequency (ELF) spectrum, with peaks at frequencies between 8 and 32 Hz, the Schumann resonances (1957). Their electric components are around 0.01 V/m, with magnetic fields of 1-10 nanotesla. The earth's *static* magnetic field at 50 microtesla is 5000 times larger than the natural oscillations, but still substantially less than a wide range of daily human exposures to static and oscillating fields in domestic and occupational environments. Importantly, there is no significant level of natural energy at RF/microwave frequencies

Generation and distribution of electric power has spawned a vast and ever growing vista of new electronic devices and systems. They overwhelm the natural electromagnetic environment with more intense fields. They include oscillations far into the microwave spectrum, many octaves higher than the Schumann resonances. This growth of RF/microwave fields is further complicated by the advent of digital communication techniques. In many applications, these microwave fields, oscillating billions of times per second, are systematically interrupted (pulsed) at low frequencies. This has raised important biological and biomedical questions, still incompletely answered, about possible tissue mechanisms in detection of amplitude- and pulse-modulation of RF/microwave fields (Adey, 1999).

2.1.2 Electric Power and Electronic Systems in Homes and Offices: Historical and Environmental Evidence

2.1.2.1 The Growth of Fields at Electric Power Frequenciess

A peak in childhood leukaemia at ages 2 through 4 emerged *de novo* in the 1920s in the United Kingdom and slightly later in the United States. Electrification in U.S. farms lagged behind urban areas until 1956. Using U.S. census data for 1930, 1940 and 1950, Milham and Osslander (2001) found that during 1928-1932, in states with above 75% of residences served by electricity, leukaemia mortality increased with age for single years 0-4, whereas states with electrification levels below 75%showed a decreasing trend with age ($P = 0.009$). During 1949-1951, all states showed a peak in leukaemia mortality at ages 2-4. At age 0-1, leukaemia mortality was not related to electrification levels. At ages 2-4, there was a 24% (95% CI, 8%-4!%) increase in leukaemia for a 10% increase in percentage of homes served by electricity. They conclude that the peak in the common childhood acute lymphoblastic leukaemia (ALL) may be attributable to electrification.

Design of modern office buildings has led to their electrification through one or more large distribution transformers that may be located in basement vaults, or in some cases, located on each floor of the building. Milham (1996) has examined cancer incidence in such a building over a 15-year period. A small cohort of 410 office workers (263 men and 147 women) were exposed to strong magnetic fields from three 12 kV transformers located beneath their ground-floor office. Cancers were detected in eight subjects over the 15-year period. There were indications of *cumulative risks*. Only one cancer was detected in 254 workers employed for less than 2 years, compared to seven cancers detected in 156 workers employed for 2 years or more ($P = 0.0057$; Fisher's exact test). An analysis of linear trend in cancer incidence, using average years employed as an exposure score, was positive ($P = 0.00337$), with an odds ratio of 15.1 in workers employed longer than 5 years. This positive correlation of cancer cases with duration of employment was seen for males and females separately and together ($P < 0.05$). For workers employed more than 2 years, the standardized cancer incidence ratio was 389 (95% confidence interval, 156-801).

None of these studies support tissue heating as an adequate model for the observed bioeffects.

2.1.2.2 The Appearance of Office Wireless Networks

In large modern offices, the electromagnetic environment has been further complicated by introduction of local-area-networks (LANs) for local telephonic and data transmission. Operating at gigahertz microwave frequencies, workers are continuously exposed to fields from a plethora of sources located on each computer and on network hubs. Many of these sources utilise some form of low-frequency pulsed modulation. Their power output is typically in the low milliwatt to microwatt range - so low that significant heating of workers' tissues is improbable. Any bioeffects attributable to their operation strongly suggest *nonthermal* mechanisms of interaction, and raise further important questions about mechanisms mediating a cumulative dose from repeated, intermittent exposures, possibly over months and years.

Tissue components of environmental RF/microwave fields are consistent with two basic models. Sources close to the body surface produce *near-field exposures,*as with users of mobile phones. The emitted field is magnetically coupled directly from the antenna into the tissues. At increasing distances from the source, the human body progressively takes on properties of a radio antenna, with absorption of radiated energy determined by physical dimensions of the trunk and limbs. This is a *far-field exposure,* defined as fully developed at 30 wavelengths from the source. Permissible exposure limits (PELs) have rested on measurement of microwave field energy absorbed as heat, expressed as the *Specific Absorption Rate (SAR)* in W/kg.

The American National Standards Institute (1992) first recognised a tissue dose of 4.0 W/kg as a *thermal* (heating) tissue threshold likely to be associated with adverse effects and proposed an exposure limit in Controlled Environments (occupational) at 0.4 W/kg, thus creating a supposed "safety margin" of 10. For Uncontrolled Environments (civilian), a larger safety margin was set with a PEL 50 times lower at 0.08 W/kg. Since actual measurement of tissue SARs under environmental conditions is not a practical technique, PELs are typically expressed as a function of *incident field power density*, the amount of energy falling on a surface/unit area, and expressed in mW/cm^2. A simple yardstick relating incident field energy to its electric component relates an incident field of 1.0 mW/cm^2 to an electric field of 61V/m.

In the interim since formulation of these ANSI guidelines, there have been extensive reviews and proposed revisions. The U.S. government Interagency Radio Frequency Working Group (1999) has emphasised the need for revisions recognising nonthermal tissue sensitivities: "Studies continue to be published describing biological responses to nonthermal ELF-modulated RF radiation exposures that are not produced by CW (unmodulated) radiation." These studies have resulted in concern that "exposure guidelines based on thermal effects, and using information and concepts (time-averaged dosimetry, uncertainty factors) that mask any differences between intensity-modulated RF radiation exposure and CW exposure, do not directly address public exposures, and therefore may not adequately protect the public."

2.2   SENSITIVITIES TO NONTHERMAL STIMULI: TISSUE STRUCTURAL AND FUNCTIONAL IMPLICATIONS

2.2.1 Conductance Pathways in Multicellular Tissues

In its earliest forms, life on earth may have existed in the absence of cells, simply as a "soup" of unconstrained biomolecules at the surface of primitive oceans. It is a reasonable assumption that the first living organisms existed as single cells floating or swimming in these primordial seas. Concepts of a cell emphasise the role of a bounding membrane, surrounding an organised interior that participates in the chemistry of processes essential for all terrestrial life. This enclosing membrane is the organism's window on the world around it.

For unicellular organisms that swim through large fluid volumes, the cell membrane is both a sensor and an effector. As a sensor, it detects altered chemistry in the surrounding fluid, and provides a pathway for inward signals generated on its surface by a wide variety of stimulating ions and molecules, including hormones, antibodies and neurotransmitters. These most elemental inward signals are susceptible to manipulation by a wide variety of naturals or imposed electromagnetic fields that may also pervade the pericellular field. As effectors, cell membranes may also transmit a variety of electrical and chemical signals across intervening intercellular fluid to neighbouring cells, thus creating a domain or ensemble of cells, often able to "whisper together" in a faint and private language. Experimental evidence suggests that these outward effector signals may also be sensitive to intrinsic and imposed electromagnetic fields.

Rather than being separated in a virtually limitless ocean, cellular aggregates that form tissues of higher animals are separated by narrow fluid channels that take on special importance in signalling from cell to cell. Biomolecules travel in these tiny "gutters", typically not more than 150 wide, to reach binding sites on cell membrane receptors. These gutters form the *intercellular space* (ICS). It is a preferred pathway for induced currents of intrinsic and environmental electromagnetic fields. Although it occupies only ~ 10% of the tissue cross-section, it carries at least 90% of any imposed or intrinsic current, directing it along cell membrane surfaces. Whereas the ICS may have a typical impedance of ~ 4-50 ohm cm^{-1}, transmembrane impedances are ~ 10^4 - 10^6 ohm cm^{-2}.

2.2.2 Structural and Functional Organization of the Extracellular Space

Spaces in the ICS are not simple saline filled channels. Numerous stranded protein molecules protrude into these spaces from the cell interior and form a *glycocalyx* with specialised receptor sites that sense chemical and electrical stimuli in surrounding fluid. Their amino sugar tips are highly negatively charged. They offer an anatomical substrate for the first detection of weak electrochemical oscillations in pericellular fluid, including field potentials arising in activity of adjoining cells, or as tissue components of environmental fields. Research in molecular biology has increasingly emphasised essentially direct communication between cells due to their mutual proximity. Bands of *connexin* proteins form *gap-junctions* directly uniting adjoining cell membranes. Experimental evidence supports their role in intercellular signalling

2.2.3 Tissue Detection of Low Frequency Fields and RF/Microwave Fields Amplitude-Modulated at Low Frequencies: Structural and Functional Options

Differential bioeffects, to be discussed below, have been reported between certain nonthermal RF/microwave fields with low-frequency amplitude- or pulse-modulation when compared to exposures to unmodulated continuous wave (CW) fields at similar power levels. The findings suggest, but do not yet establish unequivocally, that this frequency dependence may be a system property in a sequence of molecular hierarchies beyond the first transductive step. If the concept of modulation frequency-dependence continues to gain support in further research, answers must be sought as to the manner of its detection.

For ELF fields, models based on joint static/oscillating magnetic fields have been hypothesised. They include ion cyclotron resonance (Liboff, 1992), where mono- and divalent cations, such as potassium and calcium, abundant in the cellular environment, may exhibit cyclotron resonance at ELF frequencies in the presence of ambient static fields of less than 100 μT, such as the geomagnetic field. Other models describing ELF frequency dependence have considered phase transitions (Lednev, 1991) and ion parametric resonance (Blackman *et al.*, 1994), but interpretation of this frequency dependence based on ion parametric resonance remains unclear (Adair, 1998).

For amplitude or pulse-modulated RF/microwave fields, there is the implication that some form of *envelope demodulation* occurs in tissue recognition of ELF modulation components, but the tissue may remain essentially transparent to the same signal presented as an unmodulated carrier wave (Adey, 1981, 1999). However, crucial questions remain unanswered. It is not known whether biological low frequency dependence is established at the transductive step in the first tissue detection of the field, or whether it resides at some higher level in an hierarchical sequence of signal coupling to the biological detection system (Engstrom, 1997). For ELF magnetic fields, experimental evidence points to a slow time scale in inhibition of tamoxifen's antiproliferative action in human breast cancer cells (Harland *et al.*, 1999).

It is a principle of radio physics that extraction of ELF modulation information from an amplitude-modulated signal requires a *nonlinear element* in the detection system. This required nonlinearity may involve a spatial component, such as differential conduction in certain directions along the signal path; or the path itself may exhibit nonlinearities with respect to such factors as spatial distribution of electric charges at fixed molecular sites (so called *fixed charges*); or conduction itself may involve a nonlinear quantum process, as in electron tunnelling across the transverse dimensions of the cell membrane.

These constraints impose a further essential condition for demodulation to occur in the multicellular tissues of living organisms. There must be a *site for demodulation* to occur. Evidence supports a role for cell membranes to act in this way, based not only on their intrinsic structure, but also on their proximity to neighbouring cells in the typical organisation of tissues of the body. Typical tissue organisation meets the three criteria outlined above, but as cautionary note, does not allow calculation of possible detection efficiency. Direct neighbour-neighbour cellular interactions will invite our further consideration of properties of cellular ensembles or domains in determining tissue threshold sensitivities.

2.2.3.1 Directional Differences in Tissue Signal Paths

As already noted, the narrow gutters of the intercellular spaces offer preferred conduction pathways, with conductivity 10^2-10^4 higher through extracellular spaces than through cell membranes (Adey *et al.*, 1963). Thus, the intercellular spaces become preferred pathways for *conduction along (parallel to) cell membrane surfaces,* and will reflect the changing directions and cross sections of myriad channels. Although predominantly an ionic (resistive) conduction pathway, it may also exhibit reactive components, due to the presence of protein molecules in solution.

2.2.3.2 Nonlinearities Related to Electric Charge Distribution

A suggested basis for envelope demodulation at cell surfaces may reside in the intensely anionic charge distribution on strands of glycoprotein that protrude from the cell interior, forming the glycocalyx (Bawin and Adey, 1976; Adey, 1999). As already noted, they provide the structural basis for specific receptor sites and they attract a surrounding cationic atmosphere, composed largely of calcium and hydrogen ions. This charge separation creates a Debye layer. In models and experimental data from resin particles, Einolf and Carstensen (1971) concluded that this physical separation creates a large virtual surface capacitance, with dielectric constants as high as 10^6 at frequencies below 1 kHz. Displacement currents induced in this region by ELF modulation of an RF field may then result in demodulation.

2.2.3.3 Electron Tunnelling in Transmembrane Conduction: Nonlinearities in Space and Time

Experimental studies of transmembrane charge tunnelling by DeVault and Chance (1966) and their more recent theoretical development by Moser *et al.* (1992) offer an example of extreme functional nonlinearity within the cell membrane. Chance described temperature-independent millisecond electron transfer over a temperature range from 120 K to 4 K. Considering a cell membrane transverse dimension of 40, Moser *et al.* noted that a variation of 20 in the distance between donors and acceptors in a protein changes the electron transfer rate by 10^{12}-fold. Concurrently in the time domain, the electron transfer rate is pushed from seconds to days, or a 10-fold change in rate for a 1.7 change in distance.

2.2.3.4 Issues of Comparability Between Bioeffects of ELF Fields, ELF-Modulated RF Fields, and Unmodulated (CW) RF Fields

From the beginning of these studies in the 1970s, it was noted that there were similarities in responses of tissues and cultured cells to environmental fields that were either in the ELF spectrum, or were RF/microwave fields modulated at ELF frequencies. Available evidence has indicated similarities between certain cell ionic and biochemical responses to ELF fields, and to RF/microwave fields amplitude-modulated at these same ELF frequencies, suggesting that tissue demodulation of RF/microwave fields may be a critical determinant in ensuing biological responses.

These findings have been reviewed in detail elsewhere (Adey, 1997, 1999). They are briefly summarized here in experiments at progressively more complex levels in the hierarchies of cellular organization. Early studies described calcium efflux from brain tissue in response to ELF exposures (Bawin and Adey, 1976; Blackman *et al.*, 1985), and to ELF-modulated RF fields (Bawin and Adey, 1975; Blackman *et al.*, 1979, 1985; Dutta *et al.*, 1984). Calcium efflux from isolated brain subcellular particles (synaptosomes) with dimensions under 1.0 μm also exhibit an ELF modulation frequency-dependence in calcium efflux, responding to 16 Hz sinusoidal modulation, but not to 50 Hz modulation, nor to an unmodulated RF carrier (Lin-Liu and Adey, 1982). In the same and different cell culture lines, the growth regulating and stress responsive enzyme ornithine decarboxylase (ODC) responds to ELF fields (Byus *et al.*, 1988; Litovitz *et al.*, 1993) and to ELF-modulated RF fields (Byus *et al.*, 1987; Litovitz *et al.*, 1993; Penafiel *et al.*, 1996).

In more recent studies also related to cellular stress responses, Goodman and Blank and their colleagues have reported rapid, transitory induction of heat-shock proteins by microtesla level 60 Hz magnetic fields (Lin *et al.*, 1997). In human HL60 promyelocytic cells these exposures at normal growth temperatures activated heat shock factor1 and heat shock element binding, a sequence of events that mediates stress-induced transcription of the stress gene HSP70 and increased synthesis of the stress response protein hsp70kd. Thus, the events mediating the field-stimulated response appeared similar to those reported for other physiological stressors (hyperthermia, heavy metals, oxidative stress), suggesting to the authors a general mechanism of electromagnetic field interaction with cells. Their further studies have identified endogenous levels of c-*myc* protein as a contributor to the induction of HSP70 in response to magnetic field stimulation (Lin *et al.*, 1998), with the hypothesis that magnetic fields may interact directly with moving electrons in DNA (Blank and Goodman, 1999, 2001; Blank and Soo, 2001).

Immune responses of lymphocytes targeted against human lymphoma tumour cells (allogeneic cytotoxicity) are sensitive to both ELF exposures (Lyle *et al.*, 1988) and to ELF-modulated fields, but not to unmodulated fields (Lyle *et al.*, 1983).

Communication between brain cells is mediated by a spectrum of chemical substances that both excite and inhibit transaction and transmission of information between them. Cerebral amino acid neurotransmitter mechanisms (glutamate, GABA and taurine) are influenced by ELF fields (Kaczmarek and Adey, 1974; Bawin *et al.*, 1996), and also by ELF-modulated microwave fields, but not by unmodulated fields. Kolomytkin *et al.* (1994) examined specific receptor binding of three neurotransmitters to rat brain synaptosomes exposed to either 880 or 915 MHz fields at maximum densities of 1.5 mW cm^{-2}. Binding to inhibitory gamma-aminobutyric acid (GABA) receptors decreased 30% at 16 pulses/sec, but was not significantly altered at higher or lower pulse frequencies. Conversely, 16 pulses/sec modulation significantly increased excitatory glutamate receptor binding. Binding to excitatory acetylcholine receptors increased 25% at 16 pulses/sec, with similar trends at higher and lower frequencies. Sensitivities of GABA and glutamate receptors persisted at field densities as low as 50 μW cm^{-2}.

A selective absence of responses to unmodulated (CW) RF/microwave fields reported in many of these earlier studies has focused attention on establishment of threshold sensitivities to CW field exposures. De Pomerai *et al.* (2000) has reported cellular stress responses in a nematode worm as a biosensor of prolonged CW microwave exposures at athermal levels. Tattersall *et al.* (2001) exposed slices of rat hippocampal cerebral tissue to 700 MHz CW fields for 5-15 minutes at extremely low SARs in the range 0.0016-0.0044W kg^{-1}. No detectable temperature changes (+/- 0.1C) were noted during 15minute exposures. At low field intensities, a 20% potentiation of electrically evoked population potentials occurred, but higher field intensities evoked either increased or decreased responses. The exposures reduced or abolished chemically induced spontaneous epileptiform activity. Bawin *et al.* (1996) also tested the rat hippocampal slice, using ELF magnetic fields. At 56 μT (0.35-3.5 nV mm^{-1}), magnetic fields destabilised rhythmic electrical oscillations via as yet unidentified nitric oxide mechanisms involving free radicals.

2.3 INITIAL TRANSDUCTION OF IMPOSED ELECTROMAGNETIC FIELDS AT NONTHERMAL ENERGY LEVELS

There is an initial dichotomy in possible modes of interaction of cells in tissue with environmental microwave fields. It is principally determined by the separation of responses attributed to tissue heating from those elicited by certain fields at levels where frank heating is not the basis of an observed interaction. Their interpretation and possible significance has required caution in both biological and biophysical perspectives. Many of these biological sensitivities run counter to accepted models of physiological thresholds based in equilibrium thermodynamics of kT thermal collision energies. In a physical perspective, the search also continues for biological systems compatible with a first transductive step in a range of functionally effective vibrational and electromagnetic stimuli that are orders of magnitude weaker than kT. Their occurrence invites hypotheses on directions of future research.

2.3.1 Cell Membranes as the Site of Initial Field Transductive Coupling

Collective evidence points to cell membrane receptors as the probable site of first tissue interactions with both ELF and microwave fields for many neurotransmitters (Kolomyttkin *et al.*, 1994), hormones (Liburdy, 1995; Ishido *et al.*, 2001), growth-regulating enzyme expression (Byus *et al.*, 1987; Chen *et al.*, 2000; Litovitz *et al.*, 1993; Penafiel *et al.*, 1997), and cancer-promoting chemicals (Cain *et al.*, 1993; Mevissen *et al.*, 1999). In none of these studies does tissue heating appear involved causally in the responses. Physicists and engineers have continued to offer microthermal, rather than athermal, models for these phenomena (Barnes, 1996; Astumian *et al.*, 1995), with views that exclude consideration of cooperative organisation and coherent charge states, but it is difficult to reconcile experimental evidence for factors such as modulation frequency-dependence and required duration of an amplitude-modulated signal to elicit a response (*coherence time*) (Litovitz *et al.*, 1993) with models based on the equilibrium dynamics of tissue heating.

2.3.2 Evidence for Role of Free Radicals in Electromagnetic Field Bioeffects

Examination of vibration modes in biomolecules, or portions of these molecules, (Illinger, 1981) has suggested that resonant microwave interactions with these molecules, or with portions of their structure, is unlikely at frequencies below higher gigahertz spectral regions. This has been confirmed in studies showing collision-broadened spectra, typical of a heating stimulus, as the first discernible response of many of these molecules in aqueous solutions to microwave exposures at frequencies below 10 GHz.

However, there is an important option for biomolecular interactions with static and oscillating magnetic fields through the medium of *free radicals* (see Adey, 1993, 1997 for summaries). Chemical bonds are magnetic bonds, formed between adjacent atoms through paired electrons having opposite spins and thus magnetically attracted. Breaking of chemical bonds is an essential step in virtually all chemical reactions, each atomic partner reclaiming its electron, and moving away as a free radical to seek another partner with an opposite electron spin. The brief life time of a free radical is about a nanosecond or less, before once again forming a *singlet pair* with a partner having an opposite spin, or for electrons with similar spins, having options to unite in three ways, forming *triplet pairs* (reviewed in Adey, 1999).

During this brief lifetime, imposed magnetic fields may delay the return to the singlet pair condition, thus influencing the *rate* and the *amount of product* of an on-going chemical reaction (McLauchlan, 1992). McLauchlan points out that this model predicts a potentially "enormous effect" on chemical reactions for static fields in the low mT range. For oscillating fields, the evidence is less clear on their possible role as direct mediators in detection of ELF frequency-dependent bioeffects. *Spin-mixing* of orbital electrons and nuclear spins in adjacent nuclei is a possible mechanism for biosensitivities at extremely low magnetic field levels, but these interactions are multiple, complex and incompletely understood (McLauchlan and Steiner, 1991). The highest level of free radical sensitivity may reside in hyperfine-dependent singlet-triplet state mixing in radical pairs with a small number of hyperfine states that describe their coupling to nearby nuclei (Till *et al.*, 1998; Timmel *et al.*, 1998). Singlet-triplet interconversion would need to be sufficiently fast to occur before diffusion reduced the probability of radical re-encounter to negligible levels.

Lander (1997) has emphasised that we are at an early stage of understanding free radical signal transduction. "Future work may place free radical signalling beside classical intra- and intercellular messengers and uncover a woven fabric of communication that has evolved to yield exquisite specificity." A broadening perspective on actions of free radicals in all living systems emphasises a dual role: first as messengers and mediators in many key processes that regulate cell functions throughout life; and second, in the pathophysiology of *oxidative stress diseases*.

At cell membranes, free radicals may play an essential role in regulation of receptor specificity, but not necessarily through a "lock and key" mechanism. As an example, Lander cites the location of cysteine molecules on the surface of P21-*ras* proteins at cell membranes. They may act as selective targets for nitrogen and oxygen free radicals, thereby inducing covalent modifications; and thus setting the *redox potential* of this target protein molecule as the critical determinant for its highly specific interactions with antibodies, hormones, etc. Magnetochemistry studies have suggested a form of cooperative behaviour in populations of free radicals that remain *spin-correlated* after initial separation of a singlet pair (Grundler *et al.*, 1992). Magnetic fields at 1 and 60 Hz destabilise rhythmic oscillations in brain hippocampal slices at 56 μT (0.35 to 3.5 nV mm^{-1}, via as yet unidentified nitric oxide mechanisms involving free radicals (Bawin *et al.*, 1996). In a general biological context, these are some of the unanswered questions that limit free radical models as general descriptors of threshold events.

2.4 THE ROLE OF CELLULAR ENSEMBLES IN SETTING TISSUE THRESHOLDS FOR INTRINSIC AND ENVIRONMENTAL STIMULI

Our pursuit of mechanisms mediating tissue electromagnetic sensitivities at nonthermal levels raises questions about the relevance of observed thresholds in the sensory physiology of other modalities. By extrapolation, do these data suggest the need to explore collective properties of populations of cells in setting thresholds by forms of intercellular communication? Do cooperative processes yield one or more faint and private languages that allow ensembles of cells to "whisper together" in one or more faint and private languages? Do observed *tissue* sensory thresholds differ significantly from thresholds measured in single cells in isolation from their neighbours?

2.4.1 Evidence for Domain Functions as a General Biological Property in Tissues

Research in sensory physiology supports this concept that some threshold properties may reside in highly cooperative properties of populations of elements, rather than in a single detector (Adey, 1998). Seminal observations in the human auditory system point to a receptor vibrational displacement of 10^{-11} m, or approximately the diameter of a single hydrogen atom (Bialek, 1983; Bialek and Wit, 1984); human olfactory thresholds for musk occur at 10^{-13} M, with odorant molecules distributed over 240 mm^2 (Adey, 1959); and human detection of single photons of blue-green light occurs at energies of 2.5 eV (Hagins, 1979). In another context, pathogenic bacteria, long thought to operate independently, exhibit ensemble properties by communication through a system recognizing colony numbers as an essential step preceding release of toxins. These *quorum sensing* systems may control expression of virulence factors in the lungs of patients with cystic fibrosis (Erickson *et al.*, 2002).

2.4.1.1 Domain Properties in Systems of Excitable Cells

Bialek addressed the problem of the auditory receptor in quantum mechanical terms. He evaluated two distinct classes of quantum effects: a *macroquantum effect* typified by the ability of the sensory system to detect signals near the quantum limits to measurement; and a *microquantum effect*, in which "the dynamics of individual biological macromolecules depart from predictions of a semi-classical theory." Bialek concluded that quantum-limited sensitivity occurs in several biological systems, including displacements of sensory hair cells of the inner ear. Remarkably, quantum limits to detection are reached in the ear in spite of seemingly insurmountable levels of thermal noise.

To reach this quantum limit, these receptor cells must possess amplifiers with noise performance approaching limits set by the uncertainty principle. It is equally impressive that suppression of intrinsic thermal noise allows the ear to function as though close to $0°K$. Again, this suggests system properties inherent in the detection sequence. These "perfect" amplifiers could not be described by any chemical kinetic model, nor by any quantum mechanical theory in which the random phase approximation is valid. The molecular dynamics of amplifiers in Bialek's models would require preservation of quantum mechanical coherence for times comparable to integration times of the detector. It is not known whether comparable mechanisms may determine electromagnetic sensitivities as a more general tissue property at cellular and subcellular levels.

Behavioural electrosensitivity in sharks and rays may be as low as 0.5 nV mm^{-1} for tissue components of electrical fields in the surrounding ocean (Kalmijn, 1971). These marine vertebrates sense these fields through specialized jelly-filled tubular receptors (ampullae of Lorenzini) up to 10 cm in length, located near the snout and opening on the skin surface through minute pores. Sensing nerve cells lie in the wall of this ampullary tube. In support of a cooperative model of organisation of these neurons, behavioural electrosensitivity in sharks and rays is

100 times below measurable thresholds of individual electroreceptor neurons (Valberg *et al.*, 1997).

2.4.1.2 Domain Properties in System of Non-Excitable Cells: Culture Dimensions and "Bystander" Effects

Jessup *et al.* (2000) have pioneered studies on the role of gravitational fields in determining trends towards either apoptosis (programmed cell death) or towards cell proliferation. Concurrently, they tested the physical configuration of cell cultures in their influence on these same trends. Based on a colorectal cancer cell line, they compared cells cultured in adherent monolayers with three-dimensional ("3D") cultures. Biochemical measures of apoptosis and cell proliferation were tested in (1) static cultures, (2) in cultures subjected to slow rotation, and (3) in cultures exposed to the microgravity of low earth-orbital space flight. Over the course of 6 days on earth, static 3D cultures displayed the highest rates of proliferation and lowest apoptosis. Rotation appeared to increase apoptosis and decrease proliferation, whereas static 3D cultures in either unit-gravity or microgravity had less apoptosis. Expression of the carcinoembryonic antigen (CEA) as a marker of cell differentiation was increased in microgravity.

For ionising radiation, the U.S. National Council on Radiation Protection (NCRP) has recommended that estimates of cancer risk be extrapolated from higher doses by using a linear, no-threshold model. This recommendation is based on the dogma that the DNA of the nucleus is the main target of radiation-induced genotoxicity and, as fewer cells are directly damaged, the deleterious effects of ionising radiation proportionally decline. Experimental evidence seriously challenges this concept (Zhou *et al.*, 2001). They used a precision microbeam of α-particles to target an exact fraction (either 100% or $\leq$ 20%) of the cells in a confluent cell population and irradiated their nuclei with exactly one α-particle each. The findings were consistent with non-hit cells contributing significantly to the response, designated *the bystander effect*. Indeed, irradiation of 10% of a confluent mammalian cell population with a single α particle resulted in a mutant yield similar to that observed when all the cells in the population were irradiated. Importantly, this effect was eliminated in cells pretreated with 1mM octanol, which inhibits intercellular communication mediated by gap-junction proteins. "The data imply that the relevant target for ionising radiation mutagenesis is larger than an individual cell."

2.5 CONCLUSIONS

Epidemiological studies have evauated ELF and radiofrequency fields as possible risk factors for human health, with historical evidence relating rising risks of such factors as progressive rural electrification, and more recently, to methods of electric power distribution and utilisation in commercial buildings Appropriate models describing these bioeffects are based in nonequilibrium thermodynamics, with nonlinear electrodynamics as an integral feature. Heating models, based in equilibrium thermodynamics, fail to explain an impressive spectrum of observed electromagnetic bioeffects at nonthermal exposure levels. We face a new frontier of much greater significance.

In little more than a century, our biological vista has moved from organs to tissues, to cells, and most recently, to the molecules that form the exquisite fabric of living systems. We discern a biological organization based in physical processes at the atomic level, beyond the realm of chemical reactions between biomolecules. Much of this signalling within and between cells may be mediated by free radicals of the oxygen and nitrogen species. In their brief lifetimes, free radicals are sensitive to imposed magnetic fields, including microwave fields. Free radicals are involved in normal regulatory mechanisms in many tissues. Disordered free radical regulation is associated with oxidative stress diseases, including Parkinson's and Alzheimer's diseases, coronary heart disease and cancer.

Although incompletely understood, tissue free radical interactions with magnetic fields may extend to zero field levels. Emergent concepts of tissue thresholds to imposed and intrinsic magnetic fields address ensemble or domain functions of populations of cells, cooperatively "whispering together" in intercellular communication, and organized hierarchically at atomic and molecular levels.

2.6 REFERENCES

Adair, R.K. 1998. A physical analysis of the ion parametric resonance model. *Bioelectromagnetics*, **19**, pp. 181-191.

Adey, W.R. 1959. The sense of smell. In *Handbook of Physiology*, Volume 1, edited by J. Field, H.W. Magoun and V.E. Hall (American Physiological Society, Washington, DC. pp. 535-548).

Adey, W.R. 1981. Tissue interactions with nonionizing electromagnetic fields. *Physiological Reviews*, **61**, pp. 435-514.

Adey, W.R. 1993. Electromagnetics in biology and medicine. In *Modern Radio Science*, edited by H. Matsumoto (Oxford University Press), pp. 231-249.

Adey, W.R. 1997. Bioeffects of mobile communication fields; possible mechanisms for cumulative dose. In *Mobile Communications Safety*, edited by N. Kuster, Q. Balzano and J.C. Lin. (New York, Chapman and Hall), pp. 103-140.

Adey, W.R. 1998. Horizons in science: physical regulation of living matter as an emergent concept in health and disease. In *Electricity and Magnetism in Biology and Medicine*, edited by F. Bersani (New York, Kluwer/Plenum Publishers), pp. 53-57.

Adey, W.R. 1999. Cell and molecular biology associated with radiation fields of mobile telephones In *Review of Radio Science 1996-1999*, edited by W.R. Stone and S. Ueno (Oxford University Press), pp. 845-872.

Adey, W.R., Kado, R.T., Didio, J. *et al.* 1963 Impedance changes in cerebral tissue accompanying a learned discriminative performance in the cat. *Experimental Neurology*, **7**, pp. 259-281.

Astumian, R.D., Weaver, J.C. and Adair, R.K. 1995. Rectification and signal averaging of weak electric fields by biological cells. *Proceedings of National Academy of Sciences USA*, **92**, pp. 3740-3743.

Barnes, F.S. 1996. The effects of ELF on chemical reaction rates in biological systems. In *Biological Effects of Magnetic and Electromagnetic Fields*, edited by S. Ueno (New York, Plenum Press), pp. 37-44.

Bawin, S.M. and Adey, W.R. 1976. Sensitivity of calcium binding in cerebral tissue To weak environmental electric fields at low frequency. *Proceedings of National Academy of Sciences USA*, **73**, 1999-2003.

Bawin, S.M., Satmary, W.M., Jones, R.A., *et al.* 1996. Extremely low frequency magnetic fields disrupt rhythmic slow activity in rat Hippocampal slices. *Bioelectromagnetics*, **17**, pp. 388-395.

Bialek, W. 1983. Macroquantum effects in biology: the evidence. Ph.D. thesis, Department of Chemistry, University of California, Berkeley. 250 pp.

Bialek, W. and Wit, H.P. 1984. Quantum limits to oscillator stability: theory and experiments on emissions from the human ear. *Physics Letters*, **104A**, pp. 173-178.

Blackman, C.F., Elder, J.A., Weil, C.M, *et al.* 1979. Induction of calcium efflux from brain tissue by radio frequency radiation. *Radio Science*, **14**, pp. 93-98.

Blackman, C.F., Benane, S.G., House, D.E. *et al.* 1985. Effects of ELF (1-120 Hz) and modulated (50 Hz) RF fields on the efflux of calcium ions from rain tissue *in vitro*. *Bioelectromagnetics*, **6**, pp. 327-338.

Blackman, C.F., Blanchard, J.P., Benane S.G. *et al.* 1994. Empirical test of an ion parametric resonance model for magnetic field interactions with PC-12 cells. *Bioelectromagnetics*, **15**, pp. 239-260.

Blank, M. and Goodman, R. 1999. Electromagnetic fields may act directly on DNA. *Journal of Cellular Biochemistry*, **75**, pp. 369-374.

Blank, M. and Goodman, R. 2001. Electromagnetic initiation of transcription at specific DNA sites. *Journal of Cellular Biochemistry*, **81**, pp. 689-692.

Blank, M. and Soo, L. 2001. Electromagnetic acceleration of electron transfer reactions. *Journal of Cellular Biochemistry*, **81**, pp. 278-283.

Byus, C.V., Pieper, S.E. and Adey, W.R. 1987. The effects of low-energy 60 Hz environmental electromagnetic fields upon the growth-related enzyme ornithine decarboxylase. *Carcinogenesis*, **8**, pp. 1385-1389.

Byus, C.V., Kartun, K., Pieper, S. *et al.* 1988. Increased ornithine decarboxylase activity in cultured cells exposed to low energy modulated microwave fields. *Cancer Research*, **48**, pp. 4222-4226.

Cain, C.D., Thomas, D.L. and Adey, W.R. 1993. 60-Hz magnetic field acts as co-promoter in focus formation of C3H10T $^1/_2$ cells. *Carcinogenesis*, **14**, pp. 955-960.

Chen, G., Upham, B.L., Sun, W. *et al.* 2000. Effect of electromagnetic field exposure on chemically induced differentiation of friend erythroleukemia cells. *Environmental Health Perspectives*, **108**, pp. 967-972.

Di Pomerai, D., Daniells, C., David, H. *et al.* 2000. Non-thermal heat-shock response to microwaves. *Nature*, **405**, pp. 417-418.

DeVault, D. and Chance, B. (1966). Studies of photosynthesis using a pulsed laser. I. temperature dependence of cytochrome oxidation rate in chromatin. Evidence of tunneling. *Biophysical Journal*, **6**, pp. 825-847.

Dutta, S.K., Ghosh, B. and Blackman, C.F. 1989. Radiofrequency radiation-induced calcium ion efflux enhancement form human and other neuroblastoma cells in culture. *Bioelectromagnetics*, **10**, pp. 7-20.

Einolf, C.W. and Carstensen, E.L. 1971 Low-frequency dielectric dispersion in suspensions of ión-exchange resins. *Journal of Physical Chemistry*, **75**, pp. 1091-1099.

Engstrom, S. 1997. What is the time scale of magnetic field interaction in biological systems? *Bioelectromagnetics*, **18**, pp. 244-249.

Erickson, D.L., Endersby, R., Kirkham, A., *et al.* 2002. Pseudomonas aeruginosa quorum-sensing systems may control virulence factor expression in the lungs of patients with cystic fibrosis. *Infectious Immunology*, **70**, pp. 1783-1790.

Grundler, W., Keilmann, F., Putterlik, V. *et al.* 1992. Mechanics of electromagnetic interaction with cellular systems. *Naturwissenschaften*, **79**, pp. 551-559.

Hagins, W.A. 1979. Excitaion in vertebrate photoreceptors. In *The Neurosciences: Fourth Study Program*, edited by F.O. Schmitt and F.G. Worden (Cambridge, MIT Press), pp. 183-192.

Harland, J.D. and Liburdy, R.P. 1997. Environmental magnetic fields inhibit the antiproliferative action of tamoxifen and melatonin in a human breast cancer cell line. *Bioelectromagnetics*, **18**, pp. 555-562.

Harland, J.D., Engstrom, S. and Liburdy, R. 1999. Evidence for a slow time-scale of interaction for magnetic fields inhibiting tamoxifen's antiproliferative action in human breast cancer cells. *Cellular and Biochemical Physics*, **31**, pp. 295-306.

Illinger, K.H. (ed.). 1981. *Biological Effects of Nonionizing Radiation*, (Washington DC. American Chemical Society Symposium Series, No. 157) 342 pp.

Ishido, M., Nitta, H. and Kabuto, M. 2001. Magnetic fields (MF) of 50 Hz at 1.2 microT cause uncoupling of inhibitory pathways of adenylyl cyclase mediated by melatonin 1a receptor in MF-sensitive MCF-7 cells. *Carcinogenesis,* **22**, pp. 1043-1048.

Jessup, J.M., Frantz, M., Sonmez-Alpin, E. *et al.* 2000. Microgravity culture reduces apoptosis and increases the differentiation of a human colorectal carcinoma cell line. *In Vitro Cell Development and Biology-Animal*, **36**, pp. 367-373.

Kaczmarek, L.K. and Adey, W.R. 1974. Some chemical and electrophysiological effects of glutamate in cerebral cortex. *Journal of Neurobiology*, **5**, pp. 231-241.

Kalmijn, A.J. 1971. The electric sense of sharks and rays. *Journal of Experimental Biology*, **55**, 371-382.

Kolomytkin, O., Yurinska, M., Zharikov, S. *et al.* 1994. Response of brain receptor systems to microwave energy exposure. In *On the Nature of Electromagnetic Field Interactions with Biological Systems*, edited by A.H. Frey (Austin, Texas, R.G. Landes), pp. 195-206.

Lander, H.M. 1997.An essential role for free radicals and derived species in signal transduction. *FASEB Journal*, **11**, pp. 118-124.

Liboff, A.R. 1992. The "cyclotron resonance" hypothesis: experimental evidence and theoretical constraints. In *Interaction Mechanisms of Low-Level Electromagnetic Fields and Living Systems,* edited by B. Norden and C. Ramel (Oxford, University Press), pp. 130-147.

Liburdy, R.P. 1995.Cellular studies and interaction mechanisms of extremely low frequency fields. *Radio Science*, **30**, pp. 179-203

Lednev, V.V. 1991. Possible mechanism for the influence of weak magnetic fields on biological systems. *Bioelectromagnetics*, **12**, pp. 71-75.

Lin, H, Opler, M., Head. M. *et al.* 1997. Electromagnetic field exposure induces rapid transitory heat shock factor activation in human cells. *Journal of Cellular Biochemistry*, **66**, pp. 482-488.

Lin, H., Head, M., Blank, M. *et al.* 1998. Myc-activated transactivation of HSP70 expression following exposure to magnetic fields. *Journal of Cellular Biochemistry*, **69**, pp. 181-188.

Lin-Liu, S. and Adey, W.R. 1982. Low frequency amplitude-modulated microwave fields change calcium efflux rates from synaptosomes. *Bioelectromagnetics*, **3**, pp. 309-322.

Litovitz, T.A, Krause, D., Penafiel, M. *et al.* 1993. The role of coherence time in the effect of microwaves on ornithine decarboxylase activity. *Bioelectromagnetics*, **14**, pp. 395-403.

Lyle, D.B., Schechter, P., Adey, W.R. *et al.* 1983. Suppression of T lymphocyte cytotoxicity following expossure to sinusoidally amplitude-modulated fields. *Bioelectromagnetics*, **4**, pp. 281-292.

Lyle, D.B., Ayotte, R.D., Sheppard, A.R. *et al.* 1988. Suppression of T lymphocyte cytotoxicity following exposure to 60 Hz sinusoidal electric fields. *Bioelectromagnetics*, **9**, pp. 303-313.

McLauchlan, K.A. 1992. Are environmental electromagnetic fields dangerous? *Physics World*, pp. 41-45, January.

McLauchlqan, K.A. and Steiner, U.E. 1994. The spin-correlated radical pairs a reaction intermediate. *Molecular Physics*, **73**, pp. 241-263.

Milham, S. 1996.Increased incidence of cancer in a cohort of office workers exposed to strong magnetic fields. *American Journal of Industrial Medicine,* **30**, pp. 702-704.

Milham, S. and Osslander, E.M. 2001. Historical evidence that residential electrification caused the emergence of the childhood leukemia peak. *Medical Hypotheses,* **56**, pp. 290-295.

Moser, C.C., Keske, J.M., Warncke, K. *et al.* 1992. Nature of biological electron transfer. *Nature*, **355**, pp. 796-802.

Penafiel, L.M., Litovitz, T., Krause, D. *et al.* 1996. Role of modulation effects on the effects of microwaves on ornithine decarboxylase activity in L929 cells. *Bioelectromagnetics*, **18**, pp. 132-141.

Schumann, W.G. 1957. Uber elektrische Eigenschwindungen des Hohlraumes Erd-Luft-Ionosphare, erregtdurch Blitzentzladungen. *Zeitschrift Angewiss. Journal Physik,* **9**, pp. 373-378.

Tattersall, J.E., Scott, I.R., Wood, S.J. *et al.* 2001. Effects of low intensity radiofrequency electromagnetic fields on electrical activity in rat hippocampal slices. *Brain Research*, **904**, pp. 43-53.

Till, U., Timmel C.R., Brocklehurst, B. *et al.* 1998. The influence of very small magnetic fields on radical recombination reactions in the limit of slow recombination. *Chemical Physics Letters*, **298**, pp. 7-14.

Timmel, C.R. Till, U., Brocklehurst, B. *et al.* 1998. Effects of weak magnetic fields on free radical recombination reactions. *Molecular Physics*, **95**, pp. 71-89.

Valberg, P.A., Kavet, R. and Rafferty, C.N. 1997. Can low level 50/60 Hz electric and magnetic fields cause biological effects? *Radiation Research*, **148**, pp. 2-21.

Zhou, H., Suzuki, M., Randers-Pehrson, G. *et al.* 2001. Radiation risk to low fluences of α particles may be greater than we thought. *Proceedings of National Academy of Sciences USA*, **98**, pp. 14410-14415.

CHAPTER 3

Effects of Electromagnetic Fields in the Living Environment

Cyril W. Smith

3.1 INTRODUCTION

The title of this Chapter is deliberately ambiguous. It concerns both the role of electromagnetic fields in that part of our environment which is alive and the electromagnetic effects from those environments in which we live.

The writer has been involved since 1974 in experimental research on the interactions of electromagnetic fields with bio-materials and living systems. He cooperated in this with the late Professor Herbert Fröhlich FRS, a theoretical physicist. An early conclusion from this work was that there were anomalous magnetic field effects in water and in living biological systems and that these were only explicable in terms of coherence phenomena.

He first became involved in the diagnosis and therapy of patients hypersensitive to their electromagnetic environment in 1982 at the request of Dr. Jean Monro in London. Working with her electrically hypersensitive patients and those of Dr. W.J. Rea in Dallas, Texas, has given the writer an insight into the extremes of sensitivity and speed of reaction as evidenced by living systems exposed to electromagnetic stress when their dynamic homeostatic regulation has failed (Smith and Best, 1989).

These patients have a long history of existing hypersensitivities to many chemicals, and/or foods and particulates. They may react within seconds to something in their environment and they can readily distinguish *verum* from *placebo*. The frequency of an electromagnetic stress and its coherence (precision) is the clinically important parameter. There is a threshold for the intensity or amplitude of the field at the patient for the onset of any effects but, once this is exceeded its value usually matters little until the onset of thermal effects; it is the frequency which is important. The effects of frequencies are unique to each individual. As the frequency is scanned, therapeutic frequencies usually alternate with stressful frequencies. The clinically effective frequencies range from below 1 milliHertz (1000 sec/cycle) to hundreds of GigaHertz (10^9 Hz). Identical (false anticipatory) reactions can be triggered in a patient by chemical means and neutralised electrically, or triggered electrically and neutralised chemically; their electromagnetic and chemical environments interact. Frequency is the common factor linking the electromagnetic effects in these patients with the therapies acupuncture and homeopathy.

The first important result from this work was the finding that the clinical effects on patients of environmental frequencies, or chemicals, or homeopathic potencies could be reproduced with water contained in flame-sealed glass ampoules after its imprinting with coherent frequencies. This was without any chemical contact possible through the sealed glass. The unexposed water in these ampoules produced no clinical effects.

It was subsequently observed that certain specific and highly coherent frequencies would stimulåte the chakras, the Ting acupuncture points on the hands and feet as used in electroacupuncture as well as the other classical meridians which do not terminate at the Ting acupuncture points.

The acupuncture meridians are therefore envisaged as communication paths based on endogenous coherent frequencies. These are postulated to originate from coherence established as the organism develops from its embryo where the ectoderm, endoderm and mesoderm cell layers are in close proximity. Meridians of coherence persisting and growing as the organism develops would eventually link the acupuncture points to the target organs in the mature organism. This common frequency link can be followed from an acupuncture point along a meridian to its target organ.

There is enough water remaining in a histological specimen of an organ tissue even after staining and fixing, for a frequency measurement to be made through the glass of the microscope slide. These measured frequencies correlate well with the endogenous frequencies on the corresponding meridians.

Chemicals have characteristic frequency signatures resulting from their interaction with hydrogen-bonded water. In contact with the body, these are equally effective in producing an entrainment of the endogenous frequencies at appropriate meridians. Holding a glass bottle containing a chemical for just one minute is sufficient to entrain an acupuncture point to the signature; it takes about 10 minutes to relax back to the endogenous frequency. Failure to relax following chronic exposure would be equivalent to a chemical toxicity. The presence of frequencies, which fluctuate over a limited range, is a sign of appropriate dynamic responses in a normal healthy biological system.

Chemical toxicity in these patients and also in cells cultured *in vitro* is manifest through the imposition of the chemicals' fixed frequency signatures on the endogenous frequencies living systems. It has been found possible to erase and then re-program the frequency imprints of a cell culture and have these re-programmed frequencies were transmitted correctly to cultured daughter cells. This demonstrates that lasting effects, toxic or therapeutic, are possible.

Following from this work on the effects of electromagnetic fields on hypersensitive persons and of frequencies on acupuncture meridians, it is but a short step to examine of the effects of frequencies in the Living Environment on its inhabitants. This leads to the ancient Chinese discipline of Feng Shui which shares its basic concepts with those of acupuncture and thence relates to the electromagnetic environment and the impact of a building on the health of its occupants.

3.2 BACKGROUND SCIENCE

3.2.1 Fields

A field is 'something' capable of producing a mechanical force on an object located within its area or volume of action. A force gives rise to motion and these results in work being done. The rate at which work is done is power.

There are four kinds of these fields. The electric field which acts on electric charges which are atomic constituents associated with chemical bonds. The electric charges themselves generate a self-field and through this they can interact. Static electricity is the name given to the effects of electric charge when stationary. Electric charge in motion is an electric current. This current may be the bombardment of a TV/VDT screen by electrons moving in a vacuum or their flow along a wire. A current gives rise to a magnetic field. If electric charge is accelerated, the electric and magnetic effects radiate away at the speed of light. This is electromagnetic radiation.

There is a fourth field component, the magnetic vector potential which is in the direction of the current producing it and is proportional to it. Mathematically, it is a consequence of the magnetic field occurring only in closed loops with no free ends or poles. A toroid retains the magnetic field within the ring, the magnetic vector potential appears in the surrounding space.

Two sorts of electric charge motion may occur - random and oscillatory. Random motion of charges occurs as a background at all temperatures (other than absolute zero), it is called 'noise' from analogy with the sound of waves breaking on the beach. Oscillatory motion is analogous to the swing of a clock pendulum. It has some well-defined frequency {cycles per second or Hertz (Hz)} which may or may not be stable over time. It will also have phase. By analogy, the phases of the moon describe its position within the monthly cycle relative to new-moon. The degree to which various oscillations of the same frequency are in a constant phase is called their coherence. Coherent frequencies are associated with electromagnetic fields originating from technological sources, spatial structures, and chemical and particle interactions with water. All can interact with living systems and living systems with them. All may affect the status of a building as a healthy living environment.

The "classical" physics description of electromagnetism is adequate if the only forces are mechanical forces, which obey Newton's Laws of Motion. From the latter part of the 19th Century, effects started to be discovered which showed this to be an inadequate description of the natural world. Electric charges were found to be ejected by radioactive materials and by the action of light. Although radiation from a hot-body like the sun forms a continuous spectrum or rainbow, the radiation from individual atoms only comes in single colours (like sodium street lighting). This can only occur if the changes in the atom take place in whole number (integer) amounts called 'quanta', just as chemical changes take place involving integer numbers of atoms.

Only 'quantum physics' can describe such events, there is no chemistry in 'classical physics'. The positions of atoms and their component electrons are not precise; they have a certain probability of being in some given place, with the balance of probability for them being elsewhere; since they must be somewhere the total probability equals unity. This probability of finding an electron is described mathematically by a 'wave equation', which has solutions looking like oscillations or waves. This quantum state is called a 'qubit' and is the basis of quantum computing.

If the temperature is brought close to absolute zero, random motion becomes less than the oscillatory motions. Under these conditions, a long-range order of common phase-coherence may become established giving rise to a number of peculiar effects lumped under the term 'superconductivity' - zero electrical resistance - zero viscosity - zero penetration by magnetic fields. These effects are not due specifically to low temperature but to long-range order. If Nature can establish an equivalent long-range order by other means then it has the phenomena of superconductivity available for its use. A test for whether this actually has happened in Nature is whether the component of the magnetic field called the 'magnetic vector potential' has any biological effects. This 'magnetic vector potential' was for a long time though to be a mathematical fiction arising from the fact that magnetic fields only occur as closed loops. It can affect the phase of the wave equation's description of the probability of finding an electron. Its physical existence was demonstrated by its effects on interference between two electron beams. If any bioelectromagnetic effects or hazards are due to the magnetic vector potential, then no experiments in 'classical physics' can evoke or explain these.

The writer encourages experimental biologists to replace their ordinary coil with a toroid for magnetic field experiments. One who tried this wrote, "The results with the toroidal coil are quite tantalizing. Despite the fact that the magnetic field is negligible, significant increases in abnormalities (in pattern formation in *Drosophila* embryos) are found over matched controls, both when the embryos are in place before or after the power supply is switched on" (Ho *et al.*, 1994).

3.2.2 Coherence

3.2.2.1 Fröhlich and Coherence in Biological Systems

"A Voyager of Discoveries" was the title I gave to my appreciation of Herbert Fröhlich written for Electromagnetics News (Best, 2001). Professor Dr. Herbert Fröhlich FRS, Professor Emeritus of Theoretical Physics at Liverpool University, was born at Rexingen in the Black Forest of Germany on the 9th December 1905 and died in Liverpool on the 23rd January 1991.

He belonged to the generation of the founding fathers of theoretical physics all of whom he knew well. He left school at the age of 15 and for some time travelled the German countryside collecting folk songs. After a short while in the family business he became interested in physics through building his own wireless-set and taught himself enough mathematics to gain admission to Munich

University. Here he studied under the great teacher, Arnold Sommerfeld who set him a problem concerning the absorption of light by metals. In just over two years, Fröhlich produced a thesis containing a solution for which he received his D.Phil. from Munich without ever doing a bachelors degree.

He was then offered and accepted a teaching post at Freiburg University and started what should have become a distinguished academic career in Germany by commencing his first book, "Elektronentheorie der Metalle". But, as early as 1933, the Nazis had organized his dismissal from Freiburg. He was forced to emigrate and went to work with Frenkel at the Physico-Technical Institute in Leningrad until two years later Stalinism forced his hurried departure to Bristol University. Here, he worked under Professor Sir Nevill Mott in the H.H. Wills Physical Laboratory becoming interested in cosmic rays, and dielectrics and through cooperating with Heitler he became familiar with quantum field theory.

Fröhlich remained at Bristol University until 1948 when Professor Chadwick invited him to take up the first chair of Theoretical Physics at Liverpool University. Here he made major advances in the theoretical physics of metals, dielectrics and semiconductors, mesons and superconductivity, and the application of theoretical physics to biology. His tenure of this Chair lasted from 1948 until 1973 when he retired and became Emeritus Professor. Throughout this time, he taught and formed two generations of theoretical physicists at his coffee table discussions and weekly seminars. His kindness and informality is remembered by all, students, collaborators and colleagues alike. His incisive insight and immediate grasp of the essentials of physical problems was legendary.

He had already considered biological problems in relation to theoretical physics in the 1930's. His friend, the endocrinologist Max Reiss told him that biological membranes maintain a small electric potential of a tenth of a volt, he asked their thickness, and on hearing that it was a millionth of a centimetre, realised that this corresponded to the enormous electric field of ten million volts per metre which ordinary electrical insulators (dielectrics) will sustain only when special precautions are taken. An elastic constant corresponding to the velocity of sound gave him a likely resonant frequency of the order of a hundred gigaherz. Such frequencies were not available for experimentation at the time but, he sought the help of Willis Jackson to see whether they could be generated to investigate the millimetre wavelength properties of bio-systems as had been proposed by Victor Rothschild. But, the Second World War intervened and he had no opportunity to develop these ideas until in 1967 he was able to present them in his paper "Quantum Mechanical Concepts in Biology" to a conference on Theoretical Physics and Biology arranged by the Institut de la Vie and held in Versailles, France. Here, he compared the collective (co-operative) behaviour biological systems with the long-range phase correlation (coherence) found in the Einstein condensation of a Bose gas to a single quantum state. This occurs in superfluidity and superconductivity phenomena at very low (liquid helium) temperatures. It was at that same meeting that Prigogine presented his ideas about "Dissipative Structures".

The subsequent development of Fröhlich's ideas and the work of his collaborators around the world in the validation of his theoretical predictions are summarised in his two "Green Books" ([Fröhlich *et al.*, 1983; Fröhlich *et al.*, 1988). It must not be forgotten that concurrently he also was involved in major contributions to three other areas of theoretical physics. A complete picture of his work is to be found in the *Festschrift* for his eightieth birthday [Barrett and Pohl, 1987). Pauli once remarked of Fröhlich, "Ah yes! There we have someone who can not only calculate but can also think".

In general, Fröhlich considered that from the point of view of physics, active biological systems could be characterised by three properties: they exhibit a non-trivial order, they have extraordinary dielectric properties and they are relatively stable but far from equilibrium.

In 1988 his *Second Green Book* was published under the title, "Biological Coherence and Response to External Stimuli". This was his last contribution and in it he remarked that a multi-causal approach is required in dealing with the causation of such activities citing as an example, an enzyme and substrate which have to be brought together before they can interact.

The following summarises Fröhlich's work on biological systems:

In 1972, he published a paper in which he considered the resonance interaction between two molecules one within, the other outside a membrane. He showed that this resulted in long range interactions and the possibility of coupling to the natural membrane resonance which he predicted to be of the order of 100 GHz.

In 1973, he was considering the connection between Macro- and Micro-Physics based on an analogy with macroscopic wave functions in superconductivity's long-range coherence. He remarked that the organisation of biological systems is not based on spatial order and that the supply of metabolic energy could give rise to a long-range coherence as in the case of a laser.

In 1977 he showed that coherent excitations might give rise either to a single polar mode, to a metastable state, or to limit cycles (Lotka-Volterra oscillations) and considered the possibility of resonance effects under externally applied radiation. He surveyed the likelihood of macroscopic quantum effects in physics and biology in general.

In 1978 he was dealing with multi-component systems and the cancer problem. He pointed out that an absence of bio-control suggests loss of the long-range order and thereby, the loss of one or more 'Cellular Communication Systems'. If a cancer cell does not contribute to the normal collective excitation it loses coherence of frequency and phase. He also pointed out that the disordered cancer state might be restored to normality by externally applied radiation of the correct frequency.

In 1980, he discussed the biological effects of microwaves A 1974 report of a meeting of the USSR Academy of Sciences had detailed nine papers reporting the influence of millimetre waves on various biological systems and this was the first experimental evidence in support of his 1936 calculation. In 1975, he had considered the extraordinary dielectric properties of biological materials and the action of enzymes. In 1980, he followed this with a mathematical analysis of periodic enzyme reaction rates and examined the interaction between highly polar dielectric states and elastic displacements leading to metastable ferroelectric states.

In 1982 he discussed the conditions for coherent excitations in biological systems and in 1983 he presented evidence for coherent excitations in biological systems in his first *Green Book*, "Coherent Excitations in Biological Systems". In 1986 he was again considering coherent excitations in active biological systems and made the prophetic remark,

"Any future theory of biology will make use of coherent excitations in active biological systems".

3.2.2.2 A QED Theory for Coherence in Water

The late Professor Giuliano Preparata and his co-workers in the University of Milan, began to investigate theoretically the phenomena of coherence in condensed matter (i.e. anything not a gas) with the tool of Quantum Electrodynamics (QED). Del Giudice and co-workers remarked that, "The proposal of Herbert Fröhlich to assume the density of electric polarization as the 'order parameter' relevant to biological systems was then a fundamental step. Living systems can be affected by many agents in many different ways, but these add up to modifications of this basic parameter, the density of electric polarization".

Del Giudice, Preparata and colleagues went on to show that the usually neglected interaction between the electric dipole of the water molecule and the quantized electromagnetic radiation field if treated by QED gave collective modes and the appearance of permanent electric polarization around any electrically polarized impurity (Arani *et al.*,1995).

Preparata, in his book "QED Coherence in Condensed Matter" (Preparata, 1995) considered the application of QED coherence techniques to liquid helium, electrons in metals and the Mössbauer effect, superconductivity, gravitational waves, cold fusion, and ferromagnetism before coming to the more difficult case of water. The model that QED coherence presents for water is the same as for all other QED coherent condensed matter systems namely, two distinct interspersed phases.

One phase is an incoherent phase comprising water molecules as in the gas phase but more densely packed. The other phase is a coherent phase consisting a collection of "coherent domains" 75 nm in size within which the water molecules interact coherently with a very large classical electromagnetic field. Once water vapour exceeds a critical density, the coherent phase arises from naturally excited coherent oscillations involving the 5d-transition between the molecular ground energy state. The 12.06eV excited state corresponds to the lowest critical vapour density and this means that 75 nm domains are those most likely to form as water vapour condenses. These domains are stable against thermal fluctuations. At 300°K the fraction of coherent water is calculated to be 28%.

This theory of water is the only one to give a complete description of the dynamics of the coherent phase and the thermodynamics of the incoherent phase. Its predictions are the only ones in good agreement with experimentally determined values for the physical constants of liquid water - relative density - critical density - specific heat - the compressibility of super-cooled water - the 230°K anomaly - boiling temperature - latent heat of vaporisation - the low frequency dielectric constant. It offers qualitative explanations for hydrogen-bonding - the density anomaly at 4°C - the boiling process - the formation of biological membranes (Del Giudice and Preparata, 1994) and also gives a good account of like-charges attraction in colloids.

3.2.2.3 Diamagnetism and Order in Water

It is possible to remain within accepted "classical physics" and yet to see that there must be something very unusual going on in water. Energy calculations do not involve the details of any process by which they take place, only the initial and final states are involved. Water is a diamagnetic substance even in "classical physics". That is, it permanently reduces the strength of any applied magnetic field. This can only happen through the molecular equivalent of a short-circuited loop (magnetic dipole) of induced superconducting current opposing the magnetic field giving rise to it. The magnetic susceptibility χ_m (magnetisation $\mathbf{M}$ / magnetic field $\mathbf{H}$) for water is $\chi_m ?= -0.88 \times 10^5$ and its relative permeability $\mu_w = 1 + \chi_m$.

The energy per unit volume in a magnetic flux density B is given by

$$U = \tfrac{1}{2}\, B2 / \mu_0 \mu_r \quad \text{J/m}^3 \tag{3.1}$$

The difference between this energy in water and the energy in free space of same volume in the nominal geomagnetic field of 50 µT, is given by

$$\Delta U = \tfrac{1}{2}\,(B^2 / \mu_0)\, \{1/(1 + \chi_m) - 1\} \tag{3.2}$$

$$= \tfrac{1}{2}\,(B^2 / \mu_0)\, \{-\chi_m/(1 + \chi_m)\} \tag{3.3}$$

$$\simeq \tfrac{1}{2}\, B2 / \mu_0 \{-\chi_m\} = 8.73 \times 10^9 \ \text{J/m}^3 \tag{3.4}$$

The thermal energy at ambient temperature is $kT_{20}°\,C = 4.04 \times 10^{-21}$ J. The minimum volume within which the energy change due to water diamagnetism equals the thermal energy is 4.63×10^{-13} m^3 or 77 (µm)^3. For domains of this size or greater, the diamagnetism of water is able to produce an ordering which is stable against thermal fluctuations.

The experimental threshold for potentising a water film between the jaws of a micrometer is 88 µm (with the micrometer axis in the N-S direction). The wavelengths corresponding to transitions between the FIR water lines 357 – 213 cm^{-1} = 144 cm^{-1} = 68 µm are consistent with this diamagnetic energy calculation.

Taking the 50 cm^{-1} ($\omega_0 = ?1.5 \times 10^{12}$ Hz) resonance in water instead of 12.06 eV, Del Giudice (Frontier Sciences - Maria Laach Symposium, 1989) showed that there would be 100 µm size domains of coherence which would contain N ~10^{17} molecules. If quantum fluctuations must not break the coherence, their frequency v must be given by $v \le ??1.5 \times 10^{12} / N^{\frac{1}{2}} = 5$ kHz. The velocity V of the coherence would be given by $v\,\lambda$ where the coherence length λ is twice the domain size. This gives V ~ 1 m/s which is of the same order as the measured velocity of propagation of coherence in water. The 75 nm domains are those likely to form first on condensation, they may not represent the only level of coherence in liquid water.

3.2.3 Coherence

There are several very important effects of particular relevance to water and bio-systems, which can take place within a coherent system. Firstly, the coherence length becomes the constant parameter instead of the velocity of propagation. Frequency then becomes proportional to the velocity and apparently without restriction so long as one remains within the coherent system. Energy only becomes involved in setting up the coherence in the first place. There can be many velocities (even greater than the vacuum velocity of light) with corresponding frequencies. Interactions between widely different parts of the spectrum can occur, as has been found experimentally.

Two important velocities which occur are firstly, the free-space velocity of light [300 Mm/s] representing no interaction between the radiation and the individual molecules of the coherent system, that is there is unity refractive index. Secondly, the velocity of coherence propagation [~m/s] which represents interaction of the radiation with the whole of the coherent system.

Coherence propagates as *order* by a diffusion process, rather like heat travelling along a saucepan handle although heat represents propagation of disorder. The soliton equation is a non-linear diffusion equation. It has been conjectured that coherence might propagate as solitons, which are solutions representing spatially localised waves of constant shape and velocity.

The lack of characteristic frequency scales is a consequence of coherence and implies a fractal system with self-similarity, scale invariance and power laws. Power laws involving frequency have been widely observed in work on biological systems. They are implicit whenever a log-log plot is required to display the data, the power is its slope.

3.3 COHERENT FREQUENCIES IN THE LIVING ENVIRONMENT

Considerable effort is directed towards what is called 'Electromagnetic Compatibility'. This involves the application of techniques for ensuring that one piece of electrical equipment does not interfere with another.

Two examples of what can happen when there is interference - there was a report of a nuclear weapon having become armed inside a parked aircraft due to a particular combination of signals being picked-up. These actually originated as pop-music from a local radio station and transmissions from a police car on a road nearby. The other example is that of a premature infants' incubator developed in a workshop in the basement of a hospital which overheated when taken into wards high in a tower building overlooked by another building with a radio transmitter on top of it. These radio signals paralysed the temperature control circuits of the incubator.

So far, no effective consideration has been paid to making electrical equipment compatible with human regulatory circuits which may be equally electronic, nor to making electrical equipment compatible with radiation emitted by living systems. The following Sections describe various sources from which coherent signals can enter the Living Environment and possible ways in which these could affect persons exposed to them.

3.3.1 Technical Oscillations

The past 150 years has seen the widespread development of the supply of electrical power to the living environment. Originally this was direct current (DC) from lead-acid storage batteries recharged at intervals by dynamos. The DC gave rise to the same corrosion problems that used to occur at car battery terminals. In the damp English climate, wallpaper used to become stained green/black near light switches.

The invention of the transformer enabled large amounts of electric power to be generated centrally and transmitted efficiently over considerable distances. This required an alternating current. The frequency adopted in Europe was 50 Hz while America adopted 60 Hz. There are other frequencies around, the railways in Germany and some other countries use 16.6 Hz (one-third of 50 Hz) as it makes starting the trains easier. With very long transmission lines their length becomes significant in relation to the wavelength of 50 Hz, this causes problems which may be avoided by going to 25 Hz or DC. The frequency 400 Hz may be encountered in ships and aircraft as it makes for lighter-weight transformers. Much of the world in now permeated with 50 Hz or 60 Hz the closely controlled frequencies of the supply voltage oscillations. Harmonics may be present, and the currents and hence the magnetic fields may contain frequency components due to load fluctuations. For example, a major power line circuit when measured had a considerable 300 Hz magnetic field component originating from high power 6-phase rectifiers supplying a DC cable.

Electrical power supplies and electromagnetic transmissions are brought into almost every living environment to feed a wealth of electrical and electronic apparatus: heaters, motors, cookers, microwaves, radio, TV, recorders, satellite transmissions, portable and mobile phones and their base stations, computers, security systems, etc. In hospitals and industrial places there may be radiofrequency heaters (diathermy) and pagers as well as specialist apparatus such as magnetic resonance and X-ray imaging systems. The whole of the available electromagnetic spectrum with a few exceptions for the benefit of radio astronomy has been carved up and allocated for some purpose by international agreement.

3.3.2 Electromagnetic Resonators

Electrical oscillations are waves just as much as sound although they do not involve oscillations of air-mass. The size of a musical instrument determines its note or pitch. It is the amplitude of oscillations, the length of a string or pipe and the size and shape of an amplifying resonator which gives a musical sound its characteristic loudness, pitch and timbre. Likewise, the dimensions of any object, which has electrical properties different from the general environment, can give it similar resonance characteristics. The interior of a car can act as a resonant cavity for mobile phone frequencies. Railings (any array of metal posts) can act as a resonator as well as a diffraction grating which EM sensitive persons can sense. In effect, any metal object will have characteristic electromagnetic resonances within some part of the electromagnetic spectrum now in use.

3.3.3 Nature's Resonators

The layer of electrically conducting air in the upper atmosphere called the ionosphere, which bounds the earth-ionosphere resonant cavity gives natural frequencies of resonance to oscillations excited by distant thunderstorms. These are the Schumann Bands of radiation, which extend from about 7.8 Hz to about 30 Hz and come within the range of natural brain-wave frequencies.

Experiments carried out by Wever in which volunteers lived in underground steel bunkers for extended periods showed that removal of the Schumann radiation resulted in the body-clock going into a free-running mode with a 25-hour day or longer. The provision of a weak field at 10 Hz would maintain synchronism and re-establish lost synchronicity. It was even possible to force a 23-hour day with this (Wever, 1973).

There are also naturally occurring electromagnetic impulses called atmospherics or "sferics". These are in the range 10-50 kHz. They arise from lightning discharges and counter movements of air masses and can have biological as well as other effects such as those found by Hans Baumer on the bichromate-gelatin colour printing process. The ionosphere is itself affected by atmospheric tides, solar radiation and other incoming radiation and thus can transmit these changes to living systems through its frequencies

3.4 COHERENT FREQUENCIES IN WATER

There is a duality between the chemical bond and frequency – otherwise chemical analysis by spectroscopy would not be possible. The phenomenon of multiple frequencies within coherent systems can bring the frequencies corresponding to chemical spectra within the range of biological effectiveness. In general, any chemical, which can hydrogen-bond even to traces of water will have a characteristic frequency signature. Chemical elements have a single frequency, simple compounds like sodium chloride have three frequencies, more complicated molecules have more frequencies. Molecules such as 100% halogen saturated hydrocarbons appear to have no characteristic frequency signatures.

In addition to these characteristic frequencies associated with water, it is possible to imprint frequency resonances into water, and therefore into living systems which are mostly water.

3.4.1 Techniques for Imprinting Frequencies into Water

The following techniques are available to imprint a frequency into water from a source such as a solenoid or toroid, fed with an oscillating current or to copy an existing water imprint into "clean" water (Smith, 1994).

3.4.1.1 Clinical Aspects

Clinically significant information can be imprinted into a vial of water by succussion in contact with a biological system, a chemical or a homoeopathic potency. It is the *succussion* or sharp banging process of homoeopathic remedy preparation that creates the *potency*. Body frequency information from a patient can be collected similarly. If the patient holds and succusses a vial of water, this will contain an imprint of the body frequencies; it can be wrapped in aluminium foil and mailed for away measurement. Information imprinted in water seems to be retained indefinitely under normal conditions. A homoeopathic potency prepared by Samuel Hahnemann was found to be clinically effective 150 years later.

3.4.1.2 Imprinting by Contact

Frequency transfer takes place through the glass of a vial containing "clean" water if immersed in water imprinted with the required frequencies. The high frequencies potentise quickly, low frequencies may take hours or even days to imprint (Smith, 1995).

3.4.1.3 Proximity and Succussion

A vial of "clean" water is placed near to a source of the frequencies (e.g. chemical, potency or coil). It is potentised or imprinted either by mechanically succussion or, by briefly applying the field from a strong permanent magnet. For this purpose it is sufficient to effect mechanical succussion by a single sharp bang on a resilient surface such as wood, this is so as not to break the glass. The Hahnemannian method uses more impacts. Each subsequent dilution and succussion will in general introduce more frequencies, certainly different ones.

3.4.1.4 Toroids

The toroid contains the magnetic field within itself but radiates the magnetic vector potential into the surrounding space. The magnetic vector potential may appear as a chemical potential term in the wave function.

Placing a ferrite toroid between the source of frequencies and a vial of "clean" water will couple the frequencies to the water. Succussion will imprint them. Strangely imprinting also occurs if a toroid is succussed instead of the water. Clearly, something is going on in time and space which is not understood. Using a single toroid interchanges the stimulatory and depressive phases of the biological effects imprinted into the water. A second toroid gives a second phase reversal and hence makes an exact copy of the source frequencies. A small remanence magnetic field is required inside the toroid for bias as in a tape-recorder, otherwise twice the frequency is copied.

This has clinical applications in that the effects of potencies can be directly applied to the body using two toroids. Allergens can be neutralised using a single toroid to imprint anti-allergen bio-information from a sample of the allergen.

3.4.1.5 Voltage Pulses

A sequence of 7-unidirectional pulses of electric potential applied to a metal beaker containing a frequency source which may be an ampoule of frequency imprinted water or a toroidal coil connected to an electrical oscillator set to the required frequency will make a copy into a nearby ampoule of "clean" water which need not be inside a metal beaker. Reversing the polarity of the pulses reverses the phase of the effect. The need for the beaker implies that this is a change of electric potential effect rather than an electric field effect. The more so as an experiment using the very high electric field produced at a sharp metal point would not imprint its frequency intro water. Such pulses, even at the level of nerve impulses in living systems should be able to perform mathematical operations on endogenous frequencies or frequency imprints (Smith, 2001).

3.4.1.6 Magnetic Fields

This method of imprinting water requires an alternating current in a coil (e.g. solenoid) at the required frequency with sufficient current to give a magnetic field above the threshold required for imprinting. Alternatively, a toroidal coil may be used to generate an alternating magnetic vector potential. In this case, an additional magnetic field or succussion is required to effect the imprinting. Such a magnetic field may be from a permanent magnet or from a coil using any frequency, which is not greater than that being imprinted from the alternating magnetic vector potential.

3.4.1.7 Chemical Potentising

Chemical signatures can be used as sources of frequencies. The addition of a drop of dilute solution of hydrogen peroxide to water in direct contact or through glass, with a chemical signature is equivalent to succussion, and effects an imprint. Succussion of a dilute solution of hydrogen peroxide produces a continuum of stimulatory phase frequencies. This is the nearest approximation to water imprinted by a 'Healer'. Formaldehyde gives the opposite on succussion, a continuum of stressful or depressive frequencies.

3.4.2 Techniques for Erasing Frequency Imprints in Water

3.4.2.1 De-Gaussing

Electrically "clean" water is chemically pure water which has been "erased" of its frequency imprints by briefly shielding it from the geomagnetic field. That is by placing it inside a suitable steel enclosure giving less than about 380 nT. This erasure seems permanent and imprints thus erased cannot be recovered by any method found so far. Chemical frequency signatures are unaffected by this and appear to be permanent so long as at least a trace of water is present.

3.4.2.2 Heating

Heating imprinted water to between 70°C and 90°C alters the imprint. For example, 1 kHz was imprinted into water. At 70°C additional resonances appeared at 100 Hz and 10 kHz. Succussion of an imprint heated above 70°C but below 90°C restored the original 1 kHz and the 100 Hz and 10 kHz disappeared. From 90°C to boiling point no imprint of 1 kHz remained. However, the original 1 kHz imprint could be "recovered" by succussing in the presence of 7.8 Hz, 384 MHz; 1.42 GHz or 2.65 GHz. The 1.42 GHz corresponds to the 21-cm wavelength resonance of molecular hydrogen and water will not take up a frequency imprint in a 21 cm open-ended tube nor in a closed 10 cm cuvette.

Water imprinted at 1 kHz was heated in an autoclave to 121°C for 3 minutes and allowed to cool. It showed no imprint when tested with a toroid but the imprint could be detected with a 'Caduceus' (bi-filar) coil. Succussion at any of the above frequencies fully restored the 1 kHz.

3.4.2.3 "Hiding" Frequency Imprints

The frequencies 2.65 GHz, 1.42 GHz and 384 MHz have further unexpected properties. It is possible to "hide" an imprint by succussing it on one side (depending on the relative direction of the geomagnetic field) of an oscillator output coil at one of the above peculiar frequencies. Similarly, a dilute solution of a chemical (e.g. NaCl at 6 mM) can have its chemical frequency signature "hidden". The "hiding" of a chemical signature only works for a dilute solution, this suggests that the molecules of the chemical may have become enclosed within coherence domains which protects them from external measurement fields. If so, this would enable the concentration of domains to be determined.

The frequency 384 MHz is the high frequency branch of the heart meridian and chakra endogenous frequency, the low frequency branch is 7.8 Hz. This frequency also "restores" a hidden imprint as does holding a hidden imprint near the heart chakra. That 7.8 Hz and 384 MHz have unusual effects on water is no more remarkable than the heart meridian and chakra having their endogenous frequencies on a Schumann (geophysical) resonance.

For clinical and environmental purposes, it may be sufficient to hide frequencies so that the body does not recognise them although there remains the possibility that the heart frequencies will be able to un-scramble the hidden bio-information. Some protective devices make use of this phenomenon.

3.4.3 Techniques for Reading a Water Frequency Imprint

This presents a very great measurement problem because coherent endogenous frequencies in living systems may be anywhere in the spectrum from below milliHertz to beyond GigaHertz. These frequencies do not seem to be frequencies of "classical" electric or magnetic fields but rather frequencies of magnetic vector potentials or quantum fields which for instrumental measurement must be converted to the former. The following paragraphs list the presently available techniques for the detection and measurement of such frequencies.

3.4.3.1 Clinical

Homeopathic "Proving-Symptoms" and results of clinical tests on electrically hypersensitive patients make use of this biological detection. A century ago, the American physician, Dr. Albert Abrams, was using percussion of the abdomen for clinical diagnosis; he also noted the geo-directional effects.

3.4.3.2 Dowsing

Dowsing (Rutengänger, Radiesthesie) techniques have been developed by the Writer for the detection of resonances in water, allergen dilutions and homoeopathic potencies and for the diagnosis of reactions in the most highly hypersensitive patients. Imprinted and endogenous frequencies are correctly detected. Characteristic coherent frequency signatures can be measured in most chemicals containing traces of water. These disappear after thorough drying but return when traces of water are added. No chemical signatures were found in the case of 100% halogen saturated molecules. The Writer's personal sensitivity threshold for detecting the chemical frequency signature of sodium chloride in solution as it is successively diluted occurs at a concentration of about 0.3 ppm by weight. Although the internal consistency of the measurements is good, mind-body interactions make 'double-blinding' difficult. Computer generated blinding seems to offer the best prospects.

3.4.3.3 Electrodes

Electrodes immersed in frequency imprinted water and connected to the input of a sensitive, low-noise and narrow band amplifier, or a signal analyser, can be made to detect the imprinted frequencies in the kilohertz region when the water is excited by that frequency. This technique is very difficult to implement consistently although imprinted frequencies have been correctly detected. A possible physical mechanism is one whereby charges entering coherent water domains must do so as electron-pairs; this depletes the charge density at the water/electrode interface resulting in an increased electrical resistance which in turn is converted to a differential input voltage by the small DC off-set current of the head amplifier (Smith, 1994).

3.4.3.4 Laser

Dr. Peter Gariaev and Dr. George Tertishny, from the Institute of Control Science, Russian Academy of Science (Prangishvili *et al.*, 2000) have developed a special 2-beam laser using a non-standard He-Ne laser (632.8 nm) in which the beams are polarised orthogonally. Each beam interacts with the other giving entangled states to the photons. The beams can be separated with polaroids or combined confocally. With this, they can measure structures producing optical rotation. If liquid crystalline DNA (between glass slides) is placed in this laser beam and an aligned mirror is placed about 80cm further away, radio-frequencies can be picked up on an ordinary AM radio in the vicinity but only when the laser is on. These measurements are made at radio-frequency and A/D converted so a Fourier spectrum can be generated. The orthogonally polarised laser beams cannot interfere unless rotated by the DNA.

In experiments using this laser system with homeopathic potencies, a control sugar solution showed no bio-information, an active potency of Pt-D12 gave 2-peaks, at 2.2 kHz and 4.5 kHz with some fine structure.

Subsequently, the Writer prepared a D12 potency of platinum and independently measured the following three frequencies: 2.301 kHz, 4.455 kHz and 2.57 MHz. The latter was obviously the above radio-frequency and the others its modulation. This laser system does seem to be able to measure frequencies imprinted into water.

The writer has detected resonances imprinted in water using two orthogonally polarised beams of light from an LED, detected with a photodiode.

3.4.3.5 Microcalorimetry

Professor V. Elia (Elia and Niccoli, 1999) and his co-workers have carried out an extensive thermodynamic microcalorimetric study on aqueous solutions obtained through successive dilutions and succussions. The exothermic heat of mixing with acids or bases differs between the dilute solutions, homoeopathic potencies and frequency imprinted water when compared with the control untreated solvent water. It was considered that successive dilutions and succussions may permanently alter the physical-chemical properties of the solvent water for which an hypothesis of a disorder-order transition could be proposed.

3.5 WATER MEMORY

3.5.1 Trace Water in n-alkanes Points to a Water Memory Mechanism

Following retirement, the Writer had to close his university laboratory in 1991. Before this, he measured the signature frequencies of all the chemicals in stock. In the course of this, ELF resonances were found in the n-alkanes but only when there was a (>14ppm) trace of water present (Smith, 2000). In general, the highest frequencies increased with the n-alkane chain length.

If there are to be interactions involving the spectra of coherence domains in water and the characteristic molecular spectra of n-alkanes, these must be in the far-infra-red (FIR) rotational spectrum because this is the only region where n-hexane has a spectrum. It is widely used as a solvent in spectroscopy because of its clean spectrum.

There needed to be some arbitrary restriction on the choice of far-infra-red water lines because there are the hundreds of rotational water lines which might otherwise have had to be considered. It was noted that the water lines 28 μm (357 cm⁻¹), 47 μm (213 cm⁻¹), 78 μm (128 cm⁻¹) and 119 μm (84 cm⁻¹) can become coherent enough for use in a water vapour laser, hence it was conjectured that these might also provide the necessary coherence for water memory.

The FIR wave numbers of the first three of these water lines and the tabulated spectra for the n-alkanes were taken and the energy gaps taken as their wave number differences. Figure 3.1 shows the result for n-hexane. It is emphasised that the ordinate is derived from spectral tables and only the abscissa is a measurement. The overall ratio for the combined n-pentane and n-hexane lines was 1.97(± 0.16) × 10¹¹ Hz_{FIR}/Hz_{ELF} or 6.57 cm⁻¹ per Hz_{ELF}.

These resonances could be simulated using 3cm/Å plastic molecular models of the n-alkanes immersed in saline as described in Section 3.9.3.11. This is the ratio of the velocity of light to the velocity of coherence propagation. The ELF frequencies measured were the same as those frequencies measured on the actual

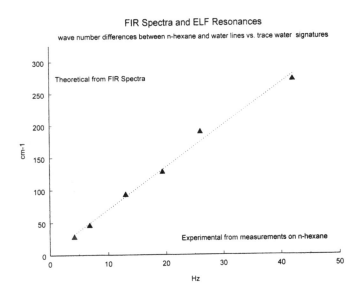

Figure 3.1 Wave number differences between the spectra of n-hexane and the water laser lines with the measured ELF resonances for trace water in n-hexane.

chemicals when the whole length of the chain was bridged with water molecules. This technique is possible because of the multiple-frequency phenomenon in coherent systems. Since insulating molecular models immersed in a conducting medium can be used, Babinet's Principle must be applicable.

3.5.2 Frequency Ratios in Bulk Water

The previous section relates to frequency effects in trace water in n-alkanes. The next question was whether the same arguments could be applied to water alone even if there was no n-alkane present. The energy level differences would have to be those between the water lines themselves.

A preliminary test was made to determine whether the above high-to-low frequency ratio could be found in water. Accordingly, water was imprinted at frequencies between 0.001 Hz and 0.01 Hz which were chosen for reasons of available frequency coverage. This water showed corresponding resonances between 200 MHz and 2GHz giving a mean frequency ratio of $1.98 (\pm 0.07) \times 10^{11}$ Hz_{FIR}/Hz_{ELF}, which is within the standard deviation for water in n-alkanes.

The converse experiment required water to be imprinted at frequencies between 200 MHz and 2GHz. These showed resonances between 0.001 Hz and 0.01 Hz with a mean frequency ratio of $2.09 (\pm 0.43) \times 10^{11}$ $HzFIR/Hz_{ELF}$. The worse standard deviation reflects the greater difficulties associated with the measurement of the very low frequencies. In carrying out such measurements by dowsing a phase comparison must be being made since at periods between 1000 sec/cycle and 100 sec/cycle each measurement takes less than a single cycle of the oscillation.

3.5.3 A Possible Mechanism for "Water Memory"

The next question to be settled was whether there were any measurable ELF frequencies corresponding to differences between the FIR water lines alone and in the absence of n-hexane. The difference between the water laser lines, 357 cm^{-1} - 213 cm^{-1} = 144 cm^{-1} and the ratio 6.57 cm^{-1}/Hz$_{ELF}$ gives 22.6 Hz (pentagonal water helix resonance). Likewise, 213 cm^{-1} - 128 cm^{-1} = 85 cm^{-1} giving 13.3 Hz (random chain water resonance). With 357cm^{-1} - 128 cm^{-1} = 229 cm^{-1} it gives 34.8 Hz (Caduceus resonance). All these ELF frequencies were detected.

3.5.4 Effect of a Frequency Imprint in Water

Then, it was necessary to determine what happens to all these frequencies if an ELF frequency is imprinted into water by succussion. Table 3.1 shows that imprinting at ELF (10 Hz) splits the frequencies in the ELF, RF and FIR. For this to happen, all these lines must be extremely coherent. The limited accuracy of measurements in the FIR (calibration at 2 microns intervals) makes it difficult to assign the measured wave numbers. What was measured as 364 cm^{-1} was probably the 357cm^{-1} line and that measured as 239 cm^{-1} was probably the 213 cm^{-1} line. If the imprint frequency was greater than the endogenous frequency, only the sum frequency was detected.

Table 3.1 Effect of imprinting frequencies into water by succussion

Frequency Imprinted	10 Hz
Frequencies Measured with Oscillators	
32.15 Hz and 12.78 Hz	= 22.6 ± 10 Hz
22.21 Hz and 3.196 Hz	= 13.3 ± 10 Hz
2.17 GHz and 0.33 GHz	= 1.25 ± 0.92 GHz
3.21 GHz and 1.25 GHz	= 2.23 ± 0.98 GHz
Frequencies Measured with Resonators	
24 μm (416 cm^{-1}) and 32 μm (312 cm^{-1})	= 364 ± 52 cm^{-1}
32 μm (312 cm^{-1}) and 60 μm (166 cm^{-1})	= 239 ± 73 cm^{-1}

3.5.5 Frequency Effects on Dilution and Succussion

If potentisation is a quantum phenomenon, it is not likely to be linear. The following tests were carried out to determine whether there was linearity or discontinuity.

Water was imprinted by succussion at 1 Hz. It was then serially diluted tenfold (1+9). The 1 Hz remained. When it was succussed, the 1 Hz disappeared and was replaced by 10 Hz. In general, what happened for the many frequencies and dilutions tested was that the original 1 Hz disappeared to be replaced after succussion by a frequency = 1 Hz × the dilution factor.

To further investigate whether potentisation is a linear or discontinuous process, water was imprinted by succussion at 1 Hz. Then, 10ml aliquots were diluted in 10% steps and succussed individually. The result was that the imprinted frequency remained at 1 Hz until a dilution ratio of 1.4 was reached. When the dilution was 1.5, the frequency jumped to 1.5 Hz. For dilutions from here to 1.9, the frequency remained at 1.5 Hz. At dilution 2.0, the frequency jumped to 2 Hz. All this supports the idea of discontinuities with selection rules. This was the general pattern at other dilution ratios.

However, there were some exceptions. Although on succussion the 4-fold dilution gave 4 Hz so did a 5-fold dilution; the 6-fold dilution gave 6 Hz but so did a 7-fold dilution; the 11-fold and 13- to 19-fold dilutions did not imprint any frequency at all; the 20- to 23-fold dilutions all gave 20 Hz; the 24- to 29-fold dilutions all gave 24 Hz; the 30-fold dilution gave 30 Hz. There were similar integer results for the dilution ratios in the 10's, 100's and 1000's and with similar exceptions.

3.6 ENDOGENOUS COHERENT FREQUENCIES ON CHAKRAS AND ACUPUNCTURE POINTS

3.6.1 Coherent Frequencies on the Chakras

So far, the presence of characteristic frequencies for chemicals has been described. The next step concerns living systems. As a result of investigating various devices claimed to provide therapeutic benefits to the user, it was realised that many generated the Schumann Band frequency 7.8 Hz. If these devices were genuine, then 7.8 Hz ought to be able to produce some measurable effect on the body. The body chakras were the most accessible points on which to check this out. It transpired that 7.8 Hz stimulated the heart chakra. This being so, then it was clear that there ought to be frequencies which could stimulate the other chakras. These were quickly found and the frequencies are given in Table 3.2.

Table 3.2 Frequencies stimulating the chakras

Chakras		Low Band	High Band
Crown of Head	(Sahasrara)	0.26 Hz	12.3 MHz
Forehead	(Ajna)	3.0 Hz	148 MHz
Thyroid	(Vishudda)	81 Hz	3.9 GHz
Heart	(Anahata)	7.4 Hz	384 MHz
Umbilical	(Manipura)	23 Hz	1.13 GHz
Pubic	(Svadhisthana)	81 Hz	3.9 GHz
Coccyx	(Maladhara)	81 Hz	3.9 GHz

3.6.2 Coherent Frequencies on the Ting Acupuncture Meridians

The meridian system of acupuncture also has heart channel and this too was found to be stimulated by 7.8 Hz. Thereafter, the endogenous frequencies on other acupuncture meridians were measured. These included the Ting points as used in electroacupuncture and other meridians of Classical Chinese Acupuncture.

The Ting acupuncture points on the hands and feet as used in electroacupuncture are listed in Table 3.3.

Table 3.3 Ting acupuncture points (after Voll). These points are located on the skin at either corner of the nail bed

Acupuncture Points		Target Organs
Points on Hands		
Thumb		
Outside	Ly1	Lymphatic tissue, Lungs
Inside	Lu1	Lungs
Index Finger		
Outside	LI1	Large intestine
Inside	ND1	Nerve degeneration
3rd. Finger		
Outside	Ci9	Circulation, Pericardium
Inside	AD1	Allergy
4th. Finger		
Inside	Or1	Organ degeneration
Outside	TW1	Triple Warmer, Endocrine
Little Finger		
Inside	He9	Heart
Outside	SI1	Small intestine
Points on Feet		
Big Toe		
Inside	Pn1	Spleen, Pancreas
Outside	Liv1	Liver
2nd. Toe		
Inside	JD1	Joint degeneration
Outside	St45	Stomach
3rd. Toe		
Inside	FibD1	Fibroid degeneration
Outside	Sk1	Skin degeneration
4th. Toe		
Inside	FatD1	Fatty degeneration
Outside	GB44	Gall bladder
Little toe		
Inside	Ki1	Kidney
Outside	BL67	Bladder (urinary)

The frequencies, which stimulate these points, were measured on the Writer, by the Writer, and are listed in Table 3.4.

Table 3.4 Frequencies stimulating Ting acupuncture points

Acupuncture Points	Nominal Frequencies Hz	
	Low Band	High Band
Hands		
Lymphatic tissue, Lungs, Ly1	6.0×10^{-2}	$3.0 \times 10^{+6}$
Lungs, Lu1	4.8×10^{-1}	$2.4 \times 10^{+7}$
Large intestine, LI1	5.5×10^{-2}	$2.7 \times 10^{+6}$
Nerve degeneration, ND1	5.5×10^{-4}	$2.7 \times 10^{+4}$
Circulation, Pericardium, Ci9	5.0×10^{-2}	$2.5 \times 10^{+6}$
Allergy, AD1	2.0×10^{0}	$9.5 \times 10^{+7}$
Organ degeneration, Or1	7.8×10^{-2}	$3.9 \times 10^{+6}$
Triple Warmer, TW1	$6.0 \times 10^{+3}$	$3.0 \times 10^{+11}$
Heart, He9	7.8×10^{0}	$3.8 \times 10^{+8}$
Small intestine, SI1	2.5×10^{-2}	$1.2 \times 10^{+6}$
Feet		
Bladder (urinary), BL67	5.5×10^{0}	$2.7 \times 10^{+8}$
Kidney, Ki1	9.5×10^{-4}	$4.7 \times 10^{+4}$
Gall bladder, GB44	5.0×10^{-2}	$2.5 \times 10^{+6}$
Fatty degeneration, FatD1	7.4×10^{-1}	$3.6 \times 10^{+7}$
Skin degeneration, Sk1	3.5×10^{-3}	$1.7 \times 10^{+5}$
Fibroid degeneration, FibD1	$8.0 \times 10^{+2}$	$3.9 \times 10^{+10}$
Stomach (right foot), St45	4.4×10^{-2}	$2.2 \times 10^{+7}$
Stomach (left foot), St45	4.4×10^{-1}	$2.2 \times 10^{+6}$
Joint degeneration, JD1	3.0×10^{-1}	$1.5 \times 10^{+7}$
Liver, Liv1	4.8×10^{0}	$2.4 \times 10^{+8}$
Spleen, Pancreas, Pn1	5.5×10^{-2}	$2.7 \times 10^{+6}$

Mean Ratio: High /Low Band $49.185 \pm 0.075 \times 10^{+6}$ (SD $\pm 0.15\%$)

The endogenous frequencies present on these acupuncture points are close to the nominal frequencies which stimulate them. The paired-values correlation coefficients for between the endogenous and exogenous frequencies are 0.99999920 for the low band and 0.99977634 for the high band. The paired-values correlation coefficients for the endogenous frequencies between two different persons are 0.99999806 for the low band and 0.99989970 for the high band. These statistics are more like those to be expected in physics than biology or medicine.

The acupuncture meridians are envisaged as communication paths or lines along which there are coherent endogenous frequencies. These are postulated to originate from coherence established as the organism develops from the embryo where the ectoderm, endoderm and mesoderm are in close proximity. Meridians of coherence could persist and grow as the organism develops until they link the acupuncture points to the target organs in the mature organism (Smith, 1990).

3.6.3 Frequencies on other Acupuncture Meridians

There are acupuncture meridians which do not terminate at the Ting points on the nail beds of the fingers and toes. The Pericardium (Pe) channel terminates at the tip of the middle finger (Pe9). The Du Mai (Governing Vessel – GV) and Ren Mai (Ren) channels do not correspond directly to any internal organ. The frequencies measured on these meridians are listed in Table 3.5.

Table 3.5 Frequencies at pericardium, Du Mai and Ren Mai meridians

Pe 9	0.25 Hz and 13.4 MHz
GV14	4.3 Hz and 149 MHz
Entrainment	>4.2GHz to 10 Hz and 4.080 Hz to 0.1760 mHz.
Ren24	14.3 Hz and 730 MHz
Synchronises:	11.4-17.6 Hz and 590-920 MHz
Entrains:	7.9-18.6 Hz and 460-1070 MHz

The acupuncture point GV14 is at the back of the neck, below the spinous process of the *vertebra prominens*. It reflects the status of the cerebro-spinal fluid and appears to be very sensitive to environmental frequencies. The endogenous to the highest frequency tested 4.2 GHz, when the low frequency band has moved to 10.65 Hz. It was confirmed that GV14 would entrain mobile phone frequencies from an actual handset. However, GV14 will only entrain a single frequency; if two frequencies are presented by the environment, only the stronger signal is entrained. This suggests that a Bose Condensation is involved in this entrainment whereby all the Fröhlich modes condense to a single giant model.

3.6.4 Energy Medicine

Energy practitioners have been said to consider that the intensity of the energy emitted from the middle finger is greater than that from the other digits. Accordingly, the Writer measured his own endogenous frequency at acupuncture point Pe9 which is located at the end of the pericardium meridian on the tip of the middle finger. He then concentrated on healing energy for one minute and re-measured. I relaxed for one minute and re-measured again. He then concentrated again on healing and on trying to push the frequency of Pe9 from its 0.25 Hz up to the 7.8 Hz of the heart chakra and heart meridian (He9). To measure the frequencies on Pe9, a pipette containing "erased" water was placed so the water was in contact with acupuncture point Pe9, the hand and the pipette were moved close to a strong permanent magnet to imprint the frequency into the water which was then measured. The results in Table 3.6 show that it would be possible for energy practitioners to change the frequencies on their Pe9 acupuncture points. This might be done in response to what is sensed as the need of the patient, remembering that 7.8 Hz is generally regarded as a therapeutic frequency.

Table 3.6 Results of self-test on Pe9

1.	Initial endogenous frequency on Pe9	0.2542 Hz
2.	Frequency on Pe9 after concentrating on healing for 1 min	6.732 Hz
3.	Frequency on Pe9 after relaxing for 1 min	0.2012 Hz
4.	Frequency on Pe9 after concentrating on healing for 1 min	7.575 Hz
5.	Frequency on Pe9 after relaxing for 1 min	0.2762 Hz

3.6.5 Entrainment of Environmental Frequencies

There is a surprising degree of interaction between the acupuncture meridians and endogenous frequencies. Although the bandwidth on a meridian is only about ±2% of its mean frequency, it can be 'entrained' or 'pulled' by an external oscillation. This may be from an electrical oscillator or other environmental source of radiation. Such as a computer, TV, mobile phone, etc., by the frequency signature of a chemical or, by a homeopathic potency.

This entrainment may be up to ± 30% before the acupuncture point frequency jumps back to its normal endogenous value. Table 3.6 shows entrainment at the heart acupuncture meridian (He9) for which the endogenous frequencies before the measurements were 7.768 Hz and 382 MHz. For this experiment, the subject was exposed only to the high frequency by sitting in front of the output loop of a microwave oscillator for 3 minutes. After this, the frequencies on acupuncture point He9 were immediately imprinted into water and measured as described above. The microwave power density at the subject was estimated to be of the order of mW/m². The frequency measurements took about 5 minutes following the exposure. By this time the acupuncture point frequency had reverted to the unexposed value.

Table 3.7 shows that at 260 MHz and at 500 MHz there was no entrainment but that from 270 MHz to 480 MHz, the frequencies as measured on He9 had entrained to the exposure frequency. The corresponding low band frequencies shifted in proportion.

Table 3.8 lists the ranges of entrainment for the high and low frequency bands at all the Ting acupuncture points. The frequencies on the St45 points differ by a factor of about ten between the left and right foot. This effect still unexplained.

Table 3.7 Exogenous frequencies entrained at heart meridian He9. Entrainment is shown in **bold** type

Exposure Frequency High Band	Endogenous Low Band	Endogenous
MHz	MHz	Hz
None	382	7.768
260	382	7.718
270	**270**	**5.245**
370	**370**	**7.652**
390	**390**	**7.864**
400	**400**	**7.933**
450	**450**	**9.830**
480	**480**	**9.657**
500	382	7.660

Table 3.8 Range of synchronisation of acupuncture points to exogenous frequency (Hz)

Hands				
Lymphatic tissue, Lungs, Ly1	5.80E-02	6.60E-02	2.85E+06	3.15E+06
Lungs, Lu1	4.20E-01	5.30E-01	2.20E+07	2.40E+07
Large intestine, LI1	5.10E-02	6.80E-02	2.50E+06	2.90E+06
Nerve degeneration, ND1	4.90E-04	5.20E-04	2.50E+04	2.90E+04
Circulation, Pericardium, Ci9	4.90E-02	5.30E-02	2.33E+06	2.60E+06
Allergy, AD1	1.90E+00	2.40E+00	9.40E+07	1.03E+08
Organ degeneration, Or1	7.60E-02	7.90E-02	3.60E+06	4.05E+06
Triple Warmer, Endocrine, TW1	5.70E+03	6.40E+03	>>	>>
Heart, He9	7.40E+00	8.20E+00	3.75E+08	3.95E+08
Small intestine, SI1	2.20E-03	2.70E-02	1.15E+06	1.29E+06
Feet				
Bladder (urinary), BL67	5.30E+00	5.80E+00	2.65E+08	2.75E+08
Kidney, Ki1	9.30E-04	9.80E-04	4.20E+04	5.00E+04
Gall bladder, GB44	4.60E-02	5.70E-02	2.33E+06	2.70E+06
Fatty degeneration, FatD1	7.30E-01	7.80E-01	3.45E+07	3.80E+07
Skin degeneration, Sk1	3.20E-03	3.60E-03	1.72E+05	1.90E+05
Fibroid degeneration, FibD1	7.80E+02	8.20E+02	>>	>>
Stomach (right foot), St45/R	4.20E-02	4.70E-02	2.02E+07	2.24E+07
Stomach (left foot), St45/L	4.30E-01	4.70E-01	2.00E+06	2.20E+06
Joint degeneration, JD1	2.80E-01	3.30E-01	1.40E+07	1.55E+07
Liver, Liv1	4.50E+00	5.20E+00	2.30E+08	2.40E+08
Spleen, Pancreas, Pn1	5.20E-02	5.50E-02	2.60E+06	2.85E+06

3.7 EFFECTS OF COHERENT FREQUENCIES ON PEOPLE

3.7.1 Frequency Entrainment by EM Sensitive Patients

The frequencies which affect electromagnetically hypersensitive patients extend from below mHz to above GHz. Apart from a correlation with the power supply frequency of the country in which the patient was resident there was no explanation of the observed patterns of alternating stress and neutralisation until the endogenous frequencies on the acupuncture meridians had been found and measured. It then became clear that these patients were responding to the effects of the environmental frequencies on their sensitive meridians which happened have an endogenous frequency close by.

For ten patients who lived in the EU, 19 out of 54 of their imprinted tubes contained their 50 Hz power supply frequency. Two patients who lived in N. America imprinted 3 out of 5 of their tubes with their 60 Hz power supply frequency, but nothing at 50 Hz. Power supply frequency entrainment appears to be quite common among electromagnetically sensitive patients.

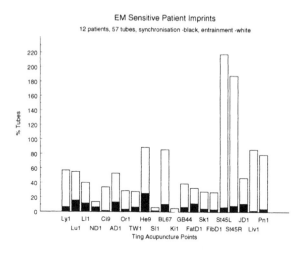

Figure 3.2 Frequency entrainment by EM sensitive patients.

Figure 3.2 summarises the frequency imprinting by 12 electrically sensitive patients who during the course of their therapy had imprinted a total of 57 tubes of water with a total of 726 frequencies. Of these, 167 would have been capable of synchronisation at one of the Ting acupuncture points, and 655 would have been capable of entrainment at one or more of these points. Many tubes had more than one frequency capable of entraining a particular meridian. This is seen particularly at stomach meridians St45 for which the values are greater than 100%. Only 49 out of 726 frequencies were outside any entrainment ranges of the acupuncture points examined.

These results show that the patients' sensitive frequencies and water imprinted by patients holding and succussing a tube gives very good coverage of the Ting meridians and can be used to indicate which meridians are under stress and which need stimulation.

3.7.2 Entrainment of Frequency Signatures of Chemicals

3.7.2.1 Entrainment by Persons

The frequency signatures of chemicals in contact with the body are as effective in producing frequency entrainment at an acupuncture point as frequencies from an external oscillator, should they happen to come within its entrainment range. Holding a glass bottle containing a chemical for just one minute is sufficient to entrain an acupuncture point from its endogenous frequency to that of the chemical. It takes about 10 minutes for the point to relax back again to its endogenous frequency.

Homeopathic potencies have characteristic frequency imprints of their own, besides the frequency signatures of chemicals used for the 'Mother Tincture'. These arise from the dilution and succussion process of potentisation. These frequencies remain even after all chemicals have been diluted away. Dilution to Avogadro's number $\sim 10^{-24}$ or C12 potency is usually regarded as the threshold for this. Frequency appears as a common factor linking acupuncture, homeopathy and the living environment.

Table 3.9 lists the acupuncture points within the entrainment range of the chemical frequency signatures of sodium chloride and the endogenous frequencies at those acupuncture points. When holding a tube containing sodium chloride solution the frequencies on these meridians become entrained to those for sodium chloride.

It is possible to "hide" the chemical frequency signature of a dilute solution so that it no longer entrains. As mentioned in Section 3.5.3, one method is to succuss it on one side (depending on the relative direction of the geomagnetic field) of an oscillator output coil at a particular frequency. One frequency having this property is 1.42 GHz, the 21cm resonance of the hydrogen molecule. Succussion at the opposite side of the coil "recovers" the frequency signature and its entrainment effects return. Since this effect only happens with dilute solutions, it is possible that the sodium chloride molecules move inside coherence domains where they are screened from external fields. Relaxation occurs when holding the "hidden" NaCl and the entrainment returns when its activity is restored. Similar entrainment effects are to be found in cell cultures prepared in the presence of low concentrations of toxic environmental chemicals (Smith, 2000).

Table 3.9 Entrainment of acupuncture points by chemical signatures of NaCl. Entrainment is shown in **bold** type. All frequencies in MHz

Acupuncture points	SI1	Or1	FatD1
Endogenous frequencies	1.23	3.80	36.5
Chemical signature of NaCl	**1.24**	**5.10**	**40.0**
Holding NaCl	**1.24**	**5.10**	**40.0**
Holding "hidden" NaCl	1.23	3.80	36.5
Holding "recovered" NaCl	**1.24**	**5.10**	**40.0**

3.7.2.2 Entrainment by T-lymphocytes

The frequency pattern of the normal living cells is one of a quasi-periodic variation asynchronously over a number of frequencies and to a limited extent over a few hours duration. This is typical of a normal healthy living system, whether a single cell or the human cell complex.

Frequency signatures were measured at the Environmental Health Center in Dallas, Texas, for some common environmental pollutant chemicals being investigated for tests on T-lymphocyte cultures. These were toluene, xylene, phenol, methyl ethyl ketone, ethanol, pyrethrin, sodium hypochlorite, 1,2,4-trimethylbenzene, 1,1,1-trichloroethane, trichloroethylene, 1,5-diaminopentane. The *in vitro* T-lymphocyte cultures were challenged with the organo-chemicals during incubation at concentrations of the chemicals corresponding to their 'Threshold Limit Values - Time Weighted Average' (TLV-TWA). This is the officially regarded concentration for a normal 8-hour work-day and 40-hour work-week to which workers may be repeatedly exposed, day after day, without adverse effect.

The chemically exposed cultures had certain frequencies in their patterns entrained by the chemical signatures corresponding to the challenging chemical so that these frequencies were no longer free to fluctuate as normal. This entrainment restricts the typical normal frequency fluctuation pattern and has the same effect as a "depressing" electrical frequency has on a living system. It represents a biological stress because the number of degrees of freedom available to the living system in which it can react to the demands of metabolism and the environment, are correspondingly reduced.

All the organo-chemicals tested were found to modify the normal cell cycle profiles of T-lymphocytes which were subsequently measured by flow-cytometry. Therefore, it must be concluded that these incitants are capable of interrupting the ordered and orderly progression of the cell cycle are firstly are capable of the destruction of specific proteins (cyclines) and enzymes, secondly, they prevent apoptosis from taking place with the result that wrong translations are made from the DNA and wrong signals are sent for the control of cell progression and thirdly, they compromise the immune system leading to multiple manifestations including cancer.

3.7.3 Electrical and Chemical Hypersensitivities

3.7.3.1 Electrical Hypersensitivity

In the predator-prey situation, the ultimate survivors will be those organisms whose bio-sensors have a sensitivity only limited by the fundamental laws of quantum physics when the predator will feed or the prey will escape. Such sensitivities do not give linear responses. Weak stimuli below some threshold will evoke no response; stronger stimuli above this threshold can evoke the maximum or "panic response", which if out of control is the characteristic of hyper-sensitivity. The normal eye when dark-adapted is sensitive to single photons, it can accommodate to a sunlit snow scene.

Allergy used to be concerned solely with skins and respiration. In recent years, allergic type responses have been found to occur so widely that they must now be thought of as 'hypersensitivity' or the 'failure of a regulatory system'. Severely ill hypersensitive patients have acquired inappropriate reactions to many chemical, environmental and nutritional substances at very low concentrations. About 15% of any given population function to some extent below their best possible performance through some regulatory system disfunction due to chemical, particulate or nutritional sensitivity. Of these, 1% will have added electrical hypersensitivity to their package of sensitivities.

Several hundred such patients have been tested since 1982 when Dr. Jean Monro first asked the Writer to help with her electrically sensitive patients. Most of their reactions were critically dependent on specific frequencies which might be anywhere in a range extending from below milliHertz (mHz) to beyond GigaHertz (GHz). Their reactions were little dependent upon the field strength so long as this exceeded some patient specific threshold value. The most common symptoms of electrically sensitive patients showed a similarity with the proving symptoms for the homoeopathic potencies of electricity, magnetism and X-ray.

New sensitivities can be readily acquired, or reactions transferred to another "trigger" if a patient becomes exposed to some hitherto innocuous substance, or frequency, while reacting strongly to an existing "trigger". An electromagnetic frequency in the environment can thus become a "trigger" for the same specific pattern of responses on each and every subsequent encounter.

Electromagnetic emissions from a reacting patient in the audio-frequency part of the spectrum may be demonstrated by getting the patient to hold a tape recorder having a radiation-penetrating plastic body, with the tape running and the recorder set to "record" but, with no microphone connected. If the reaction is strong, there will be sufficient interference passing through the plastic case to be picked up by the amplifier circuits, recorded and heard on re-play. The frequencies can be measured this way only cover the limited audio frequency range of a tape recorder.

Electrically hypersensitive patients are very well aware of their problems which are wide ranging and they know what "triggers" them. Besides upsetting other sensitive patients, their radiation may interfere with electronic equipment, this is human "electromagnetic incompatibility". Many patients are not compatible with modern technology-period! One example is of a hypersensitive patient who had a whole robotic system in a factory malfunction when standing near it; another had the electronic ignition system on successive new cars fail as soon as a "reaction" was "triggered" by diesel fumes in traffic.

3.7.3.2 Diagnosis of Electrical Hypersensitivity

This procedure for testing and treating electrically hypersensitive patients is based on the therapy for chemical and nutritional allergic responses developed in America by Dr. Joseph Miller who showed that the diameter of the wheal which developed on the skin of a patient following a 'skin-prick' (intradermal injection) test with an allergen depended upon the dilution of the allergen used. Furthermore, if a whole sequence of serially diluted allergens was applied successively, certain dilutions gave large wheals but, after further serial dilutions, one would eventually attain a dilution of the allergen at which no wheal was produced. Still further serial dilutions would recycle through similar patterns of response and no-response. The dilution at which no wheal resulted from the 'skin-prick' test was termed that patient's 'neutralising dilution'. This dilution could be injected to give neutralisation of the symptoms provoked by that particular allergen and could provide a prophylactic protection against subsequent environmental or nutritional exposure.

The following technique may be used to screen patients for electrical sensitivities. However, if it is suspected from the patient's history that there is a severe degree of hypersensitivity, then any testing must be regarded as a clinical procedure. It should not to be attempted without the immediate availability of facilities and staff medically competent to treat the remote but real risk of anaphylaxis (a serious shock reaction arising from a hypersensitive condition of the body). The most effective and rapid therapy for anaphylaxis in these patients is to place a drop of the patient's neutralising dilution of chemical allergen (found from previous testing) anywhere on the skin or sub-lingually. The patient's neutralising dilution for reactions to histamine should also be available.

No electrical connection to the patient is necessary, nor should it be contemplated for reasons of electrical safety and the risk of over-stimulation of the patient's responses from which it could take days, weeks or months to recover. Delayed reactions are always a possibility, even 24 hours after the original challenge; nurses need to be on the alert for this. In general, the symptoms to be expected on electrical testing are the same as those provoked by previous chemical or nutritional tests on that same patient.

The room used for electrical testing must be chemically and particulate as well as electrically "clean". One is seeking to produce clinical effects by merely having the patient in a clean environment containing only the sort of electromagnetic fields that leak from electrical equipment. A controlled electrical frequency is radiated from a typical commercial laboratory oscillator (waveform generator). The frequency is scanned slowly, beginning at less than 1 mHz and continuing to as high a frequency as appears necessary (or is available). Increasing frequency has the same clinical effect as increasing allergen dilution or homoeopathic potency. Because of the multiple frequency effect in coherent systems it is usually sufficient to test over a representative part of the electromagnetic spectrum, as this will challenge other regions through coherence interactions. With multiple sensitivity, this may not be possible because there may be more than one independent coherent system involved. It may be possible to "trigger" reactions to electromagnetic fields only when the patient's sensitivity has been enhanced by prior chemical exposure.

During a testing, the patient's reactions and the frequencies at which they occur are noted. It is usually possible to find certain frequencies at which all the patient's adverse reactions cease. With a sensitive patient this can take several hours. The test may even have to be abandoned or postponed if the reactions become too severe. It only takes about 15 minutes to demonstrate to normal patients that they have no abnormal sensitivity to their electrical environment. Because the above technique can only be used if the patient reacts within seconds and is not too electrically sensitive, it soon became essential to devise alternative techniques to be able to help the more highly sensitive patients.

A patient's reactions to frequencies can be detected with kinesiology or radiesthesia before any symptoms appear. This avoids having to wait for the symptoms and avoids provoking painful reactions, which in turn improves patient cooperation. It also enables one to test patients who cannot even be in the same building as a frequency generator, or who live at a distance and are unfit to travel. All that is necessary is for the patient to hold a strong glass tube of water and bang (succuss) the base of the tube on a wooden surface (but not so hard as to break the glass!). This 'succussion' imprints the principal body frequencies into the water. These can subsequently be measured by radiesthesia. Such imprinted tubes can be sent through the mail (or air mail) if wrapped in aluminium foil without the imprinted frequencies becoming corrupted or erased. However, transporting on electrified railways may cause problems.

3.7.3.3 Therapies for Electrical Hypersensitivity

While it is possible for patients to gain relief from their sensitivities by having an oscillator set to a neutralising frequency and left-on in their room, this is not a satisfactory or cost effective solution particularly at microwave frequencies. Furthermore, one patient's neutralising frequency is another's "trigger" and in a crowded hospital or clinic, a multiplicity of personal oscillators would be as destructive as the presence of perfumes.

When first faced with the problem of a patient, who could only be neutralised at a microwave frequency, it was remembered that the homeopathic "Materia Medica" includes potencies of prepared from electric currents, magnetic fields and

X-rays. Accordingly, a sample of water known to be tolerated by the patient was placed within a coil and exposed to an alternating magnetic field at the frequency, which neutralised the patient's reactions. This water then became clinically as effective when in skin contact with the patient as did having the microwave oscillator switched on at the same frequency in the room.

Tubes of water can be imprinted with the ('neutralising') frequencies that the patient needs to have stimulated. These are to be treated as "Master Copies", handled by holding above the water-line so as not to imprint any "toxic" body frequencies from on-going reactions and kept wrapped in aluminium foil when not in use. The door of a refrigerator is a convenient safe-place. No one else must handle the tubes because of the risk of overwriting them.

To make working copies, the "Master" tube is placed in a glass of water for 24-hours, after which time most of the frequencies will have imprinted into the water. Alternatively, the base of a strong glass containing the tube and the water may be succussed against a wooden surface with a single sharp impact, taking care not to break the glass or spill the water. Then, all the imprinted frequencies transfer instantaneously. Water exposed to magnetic fields at a neutralising frequency can be used in the same way as a neutralising dilution of allergen and resembles a "potentised" homoeopathic preparation. This process must involve the physics of water because a sealed ampoule can acquire a clinical effectiveness equivalent to that of a homoeopathic potency or allergen dilution, by imprinting the appropriate frequency into it with no chemical precursor what-so-ever.

This neutralisation will help to stabilise a patient so that meaningful testing for chemicals can be carried out. It will not hold for long if there is a high body load of toxic chemicals as their chemical frequency signatures will overwrite the electromagnetic bio-information supplied.

3.7.4 Chemically and Electrically Sensitive Patients

It is very rare to find a patient with electrical hypersensitivity who does not also have many chemical sensitivities. Even the exception may be just a case of inadequate chemical sensitivity diagnosis.

Both chemically and electrically sensitive patients are very assertive and highly vocal in complaining that nobody is doing anything about the environment to relieve their symptoms. Chemically sensitive patients are very well aware of their reactions and can usually pinpoint accurately and immediately when they are reacting and what "triggered" it. They are able to accurately 'test' by just holding glass tubes of diluted antigen (even if frozen) or frequency imprinted water and say which they can tolerate. They are acutely aware of traces of substance toxic to them in the air, water and foods.

The electrically sensitive are equally acutely aware of the "triggering" of their symptoms by specific electromagnetic sources in their environment. Often, they are unable to tolerate metals (e.g. needles) and so cannot be allergy tested by skin-prick or sub-dermal injection but need to be tested either by holding the tubes of allergen dilutions or by drops of allergen applied sublingually.

The chemically sensitive patients have problems with the common disinfectants. The phenol sensitive, can only tolerate injections which are phenolic preservative-free and need freezer storage. A further hazard for the electrically sensitive is that their medications, injections and drips may have acquired an electromagnetic frequency imprint during transport or storage to which they react. This may be hidden by heating above 70°C if the substance will tolerate this otherwise, erasure is possible by placing the substance briefly inside a closed steel box to shield it from the Earth's magnetic field as already described. If the chemical signatures of the medication are a problem, a patient can be neutralised to these frequencies.

If the patient's chemical sensitivities can be resolved, the electrical ones usually go away. This is due to the reduction of the total body load of allergens for which electromagnetic frequencies mimic chemical signatures. It is possible to speed up the elimination of toxic substances when all or most of the detoxification pathways are inoperative by medication, exercise and massage, and saunas. The young, once fit and still highly motivated do particularly well at this.

Visitors need to be convinced that they must not enter the patient area wearing perfume or after-shave, after having used chewing gum, after having used scented soap or detergent, after re-fuelling the car or, having been exposed: to traffic fumes, moulds or tobacco smoke. Odours linger, particularly on clothing and hair. With no offence intended, patients hypersensitive to odours may say spontaneously, "You are smelly" and indicate firmly that they cannot tolerate it. These patients may react to the body fields of the nurse testing them and therefore, they tend to prefer a particular nurse whom they know does not affect them. Nurses need to realise that there is nothing personal in this and should not be offended. Such is hypersensitivity!

3.7.5 Chemical Frequency Signatures in EM Sensitive Patients

Figure 3.3 shows that the signatures of a number of toxic environmental chemicals present in the blood of 22 EM sensitive patients were entrained. The frequencies entrained by the patients correlate exactly with those of the chemicals. The blood levels (where available) are the average of values of measured concentrations which varied by up to a factor of ten between patients.

Figure 3.4 relates to an extremely chemically and electromagnetically hypersensitive patient who was one of about 200 persons who suffered as the result of chronic exposure to a toxic chemical in their living environment. This was acquired in following manner. In a new building, the waste steam from the heating boiler was used to humidify the air in the air-conditioning system. Unfortunately, an anti-corrosion and de-scaling chemical product was put into the boiler, this vaporised into the steam and was circulated throughout the building. The chemicals' signature contained 31 frequencies. That particular patient had 21 frequencies needing treatment of which 10 corresponded exactly to those of the chemicals as shown in Figure 3.4. This represents an almost 50% entrainment of the patients endogenous frequency activity by identifiable chemical contamination of the building, namely by cyclohexylamine and morpholine. Of the 10 frequencies common to the chemicals and the patient, any of the lowest three would entrain theallergy meridian at AD1, the next three the heart meridian at He9 and the heart chakra, all but one would entrain the Du Mai meridian at GV14 which relates to the cerebro-spinal fluid, all are potentially able to entrain the frequency activity of T-lymphocytes.

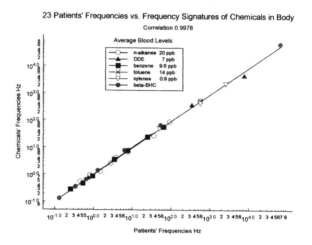

Figure 3.3 Chemical signatures entrained by EM sensitive patients

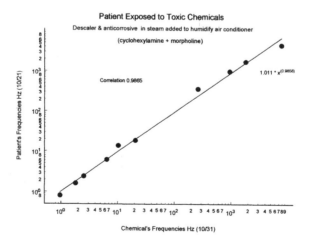

Figure 3.4 Patient exposed to toxic chemicals

3.8 ACUPUNCTURE, FENG SHUI AND FREQUENCIES

3.8.1 Background to Acupuncture and Feng Shui

Acupuncture and Feng Shui are a part of the ancient Chinese culture of being in harmony with nature. Feng Shui means 'wind and water' and seeks to explain how nature affects human life and living conditions, see for example, "Feng Shui goes to the Office" (Wydra, 2000). Acupuncture in Chinese medicine is based on the same philosophy of nature and a natural law called Tao. This is seen as an abstract creative force-giving rise to a life force or energy called Qi. This in turn has a complementary pair of opposites, Yang and Yin. It is the tension between these which is considered to give rise to life processes in the human body and if unbalanced leads to medical conditions (Stux and Pomeranz, 1991).

In the body, the Qi is considered to accumulate in the body organs and flow in channels or meridians. It is part inherited, but also acquired from food and respiration. It is expressed in body growth warmth, movement, thoughts, consciousness and spiritual activity. It also has a protective function against illness. Illnesses are regarded as excesses or deficiencies of Qi. They may be caused by external pathological climatic influences such as cold, heat damp, dryness or wind. This is where Feng Shui becomes involved.

Table 3.10 Classification of body organs according to the 'Five Elements' showing the endogenous (low band) frequencies on their meridians

Element	Internal Organs (Hz)		Hollow Organs (Hz)	
FIRE	Heart	7.8×10^{0}	Small Intestine	2.5×10^{-2}
EARTH	Spleen	5.5×10^{-2}	Stomach (LHS)	4.4×10^{-1}
METAL	Lung	4.8×10^{-1}	Large Intestine	5.5×10^{-2}
WATER	Kidney	9.5×10^{-4}	Urinary Bladder	5.5×10^{0}
WOOD	Liver	4.8×10^{0}	Gall Bladder	5.0×10^{2}

In addition to the Yang and Yin used for describing polar processes, there are the so-called 'Five Elements' which may be regarded as labels showing that something fundamental is going on in the environment. These 'elements' are labelled: WOOD, FIRE, METAL EARTH, and WATER. They are interlinked so that each regulates another while itself is being regulated by a third. Table 3.10 shows the classification of generalised body organs according to these 'Five Elements'. Their endogenous frequencies have been added.

3.8.2 Organs and Meridians

Chinese acupuncture recognises 11 organs in the sense of them being general structural and functional entities. There are 6 Yang organs (Fu) and 5 Yin organs (Zang) which interact closely with the channels or meridians serving them. There are 12 channels running parallel to each other in the limbs and these are paired, one is Yang and the other Yin (Stux and Pomeranz, 1991).

Additionally, the pericardium is given a channel and there are some other channel systems. These include the meridian Ren Mai (Yin) which runs up the ventral mid-line of the body and the meridian Du Mai (Yang) which runs up the dorsal mid-line. Together these make up the 14 channels or meridians on which the 361 'Classical Chinese Acupuncture Points' are located. Knowledge of the frequency to expect on any given meridian is of considerable practical use when trying to locate specific meridians which run close to other meridians. These channels or meridians are divided into 'Three Courses' as shown in Figure 3.5. Surprisingly, certain meridians seem to have the same, or very near, frequencies and these are shown linked together by a diagonal.

3.8.3 Not to Reason Why!

Tables 3.11 & 3.12 show the 'Five Elements' related to the environmental factors involved in Feng Shui (Wydra, 2000). This is to all appearances completely unscientific but, notwithstanding this the Writer has tried to apply dowsing techniques to the listed quantities.

Table 3.11 Relation the 'Five Elements' to factors of importance to Feng Shui

Element	Colour	Shape	View from	Taste	Sensor
FIRE	Red	Triangle	South	Bitter	Tongue
EARTH	Yellow/ Brown	Square	Centre (upwards)	Sweet	Lips
METAL	White/ Shiny	Circle	West	Spicy	Nose
WATER	Black/ Blue	Waves	North	Salty	Ear
WOOD	Green	Rectangle	East	Sour	Eye

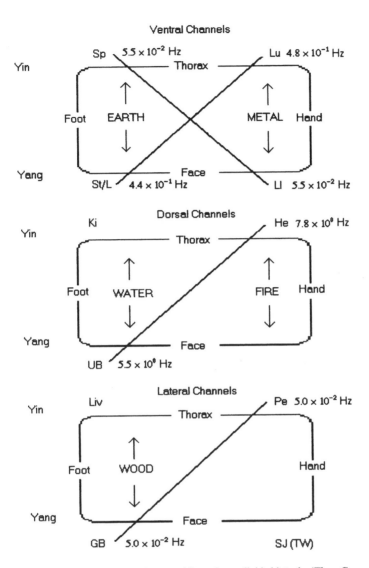

Figure 3.5 The acupuncture channels or meridians shown divided into the 'Three Courses' with their endogenous frequencies. Those sharing the same or nearby frequencies are joined by a diagonal

Table 3.12 Relation of colours from Table 3.11 to other factors of importance to Feng Shui

Colours	Sounds	Texture	Effect	Emotion
RED	Cymbals	Rough	Energy	Joy
YELLOW BROWN	Strong Beat	Solid	Clarification Stabilisation	Anxiety
WHITE	Tin Whistle	Smooth	Mental Stimulus	Sadness
BLACK/ BLUE	Mantra	Irregular	Goal Visualisation Relaxation	Fear
GREEN	Heart Sounds	Prickly	Generates Ideas	Anger

Starting from the knowledge of the frequencies which stimulated the acupuncture points, it was then found that additional exposure to any one other of the factors listed in Tables 3.11&3.12 would produce a whole body stimulation reaction.

For example, 7.8 Hz stimulated the heart chakra and heart meridian but, simultaneously looking at a red card while exposed to 7.8 Hz would produce a whole body stimulation reaction. Looking at a red card with a triangular shape superimposed on it would give whole body stimulation reaction. Looking at a yellow card while there was a taste of sugar (or sugar substitute) in the mouth would give whole body stimulation reaction. Viewing rectangular shapes from the East would give a whole body stimulation reaction. Similarly for the other pairs of stimuli.

The alleged effects of sounds were checked by looking at coloured cards while listening to music. The particular record was chosen to provide sounds resembling those indicated by Fen Shui. The introductions and finales of marches with much clashing of cymbals gave whole body stimulation when viewing a red card, as did the finale of Tchaikovsky's 5th. Symphony and Walton's Belshazzar's Feast. The yellow card gave whole body stimulation when complemented with Bavarian 'oomph-ah' folk music. The white card gave whole body stimulation when accompanied by the Spinners' tin whistle rendering of the 'Skye Boat Song'. The blue/black cards gave whole body stimulation when listening to Taizé Music (mantra-style chanting). With the green card, whole body stimulation was best achieved by listening to a recording of heart sounds but, Elgar's 'Introduction and Allegro for Strings' or Barber's Adagio for Strings also gave stimulation. Military bands with pipes and drums stimulated more than one meridian.

To get any reaction from the 'texture' factors, it was necessary first to provide some whole body stimulation such as by looking at the combination of a colour and a geometrical shape. Only then would hold the corresponding texture produce any effect, which was to cancel that stimulation. Not having been to drama school, the Writer did not attempt to check out the 'Effect' and 'Emotion' columns!

While these preliminary tests only relate to subjective effects on the Writer, they are given as an indication of ways in which such factors in the living environment might be investigated and to show that environmental frequencies can have far ranging effects within the living space. There is no doubt about the power of visual and dramatic art and music to influence us.

Electromagnetically sensitive persons are particularly good at spotting frequencies hidden in the environment. For example, they have pointed out to the Writer that one of the 'soap-operas' has had a 'feel-good' frequency as a sub-liminal on its sound channel for years. Again, a certain party political broadcast had different sub-liminals applied to each political party's presentation. Robert Beck during a broadcast in America remarked on a certain restaurant which had relaxing sub-liminal frequencies in the bar muzak to promote drink sales, but when the diners were at table there was a 'fight-or-flee' frequency as the sub-liminal to speed up the table availability. This technology is around, someone out there has it!

3.9 BUILDINGS AND COHERENT FREQUENCIES

3.9.1 Coherence perception

Buildings are structures created in space for the purpose of providing a Living Environment for dwelling or working. Structures have frequency implications. Even a cave dwelling has its resonances and echoes. In the above sections the Writer has attempted to show how endogenous frequencies are involved in the living system and how exogenous frequencies may interact with it. In this the duality between the electromagnetic frequency and the chemical signature has been particularly emphasised. Without this duality, chemical analysis by spectroscopy would not be possible. Living systems are also coherent systems and the most important consequence of this is the possibility of multiple-frequency effects which are interactions between effects in many widely different parts of the spectrum. Furthermore, the whole of the electromagnetic spectrum is effectively within the man-made environment complementing the chemical signatures of the tens of thousands of man-made chemicals in the environment. If some chemicals are so toxic that they are never synthesised, there must be equivalent patterns of frequencies which are equally toxic.

Because electrically polarised membranes form an important part of biological structures, it must be expected that these will act as acoustic transducers between the acoustic and the electrical. Very occasionally, one comes across a patient whose sensitivity is to acoustic frequencies and not to the same frequencies in the electromagnetic spectrum.

The perception of frequencies in the environment of the human living system involves the five-sense and a sixth sense of direct interaction with the body-fields or aura. The techniques developed by the Writer for the measurement of frequencies by dowsing inform a major part of this Chapter.

3.9.2 Perception of Endogenous Frequencies

3.9.2.1 Stimulation by Geopathic Stress-line Frequencies

Stress lines are primarily found by dowsing although, there have been attempts at instrumentation. There are lines on the Earth's surface above which living systems experience stress and illnesses develop following prolonged exposures and therapies fail. These are called geopathic lines or zones. In a few cases the lines are geo-positive and give a healthy stimulation. Käthe Bachler has written a book (Bachler, 1989) in which she records the results of dowsing over 11,000 cases in over 3,000 homes in 14 countries and correlates them with illnesses found there.

The Writer once demonstrated to a Conference of Dowsers how a toroid placed on a stress line and fed with a current of an appropriate frequency could render it 'invisible' to detection by dowsing. It is possible that what the dowser detects as so-called 'water lines' are the frequency patterns of chemical signatures and imprinted frequencies in ground water. Somewhere in all this the mind-body interaction must take effect because of the possibility of map-dowsing for example.

The Writer has found that a geopathic line is a source of frequencies which can be measured by dowsing. These frequency patterns may be stressful to the meridian and chakra systems of a person living and particularly sleeping in that location. Certainly, an electrically sensitive person would be acutely aware of many of these frequencies. Figure 3.6 shows the results of measuring the frequencies over a stress line in the Writer's house for fortnight in 1994. The frequencies over the line were not stable. Within even this short time there was an enormous peak which occurred early on May 19th coinciding with an earthquake on the other side of the world.

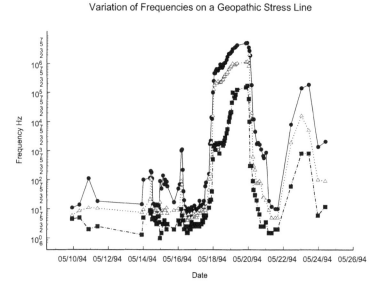

Figure 3.6 Variation of frequencies over a geopathic stress line.

3.9.2.2 Multiple Frequency Effects Involving Ultraviolet and Microwaves

Simon Best (Best, 2001) recently reported on the pioneering work of the American photobiologist, Dr John Ott (Ott, 1973) who drew on a range of plant, animal and human evidence in his classic work Health and Light to demonstrate how important natural light is for the health of the body and functioning of the brain, strongly endorsing the fact that light is an essential nutrient. Others have strongly supported this premise with a wealth of further evidence and research. Ott observed that lack of the full spectrum of natural light frequencies in offices and classrooms produced many adverse effects, including hyperactivity in children. Such 'mal-illumination' was often caused by fluorescent lighting, which lacks the full spectrum and proper balance, especially in the UV and blue/green frequencies.

When fluorescent lighting was replaced by full-spectrum lighting (FSL) in classrooms, Ott reported that children's previous misbehaviour and hyperactivity were replaced by much calmer and more attentive behaviour. Interestingly, Ott also reports that local dentists observed a 67% drop in cavities in children under similar conditions which findings were replicated by a Professor of Dentistry at the University of Alberta, Canada, who found that full-spectrum lighting even reversed the development of cavities.

To underline the stress that fluorescent lighting probably causes to young children's developing bodies and biological systems, the research by Professor Fritz Hollwich in Munich, as far back as 1980, found significantly higher levels of the stress hormones cortisol and ACTH in those working under standard fluorescent lighting compared with full-spectrum. His findings led the German Government to ban the use of standard fluorescent lighting in hospitals and medical facilities.

Figure 3.7 shows how the multiple frequency effect in ultraviolet imprinted water can give microwave frequencies from 276 MHz to 935 MHz which were

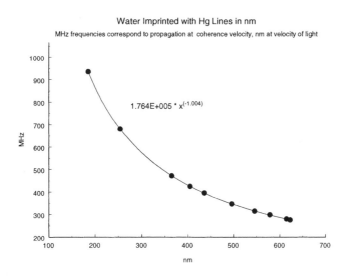

Figure 3.7 Multiple frequency effects in UV imprinted water.

found to correspond to the mercury spectral lines from 623 nm to 185 nm. The ratio of the optical frequencies to the microwave frequencies is $1.734 \pm 0.006 \times 10^6$. This represents a standard deviation of 0.34%.

Among these microwave frequencies there would be stimulators for the heart meridian at 384 MHz corresponding to 451 nm (blue) and the Ren Mai meridian at 730 MHz. corresponding to 237 nm (ultraviolet). This stimulation would be missing in limited spectrum artificial lighting. This does suggest that bio-photon effects are but one aspect of the more general phenomena of multiple-frequencies in coherent systems.

3.9.2.3 Effects of Exogenous Frequencies on Endogenous Frequency Activity

The following are examples of what seems to be the general effect of exogenous frequencies on the endogenous frequency activity of living systems. The un-stimulated endogenous frequency activity is given for *acetabularia* in Figure 3.8. It is seen that there is a quasi-periodic variation of between a half and one hour which extends over a two-to-one range of frequencies. When an exogenous frequency is stimulatory, its effect is to speed up these variations by an order of magnitude as shown in Figure 3.9. These measurements were made in March 1993 by the Writer in Kaiserslautern by courtesy of Dr. Fritz Popp. As the frequency changes, it is tracked by adjustment of the oscillator to maintain stimulation. If this is applied to a pair of living systems, their endogenous frequencies will become synchronised.

This synchronisation can be made to persist as described in the following section. If the applied frequency is depressive, all electrical variation is paralysed just as if a toxic chemical was present with its chemical signature at that frequency.

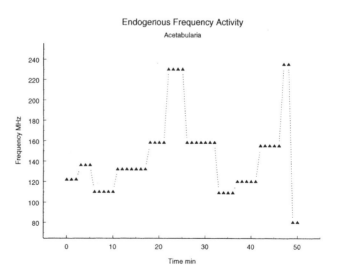

Figure 3.8 Endogenous frequency activity for *acetabularia*.

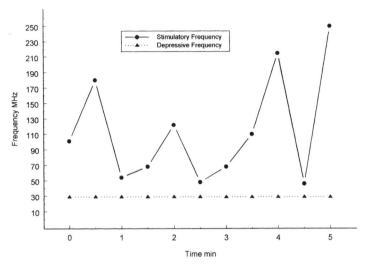

Figure 3.9 *Acetabularia* in the presence of exogenous frequencies.

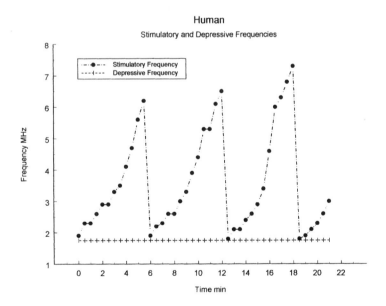

Figure 3.10 Human subject in the presence of exogenous frequencies.

This will persist (at least over-night) so long as the depressive frequency is being applied. It shows the complete analogy between the properties of absorbed traces of toxic chemicals an the effects of exogenous frequencies which happen to be depressive of endogenous frequency activity. Similar effects occur in the human subject as shown in Figure 3.10. Subjectively, it is most unpleasant to be in the presence of one's own depressive frequencies, there is a very strong urge to switch-off or get out.

3.9.3 Perception of Spatial Frequencies

An example of spatial frequency is to be seen in the bar patterns in television 'test cards'. These are generated by temporal frequencies modulating the picture scan which converts them into a spatial pattern. A number of simple basic spatial configurations have been investigated. The standard conditions adopted for the measurement of the frequencies by dowsing were to have the main axis aligned North-South. The effect of other orientations is shown in Sections 3.9.3.4&5.

3.9.3.1 Range of Coherence Interactions

This section links together the temporal, spatial aspects of environmental frequencies and bio-photon activity. Experimentally, whenever one obtains 'action at a distance' there is the possibility of inserting filters to determine the frequencies involved and where any cut-off comes.

If an endogenous stimulating frequency for one of a pair of tadpoles in separate dishes is determined then as shown in the previous section, applying an exogenous frequency will speed up its frequency variation activity. This usually results in the frequency crossing the endogenous frequency of the other tadpole, at which point both tadpoles' frequencies become synchronised to the stimulatory exogenous frequency. They will then remain synchronised even if the exogenous frequency is turned off. This synchronisation is broken if optical contact between the tadpoles is broken. This can be done by placing a card between their dishes or by placing the dishes one above the other as the interaction only seems to be in their horizontal plane, not dorsal to ventral. There is a critical distance of separation of their dishes at which the synchronisation is lost. This as shown in Figure 3.11 is about 13 metres but it is reduced if optical filters are placed in-between. With filters with cut-off greater than 530 nm all synchronisation is lost and will not spontaneously re-establish.

Similar effects are obtained between aliquots of water or ethanol imprinted with a common frequency. In this case, the coherence returns on coming back within the critical distance of a few metres separation. Again, there are similar distance effects with optical filters interposed and no coherence is detected with cut-off wavelengths greater than 550 nm.

The critical cut-off for re-establishing coherence is less than 395 nm for water, 375 nm for ethanol and 360 nm for the tadpoles. This shows the involvement of bio-photon effects as a general phenomenon associated with coherent systems and multiple frequency effects. The tadpole experiments were done in Graz by courtesy of Dr. Christian Endler.

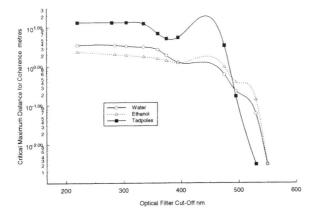

Figure 3.11 Spatial range of coherence interactions for water, ethanol and tadpoles.

3.9.3.2 Frequencies of Resonant Structures

Buildings are essentially geometrical structures, coherent in space and having a long-range order, otherwise they would be sand dunes. Any structure will have one or more characteristic resonances for acoustic and electromagnetic waves as well as the frequencies of the chemical signatures of its constituents. A cave has its echoes, railings have the resonances of a single rod multiplied many-fold and the diffraction grating directional effects of the extended structure. The inside of a car is an electromagnetic resonator for mobile phone transmissions. All such resonances can be sensed by living systems and particularly by hypersensitive persons. The Writer uses dowsing techniques to keep the sensitivity under control.

The most basic electrical resonator is the inductance-capacitance (LC) circuit for which a coil (inductance) is connected in series or in parallel with a pair of metal plates (capacitance). Dowsing this combination to measure the resonant frequencies gives the precisely same frequencies as are detected with electrical measuring apparatus and are calculated from the inductance and capacitance values. These in turn depend on the geometry and the fundamental electrical constants. The fundamental frequency and the odd harmonics appear to be stimulatory to living systems and the even harmonics are depressive.

As frequency is increased, the coil eventually becomes a straight piece of wire, the capacitance is its electric charge storage capability to ground. It has become a microwave resonant transmission line. The velocity with which an electromagnetic wave will travel along any structure depends on the geometry, the materials involved and their electrical properties. There may be resonances at lengths corresponding to a quarter, half, three-quarters or full wavelength (and higher harmonics). As with musical instruments and electromagnetic resonators, the larger the resonating dimension the lower the resonant frequency. The equation: *constant velocity equals frequency times wavelength* $(V = f \times \lambda)$ governs this.

In the measurement of the characteristic frequencies of spatial objects by dowsing, is it unusual to find something for which the frequency decreases as the resonator gets larger. The norm seems to be the opposite. This first appears by implication in Section 3.5.1 where the longer the alkane chain, the higher the top frequency.

The graph in Figure 3.12 shows the lowest frequency that a given volume of water will take up as an imprint using the magnetic field of a permanent magnet to potentise into the water the frequency information from the field of a toroid fed from an oscillator. The frequency decreases linearly as the volume is increased over a range of 10^4. When this experiment is done in terms of length using water in a uniform tube rather than a volume of water it gives a cube power law except for very small lengths where it is linear.

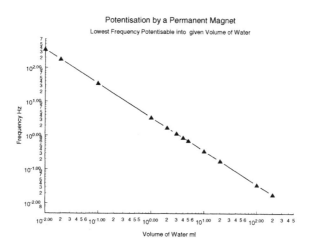

Figure 3.12 Potentisation of a given volume of water by a permanent magnet.

Although this frequency trend is in the expected direction, it still not consistent with the $V = f \times \lambda$ equation. A coherent system needs to be considered to understand this anomaly.

If it is assumed that the theoretical 75 nm domains of coherence are involved (see Section 3.2.2), then there would be 2.37×10^{21} domains/m^3. From Figure 3.12, the lowest frequency potentisable into 100 ml is 0.035 Hz and this 100 ml would contain 2.37×10^{17} domains. Therefore the lowest frequency for a single domain would be 8.3×10^{15} Hz or a wavelength of 36 nm which is about the radius of these assumed coherence domains. This implies that the limiting resolution for coherent water memory is determined by the number of domains involved and for 100 ml the resolution would be 4.2×10^{-18} Hz/Hz.

3.9.3.3 Velocity of Propagation of Coherence

However, the $V = f \times \lambda$ equation does work for the propagation of coherence. The velocities in a number of materials were measured using a modification of Fizeau's transit-time method and also from the critical angle at a water interface as shown in Table 3.13.

Table 3.13 Velocity of propagation of coherence

Fizeau's Method

Material	Velocity (m/s)
Specimen lengths ~1 metre	
Human leg	6.0
Tinned-copper (wire)	4.0
Gold (wire)	3.4
Normal saline	3.1
'Volvic' water	2.6
Silver (wire)	1.5
Steel (wire)	0.9
Brass (rod)	0.81
Molybdenum (wire)	0.5
Glass (pyrex, rod)	0.42
Solder (bar)	0.4
Tungsten (wire)	0.36
Aluminium (wire)	0.25
Air	0.05
Wood, cork, plastic	(not measurable)

Critical Angle Method - Total Reflection at Water Interface

Galvanized iron (sheet)	2.6	(velocity m/s)
Formica	1.5	
Roofing tile	1.3	
Wood	0.93	
Limestone	0.8	
Window glass	0.72	
Sandstone	0.41	
Pyrex glass	0.38	
Aluminium (foil)	0.21	
Quartz (rock)	0.17	
Air	0.07	

The coherence time for imprinting water was measured using 100% square wave modulation of the imprinting frequency (1 kHz). There was only imprinting if the modulation frequency was ≤ 1.8 Hz. The coherence length was measured by following the coherence along a plastic tube filled with water from a tube of frequency imprinted water pushed into one end, the coherence ceased at 1.38 m from the source. The product 1.8 Hz × 1.38 m = 2.48 m/s. The velocity for water from Table 3.13 is 2.6 m/s.

3.9.3.4 Frequencies of Simple Coherent Structures

The most basic structure is the mathematical 'straight rod'. Accordingly, the frequencies for various lengths of copper tubing both air-filled and saline-filled were measured and are shown in Figure 3.13. The frequency increases with the length. The potentising anomaly which occurs when the tube length is equal to the wavelength of the 21cm hydrogen molecule resonance also appears as an anomaly with the water-filled tube frequencies.

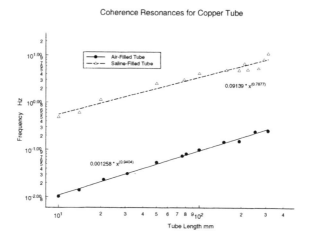

Figure 3.13 Coherence resonances in copper tube of various lengths.

The ratio of the frequencies of saline-filled to air-filled tubes (with values taken from outside the 21 cm anomaly) is 46.09 ± 2.18 (4.7% SD). The velocities of coherence in saline is 3.1 m/s and for air the mean is 0.06 m/s as given in Table 3.14 and this ratio comes to 52. The accuracy is limited by the difficulty of measuring the velocity of coherence in air.

The coherence length is given by the velocity of propagation of the coherence divided by the frequency. It is inversely proportional to the physical length as shown in Figure 3.14.

If the rod is within a coherent system, the coherence frequency becomes proportional to the spatial length L. Thus, frequency f equals velocity of coherence propagation v_{coh} times coherence length λ_{coh} which makes frequency proportional to spatial length L. However, this is not dimensionally correct unless an area A is added, thus:

$$f = v_{coh} / \lambda_{coh} = v_{coh} \times L / A \qquad (9.1)$$

Measurements on a glass cuvette (100 × 16 × 8 mm) containing water gave the results shown in Table 3.14 in terms of the constancy of f / L × F(A) where F(A) is that trigonometric function which best appears to fit the data and preserve the dimensional integrity. The angle of dip of the geomagnetic field is 60°.

Table 3.14 The first column gives the orientation of the cuvette (100 × 16 × 8 mm), f is the measured frequency, L is the length component in the North-South direction, F(A) is an empirical trigonometric function

Cuvette N-S mm	E-W mm	f Hz	f / L	F(A)	F(A) × f / L
100	28	0.4014	4.014	cot 60°	2.317
100	16	0.3914	3.914	cot 60°	2.260
16	100	0.0404	5.075	cos 60°	2.538
16	8	0.0391	2.444	1	2.444
8	100	0.0111	1.391	tan 60°	2.409
8	16	0.0221	1.383	tan 60°	2.395

Mean & S.D. 2.393 ± 0.097 (± 4%)

Relation between Coherence and Spatial Lengths

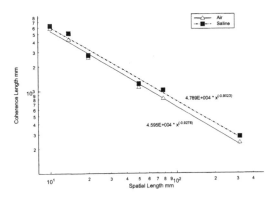

Figure 3.14 Coherence length as a function of the spatial length.

The measurements described above were standardised with an alignment along the North-South axis so as to be perpendicular to this and avoid interactions with it.

3.9.3.5 Effect of Orientations Relative to North-South Axis

It appears that the geophysical reference parameter for dowsing measurements is the magnetic vector potential of geomagnetic field which is in the East-West direction corresponding to those currents which give rise to the geomagnetic field.

Figure 3.15 shows the effect of orientations relative to North-South axis for an aluminium wire and a glass pipette containing water. The ordinate is the frequency relative to that measured with the specimen along the North-south axis (264 Hz for the water and 3.1 Hz for the aluminium) multiplied by the cosine component of the total length which is in the North-South direction at each angle of inclination. If these were the only factors involved in this effect, all points would lie on the 100% line which indicates that there is some other effect operating as well.

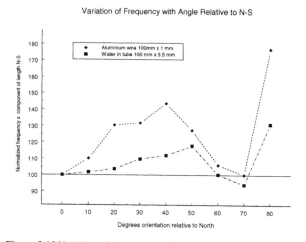

Figure 3.15 Variation of frequencies with angle relative to the North-South axis.

3.9.3.6 Frequencies of a Random Arrangement of Objects

A random arrangement of objects took the form of glass hemispherical beads about 1 cm diameter and of uniform size which were dropped one at a time into a jar of water so that they became arranged randomly. The frequencies due to this perturbation of the water were found to increase in proportion to the square root of the number of beads as is shown in Figure 3.16.

3.9.3.7 Frequencies of Simple Coherent Structures – Parallel Wires

For a combination of two wires of different metals joined at their ends the total coherence path was L = 40 cm. The velocity effective in this loop appears to be the difference between the velocities of the two metals. This is shown in Table 3.15.

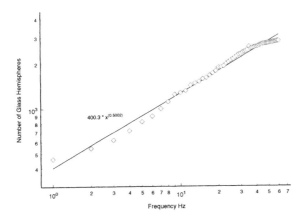

Figure 3.16 Frequencies for number of glass beads added randomly to water.

Table 3.15 Coherence frequency for pair of wires each 20 cm of different metals joined only at the ends

Metals Difference	Velocity (v_{diff}) / L	Frequency (Experimental)	Frequency
Fe – Al	0.65 m/s	1.625 Hz	1.614 Hz
Al – Cu	3.75	7.750	7.610
Cu – Fe	3.10	9.375	9.230

3.9.3.8 Frequencies for Regular Patterns of Cuvettes

Plastic cuvettes (10 mm x 10 mm x 45mm, wall 1 mm) containing water and arranged standing in a tank of water were investigated (both had water levels of 25 mm). The frequency measured for a single cuvette was 1200 Hz and was independent of its position and orientation in the tank. Table 3.16 summarises the frequencies for various regular arrangements of the cuvettes. In general, the measured frequency was the 1200 Hz for the single cuvette divided by the number of cuvettes present whether in a straight line or a square pattern. Spaces between the cuvettes of 10 mm behaved almost as if a cuvette was present. Above three cuvettes and two spaces, or two cuvettes and three spaces both of which behaved like six cuvettes, the spaces were not exactly equivalent to a cuvette.

Unusually for this work, is the result that the frequency decreases as the number of cuvettes increases? This must mean that this assembly is not functioning as a single coherent system.

Table 3.16 Frequencies for various arrangements of cuvettes

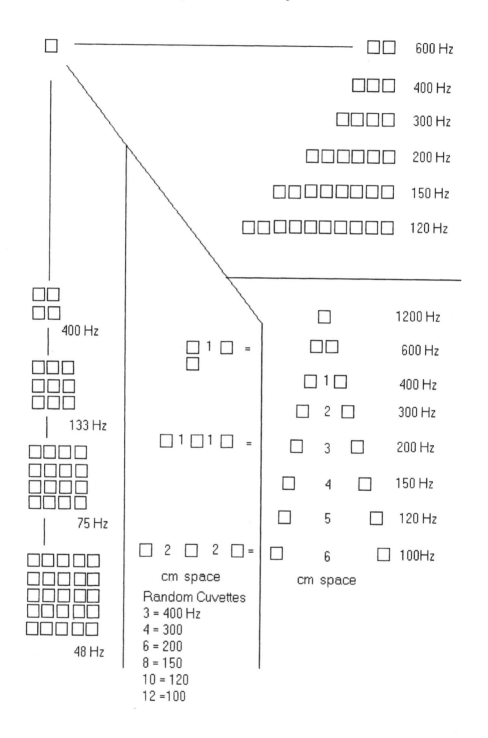

3.9.3.9 Frequencies for Waveform Patterns formed in Wire

Patterns were formed with enamelled copper wire (28 SWG) stretched against pins stuck into a corkboard. The wire was formed into triangular, saw-tooth and square wave patterns of amplitude ± 2 cm. The frequencies measured are shown in Table 3.17.

The Fourier transform of the triangular and square waveforms contain only the odd harmonics which were found. Altering the square wave from symmetrical to asymmetrical changed the 3rd harmonic from depressive to stimulatory phase. In the symmetrical arrangement, the signs alternate, in the asymmetrical arrangement, all are positive. The Fourier transform of the saw-tooth waveform contains all harmonics. All these harmonics were detected. It seems that the number of harmonics is limited by the number of cycles present in the shaped wire. However, when the square wave wire was immersed in water, the fundamental

Table 3.17 Frequencies for waveform patterns formed in wire

Waveform	Triangular 3 cycles @ 8 cm/cycle	Square 3 cycles @ 8 cm/cycle	Saw-Tooth 6 cycles @ 4 cm/cycle
Fundamental Frequency	14.31 Hz	14.31 Hz	28.62 Hz
Harmonics x2			57.24 Hz
x3	52.93 Hz	52.93 Hz	85.86 Hz
x4		114.4 Hz	
x5	71.55 Hz	52.93 Hz	143.1 Hz
x6		171.7 Hz	

3.9.3.10 Real Building Materials – Bricks

The frequencies measured for bricks (by Phillipson of Bolton) are given in Table 3.18. A block of three of these still mortared together had been conveniently prepared by the last car to impact the garden wall. There are directional effects similar to those measured for the water-filled cuvette given above (Table 3.14). The frequencies of two bricks still joined end-on with mortar is double that of the single brick; the third brick mortared across the joint gives a further increase in frequency, but not by a factor of three over that of the single brick. The important thing is that in all these cases the frequency increases with the number of bricks showing that the mortared assembly remains a coherent system.

Table 3.18 Frequencies for an assembly of builders bricks

Single brick 220 × 103 × 73 mm

N-S axis direction	Frequency Hz	Frequency / N-S dimension
220 mm	0.5416	2.464
103 mm	0.2072	2.012
73 mm	0.1452	1.989
Two bricks end-on		
2 × 220 mm + 12 mm mortar	1.087	2.405
Three bricks (2 end-on, third across join)	1.859	

3.9.3.11 Frequencies for 'Molecular Model' Patterns

The plastic straws of a molecular model building set are conveniently cut to the scale 3 cm/Å. This is the same ratio as the velocity of light in free space to the velocity of coherence propagation in saline 300 Mm/s to 3 m/s. Because of the multiple frequency effects in coherent systems, it has proved possible to model molecules with water attachments, to measure their ELF frequencies when immersed in saline and compare these measurements with the ELF measurements made on the corresponding chemicals with traces of water present. Figure 3.17 shows the frequencies for the basic 'molecular model' plastic straws as a function of their length. The line shown by triangles represents the insertion of a molecule within the length to confirm that nothing drastic results from a change of this magnitude.

Figure 3.17 shows the frequencies for molecular models of n-pentane and n-hexane at scaling lengths of 3 cm/Å. This scaling corresponds to a velocity ratio of 10^8 representing the velocity of light in free space. 3×10^8 m/s to the velocity of coherence in saline 3 m/s. Note that frequency increases in proportion to length as is usual for a coherent system. The best fits as shown in bold type correspond to the hydrogen-bonded water structure extending the whole length of the molecule in each case. This experimental technique is possible in any coherent system where multiple frequencies can occur would appear to have considerable scope for application to molecular modelling in general.

The endogenous ELF frequencies measured for water are shown in Table 3.18 for models of three possible water structures. It was found that if some frequency was imprinted into the saline containing the immersed molecular model, the measured ELF frequency was replaced by that frequency ± the imprinted frequency already found in experiments on potentisation (see Section 3.8.3).

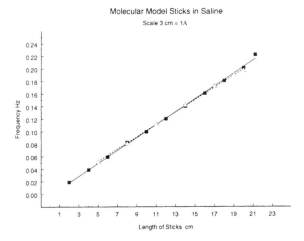

Figure 3.17 Frequencies for "molecular model" patterns.

Table 3.19 Frequencies for molecular models. Scaling lengths at 3cm Å scales velocities by 10^8 from the velocity of coherence in saline 3 m/s to the velocity of light in free space. 3×10^8 m/s. The frequency increases in proportion to the alkane chain length. The connections for the H-bonded water structures are shown

Chemicals/ Frequency	**Molecular Models**		
	C_5H_{12} only	$C_5H_{12} + 3H_2O$ $C_1 - C_4$	$C_1 - C_5$
n-pentane + trace water			
4.2 Hz	4.113 Hz	4.042 Hz	**4.012 Hz**
19.4	7.132	7.054	**19.03**
29	20.31	18.51	**28.12**
n-hexane + trace water	C_6H_{14} only	$C_6H_{14} + 4H_2O$ $C_1 - C_6$	
4.2 Hz	4.113 Hz	**4.204 Hz**	
6.8	7.132	**6.824**	
13	20.31	**13.10**	
19.4	38.11	**19.32**	
26	80.32	**25.32**	
42		**41.63**	

Table 3.20 Frequencies for molecular models of water. Scaling lengths at 3cm/ Å. The FIR to ELF conversion factor used

Water Models in water Hz	Natural Resonances levels cm^{-1}	FIR energy cm^{-1}/Hz	ELF 6.37 at
Water – random chain	13.07	213-128	13.3
Water – pentagonal spiral	22.01	357-213	22.6
Water – caduceus spirals	35.06	357-128	34.8

3.10 VISUAL STIMULATION

3.10.1 Stimulation by Flashing Light

A light-emitting-diode (LED: RS-564-015, red-3000 mcd) was viewed towards its lens at a comfortable brightness which did not give any noticeable after image when run off a digital oscillator at 0.6-0.7 V (50 ohms impedance). The chakra and acupuncture points were checked for reaction at that frequency known to stimulate

Table 3.21 Frequency range for stimulation of acupuncture meridians by a flashing light.

Chakras	Frequency
Crown	0.25 Hz
Forehead	3.0
Thyroid	80
Heart	7.8
Umbilical	23
Pubic	80
Coccyx	80
Acupuncture Points	**Frequency**
Or1	0.078 Hz
Pe9	0.25
AD1	2.0
GV14	4.3
Liv1	4.8
BL67	5.5
He9	7.8
Ren24	14.3
FibD1	800
TW1	6kHz
ND1	27kHz

(see Section 3.6). All the reactions measured were of stress, both when the LED was viewed at the frequency given below with both eyes open. There was a reaction of awareness when viewing with either eye alone. With both eyes closed there was no response. The same response was also obtained if the light from the flashing LED was viewed by reflection off, or transmission through a sheet of white paper.

The frequency thresholds for stimulation were below the 47 kHz of Ki1 and above the 0.25 Hz of Pn1. Outside that frequency range, none of the meridians responded to this visual stimulus.

3.10.2 Visual Response to Basic Patterns

It seems that in visual responses the orientation of the object relative to the geomagnetic field does not matter, unlike the responses to patterns within the body aura. The viewing distance does not matter so long as the whole of the object is within the visual field.

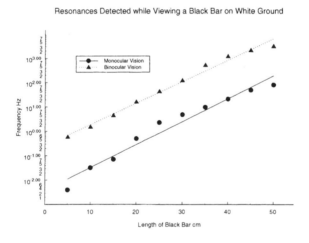

Figure 3.18 Monocular and binocular resonances viewing black bar.

A bar from 2.5 cm wide black PVC tape on a white background 40 x 60 cm was measured for lengths from 5 cm to 50 cm. The results as shown in Figure 3.18 were markedly different between viewing with monocular and binocular vision. Reversal to a white line on a black background reversed the phase of the reaction from stimulatory to that which is depressive of biological activity.

Figure 3.19 shows the same experiment carried out using different coloured A4 cards as the background. Figure 3.20 shows the frequencies for a bar pattern comprising a number of black PVC tape bars each 20 cm long arranged parallel and spaced by 2.5 cm to give alternating white bars. The frequency just continues to increase as more bars are added at least up to ten black bars.

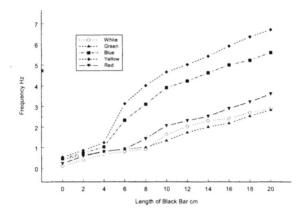

Figure 3.19　Visual perception of black bar on A4 coloured background.

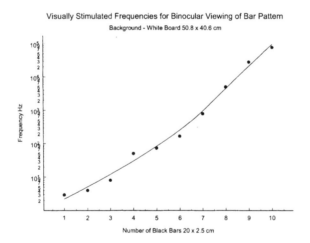

Figure 3.20 Visual perception of bar pattern.

3.10.3 Cameras and Stolen Souls

There is reluctance of peoples living close to nature to be photographed. As a concession they may allow a photograph if their eyes are, "...kept tight shut so that the soul might not be stolen by the camera..." Dowsing photographs confirmed for the Writer that there was no reaction from any person with tightly closed eyes but, that a photograph of a person with open eyes did produce a response, usually of "good will". It must remain a topic for a future consenting experiment to determine whether the map-dowsing of such photographs is a subsequent invasion of the subjects' privacy. The implication is that all photographs and images used to adorn buildings and living spaces may result in some interaction with the original subject.

In New Zealand, the Maoris have a similar reluctance to be photographed. During a visit to Rotorua, in publicity photographs for the "Maori Experience" for which the subjects were of course actors accustomed to being photographed, the Writer found a distinct dowsing reaction of "good will" could be obtained from those performers' faces looking straight at the camera.

At the Maori Art and Crafts Institute in Rotorua, there was a very strong dowsing signal from the Maori figure carvings only when standing exactly in front of the face and only if the carvings were as they say "finished" that is, with their 'shell eyes' inserted. At the carved gateway forming the entrance to the Model Pa at Whakarewarewa, the upper figures are "finished" in this way, the lower ones are not. Only the upper ones gave a dowsing reaction both at the site and subsequently from a postcard image.

A similar dowsing reaction was obtained from the carved figureheads on the stem and the carving on the stern of the canoe displayed there. There was a very strong narrow-beam signal fore-and-aft which might have been strong enough for sensitive helmsmen to be able to keep the canoes in convoy during darkness.

A more permanent navigational feature might be built into the megalithic alignments at Carnac in Brittany. This array of stones puts out a very strong dowsable beam towards the Atlantic, extending out over what is now a sandbank connecting with the Presqu'ile de Quiberon. The Writer was able to detect this beam by dowsing from on-board a Santander to Plymouth ferry while it passed through the projection of the Carnac beam about 160 km out in the Bay of Biscay. Could this have been a 'neolithic homing radar' for the prehistoric capital of Carnac? These signals were strong enough to indicate when detailed features within the beam had been crossed and there was a very strong signal change on finally leaving the beam. It would be interesting to find out if the signal would reach to the Azores. There are similar though less powerful signals coming from other stone alignments near coasts such as those at Callanish in the Hebrides.

3.11 FREQUENCIES FROM THE EXTERNAL ENVIRONMENT

3.11.1 Frequencies from Over-Head Power Lines

Each set of three conductors (the phases) of the typical overhead power line represents a three-phase transmission system. Often, there are two or more transmission circuits attached to each pylon each with three phases. The term *phase* has a double use. Not only does it denote an electrical circuit, it also denotes the fraction of a cycle of the supply frequency (50 Hz in the UK, 60 Hz in N. America) by which the voltages and currents in the conductors differ. This is expressed as an angle in degrees, 360° represents a complete cycle and is the same phase angle as 0°. This article will use *phase-angle* for the latter meaning. In a three-phase system, the currents in the respective phases have phase-angles in the same relationships as three clock faces with hour hands at 12, 4 and 8 o'clock. Mathematically, the vector resultant of any two such hour hands is equal and opposite to the third, making the vector sum of all three equals to zero. Thus, in an ideal three phase system with balanced loads (all hour-hands the same length) and symmetrical phases, the vector sum of the phase currents is zero.

The electric and magnetic fields and the magnetic vector potential all decrease in proportion to distance from a long straight wire. The magnetic vector potential has an added complication in that its absolute value is proportional to the total length of the wire.

The magnetic field (**B**) and the magnetic vector potential (**A**) are both proportional to the current in a long straight wire. When one is far enough away from a set of three-phase conductors, each conductor can be regarded as at the same distance, the vector sum of the phase currents will be zero, and **B** and **A** will be zero.

When directly beneath the centre of the conductors of a typical 400 kV line, there could be a 40% difference between the distance from the lower conductor and the mean distance of the two upper ones. Here, **B** and **A** would have values corresponding to 40% of the line current for a single conductor. The actual value obviously depends on the line configuration and the terrain.

Imprinting a frequency into water requires an alternating magnetic vector potential (**A**) at that frequency and a magnetic field at any frequency less than or equal to this.

Close to overhead lines, the effect of this unbalance of distance may be sufficient to imprint water. For a 400kV twin-conductor, double circuit transmission line, the Writer found that water would spontaneously and immediately imprint at distances of 40 to 50 metres on either side of the line-centre (see Figure 3.21). The frequencies imprinted into a tube of water so exposed were 50 Hz, 150 Hz, 16.66 Hz and 3 Hz. These represent the power supply frequency, the third and one-third harmonics and what is probably a load fluctuation at 3 Hz.

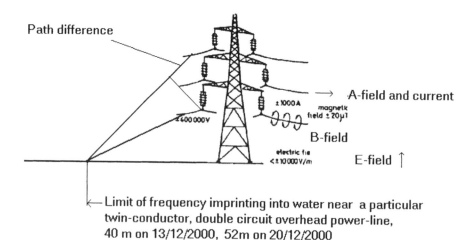

Figure 3.21 Fields near overhead power lines.

This water imprint was of the type found when imprinted or potentised water has been heated above 90°C. It is only detected with a Caduceus coil. It may be restored to its usual form by exposure to the 7.8 Hz endogenous frequency of the heart chakra by holding it against the chakra or by placing it close to a coil radiating that frequency. The body being mostly water can (for better or worse) acquire frequency imprints under power lines. These imprints will be equivalent to some uncharacterised homoeopathic potency the proving-effects of which in a healthy person will be the symptoms that it would be a remedy for in the sick. With chronic exposure, there will come a time when the proving-symptoms and the disease state become indistinguishable due to adaptation to the environmental stress. The Writer has observed 50 Hz as a stress frequency in microscope slides of myeloid leukaemia blood; also, a sample of pineal extract included both 50 Hz and 60 Hz as endogenous frequencies.

3.11.2 Distance Related Effects near Radio and Television Transmitters

A Study reported in the paper on, "Cancer Incidence near Radio and Television Transmitters in Great Britain II. All High Power Transmitters" (Dolk *et al.*, 1997) was concerned with, '...findings for adult leukemias, skin melanoma, and bladder cancer near the 20 other high power radio and TV transmitters in Great Britain...' (other than the Sutton Coldfield transmitter previously studied). It concluded that, "....while there is evidence of a decline in leukemia risk with distance from transmitters, the pattern and magnitude of risk associated with residence near the Sutton Coldfield transmitter do not appear to be replicated around other transmitters. ....". This Study and its findings appear at odds with the 'gut-feelings' of local residents communicated to the writer who re-examined it in the simplest manner so far as its published data allowed in Electromagnetic Hazard & Therapy (Best, 2001).

3.11.2.1 Population Density

In England and Wales as a whole, the average population density is 940 persons/square mile which is 363 persons per square kilometre. 'The total Study population was around 3.39 million persons'. 'The study areas were defined by circles of radius 10 km from each transmitter'. Thus, 3,390,000 persons within 20 areas each of 10 km radius (= 314.2 sq. km) gives an average population density of 539 persons per square kilometre. This is not inconsistent with the above estimate for the country as a whole since these transmitters serve major centres of population and 539 persons/sq km was adopted for the following calculations.

3.11.2.2 Cancer Incidence - Expected Number of Cases

The *Study* used cancer registration data post-coded to the address at diagnosis for the years 1974-1986 (1974-1984 in Wales, 1975-1986 in Scotland). If the overall incidence of leukaemia is approximately 5/100,000 per year (Kumar, 1990) then, over a period of 12 years the expected incidence would be 60/100,000 persons.

Table 3.22 compares the expected incidence of all leukaemias based on an average population density of 539 persons per square kilometre to the Study's observed cases as a function of distance from all transmitters taken from the data given in Table 2 on p. 14 of Dolk *et al.* (1990).

Column 1 lists the same annular bands at the various distances from the transmitters that were used in the above Study. The areas enclosed within these bands are given in Column 2. The number of persons within these areas in Column 3 for a population density of 539/sq.km, the value implied in the Study. The total expected number of cases in each annular band is given in Column 4 for a leukaemia incidence of 5/100,000 per year for 12 years (= 60/100,000) and including all 20 transmitter locations.

These expected cases mostly come out lower than the expected cases as given in Table 2 on p.14 of the Study which are quoted in Column 5. If the observed-to-expected (O/E) ratio for the Study's observed values given in Column 8 are compared to those now calculated and given in Column 7, it is seen that whereas the latter show an O/E peak which may represent a "window" for electromagnetic field effects, any peak in the Study's results is minimal.

A probability plot of Column 7 is a good fit to a 'Normal' distribution from 1.5 km to 8 km (as shown by the solid line) with the mean close to 5 km and a standard deviation of ± 1.5 km (Figure 3.22).

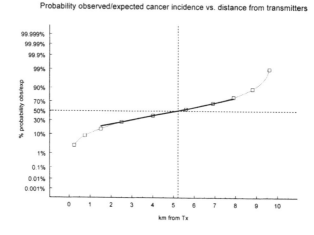

Figure 3.22 Probability of observed-to-expected ratio of cancer incidence versus distance from 20 transmitters.

Since the Study covered 20 transmitters from different parts of the UK, it is reasonable to assume that any effects related to geographical or topographical features and antenna design should average out. This only leaves the physical characteristics of the propagation of electromagnetic radiation from which to seek a mechanism.

A simple experiment involving a toroid and solenoid connected in series shows that when the A and B field vectors are in opposite directions (180° or π/2 radians phase difference) the frequency of the current is imprinted into water placed nearby. When the A and B fields are parallel (zero phase difference) the frequency imprint is erased.

The electromagnetic radiation (E- and B-fields) from a transmitter will experience refractive index and propagate at the velocity of light in air, the magnetic vector potential (A-field) does not interact with matter (it alters the phase of the electron wave-function) and so propagates at the vacuum velocity of light. At 5 km distance from the transmitter, there will be a transit time difference of 5 ns between the A and B fields. At 100 MHz this represents a 180° or $\pi/2$ radians phase difference. This is the condition for that frequency to be imprinted into any water present such as living tissues. The frequency band 70 MHz-130 MHz would cover the standard deviations of Column 7 in the Table and this would include the FM radio transmissions from the TV transmitter towers.

Thus, it seems pointless to restrict the power output in the medium-wave transmitter at Anguillara-Sabazia (Italy) in response to pressure from the government on the basis of a 'thermal effects' hypothesis from "classical physics".

Possible bio-medical effects of the FM transmissions should include stress by entrainment of the allergy acupuncture meridian (AD1 in Voll-notation) which has an endogenous frequency of 94 MHz. As with power lines, there should be stress from chronic exposure to the 'proving-symptoms' for whatever homoeopathic potencies happen contain in this case, frequencies in the region of 100 MHz.

Table 3.22 Cancer incidence with distance from 20 transmitters

Distance of Tx km	Area sq.km	Persons 539/sq.km.	Expected cases	Expected cases [1]	Observed cases [1]	Obs./Exp. cols 6/4	Obs./Exp. [1]
1	2	3	4	5	6	7	8
0.0-0.5	0.785	423	5	2.3	2	0.4	0.87
0.5-1.0	2.356	1270	15	13.8	12	0.8	0.87
1.0-2.0	9.428	5082	61	65.5	65	1.1	0.99
2.0-3.0	15.70	8462	101	135.3	155	1.5	1.15
3.0-4.9	47.16	25,419	306	494.1	539	1.8	1.09
4.9-6.3	49.27	26,557	318	589.7	623	2.0	1.06
6.3-7.4	47.30	25,495	306	518.0	547	1.8	1.05
7.4-8.3	44.40	23,932	288	453.4	434	1.5	0.96
8.3-9.2	49.50	26,681	320	493.5	497	1.6	1.01
9.2-10.0	48.30	26,034	312	427.9	431	1.4	1.01

3.12 ELECTROMAGNETIC HEALTH STATUS OF BUILDINGS

3.12.1 Healthy Living Environments

All building structures contain within them frequency implications. These arise from the chemical signatures of the building materials, the electrical resonances of metallic components, steel structural girders, concrete reinforcements, builders' hard-ware, electrical wiring and equipment, plumbing, ventilation and heating ducts, furniture and fittings, decorations, lighting, colours and textures and living systems – people, animals, insects, plants, and last but not least – herbicides and pesticides!

There will be light and sound and electromagnetic fields from internal and external sources, multiple frequency effects link many widely different parts of the electromagnetic spectrum in common effects on living systems. There may be some geopathic stress and negative air ion deficiency.

Rousseau *et al.* (1989) deal with the environmental health aspects of dwellings under the heading – 'Don't be alarmed, but be prepared!'. They include descriptions of suitable living environments for the environmentally sick and how to clean-up an existing property.

Nature gives us a high degree of apparent randomness, plant and tree growth is fractal, so is the snowflake. The rigidly geometrical is rare, the only straight line in nature is the horizon at sea.

For individuals to be able to personalise their living environment for each activity and in each part of their space, the building like a good violin should be free from "wolf notes" in its aura and visual frequencies, that is, unless the structure or its contents are intended to make some such impact statement through frequency. For example, the painting "Movement of a Bird" painted in 1916 by Fortunato Depero (1982-1960) stimulates the forehead chakra with 3.113 Hz, whether viewed in colour or in a black-and-white photocopy; there is also stimulation from the red part at 44.12 Hz and from the orange part at 7333 Hz. A trip to an art gallery may well prove as therapeutic as a visit to a physician.

When the building is occupies, there may be a need to provide for, or avoid, interactions between the aurae of the occupants, whether *aura popularis*, or *auris vitalibus vesci*.

3.13 CONCLUSION

It is never going to be possible to eliminate stress from life and living. The healthy living system can cope with stress and be strengthened by it as with athletic training. There needs to be a concept of "Environmental Stress Management" evolving as a general awareness. There will be some stresses which are optional, others are unavoidable. The individual body can only cope with a certain amount of stress whether it is chemical, electromagnetic, nutritional, personal, emotional or, social. If for example, an individual has to, or wishes to take a lot of electromagnetic stress, there must be compensatory changes elsewhere. It might even have to involve some change in life-style. Plans for increasing sources of electromagnetic stress into the environment should be compensated with a corresponding reduction in the chemical stress. This is because of the possibility of transfer of sensitivities to a new "trigger". If a chemically sensitive person is reacting to environmental chemical stress while being exposed to a frequency of an environmental electromagnetic field, the next time that frequency is encountered the chemical reactions may be "triggered".

As is often remarked, there is no profit to be made from preventative medicine hence relatively it is so little researched. It is clear that the chemical and electrical environments interact intimately. The question is how to reconcile 'equality under the law' with 'biological individuality - biochemical and bioelectromagnetic'.

3.14 REFERENCES

Arani, R., Bono, I., Del Giudice, E. and Preparata, G., 1995, QED coherence and the thermodynamics of water. *International Journal of Modern Physics B,* **9**, pp. 1833-1841.

Bachler, K., 1989, *Earth Radiation,* English translation by Gerhart, M.Z. (Manchester, Wordmasters).

Barrett, T.W. and Pohl, H. (Eds.), 1987, *Energy Transfer Dynamics,* (Berlin: Springer-Verlag).

Best, S., (Ed.) 2001, *Electromagnetic Hazard & Therapy.* (www.em-hazard-therapy.com simonbest@emhazard.dabsol.co.uk), personal communication.

Dolk, H., Elliott, P., Shaddick, G., Walls, P. and Thakrar, B, 1997, Cancer Incidence near Radio and Television Transmitters in Great Britain II. All High Power Transmitters. *American Journal of Epidemiology,* **145(1):** pp. 10-17.

Del Giudice, E. and Preparata, G., 1994, Coherent dynamics in water as a possible explanation of biological membranes formation. *Journal of Biological Physics,* **20**, pp. 105-116.

Elia V, Niccoli M. (1999) "Thermodynamics of Extremely Diluted Aqueous Solutions" *Ann. N.Y. Acad. of Sci.* **879**: 241-248.

Fröhlich, H. (Ed.), 1983, *Coherent Excitations in Biological Systems,* (Berlin: Springer-Verlag).

Fröhlich, H. (Ed.), 1988, *Biological Coherence and Response to External Stimuli,* (Berlin: Springer-Verlag).

Ho, M-W, French, A., Hafegee, J. and Saunders, P.T., 1994. In: Ho, M-W., Popp, F-A. and Warnke, U. (Eds.). *Bioelectrodynamics and Biocommunication.* (Singapore: World Scientific), p. 204.

Kumar, P.J, and Clark, M.L., 1990, *Clinical Medicine* (London: Bailière-Tindall), p. 357.

Ott, J., 1973, *Health and Light* (Columbus: Ariel Press).

Preparata, G., 1995, *QED Coherence in Matter,* (Singapore: World Scientific).

Prangishvili, I.V., Gariaev, P.P. (gariaev@aha.ru), Tertishny, G.G., Maximenko, V.V., Mologin, A.V., Leonova, E.A. and Muldashev, E.R., 2000, *Spectroscopy of Radiowave Radiations of the Localized Photons: the Path to Quantum Nonlocality of Bioinformation* Processes Sensors and Systems – 2000. Moscow: Institute of Control Science, Russian Academy of Science.

Rousseau, D., Rea, W.J. and Enwright J., 1989, *Your Home, Your Health and Well-Being.* (Vancouver BC: Hartley & Marks).

Smith, C.W., 1990, Bioluminescence, coherence and biocommunication. In *Biological Luminescence,* edited by Jezowska-Trzebiatowska, B., Kochel, B., Slawinski, J and Strek, W., 1990, (Singapore: World Scientific) pp. 3-18.

Smith, C.W., 1994, *Electromagnetic and Magnetic Vector Potential Bio-Information and Water.* In: Endler PC, Schulte J (Eds.). Ultra High Dilution: Physiology and Physics. Dordrecht: Kluwer Academic, pp. 187-202.

Smith, C.W., 1995, Measurements of the Electromagnetic Fields Generated by Biological Systems. *Neural Network World* **5**: pp. 819-829.

Smith, C.W., 2000, Notes for Presentations: "The Diagnosis and Therapy of EM Hypersensitivity" and "EM Fields in Health, in Therapies and Disease". *18th Annual Symposium on Man and His Environment*, June 8-11, 2000, Dallas, Texas.

Smith C.W., 2001, *Learning from Water, a Possible Quantum Computing Medium*, 5th. International Conference on "Computing Anticipatory Systems", HEC Liège, Belgium, 13-18 August 2001. CASYS'01 Abstract - Symposium 10, p. 19 (Proceedings to be published in Intl. J. of Computing Anticipatory Systems).

Smith, C.W. and Best, S. 1989, *Electromagnetic Man: Health and Hazard in the Electrical Environment*. (London: Dent, 1989,1990; New York: St. Martin's Press, 1989; Paris: Arys/Encre, 1995 {French edition}, Bologna: Andromeda, 1997, 1998 {Italian edition}).

Stux, G and Pomeranz, B., 1991, *Basics of Acupuncture,* 2nd ed. (Berlin: Springer-Verlag).

Wever, R.A., 1973, Human circadian rhythms under the influence of weak electric fields and the different aspects of these studies, *International Journal of Biometeorology* **17(3)**, pp. 227-232.

Wydra, N., 2000, *Feng Shui goes to the office* (Lincolnwood, Chicago IL: Contemporary Books).

CHAPTER 4

An Introduction to the Naturally Occurring and the Man Made Electromagnetic Environment

A.C. Marvin

4.1 INTRODUCTION

Over recent years, much concern has been expressed over the potential adverse health effects of electromagnetic radiation. These concerns range from the effects of low frequency magnetic fields associated with power systems and visual display units (VDU's) to the increased incidence of skin cancers attributed to excessive exposure to ultra-violet radiation from the sun.

Electromagnetic radiation is the transport of energy in the form of an electromagnetic wave. The deposition and absorption of this energy into biological tissue is the physical basis of any interaction. The absorption mechanism is determined by the type of electromagnetic radiation (see later) and the properties of the tissues.

At this point it is perhaps worth reviewing what is meant by the term energy. The Feynman Lectures on Physics (Wesley, A., 1963) state "It is important to realise that in physics today, we have no knowledge of what energy *is*." A working definition may be that energy is what is required to do something! Energy is conserved. Thus in a microwave oven, energy in the form of microwaves (a class of electromagnetic wave) travels into the chicken where it is absorbed and manifests itself as heat energy (vibrating molecules). Sufficient molecular vibration or heat is enough to change the physical and chemical nature of the chicken (a process known as cooking). When the chicken is removed from the oven and left to cool, the energy leaves the chicken as radiant heat (a class of electromagnetic radiation) and through convection in the air around the chicken. The energy transferred to the air is in the form of heat (the air is hotter) and kinetic energy (movement - the air rises). No energy is lost in this process but many things have happened. Energy is conserved and can be measured and quantified.

Ultimately, nearly all our energy (the exceptions are nuclear energy, tidal energy and geothermal energy) derives from the sun. Fossil energy sources such as gas, oil or coal are effectively stored solar energy sources. Their energy output is no more than that put in by the sunlight that was used to create them millions of years ago. Energy derived from biomass is more recently stored solar energy. Renewable energy sources such as wind energy, wave energy and solar energy are derived from the energy currently arriving from the sun. All this solar energy arrives at the earth in the form of electromagnetic radiation.

All biological systems, mankind included, have evolved in the presence of electromagnetic radiation. The basic unit of all forms of energy, including electromagnetic radiation, is the Joule, the amount of energy required to raise the temperature of a litre of water by one degree Kelvin (one degree Celsius). Although the Joule is defined in thermal terms, familiar to the engineers and scientists of the nineteenth century, the conservation and conversion of energy on which we depend means that the basic unit can be used to quantify all types of energy. The rate of energy flow is the power in Joules/second more commonly known as Watts.

Energy flow in electromagnetic radiation is described in terms of its power density in Watts per square metre, the power incident upon an area of one square metre. As all the activities of our society are reliant on our use of energy, the ultimate economic fact is that the electromagnetic energy incident upon the earth from the sun is approximately 1.4 kW/m^2 (1400 Watts per square metre). This energy powers the ecosystem of our planet including that portion associated with our human activities.

Naturally occurring electromagnetic radiation such as sunlight has been complemented during the last century by the increasing use of artificially generated electromagnetic radiation for diverse uses such as lighting, heating, communications and power transmission. Naturally occurring electromagnetic radiation facilitated our evolution and much of our culture. We now complement it with artificially generated electromagnetic radiation and our technology base and thus our culture is now completely dependent upon it.

4.2 ELECTROMAGNETIC WAVES

The following is a description of the basic physics of electromagnetic waves. As with all physics, it is a conceptual model used to account for what we observe.

An electromagnetic wave can be regarded as an oscillatory wave propagating through a volume of space. The wave has a velocity (the velocity of light) specific to the material medium in which the wave is propagating. The oscillatory wave has a frequency, the number of complete oscillations or cycles observed by a stationary observer in one second. This frequency is expressed as a value in Hertz where 1 Hertz is 1 cycle per second. The wave also has a wavelength, the distance occupied in space by one cycle. The well-known relationship between these quantities is;

$$\text{wave velocity} = \text{frequency x wavelength}$$

The classification of electromagnetic waves by their wavelength or frequency is behind the idea of the electromagnetic wave spectrum. The velocity of electromagnetic waves in a vacuum (and air) is three hundred thousand kilometres per second or 3.10^8 metres/second. The BBC radio transmitter at Droitwich used to carry the Radio Four service with a wavelength of 1500 m corresponding to a frequency of 200 kHz (200 kiloHertz or 200000 cycles per second). In recent years it has moved slightly to 198 kHz with a corresponding wavelength of 1515 m. The electromagnetic radiation used in GSM mobile phones has frequencies of either 900 MHz (mega Hertz i.e. million Hertz) or 1.8 GHz (giga Hertz i.e. thousand million Hertz) corresponding to wavelengths of 330 mm and 165 mm respectively. Green light has a wavelength of around 500 nm (nano metre i.e. one thousand millionth of a metre) and a corresponding frequency of 600 THz (tera Hertz i.e. one million million Hertz). The table below lists points in the electromagnetic spectrum over a wide frequency and wavelength range with indications of some of the artificial and naturally occurring uses of the radiation at those and similar frequencies.

Frequency (Hz)	*Wavelength (m)*	*Application*
50	6000 km	power transmission
75	4000 km	sub-sea communications
1 kHz	300 km	audio (20 Hz to 20 kHz)
100 kHz	3 km	long wave radio
1 MHz	300 m	medium wave radio
100 MHz	3 m	FM radio
600 MHz	500 mm	Terrestrial TV broadcast
900 MHz	330 mm	Mobile phones
1.42 GHz	214 mm	Hydrogen line Radio astronomy
1.5 GHz	200 mm	Global positioning system
1.8 GHz	165 mm	mobile phones
2.45 GHz	122 mm	microwave ovens short-range radio systems
5 GHz	60 mm	wireless networks radar trunk communications
30 GHz	10 mm	imaging radars
30 THz	10μm	far infra-red – radiant heat fibre-optic communications
230 THz	1.3μm	fibre-optic communications
600 THz	500 nm	green light
1200 THz	250 nm	ultra-violet light
3000 THz	100 nm	X-rays

Although electromagnetic waves are used by us to see, we cannot "see" the oscillatory nature of the waves in the same way that we can see waves on the surface of water. The oscillating components of an electromagnetic wave are described as electric and magnetic fields. A working definition of a field is that it represents some physical phenomenon that can be measured with a values associated with all points in the space where the field exists. For example, atmospheric pressure is a phenomenon that can be measured using a barometer. It varies with position on the earth's surface due to weather system activity and decreases with height above the earth's surface. It is a pressure field. The origins of electric and magnetic fields are in the observed interactions between charged particles (electrons and protons) which, along with neutrons, constitute the atoms in all matter. We observe that charged particles exert forces on each other. The force on a charged particle due to the presence of a second is described by the interaction of the first particle with the force field of the second particle existing in the space around it. All charged particles have a surrounding electric force field; charged particles in motion have a magnetic force field also. This rather abstract description has many advantages from an analytical viewpoint. It enables our description of all electrical phenomena and in particular the concept that electromagnetic interactions may exist between physically separated bodies through the medium of an electromagnetic wave.

The description of fields above leads to the concept that an electromagnetic wave has its source in moving electric charge (electric current) and that its effect can be to move other charge through its electric and magnetic force fields. Energy in the form of movement of matter can be transferred from one body to another by an electromagnetic wave. In the case of naturally occurring electromagnetic radiation, the movements of the charged particles are random and the waves have no well-defined single frequency. The random process results in a broad range of frequencies at any one time. Sunlight for example has the frequency range associated with all the colours of the spectrum. Man made electromagnetic waves often have a single, well-defined, frequency as indicated in the examples of radio services above.

One of the primary uses of man-made electromagnetic radiation is for communications. The transmission of information requires the transmission of energy, the amount of energy required being directly proportional to the amount of information. Information has an element of uncertainty and change. A regular periodic signal conveys no information. The voice or a Morse code signal conveys its information through its irregularity. This irregularity leads to a spread of frequencies, the bandwidth of the signal, which is carried on the transmission frequency through the process of modulation. Thus all communications systems have a bandwidth spread around their carrier frequency. In general, the higher the rate of information transfer, higher the bandwidth.

Communications systems have to operate in the presence of noise and interference. Generally the noise is naturally occurring electromagnetic radiation and the interference is other man-made signals occupying the same frequency range.

Naturally occurring electromagnetic radiation has many sources and is present at all frequencies. Its power density varies between the 1.4 kW/m^2 of the radiant energy from the sun to the minute amounts of electromagnetic noise which radio astronomers observe from beyond the solar system. At frequencies used by radio systems the naturally occurring noise is of very low intensity and forms a background level that determines the minimum power densities at which radio communications systems must operate. Its origin is in thermally induced vibrations of electrons, i.e. electric currents, in all matter. This includes extraterrestrial sources, the ionosphere and the material from which the earth and the structures on it are made including us. The passive infrared detectors used in security systems detect the higher levels of infrared electromagnetic waves emitted by our bodies which are slightly warmer than our surroundings.

Interference to a communications system is either somebody else's signal getting into the system by mistake or an electromagnetic wave generated by an electronic system as a by-product of its operation. An example of the latter type of interference is the noise sometimes experienced in radio systems caused by fluorescent lights or a nearby computer. Neither of these systems is intended to radiate an electromagnetic wave but, as both rely for their operation on oscillating electric currents, some electromagnetic energy is radiated from each of them. This radiated interference can never be totally eliminated in an economic design however manufacturers of such equipment are legally bound to minimise its generation.

Interference can be minimised by regulation of communications systems. In order to operate the power of the communications system needs to be sufficient to overcome the noise and interference. The higher the signal power is compared to the noise and interference the greater is the information capacity of the communication system.

Electromagnetic waves penetrate through materials with varying degrees of success. Two processes are present when a wave passes into and through a material body. The first is reflection from the surface and the second is absorption of the wave's energy in the material manifesting itself as heat. The latter effect does not always happen to any great extent. An example of this is a light wave passing through a window. Transmission through the window is done with little energy loss other than that due to reflections from the glass surfaces. Materials that are good electrical conductors such as metals are good reflectors of electromagnetic waves and do not allow penetration of the energy. Metallic enclosures are used to provide shielding against ingress and egress of electromagnetic waves in some electronic systems. Some materials, including biological tissue, are efficient absorbers of electromagnetic energy. The wave loses its energy as it penetrates the material resulting in heating of the material. The depth of penetration depends on the type of material but reduces with increasing frequency of the wave. Thus, the microwave energy used for cooking at a frequency of 2.4 GHz (wavelength 125 mm) penetrates a few tens of millimetres into the food cooking it from the inside. Infrared radiation from a grill (wavelengths of around 10 micrometres) heats only the surface layers and the heat is conducted into the food to cook it, preferably before the surface burns. Human tissue is subject to the same process. We can feel the heat from a grill and field strength or power density limits for exposure to electromagnetic radiation at frequencies used for radio services are based on the resulting heating effects.

4.3 A QUANTITATIVE DESCRIPTION OF ELECTROMAGNETIC WAVES

The description of an electromagnetic wave by means of its power density is often used in standards. Also used are descriptions of the wave in terms of its electric or magnetic field strength, both of which are related to the power density. The most commonly used method is to express the electric field in its units of Volts/metre (V/m). The electric field is related to the power density by the formula;

$$\text{Power Density} = (\text{Electric Field})^2/120\pi$$

or

$$\text{Electric Field} = (120\pi\, \text{Power Density})^{1/2}$$

The relationships with the magnetic field (units Amps/metre or A/m) are;

$$\text{Power Density} = 120\pi\,(\text{Magnetic Field})^2$$

or

$$\text{Magnetic Field} = (\text{Power Density}/120\pi)^{1/2}$$

The relationship between the electric and the magnetic field is;

$$\text{Electric Field/Magnetic Field} = 120\pi$$

The term 120π is termed the intrinsic impedance of free space, its units are Ohms and it represents the fixed ratio of electric to magnetic field intensity in an electromagnetic wave. These equations are valid for electromagnetic waves at distances from the wave source of more than about one wavelength. For shorter distances, much less than a wavelength, the equations do not hold and each field type is specified independently. The actual relationship between the two fields is dependent on the type of transmitting source.

Directly related to the magnetic field and used interchangeably is the magnetic flux density the units of which are Tesla;

$$\text{Magnetic Flux Density} = 4\pi.10^{-7}\,\text{Magnetic Field}$$

Figure 4.1, Figure 4.2 and Figure 4.3 are taken from the National Radiological Protection Board booklet "Restrictions on Exposure to Static and Time-varying Electromagnetic Fields" (1995) and show typical "Investigation Levels" for electric and magnetic fields and for power density. Note that the power density is specified only for frequencies above 10 MHz (wavelength 30 m or less). In this frequency range the equations above fit the values of field strength and power density in the figures. At lower frequencies (longer wavelengths), where the distance to the source and the size of the room or other surroundings is comparable to or smaller than a wavelength, the equations do not hold and individual electric and magnetic field strengths are specified.

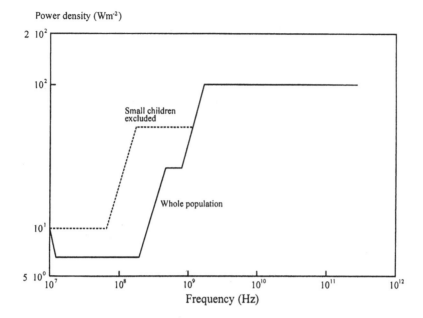

Figure 4.1 Investigation level [2] power density

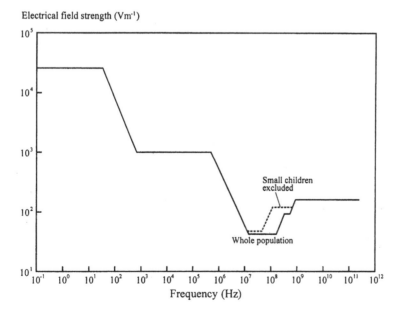

Figure 4.2 Investigation level [2] electric field

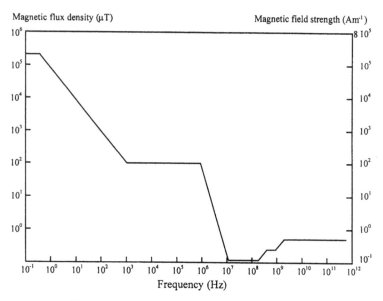

Figure 4.3 Investigation level [2] magnetic field

Estimating the field strength and power density from a distance source of electromagnetic energy is a complex process. The propagation of the waves is dependent on the structures in the intervening space between the source and the observation point. All sources have transmitting antennas with directional characteristics, that is the power density of a wave emerging from the antenna is dependent on the direction in which the wave leaves the antenna. Some antennas used for point to point communications have narrow, well-defined, beams of radiation. Others used for services such as broadcasting or mobile phones transmit energy more or less uniformly in all directions. A useful illustrative approximation for the power density and field strengths from a broadcast transmitter can be gained by assuming that the radiated energy is of equal power density in all directions and by ignoring any obstacles or structures in the vicinity of the propagation path of the wave. Under these circumstances, the power density at any distance D from the source is simply the effective radiated power of the source P divided by the area of the sphere of radius D.

$$\text{Power density} = P/4\pi D^2 \text{ Watts/m}^2$$

The field strengths can then be found from the power density. Using the equation above, the power density from a typical 800 kW TV broadcast transmitter at a distance of 55 km is 20 µW/m². The associated field strengths are 86 mV/m electric field and 230 µA/m magnetic field. A 20 W mobile phone transmitter at a distance of only 250 m gives a similar power density of 25 µW/m² (96 mV/m electric field) whilst a mobile phone transmitting on its maximum power of 2 W over a distance of 2 m gives a power density of 40 mW/m² corresponding to an electric field strength of 4 V/m and a magnetic field strength of 10 mA/m. These examples are chosen to illustrate the typical field strengths encountered in the environment. The section below gives results of a measurement survey where such field strengths are encountered.

The examples in the paragraph above illustrate the wide range of numeric values encountered in this along with many other engineering activities. The range of numbers can be conveniently reduced by using a logarithmic scale in deciBels. The logarithmic ratio of two quantities P_1 and P_2 having the dimensions of power or energy is given by;

$$\log_{10}(P_1/P_2) \text{ Bels}$$

or

$$10\log_{10}(P_1/P_2) \text{ deciBels} \quad (\text{abbreviation dB})$$

The equivalent for field quantities is expressed in the same way as;

$$10\log_{10}(P_1/P_2) = 10\log_{10}((E_1^2/120\pi)/(E_2^2/120\pi)) = 20\log_{10}(E_1/E_2)$$

Often these ratios are expressed relative to a unit quantity enabling them to be used for absolute rather than relative values of quantities. Thus, we express electric field strength in units of deciBels relative to 1 μV/m or dB μV/m as shown in the table below.

The range of a million to one from 1 μV/m to 1 V/m is reduced to a range from 0 to 120. Multiplicative operations such as doubling a numerical field strength value become additive (add 6 dB for doubling) in the logarithmic deciBel scale. Instrumentation for measuring field strength displays its results on a logarithmic (dB) scale enabling a wide range of values to be displayed. Generally small field strengths associated with radio reception are referred to on the deciBel scale whilst larger field strengths near sources such as transmitters use numerical field values.

Field Strength μV/m	*Field Strength dB μV/m*
0.1	-20
0.316	-10
1.0	0
2.0	6
3.16	10
10	20
100	40
1	60
1 V/m	120

The radio services described above make use of electromagnetic waves propagating from the transmitter to the receiver through the intervening space. Electrical energy conveyed through wiring for either power or signalling purposes also propagates along the cable as an electromagnetic wave, in this case a guided wave guided by the cable. Most lay people would assume that the energy is conveyed in the wires. In fact, it is conveyed in the electric and magnetic fields of the guided wave in the space immediately around and between the wires. Electromagnetic fields are present around wiring systems. They usually decrease in intensity with increasing distance from the cable according to an inverse cube law; i.e. the field intensity reduces by a factor eight for each doubling of the distance from the cable.

4.4 SURVEY OF ELECTROMAGNETIC FIELDS IN A MODERN OFFICE BUILDING

In order to illustrate the concepts outlined above the electric and magnetic field strengths in the offices of York EMC Services Ltd were measured. Standard calibrated antennas and a measurement receiver were used for this purpose. The offices are located on the first floor in a building with IT equipped offices on the first and second floors and a supermarket and other shops on the ground floor. The results are shown in Figure 4. The complete frequency range of Figure 4 is from 9 kHz to 2 GHz. For the range from 9 kHz to 30 MHz, both electric and magnetic field strengths are shown. Above 30 MHz only electric field strengths are given.

Figures 4.4a and 4.4b give electric and magnetic field strengths in the frequency range 9 kHz to 300 kHz. The isolated peaks (spectral lines) indicate specific frequencies at which signals are present. Both figures have a common set of such frequencies indicating that both electric and magnetic fields are present in the waves. In general, the ratio of electric to magnetic field is not 120π Ohms at these low frequencies due to perturbations by the surrounding structure. The background noise level is also more significant in the electric field plot of Figure 4a masking some of the frequencies seen in Figure 4.4b.

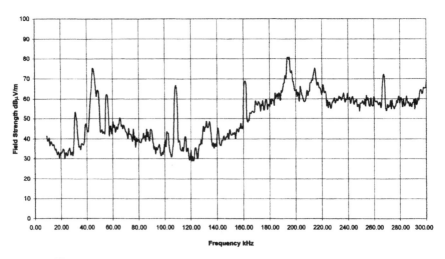

Figure 4.4a Example of electric field strength within a building 9kHz to 300kHz

Loop Antenna
RBW 1kHz VBW 1kHz

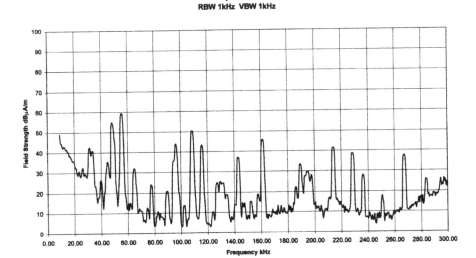

Figure 4.4b Example of magnetic field strength within a building 9kHz to 300kHz

This more significant electric field is typical of the type of noise generated by electrical equipment at low frequencies. The individual frequencies are a combination of low frequency radio stations (e.g.198 kHz and 252 kHz) and frequencies associated with interference radiated by local sources such as Visual Display Units (VDU's). Figures 4.4c and 4.4d show the field components in the frequency range 300 kHz to 30 MHz. This range encompasses all medium wave and short wave broadcasting and many signals are present. The strong signal at 1.3 MHz is the university student radio service whose transmitter is 200 m from the building. Groups of short wave broadcast signals can be seen throughout the range, particularly around 14 MHz, 18 MHz and 22 MHz. Again, the electric field appears to have the most significant noise. Figure 4e shows the electric field strength in the 30 MHz to 300 MHz frequency range. Two groups of FM broadcast signals around 90 MHz and 100 MHz can be seen. The signals from 140 MHz to 160 MHz are the local site security radio system and pagers. At 225 MHz is digital audio broadcast. Figure 4.4f shows the electric field strength from 300 MHz to 2 GHz. Notable signals include the TETRA radio system at around 420 MHz, the second local site security radio system at 460 MHz, out of area terrestrial analogue TV signals between 500 MHz and 600 MHz and local terrestrial analogue TV signals from Emley Moor near Wakefield between 600 MHz and 700 MHz. This latter group of signals is from a transmitter 55 km from the building. The typical field strength of 71 dBμV/m translates to a numerical field strength value of 4 mV/m a factor 20 lower than the simple line of sight calculation indicates. The difference can be accounted for by the level of shielding given by the building and other structures in the signal path. Mobile phone signals can be seen at around 900

MHz and around 1.8 GHz. The signal at 1.3 GHz is a local Amateur Radio transmitter at a distance of 300 m from the building.

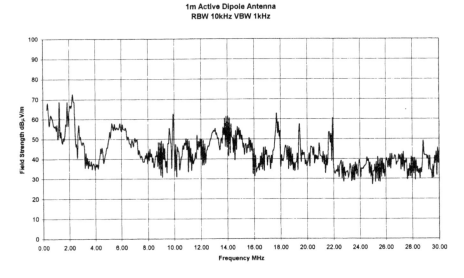

Figure 4.4c Example of electric field strength within a building 300kHz to 30MHz

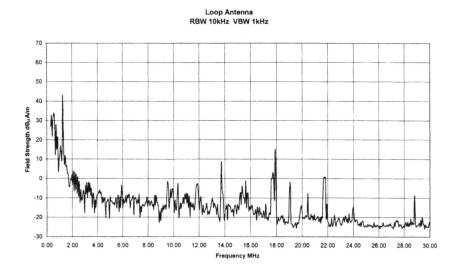

Figure 4.4d Example of magnetic field strength within a building 300kHz to 30MHz

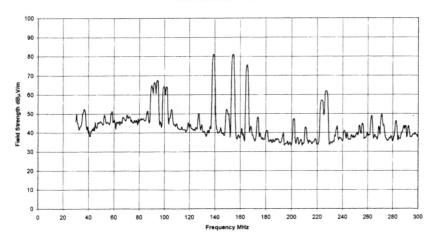

Figure 4.4e Example of electric field strength within a building 30MHz to 300MHz

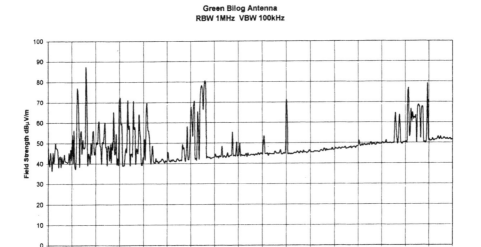

Figure 4.4f Example of electric field strength within a building 300MHz to 2GHz

More detail of the mobile phone transmitters can be seen in Figures 4.4g and 4.4h. Figure 4.4g shows the totality of signals over a two minute period while Figure 4.4h shows those present in the 20msec scan period of the receiver. The smaller number of signals is the typical occupancy of the band at any one time. The mobile phone system is very dynamic with calls coming and going and the frequency allocation of a single mobile phone changing during the call.

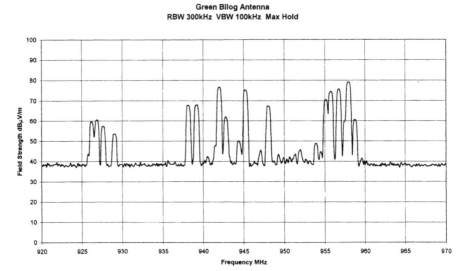

Figure 4.4g 900 MHz mobile phone electric field strength two minute time aggregation

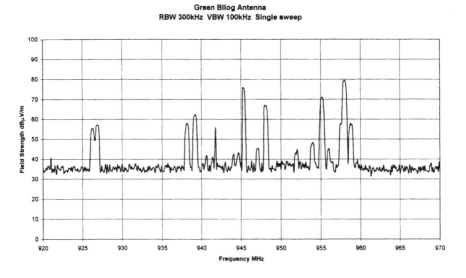

Figure 4.4h 900 MHz mobile phone electric field strength 20msec single scan

The signals shown in the figures are from the base station 250 m away. The line of sight calculation of these signal strengths is 96 mV/m. The measured field strengths of 80 dBμV/m equates to a numerical field strength of 10 mV/m. The factor ten reduction is due in part to architectural shielding and in part to the power control used by mobile phone systems which operate at reduced power where possible.

These measurements are a snap shot of the electromagnetic environment in a modern building situated in an urban area. Measurements made in another location with a different population of local and more distant radio transmitters would show a similar set of signals with levels differing from those shown. The author considers it unlikely that any increases in individual signal levels over the measurements presented here would be greater than 40 dB a factor of one hundred in field strength. This would make the highest field strength observed 126 dBμV/m or 2 V/m equivalent to a power density of 10 mW/m^2.

4.5 IONISING AND NON-IONISING ELECTROMAGNETIC RADIATION AND OPTICAL EFFECTS

The description of electromagnetic radiation above is valid across the entire spectrum. However, at the high frequency end of the spectrum a further effect needs to be considered, the process of ionisation. Here the interaction of the electromagnetic wave with the atoms in a material results in the disassociation of one or more electrons from atoms to leave electrons and positive ions (the atomic nucleus and its remaining electrons). The bio-chemical consequences of this ionisation process are known to give rise to serious health effects if exposure to the ionising electromagnetic radiation is sufficient. Ionisation is a process associated with electromagnetic radiation in the optical frequency range and beyond (ultra violet, X and gamma radiation). Electromagnetic radiation used for communications and power transmission is currently too low in frequency to exhibit ionisation effects.

Electromagnetic radiation can be focussed by lenses and mirrors to produce very high power densities over small areas. The diameter of the minimum area into which radiation can theoretically be focussed is approximately equal to the wavelength of the radiation. Focussing effects are only likely to be of significance therefore with radiation of optical frequencies. In a general indoor environment, the only systems that may require consideration of the focussing effect are infrared data links used between systems such as laptop computers and mobile phones. The power of such systems is controlled such that no dangerous power density is possible with normal installations.

4.6 CONCLUSIONS

The purpose of this paper is to describe the sources and the uses that our society makes of electromagnetic radiation in all its forms. Electromagnetic radiation is a natural phenomenon that is generally benign but which can be dangerous if misused. The energy that powers our ecosystem and our society reaches our planet by electromagnetic radiation. We have used electromagnetic radiation for communicating with each other from the earliest times. Semaphore, signal flags, bonfires and the simple act of waving to each other all rely on electromagnetic radiation in the form of light! During the last hundred years, society has made more sophisticated use of electromagnetic radiation for communications and power transmission. The electromagnetic natural electromagnetic background has been added to by numerous man-made electromagnetic waves mostly at radio frequencies; waves that penetrate biological tissue to depths much greater than infrared or optical radiation. In order for the systems employing these waves to function, the power density of the waves needs to be several orders of magnitude higher than the natural background. Safety considerations for human exposure arise and current limits are based largely on thermal effects. The measurements presented here indicate that exposure to the enhanced electromagnetic background in a typical building is still much lower than the current safety limits. The measurements did however exclude hand held and body worn mobile radio transmitters.

4.7 REFERENCES

National Radiological Protection Board, Restrictions on Exposure to Static and Time-varying Electromagnetic Fields, 1995, Chilton, Didcot, Oxon, OX11.

Wesley, A., 1963, The Feynman Lectures on Physics Section 4-1, Library of Congress card 63-20717.

Part II

Electromagnetic Fields, the Environment & the Law

CHAPTER 5

Mast Action UK – Legal Services

Mobile Phones – Mobile Networks - Safety
Local Government Responsibilities Following
The Issuing of the Revised Planning Circular
PPG8 on 23rd August 2001

Alan Meyer

5.1 PLANNING CIRCULAR PPG8 (REVISED)

1.   The Revised PPG8 Circular arises from the Government's responses since the Publication of the Stewart Report of the Independent Expert Group (IEGMP) On 11th May 2000, and was issued on 23rd August 2001 - some 15 months later and nearly 10 months after the close on the formal DETR Consultation which closed on 31st October 2000.

2.   The Revised PPG8 Circular (the Circular) follows the main announcements made by the Planning Minister in his letter to Local Planning Authorities and MP's and in his written parliamentary answer all made on 16 March 2001.

3.   The Circular repeats The Government's General Policy that it is not for Local Planning Authorities to enquire into Network Need. It also repeats the oft stated Government view that in the light of the Stewart Report's Recommendations, it is not necessary for Local Planning Authorities *'to consider further the health aspects and concerns about them"* (paragraph 98).

4.   However, having provided such Guidance based upon the Government's views the Circular sets out partially the Law as made clear by the Newport and Tandridge Decisions of the Court of Appeal (paragraph 97). However, the Circular makes neither any reference to the requirements of the Human Rights Act 1998 and the requirements of the European Convention on Human Rights which that Act embodied into English Law on 2nd October 2000, nor that the Circular has been declared by Ministers to be Human Rights Act compliant.

5.   However, in Planning matters generally the Government lays down its overall Development Policies but then leaves it to Local Planning Authorities and Development Plans to work out **where** and **how** to achieve such Governmental aims and objectives.

6.   On such a basis it remains the Local Government's responsibility to comply with the English Laws' requirements both as pronounced by the Court of Appeal, but also taking into account the requirement that **"all Public Authorities"** are required to act lawfully and **not to contravene Section' 6(l) of the Human Rights Act.**

5.2 THE HUMAN RIGHTS ACT 1998

1.   The Human Rights Act 1998 incorporating the European Convention on Human Rights came into effect in England and Wales on 2nd October 2000.

2.   The Act confers Rights to pursue claims against *a* **Public Authority** for breach of a Convention Right in the Courts of England and Wales instead of having to pursue such a Claim in Strasbourg in the European Court of Human Rights.

3.   Claims can only be brought against *a* **Public Authority**. In Planning a Local Planning Authority acting through its Planning & Development Control Committee or under delegated powers, its Director of Planning **constitutes Public Authority.**

4   Section 6(l) of the Act states:

"It is unlawful for a Public Authority to act in a way which is incompatible with a Convention Right".

5.   Generally, if a Public Authority breaches a Convention Right without any defense within the exceptions in Section 6, the Claimant can obtain a Declaration from the Court that the Public Authority is in breach of the Convention Right and has acted unlawfully, and the legal costs will be awarded against the Public Authority. There is no automatic Claim for damages or compensation.

6.   However, in certain circumstances or claims it maybe possible to show a direct financial loss, and in such circumstances the **Public Authority** may **not only have to pay the Legal Costs but also damages** for the failure to comply with a Convention Civil Right.

7.   In planning the two main Convention Rights which require to be observed are contained in **Articles 6** and **Article 8** as expanded by Article I of the First Protocol to the Convention (see Appendix).

Article 6: *The Right to a Fair and Impartial Hearing in Public within a reasonable timescale.*

Article 8: *The Right to respect for Private and Family Life, the Home, and possession.*

5.3 PLANNING APPLICATIONS AND NOTIFICATIONS

1.   In Telecommunications and Mobile Phone Masts there are two types of Planning Applications submitted to Local Planning Authorities:

Full Planning Applications for Ground Based Masts and Base Stations over 15 meters; or

Prior Approval Notifications for Ground Based Masts and Base Stations under I5 meters and for Masts and Antennae on Buildings.

2.   Since the Planning Ministers Announcement on 16th March 2001 and the issuing of the Circular PPG8, proper Consultation and Notification to the Public on both types of these Applications **are intended in the new Guidance to be the same** so as to enable the Local Community to put forward their concerns to be taken into consideration by the Local Planning Authority **before reaching its decision.** The time Limit for considering Prior Approval Notification has been extended to 56 days.

3.   However, the House of Commons Trade and Industry Select Committee in its 10th Report on 3rd April 2001 stated that the Revised PPG8 Circular has to be declared by Ministers to be **Compliant with the Human Rights Act,** and will also have to be consistent with the two Court of Appeal Decisions in the **Newport** and **Tandridge** Cases where the Court found that Public concern and anxiety was "a material Planning Consideration" to be taken into account by the Local Planning Authority in considering Planning Applications (*see 5.1 Paragraph 4 above*).

4.   For all Applications the 16th March 2001 proposed procedural changes now at last embodied in the Circular mean that the **"Public Authority" has only 56 days** within which to notify the Local Community, and to listen to and consider their objections and to come to its decision. If the Public Authority fails within the 56 days to come to a decision **"the deemed approval"** under the GPDO will apply to the Applications.

However, in such 'deemed approval" cases there may have been **two breaches of Civil Rights** breaches by the Local Planning Authority as the "Public Authority":

(1) *Under Article 6 because there has been no* **"Fair and Public Hearing"** *at which Objections were considered or listened to by a Public Authority; and*

(2) **Under Article 8** *there will have been* **no demonstration** *by The Local Planning Authority that the infringement or violation of person's Article 8(1) Civil Rights* **was permitted** *under the exceptions in Article 8(2) because the violation was* **"necessary"** *in the interest of the economic well being of the County because technically there was no other alternative possible Location*

5.   The Onus is placed on **"The Public Authority"** to demonstrate that the proposed site or location was **'necessary'**. The Public Authority will have to investigate properly that there were **no alternative sites or locations** which could have fulfilled adequately the Applicant's Network Needs.

In addition, where a **"need"** can be demonstrated which was **"necessary"** that there are nowadays no other available technological infrastructure alternative ways of achieving such need without **"unnecessarily"** infringing Article 8 Rights.

This is the aspect left to the Local Planning Authority by the Circular. **Where** *to meet the Network need, and technologically how that needs con be met.*

6.   Local Residents bringing claims under Articles 6 and Article 8 against the Local Planning Authority may be able to make claims against the Local Planning Authority for **diminution in value of their Homes** arising from "insensitive siting or location" of Ground Based Masts and Base Stations where other alternatives exist. The Courts in England and Wales when making a Judgment of unlawfulness against a Local Planning Authority, having taken into account the Doctrine of Proportionality i.e. striking a proper balance of **"need"** against **"loss of amenity harm"** in addition to awarding costs can make damages or compensation awards if a property value loss is proved.

7.   Accordingly the Local Planning Authority in considering both Full Planning Applications or Prior Approval Notifications should require the Network Operator Applicant to list the alternative possible sites they have investigated, and also to explain why there are nowadays no alternative technological infrastructure ways of achieving the Network needs e.g. by Site or Mast or Antenna Sharing with other Network Operators on other existing or proposed Communication Towers.

In this way the task of the Planning Officers of the Local Planning Authority will be made less difficult, and the Public Authority will find it easier to carry out its responsibilities **imposed under Articles 6 and 8** of the European Convention also taking into account the requirement to apply the Doctrine of Proportionality with regard to violation or infringement of Article 8 Civil Rights.

8.   Whilst Paragraph 98 of the Appendix to the Circular sets out once again the Government's views on health risks arising from the emissions from the Antennae on Masts, **Prolonged Stress and Anxiety** from **any cause** can trigger any number of **actual** Medical Conditions such ac Cancer, Leukaemia, Heart Conditions and acute depression. In the Circular, the Government repeated its view that it is not necessary "to consider Health Effects further". However, two additional words "and concerns" have been added so that unlike the earlier letters, Paragraph 98 now reads **"to consider health effects and concerns further"**.

Those additional words appear to contravene both the two Court of Appeal Decisions and also Article 6 of the European Convention. As a result, Paragraph 98 itself is not Human Rights Act Compliant as required by the House of Commons' Trade & Industry Select Committee.

In February 2001 the Government also announced the establishment of the Stewart £7.3 million Research Project into the possible biological effects of long term exposure to RF emissions from Masts. However, in its 16 March 2001 Announcement the Government stated that it would be "maintaining in full an Authority's ability to reject Applications on amenity grounds".

9. **Prolonged Stress and Anxiety** can be caused by the *"Insensitive siting, location, and presence of a Mast"* either as a result of a Planning Consent or a deemed approval.

Prolonged Stress an'd Anxiety can be caused by the mere presence of a Mast and can in some people trigger any number of Health Conditions regardless of whether the Antennae are dangerous or not. The mere fact that it has now been decided by the Government that further Research is both "necessary" and now has commissioned some can in itself contribute in many instances to such Stress and Anxiety.

10. During Summer 2001 the Industry through the coordinating efforts of the Federation of Electronics Industries (FEI) adopted across the five 3G Network License Operators **a Paper designed to address Mobile Phone concerns called the Operators 'Ten Commitments'.** The two opening Commitments relevant to the General Public's concerns are as follows:

(1) Develop, with other Stakeholders, clear standards and procedures to **deliver significantly improved Consultation with Local Communities.**

(2) Participate in obligatory pre-Roll Out and pre-Application Consultation with Local Planning Authorities.

11. During this year there have been numerous Planning Inspectorate Appeal Decisions concerning Local Planning Authority refusals to approve Mast Planning Applications or Prior Approval Notifications. Generally, on a two to one basis, such appeals by Network Operators have been allowed by the Planning inspectorate.

However, there have also been a significant number of Appeal refusals, where the Planning Inspector has indicated that there is a requirement **to achieve a proper balance between possible Network Rollout** *"needs"* **in the Public Interest against** *"local amenity harm"*, **including insensitive siting and location causing Local Residents anxiety.**

A number of such Appeal Decisions proceeded the Government's reissuing of Planning Guidance Policy Circular PPG8. However, that Revised Circular in Paragraphs 66 and 67 sets **out the Government's views currently on Mast and Site Sharing to keep the numbers down consistent with the efficient operation of the Network, and also with regard to proper investigation of alternative sites.** These views coincide with the requirements of Article 8(2) of the European Convention on Human Rights and the Legal Requirement contained In Section 6(1) of the Human Rights Act 1998.

12.   Amongst the Planning Inspectors' recent Appeal Decisions Me Those in the **London Borough of Harrow** – **Thurrock Borough Council** (since reversed by the Secretary of state) – **Leeds City Council**, where the Inspector cited the requirements of the Government Guidance in the Circular PPG3 Housing, and **Ipswich Borough Council** (currently subject to an Appeal to the Court) where the Inspectors have indicated **the need to achieve a proper balance between possible further Network Roll Out** *"needs"* **and** *"loss of local amenity"* **as part of the necessary consideration of the issues of siting and appearance.**

13.   In one of the most recent Planning Inspectorate Appeal Decisions (30th October 2001) the Inspector, in allowing the Appeal against the Decision of **Watford Borough Council** stated as follows:

'PPG8 does not say that the need for Telecommunication Development should always override environmental considerations.

Later, when considering other recent Planning Appeal Decisions, including those mentioned in Paragraph 12 above, the Inspector further stated:

"These Decisions simply illustrate that all Appeals must be decided on their merits and that those merits will vary according to the circumstances of the proposal and of the site. The concept of precedent does not apply".

Finally, with regard to Article 8 (1) of the European Convention on Human Rights, This Inspector applied the **European Court's Doctrine of Proportionality,** and found in this particular case that the violation of Article8(1) Rights was *'necessary"* in the overall Public Interest in accordance with the exemption contained in Article 8(2) of the Convention.

5.4 THE HUMAN RIGHTS ACT 1998

1.   **This Act came into force in England and Wales on 2nd October 2000. It provides for Civil Rights against Public Authorities only.**

2.   The Act incorporates in its Schedule the provisions of the European Convention on Human Rights and includes (*inter alia*) the following principal **Civil Rights** which have been incorporated into the Laws of England and Wales:

Section 6(1) states:
(1)   *It is unlawful for a Public Authority to act in a way which is incompatible with a Convention Right.*

Article 6: Right to a fair trial
(1)   *In the determination of his civil Rights and obligations, or any criminal charge against him, everyone is entitled to a fair and public hearing within a reasonable time by an independent and impartial tribunal established by law…*

Article 8: Right to respect for private and family life

(1) *Everyone has the right to respect for his private and family life, Pits home and his correspondence.*

(2) *There shall be no interference by a public authority with the exercise of this right except such as is in accordance with the law and is necessary in a democratic society in interest of national security, public safety or the economic well-being of the Country for the prevention of disorder or crime, for the protection of health or morals, or for the protection of the rights and freedoms for' others.*

The First Protocol to the Convention for the Protection of Human Rights and Fundamental Freedoms, as amended by Protocol 11 states:

Article 1 - Protection of Property

Every natural or legal person is entitled to the peaceful enjoyment of his possessions. No one shall be deprived of his possessions except in the public interest and subject to the conditions provided for by law and by the general principles of international law.

The preceding provisions shall not, however, in any way impair the right of a State to enforce such laws as it deems necessary to control the use of property in accordance with the general interest or to secure the payment of taxes or other contributions or penalties.

3.   Inherent in the whole of the European Convention is the need to find a fair balance between the protection of individual rights and the interests of the community at large. **The Principle of Proportionality** is concerned with defining that **"fair balance"**. **The Principle of Proportionality** has been the defining characteristic of the Strasbourg Court's approach to the protection of Human Rights.

4.   **These Civil Rights are only exercisable against "A Public Authority" usually the Local Planning Authority or on Appeal, the Planning Inspectorate.**

5.5 LOCAL GOVERNMENT RESPONSIBILITIES MEMORANDUM

5.5.1 Addendum

1.   *On 6th February 2002 at a meeting between the Planning Minister Lord Falconer and Members of Parliament held at the House of Commons offices at Portcullis House in response to a question Lord Falconer confirmed that the Town and Country Planning Act 1990, like all UK legislation, is subject to the requirements of the Human Rights Act 1998.*

2.   What does the existence of the Human Rights Act 1998 mean in practical terms with regard to the whole planning process? Some of its effects can be summarized as follows

> (A) Planning Guidance Circulars like tie enabling Town and Country Planning Act 1990 have to be implemented taking the requirements of HRA 1998 into account as well.

> (B) Proceedings at Planning Committee Meetings at the Local Planning Authority and at Appeals or Planning Inquiries conducted by the Planning Inspector appointed by the Secretary of State have to comply with the requirements of Article 6 of the European Convention on Human Rights. There has to be 'a *fair and impartial hearing*'.

> (C) The contents of Planning Guidance Circulars, like all UK legislation, have to be compatible with the HRA 1998 and the European Convention on Human Rights. However, it is necessary to understand now what that means in practice.

3.1 The main practical area which is affected in connection with Telecommunications Masts and Planning or Prior Approval Applications, and with Appeals, is the fact that objectors are permitted to raise health concerns. This is not new law because that was made clear in both the 'Newport' and 'Tandridge' Court of Appeal decisions which pre-date the coming into effect of HRA 1998 on 2nd October 2000.

3.2 Attempts to prevent *'health concerns'* being raised are *'unlawful'* and incompatible with HRA 1998 Section 6(1) and the Article C *'right to a fair and impartial'* hearing in public, and contrary to paragraph 97 of the Re-issued Circular PPG8. Paragraph 98 of that circular sets out only the Government's guidance and views which have to be taken into account and weighed up against the objector's fears and concerns

3.3 *Health fears and concerns'* can include the Stewart Reports concerns regarding-*"adverse impacts on the local environment may adversely impact on the public's well-being as much as any direct health effects"* – Stewart paragraphs 1.31 and 1.33.

4.1 However, it is also essential to understand properly the requirements of Article 8 of the European Convention. That right *"to respect for private and family life home and possession"* **can only be violated or infringed** if the violation or infringement can properly be brought within the Article 8(2) exemption. The vital words in Article 8(2) are -
      *"except such is in accordance with the law and is **necessary** in a democratic society in interests of national security, public safety or the economic well-being of the Country, for the prevention of disorder or crime..."*.

4.2 Whilst it is necessary to meet proven **"network needs"** both in telecommunications and also for the emergency services, it remains essential to demonstrate that before violating or infringing the Article 8(1) rights **that all possible alternative other sites** or technological means have been fully explored. Only after **the Applicant** has properly demonstrated that there are no other practical alternative sites available to meet a proven need should the **Public Authority** grant the required planning consent, prior approval notification or Appeal.

5.1 Lord Falconer invited Mast Action UK to meet with him and his Telecomms Planning Team at the DTLR on 15th March 2002.

During an hours discussion regarding the concerns of MAUK and its members, Lord Falconer firstly reiterated that " the Town and Country Planning Act 1990, and its supporting Guidance Circulars, like all UK legislation was now subject to the requirements of the Human Rights Act 1998."

5.2 In addition, secondly he made it clear in his opinion there was nothing wrong in the revised PPG8 Circular itself, and he would not be changing it. The problem was the practical one as to how in practice PPG8 was currently being interpreted within the planning system.

It was pointed out that paragraph 29 and Annexe Paragraph 97 stated the law as pronounced by the Court of Appeal in the "Newport" and "Tandridge" Decisions with regard to *"material planning considerations"*. Paragraph 30 and Annexe Paragraph 98 could not be taken in isolation, ignoring both the preceding paragraphs, and now the requirements of the human rights Act 1998.

5.5.2 Second Addendum

1. **Over the past year a series of Planning Inspector Appeal Decisions from planning refusals by Local Planning Authorities has established that generally Planning Inspectors, in either allowing or dismissing Appeals, carry out a test to balance Network Operator** *"Need"* **against local environmental** *"loss of amenity"* **or harm.**

2. The Inspectors take into account whether the Appellant has proved the Network Need **has to be met**; and **can only be met**, from the site or location chosen by the Network Operator.

3. Accordingly, to **satisfy now the requirements of the Human Rights Act,** and the Rights in Article 8(1) of the European Convention on Human Rights, so that **the exemption** contained in Article 8(2) can be operated, it is *"necessary"* for the test by the Planning Inspector, to establish the required "necessary Work Need",

to be a double test falling into two pats:-

to establish that there is a proper Network Operator Need in the locality "necessary" to meet customer requirements; and

that there are **no alternative sites, or technological means,** which can adequately satisfy that Network Need, an accordingly the selected site is "necessary" in the "public interest" to such a degree that "environmental loss of amenity" should be allowed, falling within the Article 8(2) Human Rights exemption.

4.   **Such a double test** should then satisfy the requirements of the European Courts **"Doctrine of Proportionality"** established to determine infringements or violations of the Human Right Act 2998 and its Section 6(1) setting out the requirements imposed upon *"Public Authorities"* include in planning initially the **Local Planning Authority**, and then on appeal the **Planning Inspector.**

5.5.3 Third Addendum

1.   On 23rd May 2002 the House of Lords issued its judgement in the Appeal Case R. -v- Hammersmith and Fulham London Borough Council (Times Law Report 24th May). The judgement dealt with leave to apply for judicial review. The House of Lords in the judgement stated:

"From observations of Mr. Justice in R. -v- Ceredigion Country Council ex parte McKeown (1998) 2PLR1 the inference has sometimes been drawn that the three-month time limit had been replaced by a six-week rule. That was a misconception. The legislative three-month limit could not be contracted by a judicial policy decision.

It as at least doubtful weather the obligation to apply promptly, read with the three-month limit, was successfully certain to comply with European Community law and the European Convention on Human Rights."

"Lord Hope agreed saying that he shared Lord Steyn's doubt as to whether the provision that the claim form must be filled promptly was sufficiently certain to comply with the right to a fair hearing within a reasonable time in Article 6.1 of the European Convention of Human Rights and, in that respect, also with Community Law".

2.   The House of Lords decision is relevant to telecommunications planning applications and prior approval notifications, and to appeals from Local Planning Authority decisions in that it emphasizes the need under Article 6.1 for **" a fair and impartial hearing within a reasonable time"**. Insofar as it affects decisions by Planning Inspectors appointed by the Secretary of State, or the decisions of the Secretary of State himself, the House of Lords decision in Alconbury Developments Limited and the European Courts decision in Bryan -v- UK both

made it clear that such decisions were fair and impartial because the final arbiter would be the courts on Judicial Review.

Until this House of Lords judgement it had seemed that the period in which to seek leave to apply for judicial review had been confined to six weeks - the same time limit as the period in which a party to a planning Inspectorate Appeal has the right to appeal the decision to the court.

In practice this should mean that there is now again a reasonable period during which the funding for a leave application can be raised, or possibly legal aid may be obtained.

3.   The Planning Minister's 16 March 2001 announcement that it was proposed that for prior approval notifications there should be public consultation similar to that required for full planning applications, requires not only   that there should be proper consultation with relevant stakeholders including the affected local community but also that **"genuine public fears and concerns"** are property listened to and considered by the Decision Maker. These issues were then incorporated into the revised guidance circular PPG8 issued on 23rd August 2001.

4.   As now confirmed by the present Planning Minister Lord Falconer on 6th February 2002 and 15th March 2002 (see Second Addendum) it is no longer open for Planning Officers and Local Planning Committees to advise objectors that they are not allowed to consider   *"Health effects and concerns"* objections because of PPG8 Annexe Paragraph 98. That is contrary to the law as set out in Section 6(1) of the Human Rights Act 1998, and also contrary to Article 6(1) of the European Convention.

5.   However, the days the issue or *"possible health effects"* divides itself into
      two separate issues:

(1) The *"scientific"* issue on which the jury is out, and on which the Government and the Industry have sponsored further research into non-thermal possible biological effects. Paragraph 98 is largely directed to scientific effects.

(2) The "second health effects" issue a *"medical" issue which does not* involve scientific opinion. It is an accepted medical concern that in some people "prolonged stress and anxiety" over a protracted period may induce any number of possible adverse health effects.
      This is the type of "indirect adverse impact on the public's wellbeing"
envisaged by Sir William Steward in Paragraph 1.31 of the Steward Report.
      Such stress for example may be induced by bereavement – family long term illness of a child divorce – redundancy and unemployment – or in the case of telecommunication masts and base stations by " **insensitive siting and proximity"** to peoples houses.   Here such stress may also be compounded by fear of being placed in a negative equity mortgage situation, or from finding that the property resale possibilities have been blighted.

Such stress issues fall within the Courts of Appeal's two Newport and Tandridge decisions concerning "genuine public fear and concern" being one of the material planning considerations which the Decision Maker needs to take into account when considering the planning application or prior approval notification.

6.   It has been remembered it is the "Public Authority" which is required by Section 6(1) of the Human Rights Act **"to act lawfully"** and not to **"unnecessarily"** contravene a Convention right. Article 8(1) is in place to protect **"private and family life, home and possessions"**. This Convention Right can only be violated under Article 8(2) if it is **"necessary"**. There the test is the double test of (a) "proven network need" (b) that only the requested location can meet that network need (see Section Addendum paragraph 4).

7.   **"Guidance does not fetter decision-makers discretion"**
     The Court of Appeal on Friday 17th May 2002 in the Education Guidance Circular Appeal cases issued judgements in **R. -v- Brent London Borough Council and Others (2)** and **R. -v- Oxfordshire Country Council's Exclusion Panel and another** (Times Law Reports 4 June 2002) and held in these educational appeal cases:
     *"Appeal Panels to keep the guidance issued by the Secretary of State in mind, **but it was not direction** and did not lay down rules to be strictly adhered to. The guidance had to promote the statutory purposes and since, **October 2, 2000, had to be Convention complaint"**.*
     In considering, Planning Policy Guidance circulars it would be necessary in the above Court of Appeal Judgements to **substitute** for the words "Appeal Panels" the words "Local Planning Authorities, or Planning Inspectors on Appeal".

CHAPTER 6

Environmental Impact of Electrosmog

G.J. Hyland

6.1 INTRODUCTION

Our environment is becoming progressively more polluted by the electromagnetie fields generated by modern technologies that are now an indispensable part of everyday life. Although the levels of emissions generally comply with existing exposure standards, this is not a guarantee of safety, as will be illustrated using VDUs and Mobile Phones.

A major contemporary threat to the health of Society is man-made 'electrosmog'. This non-ionising electromagnetic pollution of technological origin is particularly insidious, in that it escapes detection by the senses – a circumstance that, in general, tends to promote a rather cavalier attitude, particularly with respect to the necessity of ensuring an adequate degree of personal protection. Yet the nature of the pollution is such that there is literally 'nowhere to hide'. Furthermore, given the relatively short time for which we have been exposed to it, we have no evolutionary immunity either against any adverse effects it might directly have on our alive organism or, indirectly, against its possible interference with certain electromagnetic processes of natural origin, which appear to be essential for homeostasis, such as, for example, the Schumann resonance – a weak electromagnetic field that oscillates resonantly in the cavity between the earth's surface and the ionosphere at frequencies close to those that characterise the alpha rhythm of the human brain, isolation from which has been found to be deleterious to health.

What distinguishes technogenic electromagnetic fields from (the majority of) those of natural origin – such as sunlight, for example - is their much higher degree of *coherence*. This means that their frequencies are particularly well-defined, a feature that facilitates the discernment of such fields by living organisms, including ourselves, against the level of their (incoherent) thermal emission appropriate to physiological temperature. This greatly increases their biological potency, and 'opens the door' to the possibility of frequency-specific, *non-thermal* influences of various kinds, against which existing Safety Guidelines – such as those issued by the International Commission for Non-ionising Radiation Protection (*ICNIRP*) - afford absolutely no protection. For these Guidelines are based solely on consideration of the ability of radio frequency (*RF*) and microwave radiation to heat tissue, and of extremely low frequency (*ELF*) magnetic fields to induce circulating electric currents in the interior of the body, both of which are known to be deleterious to health if excessive. Since the severity of these effects increases with the strength (intensity) of the fields in question, it is this that the Guidelines restrict, in order to minimise any associated adverse health effects, the frequency of the fields being taken into account *only* in so far as it affects (through 'size' resonance effects) the ability of the organism to absorb energy from an irradiating *RF*/microwave field and heat up accordingly. The Guidelines thus afford no protection against adverse health effects that might be provoked *primarily* and *specifically* through influences that the *frequency* of the fields might have on the human organism.

A necessary condition for the realisation of such an influence is the existence in the organism of some means of detecting/demodulating the frequencies of the signals. One possibility would be the existence of the biological counterpart of an electrical *tuned circuit* – *i.e.* an endogenous oscillatory electrical activity. For then the organism can respond - in a way akin to a radio - if the frequency of the external field (either of the carrier wave, or of lower frequency amplitude modulations/pulsings) matches, or is close to, that of one of its tuned circuits, entailing, respectively, the possibility of either a resonant amplification of the associated endogenous biological activity – perhaps to an undesirably high level – or, deleterious interference with it, resulting in its degradation; another possibility is the entrainment of an endogenous activity to a frequency that is incompatible with homeostasis.

The possibility of such frequency-specific, non-thermal influences, unlike those addressed by existing Safety Guidelines, clearly requires, however, that the organism be *alive*, since only then does it support the endogenous electrical activities upon which its ability to detect external electromagnetic fields depends. Such influences can thus be considered to arise from a transfer, not primarily of energy but, of *information* (in a generalised sense) from the field to the alive organism, whereby the organism is able, through this kind of 'oscillatory similitude', to recognise – and in turn respond to – a feature of the external field *other than* its intensity. A necessary condition for the realisation of such influence is, of course, that the external electromagnetic fields be sufficiently coherent to be discernible by the body against the level of its own incoherent thermal emission at physiological temperatures. Whilst this is usually the case, it should be noted that since the radiation is not perfectly coherent, the occurrence of non-thermal effects is still contingent upon a certain minimum intensity threshold, the magnitude of which is, however, *well below* that at which any discernible heating occurs.

A good example of such an 'informational', frequency-specific, non-thermal electromagnetic influence on the alive human organism is the ability of a light flashing at a certain *regular* rate to trigger seizures in people suffering from photosensitive epilepsy, the coherence, in this case, being essentially due to the regularity of the flashing. The provocation of the seizure is primarily due, not to the brightness (intensity) of the light (provided, of course, that it is bright enough to be seen), but rather to the rate at which the light flashes – which, if close to the frequency of the particular electrical brain activity that is involved in epileptic seizures, can trigger their occurrence - *i.e.* the phenomenon is primarily a frequency-specific effect of *information* transfer from the light to the brain, the brain being able to 'recognise' the light through the rate at which it flashes. Existing intensity-based Safety Guidelines (relating to the visible part of the electromagnetic spectrum) afford absolutely no protection against such a (non-thermal) effect – unless, of course, they insisted on an intensity so low that the light was not detectable by the eye!

Some oscillatory endogenous electrical activities of the alive human body are quite familiar - such as those of the heart and brain, which are monitored by an electrocardiogram and electroencephalogram, respectively; also equally familiar is the circadian rhythm. Others, however - such as the coherent electrical excitations at the cellular (and sub-cellular) level, whose frequencies typically lie in the *microwave* region of the electromagnetic spectrum, and those, characterised by *ELFs*, pertaining to crucially important biochemical activities, involving, for example, the transport of calcium ions across cell membranes - are perhaps somewhat less well-known.

Accordingly, until the neglected frequency/information dimension of *non-visible* electromagnetic fields – such as (propagating) microwaves and other (non-propagating) electric and magnetic fields of technological origin, such as those from overhead power lines, for example - is recognised *in its own right*, these fields will continue to constitute a major potential threat to the living world in general, and to ourselves in particular. Since electromagnetic fields are so indispensable to the technology that Society is, understandably, reluctant to abandon, it is essential that a more comprehensive level of protection be developed, if this technology is to be used with a greater degree of safety than obtains at present. For, as has been explained, we are currently left vulnerable to any adverse health effects that might be provoked by non-thermal influences of the frequency dimension that escape regulation by the existing intensity-based Safety Guidelines, the sensitivity of different people varying enormously. Since, however, unlike intensity, the frequency aspect of the problem cannot be addressed by interventions pertaining to the aggressing field (without interfering with its frequency characteristics and informational content - the integrity of which must, of course, be maintained in communication technologies, such as *GSM* telephony), it becomes necessary to consider strategies that target the *person* being irradiated – rather than the irradiating field itself - and devise ways to ensure that a higher degree of immunity than currently obtains be realised. Such strategies are currently under development, and a number of related protection devices are already available commercially, although often their efficacy has not always been adequately demonstrated *biologically*. (There is an obvious parallel here with the pharmacological strategy of attempting to protect against bacterial infection by taking, for example, vitamin *C*, to fortify the immune system, rather than by wearing a protective mask to simply reduce the intensity of the bacterial field to which a person is exposed.)

It is thus essential that the domain of competence of existing Safety Guidelines be broadened by requiring that the familiar consideration of electromagnetic compatibility (*EMC*) between electromagnetic radiation and electronic instrumentation be extended *to include the alive human organism*, as an electromagnetic instrument itself, *par excellence*. The implementation of this ambitious programme of realising electromagnetic biocompatibility is an important task for the 21st century, and one that is shirked only at our peril.

Currently, there is much public concern over the possibility of adverse health effects provoked by both long *and* short-term exposure to electrosmog, particularly the contributions from overhead power lines and from *GSM* telephony; curiously, however, there is less overt concern over exposure to the emissions of *VDU* terminals, despite the fact that the exposed person is here subject to a much richer 'cocktail' of frequencies - involving, in particular, very penetrating *ELF* magnetic fields - the noxiousness of which has been apparent for a much longer time.

We now present a Case Study in which is considered, in some detail, two of the above examples – namely, exposure to the low intensity, pulsed microwave radiation used in *GSM* telephony and to the wide range of electromagnetic emissions from *VDU* screens.

6.2 THE NATURE OF THE ELECTROMAGNETIC EMISSIONS FROM MOBILE (CELLULAR) PHONES AND VDUs

6.2.1 Mobile Telephony

Mobile telephony is based on two-way radio communication between a mobile handset and the nearest (fixed) Base-station. The radio communication utilises (pulsed) microwaves (at either 900 or 1800 MHz), to 'carry' the voice information by means of small modulations of the wave's frequency. Whilst Base-stations radiate either directionally or isotropically, depending on their design, the antenna of a handset invariably radiates isotropically, so that a certain fraction of the radiation necessarily enters the head of the user. In addition to this isotropic emission of pulsed (*vide infra*) microwaves, there are low frequency (pulsed) magnetic fields, associated with the surges of electric current as the battery is rapidly switched on and off in order to implement the 'time division multiple access' (*TDMA*) system currently employed in *GSM* telephony to increase the number of users who can simultaneously communicate with a Base-station. To achieve this, each communication channel is divided into 8 time slots, which are transmitted as very short 576 µs bursts. Together, the 8 slots define a 'frame' of duration 4.6 ms, the repetition rate of which is 217 Hz. For certain technical reasons, however, relating to synchronisation, the frames transmitted by handsets and Base-stations (other than those in t he control channel) are effectively grouped into 'multi-frames' of 25 frames, by the *absence* of every 26th frame. This results in an additional low frequency pulsing of the signal at 8.34 Hz. In the case of the handsets equipped with the (energy saving) discontinuous transmission mode (*DTX*), there is an even lower frequency pulsing at 2 Hz, which occurs when the user is listening but not speaking.

Thus, the user of a mobile phone handset is simultaneously exposed to two distinct kinds of *ELF* fields – one that is simply a feature of the way in which the microwave signal is pulsed, whilst the other (from the battery) is a real magnetic field that induces a time-dependent electric field, giving rise to circulating electric currents in the interior of the exposed person.

The microwave emissions from a handset can entail both thermal and non-thermal consequences in the user, the severity of the former being strongly dependent on the particular model and on the conditions of use. It is to be stressed that there are not *separate* thermal and non-thermal emissions, but rather a single emission that can exert both thermal and non-thermal influences – namely tissue heating, *via* energy absorption from the field, and frequency-specific informational effects allied to the coherent *ELF* pulsing of the signal.

Associated with the *ELF* magnetic emissions from the battery of the handset, on the other hand, is the possibility of two quite different kinds of non-thermal effect, namely (*i*) a frequency non-specific influences associated with the induced circulating electric currents, which can entail adverse effects on the nervous system if their magnitude exceeds that of certain endogenous biosignals, and (*ii*) frequency-specific 'informational' influences on the electrical activity of the brain and on certain electrochemical activities at a cellular level.

By contrast, exposure to Base-station emissions can, because of their ultra-low intensity in publicly accessible areas, entail only non-thermal influences.

6.2.2 VDUs

These devices emit a much wider spectrum of electromagnetic radiation than do mobile phones, ranging from X-rays to low frequency electric and magnetic fields. Included in the latter are those associated with the vertical and horizontal sweeps that are effectively pulsed at 60Hz and between 15-20 kHz, respectively. It is to be noted that there are emissions both from the front and rear of the monitor, so that *it is not only the user who is exposed*! Filters and specially designed screens can help to reduce the intensity of the emitted X-rays and electric fields, but are impotent against low frequency magnetic fields, in consequence of their high penetrating power. Interestingly, safety guidelines governing exposure to *VDU* emission are much more stringent than those applicable to mobile phones – the maximum magnetic field 30 cm from the screen allowable under the Swedish MRP2 Radiation Standard, for example, being over 100 times lower than that originating from the current surges from the battery of a mobile phone handset, which is realised in the vicinity of the brain.

In the case of *VDU*s, thermal influences of the emitted radiation are, of course, completely negligible, so that any adverse health effects can only be attributed to non-thermal influences.

6.3 THERMAL AND NON-THERMAL EFFECTS OF EXPOSURE TO MOBILE PHONE & VDU EMISSIONS

6.3.1 Thermal Effects of Mobile Phone Exposure

As noted above, it is only in the case of mobile phones that thermal effects need to be considered, where heating of tissue and brain matter is an unavoidable consequence of the absorption - in particular, by tissue water in the head of a user - of energy from the pulsed microwave radiation emitted from the antenna. The amount of heating produced depends on the rate at which energy is absorbed per unit mass of tissue, averaged over a certain period of time (usually 6 minutes) - the so-called *specific absorption rate*, or *SAR*. The magnitude of the *SAR* is governed by the intensity of the radiation *once it has entered the head*, on certain electrical properties of tissue, and on the efficiency of the body's thermoregulatory mechanism. Above a certain intensity, temperature homeostasis can no longer be maintained, and adverse health effects ensue once the temperature rise exceeds about 1°C. Accordingly, Safety Guidelines impose upper limits on the *SAR* to ensure that this does not occur.

SAR values for different models of cell phone are determined using a so-called 'phantom' head. This is a replica of a human head, which is filled with certain synthetic fluids to simulate brain tissue, in which the values of the internal electric field and the temperature rise can be monitored and measured by implanted probes during exposure of the head to the pulsed microwave radiation from an antenna; it should be noted, however, that the antenna is not always that of a real cell phone, but rather of a 'surrogate', whose emission can differ from that of a real phone in ways that could well be important in the case of human exposure, but which probably have no influence on the *SAR*. The temperature rise is typically *tenths* of 1°C, but can be higher (of the order of 1°C) with some models. Whilst temperature rises of this magnitude can be sensed by the skin, they are not deemed to be deleterious, except, perhaps to the eyes, in view of their low blood supply and proximity to the antenna; indeed, an increase in uveal melanoma amongst cell phone users has recently reported (*see* Ref.26), which might be related to this. Another potentially deleterious thermal influence that can be envisaged is that on metallic tooth fillings.

6.3.2 Towards Non-thermal Effects

Despite the fact that the majority of mobile phones have *SARs* that are *well below* the limits of the Safety Guidelines, there are numerous reports of adverse health effects amongst users of both cell phones and *VDUs*, the origin of which must be sought in more subtle, *non-thermal* influences of the emitted radiation. Unlike tissue heating by microwave absorption and the induction of circulating electric currents, which have been rather well understood for many years within the realm of Solid State physics, *frequency-specific* non-thermal effects – which being contingent on aliveness are necessarily much more subtle - challenge the conventional wisdom. Furthermore, unlike tissue heating and induced electric currents, which can be controlled simply by restricting the intensity of the emitted microwaves and of any *ELF* magnetic fields, *frequency-specific* non-thermal effects cannot, since here it is more a matter of 'information' transfer from the aggressing field to the organism. It must be appreciated, however, that it is *only* when the organism is *alive* that it can it 'detect' a signal - in a similar way that a radio has to be turned-on - through the existence of an 'oscillatory similitude' between the frequency of the incoming signal and that of an endogenously excited electromagnetic activity. As has already been noted, this can result in resonant enhancement, potentially degrading interference, or entrainment of a particular endogenous activity.

Despite the controversy surrounding their reality, there is increasing evidence that frequency-specific non-thermal effects are a quite general feature of living organisms, an understanding of which is, however, dependent on relatively recent developments in biophysics, particularly in respect of the inherently *non-linear* nature of living systems. It should be stressed that being contingent on aliveness, the occurrence, severity and implication for health of frequency-specific non-thermal effects necessarily must quite generally be anticipated to vary from person to person, even under identical exposure conditions; this, of course, contrasts with thermal heating and frequency *non*-specific non-thermal effects which can occur irrespective of whether the irradiated organism is alive or dead. It is these more subtle, *frequency-specific*, non-thermal effects that are now considered in detail.

6.3.3 Non-thermal effects of Mobile Phone Exposure

There have been many investigations – both *in vitro* and *in vivo* – reporting non-thermal influences of exposure to microwave radiation (both pulsed and continuous) of an intensity close (or lower) to that realised near a *GSM* mobile phone handset, but sometimes at higher carrier frequencies. Amongst the *in vitro* studies should be mentioned:

Alteration in *ODC* enzyme activity (Byus *et al.* 1988, Litovitz *et al.* 1993; Penafiel *et al.* 1997)
Reduced efficiency of lymphocyte cytoxicity, (Lyle *et al.* 1983; Sri Nageswari, 1996; Coghill and Galonja-Coghill 2000).
Increased permeability of the erythrocyte membrane, (Savopol et *al.* 1995; Sajin *et al.* 1997)
Effects on brain electro-chemistry, (*Ca* efflux), (Bawin *et al.* 1975; Blackman *et al.* 1979; Dutta *et al.* 1984, 1986)
Increase of chromosome aberrations and micronuclei in human blood lymphocytes, (Garaj-Vrhovac *et al.* 1992)
Synergistic effects with cancer promoting drugs, such as phorbol ester. (Balcer-Kubiczek and Harrison 1991)

In vivo evidence of non-thermal influences, including exposure to *actual* GSM radiation - as opposed to that of an experimental 'surrogate' - includes:

Epileptiform activity in rats, in conjunction with certain drugs, (Sidorenko and Tsaryk 1999)
Increased permeability of the blood-brain barrier in rats, (Salford *et al.* 1994; Persson *et al.* 1997)
Effects on brain electro-chemistry (dopamine-opiates), (Frey (with 1994))
Increases in DNA single and double strand breaks in rat brain, (Lai Singh, 1995, 1996; Singh and Lai 1996)

Promotion of lymphomas in transgenic mice, (Repacholi *et al.* 1997)
Synergistic effects with certain psychoactive drugs, (Lai *et al.* 1987)
Stressful effects in healthy and tumour bearing mice, (Youbicer-Simo *et al.* 2001)
Neurogenetic effects and micronuclei formation in peritoneal macrophages in mice. (Youbicer-Simo *et al.* 2001)

Finally, human *in vivo* studies, under *GSM* or similar conditions, include:

Effects on the *EEG* - specifically, a delayed increase in spectral power density (particularly in the alpha band (von Klitzing 1995)), which has been corroborated (Reiser *et al.* 1995) in the awake *EEG* of adults exposed to *GSM* radiation. Effects on the asleep *EEG* have also been reported (Mann and Roschke 1996), including a shortening of *REM* sleep (with possible adverse effects on learning) - during which the power density in the alpha band again increases – and the detection of interference with *non-REM* sleep (Borbely *et al.* 1999). In addition, exposure to mobile phone radiation causes a significant decrease in the preparatory slow potentials in certain regions of the brain (Freude *et al.* 1998, Eulitz *et al.* 1998), and affects memory tasks (Krause *et al.* 2000) and cognitive processing (Preece *et al.* 1999, Koivisto *et al.* 2000).

Much controversy currently surrounds many of the above non-thermal effects, owing to the difficulties sometimes encountered in independent attempts to replicate them. Whilst this is undeniable, it is not unexpected, however, because the very existence of such effects is contingent upon the aliveness of the organism concerned, and in turn is thus dependent on the state of the organism when it is exposed, particularly *in vivo*. Even in the *in vitro* case, discrepant findings can sometimes be traced to differences (which are not always appreciated on superficial inspection) in the conditions obtaining or the protocols used, which, of course, undermine the fidelity of the intended replication.

A good example of this is provided the attempt of Malyapa *et al.* (2000) to replicate the increase in *DNA* breakage under low intensity microwave irradiation found by Lai and Singh (Lai and Singh 1995, 1996). Whilst both groups used microwave radiation of the same frequency, they irradiated different systems (live rats *vs.* a cell line), and used very different assays to assess the *DNA* damage; in addition, the replication attempt did not separate the (positively charged) bound protein from the (negatively charged) *DNA* strands, thus obtaining much less migration in the electrophoresis field, which was also applied for a much shorter time than in the original experiment. Both of these features militate against the formation of the 'comet' tails used to assess the degree of fragmentation.

Quite apart from these kinds of problems, it must be remembered that the highly non-linear nature of living systems makes them hypersensitive (*via* deterministic chaos (Kaiser 1987)) to the prevailing conditions, and thus again militates against the realisation of the identical conditions necessary for exact replication.

6.3.4 VDUs

There have been several investigations reporting adverse non-thermal impacts in animals of exposure to the emissions from *TV* and *VDU* screens, including:

Depression of chicken immune system (melatonin, corticosterone and IgG levels) (Youbicier-Simo *et al.* 1997),

Stress effects in mice indicating an alteration in haematology and a weakening of the immune system (Youbicer-Simo *et al.* 2001)

6.4 FROM NON-THERMAL BIOLOGICAL EFFECTS TO ADVERSE HEALTH EFFECTS.

6.4.1 Mobile Phones

With the exception of the reported increase in chicken mortality (Youbicer-Simo *et al.* 2001), it is essential to appreciate that the occurrence *per se* of non-thermal effects does not necessarily entail adverse health consequences. Given, however, the ways in which *GSM* radiation non-thermally affects a variety of brain functions (including the neuroendocrine system), it can surely be no coincidence (Hyland 2000) that the health problems reported are themselves predominantly of a neurological nature, although the precise causal connections remain, of course, to be established. Thus, for example, the anecdotal reports of headache are consistent with the effect that the radiation has on the dopamine–opiate system of the brain, and on the permeability of the Blood-brain barrier, (Salford *et al.* 1994; Persson *et al.* 1997), both of which have been medically connected to headache (Winkler *et al.* 1995; Barbanti *et al.* 1996). The reports of sleep disruption, on the other hand, are consistent both with the effect the radiation has on melatonin levels (admittedly, at present, only under *VDU* exposure) (Youbicier-Simo *et al.* 1997)) and on *REM* sleep (Mann and Roschke 1996). More recently, there has been an (as yet) uncorroborated report of an increase in uveal melanoma amongst phone users (Stang *et al.* 2001). Finally, the statistically significant increase (by over a factor 2) in the incidence of neuro-epithelial tumours (the laterality of which correlates with cell phone use) found in a epidemiological study (Muscat *et al.* 2000) in the *USA* as part of the Wireless Technology Research (*WTR*) Programme is consistent with the genotoxicity of *GSM* radiation, as indicated, for example, by the increased number of *DNA* strand breaks and the formation of chromosome aberrations and micronuclei in human blood (the latter being corroborated by the *WTR* Programme (Carlo 2000; Carlo and Schram 2001)), and with the promotional effect that *GSM* radiation has on tumour development (Repacholi *et al.* 1997).

When the evidence is considered *in its entirety*, a rather interconsistent and coherent picture starts to emerge, from which it is clear that the issue of non-thermal effects – with their potential to induce adverse health reactions of the kind reported, often admittedly anecdotally, such as the headache and sleep disruption already mentioned, as well as memory impairment – can no longer be responsibly ignored, although it may only be a small minority of people who are yet badly affected.

Because both the occurrence of the initial provoking non-thermal effect, *as well as* the severity of any associated adverse health effect, are contingent on aliveness, they must *necessarily depend on the physiological state of the organism when it is exposed to the radiation - i.e.* non-thermal effects are *non-linear* effects. Accordingly, it is quite possible that exposure to low intensity radiation can entail a seemingly disproportionately *large* (non-linear) response (or none at all), and *vice versa* (consistent with which is the familiar occurrence of frequency/intensity 'windows' of response), quite unlike the situation with the predictable (linear) thermal effects.

The physiological state of different people cannot, in general, be anticipated to be the same, however, since it depends, amongst other things, on the stability of an individual's brain rhythms against interference or entrainment by the radiation, their already prevailing level of stress, and the robustness of their immune system. Accordingly, *identical* exposure to exactly the *same* radiation can entail quite different (non-thermal) responses in different people (unlike the case of active electronic instruments), consistent both with the fact that not every exposed person is adversely affected (as also holds, of course, in the case of smoking, where not all smokers get lung cancer!), and with the difficulties encountered in some laboratory attempts to replicate non-thermal effects, particularly under *in vivo* conditions. For depending on a person's genetic predisposition, and the fact that stress is cumulative, it is quite possible that exposure to an electromagnetic field simply supplies the final contribution that raises a particular person's level of stress above some critical value, thereby 'triggering' the manifestation of some pathology that is already in a well advanced state, but which, in the absence of any exposure, would, for the time being, have remained latent.

Accordingly, the oft-repeated statement that ......*'There are no established adverse health effects of exposure to GSM radiation (of sub-thermal intensity)'*....... is actually quite true, but, in view of the above, this is necessarily so, thus making the statement essentially vacuous. The more relevant consideration is whether there is an established risk to human health. It must be concluded that such a risk does indeed exist, but - in view of the above considerations - the actual number and identity of those at risk are necessarily unknown, *a priori* - although, for the reasons given below, children (as well as highly stressed people, those with already compromised immune systems, and those who are on certain prescribed psychoactive drugs) must be considered more vulnerable.

Pre-adolescent children can be expected to be (potentially) more vulnerable than adults – as acknowledged by the Report of the UK Independent Expert Group on Mobile Phones – because:

i. Absorption of microwaves of the frequency used in mobile telephony is greatest (Gandhi *et al.* 1996) in an object about the size of a child's head – the so-called 'head resonance' – whilst, in consequence of the thinner skull of a child, the penetration of the radiation into the brain is greater than in an adult.

ii. The still developing nervous system and associated brain-wave activity in a child (and particularly one that is epileptic) are more vulnerable to aggression by the pulses of microwaves used in *GSM* than is the case with a mature adult. This is because the multi-frame repetition frequency of 8.34Hz and the 2Hz pulsing that characterises the signal from a handset equipped with discontinuous transmission *(DTX)*, lie in the range of the *alpha* and *delta* brain wave activities, respectively. The fact that these *two* particular electrical activities are constantly changing in a child until the age of about 12 years – when the delta-waves disappear and the alpha rhythm is finally stabilised – means that they must *both* be anticipated to be particularly vulnerable to interference from the *GSM* pulsing.

iii. The increased mitotic activity in the cells of developing children makes them more susceptible to genetic damage.

iv. A child's immune system, whose efficiency is, in any case, degraded by radiation of the kind used in mobile telephony, is generally less robust than is that of an adult, making the child less able to 'cope' with any adverse health effect provoked by exposure to such radiation.

6.4.2 VDUs

Working with *VDUs* for more than two hours per day has been found to provoke a variety of neuropsychological disorders (Loiret *et al.* 1994). It has also been shown (Smith *et al.* 1981; Johansson and Aronsson 1984) that the *VDU* user develop a higher level of stress compared with non-users. For example, many show (Amick & Smith 1992) both emotional (irritability, anxiety, depression) and psychosomatic (insomnia, lack of appetite, perspiration) disorders. Other complaints include headache, pains in the neck, hands, wrists and fingers, and back pain (Amick & Smith 1992).

Giving rise to more serious concern, however, are numerous reports indicating an increased risk of miscarriages amongst women workers - those exposed to more than 20 hours per week having more than twice as many as women doing other kinds of office work (Goldhaber *et al.* 1988); others reports (Bentham 1996) suggest that unsuccessful pregnancies are, on average, around 75% more common than amongst non-users. Of possible relevance to *VDU* exposure is a report of damage to the eyes of subjects exposed (at a somewhat greater distance) to the emissions of *TV* set - specifically, corneal injury and reduction in visual acuity/ accommodation (Miyata and Namba 1999).

Animal studies (Youbicier-Simo *et al.* 1997), on the other hand, indicate a significant increase in the mortality of chicken embryos exposed to *VDU* emissions.

Finally, mention should be made of on-going work, first reported (Clements-Croome and Jukes 1999), aimed at identifying, the contribution of adverse impacts of *VDU* exposure to 'sick building syndrome'. By means of a double blind crossover study employing a device that claims to protect against such emissions, a 33% decrease in sick building symptoms was found amongst the protected group, implying that these symptoms are indeed *partly* due to the electromagnetic emissions from *VDU*s: in a subsequent double blind crossover study, a reduction of 25% was found.

6.5 EPILOGUE

The above Case Study illustrates not only the magnitude of the problem involved, but also its subtlety. It constitutes a cautionary lesson – that is not confined solely to the microwave and *ELF* parts of the electromagnetic spectrum - of the absolute necessity, in the quest for electromagnetic biocompatibility, of going beyond the 'official' *SAR*-based mindset to embrace the currently largely neglected 'informational' dimension of the problem. It must be remembered that electromagnetic radiation is a wave, and as such has properties other than just the intensity that is addressed by existing Safety Guidelines, the neglected frequency dimension opening the exposed organism – *provided it is alive* - to non-thermal, *informational* influences of the radiation. The mainstream scientific approach to assessing the noxiousness of human exposure to electromagnetic fields is, on the other hand, principally guided by an essentially *linear* perception, which might well be adequate to deal with thermal effects, but is one that is quite inappropriate to any realistic consideration of the non-thermal, frequency-specific vulnerability of the alive human organism to the rather coherent electromagnetic fields of technological origin. For since, unlike thermal effects, the possibility of such non-thermal influences is contingent on the *aliveness* of the organism, their very occurrence, as well as any implications for health, necessarily both depend on the *state* of the organism when it is exposed, which, of course, varies not only between *different* individuals, but can also do so for the *same* individual, depending on his/her condition at the time of exposure – *i.e.* such influences are, technically speaking, of an inherently *non*-linear kind. As such, they often appear counter-intuitive, or even bizarre, from a linear standpoint - a feature that, together with difficulties sometimes experienced in attempts to independently replicate them, tends to bias their dismissal as experimental artefacts. Attempts to address a problem that is inherently non-linear from such a linear perspective only exacerbates things: outdated knowledge is worse than ignorance - at least the ignorant know what they do not know!

It is not so much the case that, in the haste to make new and valuable technology available to the public, the research necessary to establish its safety has been bypassed or compromised, but rather - and more reprehensibly – that the already existing body of research indicating that the technology is potentially less than safe is often *studiously ignored*, or at best not given adequate attention, not only by the Industry but also by national and international regulatory bodies.

Quite justifiably, the public remains sceptical of attempts by governments and industry to reassure them that all is well, particularly given the unethical way in which they often operate symbiotically so as to promote their own vested interests, usually under the brokerage of the very statutory regulatory bodies whose function it supposedly is to ensure that the security of the public is *not* compromised by electromagnetic exposure! Given the experience over recent years with official duplicity over *BSE/CJD* – with the initial assurances of no risk and subsequent revelations of cover-ups - the public is now understandably wary of safety assurances from 'official' governmental scientific sources in respect of electromagnetic pollution; this is particularly so when the voice of those with a view contrary to that of the prevailing officially perceived wisdom is at worst silenced, or, at best, studiously ignored. The situation is further exacerbated by reports relating to research supported financially by the Mobile Phone Industry of its attempts to 'persuade' those who discover findings that might prove to be potentially damaging *to actually alter their results* to make them more 'market friendly'. Also no doubt driven by market considerations is the attempt (in which the World Health Organisation is playing a leading role) to establish a global 'harmonisation' of mobile phone exposure standards, by attempting to persuade countries that currently operate more stringent limits – such as Russia and China - to relax them in favour of the higher levels tolerated in the West; it can be no coincidence that in Russia, where the frequency-specific sensitivity of living organisms to ultra-low intensity microwave radiation was first discovered over 30 years ago, that the exposure guidelines are approximately between 50 and 100 times more stringent that those of *ICNIRP*!

Furthermore, there is a regrettable tendency to attribute market–friendly (negative) results a greater significance, publicity and profile than positive ones indicative of the possibility of adverse health impacts. An example of this is provided by the above mentioned publication two years ago of an epidemiological study in the *USA* and the reaction to it – not only in the Press but also in professional scientific circles - which focused exclusively on the negative finding that there was no *overall* increase in the incidence of brain tumours (*i.e.* irrespective of their location) amongst mobile phone users; completely ignored was the fact that this was actually a *false* negative! For if, instead of the entire brain, attention is focused on the periphery - which is the only region into which radiation from the mobile phone can directly penetrate – then a statistically significant elevation is found (amongst mobile phones users) in the incidence of a rare kind of tumour (epithelial neuroma), in the laterality of which also correlates with that of phone usage; this was glossed over even in the text of the paper, every effort having been made between its submission and publication to reduce the level of statistical significance of this positive finding.

This bias towards negative results is also commonplace amongst so-called 'expert' reviews of available experimental data, where the most negative possible 'spin' is invariably put on any positive results that might be consistent with the possible noxiousness of the exposure, particularly if the results have proved difficult to replicate. Their reliability is invariably called into question, some reason being given as to why they should be rejected, without ever appreciating that the reason for non-replication can often be traced to an essential difference in experimental protocol *that effectively undermines the fidelity of the intended replication*! The veracity of negative results – upon which the tenability of existing Safety Guidelines is, of course, dependent – is, by contrast, seldom questioned: there are rarely any false negatives, only false positives!

Furthermore, in assessing the relevance to humans of findings obtained using animals, such as rats – it should not be forgotten that not only are they are often subject to exposure conditions quite different from those realised during human use of a mobile phone (*e.g.* whole body exposure to the *near* field), but also that also that exposure is not actually to the emissions of a real mobile phone, but rather to that of a 'surrogate' microwave generator, which invariably *lack* the bio-active *ELF* features.

Given the perceived lack of consensus amongst experts concerning both the significance and credibility to be attached to published research into biological effects of the kind of radiation now used in *GSM* telephony, for example, and whether such effects can actually provoke adverse health reactions in certain susceptible people (despite the existence of many consistent, anecdotal positive reports of such), it is probably true to say that if the same obtained in the case of a new drug or foodstuff they would simply never be licensed!

The genuine concerns of the public are thus not unfounded, and the irony of the situation with mobile phones and their Base-stations is that the current Safety Guidelines afford a greater level of protection to electronic instrumentation than they do to the alive human being!

Of particular concern to the public – and that which understandably generates the most outrage – is the involuntary subjection of certain groups of the population 24 hours/day, 7 days/week to the emissions of *GSM* base-stations that are insensitively sited near to homes, schools and hospitals. For the environment of these people is effectively permanently polluted - a pollution from which there is literally 'nowhere to hide'. This totally unacceptable state of affairs raises serious ethical questions, and arguably contravenes the Nuremberg Code, in that it is these people who will eventually reveal the degree to which chronic exposure to such fields is noxious – information that *is not currently available*: in other words, they are effectively involuntary subjects in a mass experiment.

6.6 REFERENCES

Amick III B.C. and Smith M.J. Stress, computer-based work monitoring and measurement system: a conceptual overview. *Applied Ergonomics* 1992; **23:** 6-16.

Balcer-Kubiczek E.K. and Harrison G.H.   Neoplastic transformation of C3H/10T1/2 cells following exposure to 120 Hz modulated 2.45 GHz microwaves and phorbol ester tumour promoter. *Radiation Res.* 1991; **126:** 65-72.

Barbanti P. *et al.* Increased density of dopamine D5 receptor in peripheral blood lymphocytes of migraineurs: a marker of migraine? *Neurosci. Letts.* 1996; **207**(2): 73-76.

Bawin S.M. *et al.* Effects of modulated vhf fields on the central nervous system. *Ann. NY Acad. Sci.* 1975; **247:** 74-81.

Bentham P. 'VDU Terminal Illness', 1996, Jon Carpenter Publishing, London.

Blackman C.F. *et al.* Induction of calcium-ion efflux from brain tissue by radio-frequency radiation: effects of modulation frequency and field strength. *Radio Sci.* 1979; **14:** 93-98.

Borbely A.A. *et al.* Pulsed high-frequency electromagnetic field affects human sleep and sleep electroencephalogram. *Neurosci. Lett.* 1999; **275**(3): 207-210.

Byus C.V. *et al.* Increased orthinine decarboxylase activity in cultured cells exposed to low energy modulated microwave fields and phorbol ester tumour promoters. *Cancer Res.* 1988; **48:** 4222-26.

Carlo G.L. 'Wireless Telephones and Health': WTR Final Report - paper presented to the French National Assembly, 19th June, 2000.

Carlo G. and Schram M., 'Cell Phones: Invisible Hazards in the Wireless Age', 2001, Carroll & Graf Publishers, New York.

Clements-Croome D. and Jukes J. Double blind crossover field trial of the effectiveness of Tecno AO: a technology for protection from the effects of low frequency magnetic fields *Prog. in Radiation Protection* 1999; FS-99-106-T, **1:** 236-240.

Coghill R.W. and Galonja-Coghill T. *Electro and Magnetobiology* 2000; **19**(1):43-56.

Dutta S.K. *et al.* Microwave radiation-induced calcium ion efflux from human neuroblastoma cells in culture. *Bioelectromagnetics* 1984; **5:** 71-78

Dutta S.K. *et al.* 'Biological Effects of Electropollution: Brain Tumours and Experimental Models', (Eds. S.K. Dutta *et al.*), Information Ventures Inc., Philadelphia, 1986, *pp.*63-69.

Eulitz C. *et al.* Mobile phones modulate response patterns of human brain activity. *Neuroreport* 1998; **9**(14): 3229-3232.

Freude G. *et al.* Effects of microwaves emitted by cellular phones on human slow brain potentials. *Bioelectromagnetics* 1998; **19:** 384-387.

Frey A.H. (Editor), 'On the Nature of Electromagnetic Field Interactions with Biological Systems', 1994, R.G. Landes Co., Austin, Texas.

Gandhi O.P. *et al.* Electromagnetic absorption in the human head and neck. *IEEE Trans. MTT* 1996; **44:** 1884-1897.

Garaj-Vrhovac V. *et al.* The correlation between the frequency of micronuclei and 6 specific aberrations in human lymphocytes exposed to microwave radiation *in vitro. Mutation Research* 1992; **281:** 181-186.

Goldhaber M. *et al.* Risk of miscarriage and birth defects among women who use video display units. *American J. Industrial Medicine* 1988; **13:** 695-706.

Hyland G.J. Physics and biology of mobile telephony. *The Lancet* 2000; **356:** 1833-1836.

Johansson G. and Aronsson G. Stress reactions in computerised administrative work. *J. Occupat. Behaviour* 1984; **5:**159-181.

Kaiser F. The role of chaos in biological systems, in 'Energy Transfer Dynamics', (Eds. Barrett T.W. and Pohl H.A.), Springer-Verlag, Berlin, 1987, Ch.21, pp. 224-236.

Krause C.M. *et al.* Effects of electromagnetic field emitted by cellular telephones on the EEG during a memory task. *Neuroreport* 2000; **11**(4): 761-764.

Lai H. and Singh N.P., Acute low-intensity microwave exposure increases DNA single strand breaks in rat brain cells. *Bioelectromagnetics* 1995; **16**: 207-210.

Lai H. and Singh N.P., Single and double-strand DNA breaks after acute exposure to radiofrequency radiation. *Int. J. Radiation Biol.* 1996; **69**: 13-521.

Lai H. *et al.* A review of microwave irradiation and actions of psychoactive drugs. *Engineering in Medicine and Biology*, 1987; **6**: 31-36.

Litovitz T. *et al.* The role of coherence time in the effect of microwaves on ornithine decarboxylase activity. *Bioelectromagnetics* 1993; **14**: 395-404.

Loiret P *et al.* 'Enquete Travail sur ecran TEC-2. Conclusions neuro-pyscologiques: plaintes neuro-pyscologiques et maux de tete. Direction Regionale du Travail et de l'Emploi de Poitou-Charentes. Ministère du Travail, de l'Emploi et de la Formation Professionelle, 1994.

Lyle D.B. *et al.* Suppression of T-lymphocyte cytotoxicity following exposure to sinusoidally amplitude-modulated fields. *Bioelectromagnetics* 1983; **4**: 281-292.

Malyapa, R.S. *et al.* DNA damage in rat brain cells after in vivo exposure to 2450MHz electromagnetic radiation and various methods of euthanasia. *Radiation Research* 1998; **149**(6): 637-645.

Mann K. and Roschke J. Effects of pulsed high-frequency electromagnetic fields on human sleep. *Neuropsychobiology* 1996; **33**: 41-47.

Microsc. Res. Tech. 1994; **27**: 535-542.

Miyata M. and Namba T. Ocular functions loading by visual display terminal and the effect of Tecno AO. *Jap. Rev. Clinical Opthalmology* 1999; **11**: 1634-1636/32-35.

'Mobile Phones and Health', Report of the Independent Expert Group on Mobile Phones, May, 2000.

Muscat J.E. *et al.* Handheld cellular telephone use and risk of brain cancer. *JAMA* 2000; **284**: 3001-3007.

Penafiel L.M. *et al.* Role of modulation on the effect of microwaves on ornithine decarboxylase activity in L929 cells. *Bioelectromagnetics* 1997; **18**: 132-141.

Persson B.R.R. *et al.* Blood-brain barrier permeability in rats exposed to electromagnetic fields used in wireless communication. *Wireless Networks* 1997; **3**: 455-461.

Preece A.W. *et al.* Int. J. Rad. Biol.1999; **75**: 447-456; Koivisto M. *et al.* Effects of 902MHz electromagnetic field emitted by cellular telephones on response times in humans. *Cognitive Neuroscience and Neuropsycology* 2000; **11**:413-415.

Reiser H-P. *et al.* The influence of electromagnetic fields on human brain activity. *Eur. J. Med. Res.* 1995; **1**: 27-32.

Repacholi M.H. *et al.* Lymphomas in Eμ-Pim 1 transgenic mice exposed to pulsed 900MHz electromagnetic fields. *Radiation Res.* 1997; **147**: 631-640.

Salford L.G. *et al.* Permeability of the blood-brain barrier induced by 915MHz Electromagnetic radiation, continuous wave and modulated at 8, 16, 50 and 200Hz.

Sajin G. *et al.* Low power microwave effects on erythrocyte membranes. Proc. 27th European microwave conference Vol. **I**, *pp.*596-599, 1997.

Savopol T. *et al.* Membrane damage of human red blood cells induced by low power microwave radiation. *Electro-and magnetobiology* 1995; **14**(2): 99-105.

Sidorenko, A.V. and Tsaryk, V.V. Electrophysiological characteristics of the epileptic activity in the rat brain upon microwave treatment. Proc. 'Electromagnetic Fields and Human Health', Moscow, Sept. 1999, *pp.* 283-4.

Singh N.P. and Lai H., Proc. Asia Pacific Microwave Conf. (Edited by Gupta R.S.), Vol. **1**(B1-4), *pp.*51-55, 1996.

Smith, M.J. *et al.* An investigation of health complaints and job stress in video display operations. *Human Factors* 1981: **23**: 389-400.

Sri Nageswari K. Immunological effects of chronic low power density and acute power density microwave radiation – a review. Proc. Asia Pacific Microwave Conf. (Edited by R.S. Gupta), Vol. **1**(B1.6), *pp.*59-61, 1996.

Stang A. *et al.* The possible role of radiofrequency radiation in the development of uveal melanoma. *Epidemiology* 2001; **12**: 7-12.

von Klitzing L. Low-frequency pulsed electromagnetic fields influence the EEG of Man. Phys. Medica 1995; **XI** (2): 77-80.

Winkler T. *et al.* Impairment of blood-brain barrier function by serotonin induces desynchronization of spontaneous cerebral cortical activity: experimental observations in the anaesthetised rat. *Neuroscience* 1995; **68**(4): 1097-1104;

Youbicier-Simo B.J. *et al.* Biological effects of continuous exposure of embryos and young chickens to electromagnetic fields emitted by video displays. *Bioelectromagnetics* 1997; **18**: 514-523.

Youbicer-Simo B.J. *et al.* Review of studies validating the compensative efficacy of a new technology designed to compensate potential adverse bioeffects caused by VDU and GSM Cell Phone radiation. *Radioprotecção* 2001; **1**(8,9): 105-123.

CHAPTER 7

Product Liability, Product Safety and the Precautionary Principle

Simon Pearl

Product liability is the term used in respect of private legal rights to seek compensation for injuries sustained as a result of exposure to a dangerous product. Product safety, on the other hand, is the remit of a regulatory regime and concerns public questions of enforcement by the state through the criminal courts. So far, there has been little evidence of "Joined Up" treatment of these differing regimes.

Whilst the ultimate purpose of both product safety regulations and product liability law may be similar, namely to achieve consumer protection from unsafe products, and it is inappropriate to treat the two regimes in entirely separate compartments, the fact remains that they are governed by separate European laws. The primary legislation of product liability is the Product Liability Directive 85/374, and of product safety is the General Product Safety Directive 92/59 (GPSD). The GPSD was revised last year. The revision has to be transposed into UK law within 2 years.

7.1 PRODUCT LIABILITY

The Product Liability Directive imposes strict liability for damages caused by a defective product. The liability is imposed on a producer, own brander, or first importer into the EU, or the supplier of the product if he is unable to identify the party up the chain of supply. The liability is imposed if the product does not provide the degree of safety which a person is entitled to expect. It was implemented in the United Kingdom by the Consumer Protection Act 1987 and in parallel legislation throughout Europe. In a number of countries, particularly the United Kingdom, prior to its introduction, compensation could only be recovered if a consumer was able to prove, on the balance of probabilities, that the manufacturer had been negligent in the manufacture or marketing of the product. The Directive, which was implemented in the UK in March 1988, shifted the focus from the behaviour of the manufacturer to the product in question. The definition of "product" is wide and specifically includes electricity by virtue of Article 2. Article 6 defines defect as follows:-

A product is defective when it does not provide the safety which a person is entitled, taking all circumstances into account including:

- the presentation of the product;
- the use to which it could reasonably be expected that the product would be put;
- the time when the product was put into circulation.

A product shall not be considered defective by the sole reason that a better product is subsequently put into circulation.

Article 7 provided the producer with a number of defences. A producer shall not be liable as a result of the Directive if he proves

- that he did not put the product into circulation;
- that, having regard to the circumstances, it is probable that the defect which caused the damage did not exist at the time when the product was put into circulation by him;
- that the product was neither manufactured by him for sale or any form of distribution for economic purposes ... or in the course of his business;
- the defect is due to compliance of the product with mandatory regulations issued by the public authorities;
- the state of scientific or technical knowledge at the time when he put the product into circulation was not such as to enable the existence of the defect to be discovered (the so called Development Risk Defence);
- in the case of a manufacturer of a component, the defect is attributable to the design of the product in which the component had been fitted or to the instructions given by the manufacturer of the product.

Until recently there has been remarkably little case law in Europe concerning the Directive. It was generally thought that the impact of the Directive was not significant and it had done little to improve the prospects of a consumer in product liability litigation.

This is because of the definition of defect which was based on the legitimate expectation of the consumer. Commentators had suggested that the Courts would conclude that the public could legitimately expect no more than that a manufacturer would act in a reasonable and non-negligent manner. Absolute safety was not a legitimate expectation.

The few Court decisions that there were seemed to confirm the commentator's cynicism. In the UK case of Richardson -v- LRC Products Limited, the Court held that the public could not legitimately expect a condom to be 100% effective in preventing pregnancy. A similar result was reached in the case of Foster -v- Biosil in relation to a silicone breast implant which leaked requiring explantation. The consumerist lobby called for the Directive to be amended to achieve the stated aim of making it easier for injured consumers to recover compensation.

The first significant judicial test of the Directive however demonstrated that the Courts, at least in the UK, do regard the Directive as a significant change over the common law negligence principles.

Mr Justice Burton in the Hepatitis C litigation (A and Others -v- National Blood Authority) decided in March 2001 conducted a comprehensive analysis of the Directive.

7.2 FACTS OF THE HEPATITIS C CASE

Damages were claimed by a group of over 100 individuals infected by Hepatitis C virus from blood transfusions between 1 March 1988 (the implementation date of the Directive) and September 1991 when the Hepatitis C screening was introduced in the UK. Isolation of the Hepatitis C virus had been announced in May 1988 and the screening test became commercially available in early 1990 but not immediately introduced pending, among other issues, deliberation of the accuracy of the test and Government approval. The Claimants' case was that they had received a defective product because it was infected by the virus, even though for the earlier parts of the period there was no test available which could have detected the virus as such considerations were not relevant in the strict liability compensation regime.

The Defendant, the National Blood Authority, argued that blood was a natural product which carried an inherent risk of infection with viruses and that the medical profession knew of the risk which, at least for part of the period, could not have been avoided. The definition of what constituted a defect was fundamental to the outcome of the case. The defence maintained that the inclusion in the definition of the expectation of the public being assessed taking all circumstances into account obliged the Judge to look at what had been done to prevent the infection, amongst other aspects, and therefore enquire, in effect, into the reasonableness of the Defendant's actions. This approach was not accepted by the Judge. He stressed the importance of applying the Directive purposefully in order to give effect to its stated intention to make it easier for the injured consumer to recover compensation by removing the concept of negligence as an element of liability. The Judge concluded that the fact that the screening test was not available for at least part of the period, or that the medical profession were well aware of the risk that blood could be contaminated by viruses, including viruses which caused hepatitis, were not relevant circumstances when assessing the legitimate expectation of the public as to the safety of blood.

Avoidability of a harmful characteristic, the practicality of taking measures to prevent such a characteristic, cost and benefit to society or utility of the product were not relevant circumstances because they were inconsistent with the purpose of the Directive. The Court held that had it been intended for these factors to be relevant, they would have been expressly included in the Directive as specific factors. Other circumstances such as the product's intended use had been specifically set out. Disregarding such factors, the Judge concluded that the public had a legitimate expectation that blood would not be infected with Hepatitis C.

Part of the Judge's reasoning was that blood infected with Hepatitis C was a "non-standard product". It differed from non-infected blood in that it contained a virus which caused injury. Its safety should be compared with blood that is not infected. This equates to what product liability lawyers had previously categorised as "manufacturing defect". The Judge rejected the NBA's contention that the small risk that blood was infected with Hepatitis C virus was a characteristic of all blood and one which was accepted by the public. In relation to a non-standard product, the Judge concluded that unless it could be demonstrated that the public accepted the non-standard nature of the product, it will be relatively easy for Claimants to demonstrate that the product was defective. The impact of the Judgment may be wider than this and will affect standard products where the contention is that there is a design defect affecting all products rather than a deviation from the norm.

One of the most surprising aspects of the decision is the Judge's conclusion that the knowledge of the medical profession was of no relevance. In a case such as blood, but equally applicable to many prescription only medicines for instance, the medical profession know of the risks and benefits of the product that they are giving to their patient and may have a lower expectation of the level of safety of that product than the ultimate consumer of the product. It had been thought that the expectation of the public would be legitimately informed by the knowledge of the medical profession. This argument was rejected by the Judge. Even for standard products, it is difficult to see how warnings (despite being expressly relevant circumstance within the Directive) which appear, for instance, on package inserts of drugs, or accompanying cell phones packaging would necessarily be understood fully and will be sufficient to shift an expectation of the public at large as to the level of safety of such products. The Judge failed to grapple with the profound implication of his reasoning. The inability of manufacturers to predict with any level of certainty whether warnings will be regarded as adequate to lower expectations of the public as to the level of safety could be a significant damper on innovation and is one of the profound impacts of this case.

The Judgment also dealt with the scope of the Development Risk Defence in Article 7e of the Directive. The Judge rejected the Defendant's contention that the defence related to the discovery of the defect in a particular product rather than products produced by manufacturers generally. As soon as a risk of infection was known the defence was not available even if the risk in a particular product could not have been avoided. The same would apply to all products. As soon as the manufacturer knew or should have known in light of assessable information of the risk of defect in question, even if the risk is not proven and only postulated by a minority, then the producer continues to supply the product at his own risk. This is certainly something which, in light of published literature relating to EMF, would make it difficult, if causation is ultimately demonstrated, for a Development Risk Defence to be run based on the level of scientific and technical knowledge at the time. The existence of the defect is generic and the lack of ability to determine whether the defect would materialise in any particular product cannot benefit from a Development Risk Defence. The defence protects the producer only in respect of the unknown and is extremely narrow in scope.

The consequences of this Judgment, if followed in future cases, is that once a Claimant can demonstrate that the product did indeed cause his injury (an issue which will remain the focus in most product liability cases, particularly in the technical field), it is likely that it will be easier for him to prove that the product was defective and less likely for the manufacturer to have the ability to escape liability by demonstrating that he did all that could reasonably be expected. Only if the Court concludes that the harmful characteristic was well known and socially accepted will the legitimate expectation of the public be regarded as less than absolute (e.g. a knife, alcohol or tobacco). For such products, the fact that the product has the capacity to cause injury informs the public's legitimate expectation, so that it cannot be defective if it has the characteristic that it is known to have and for which society has accepted the risks involved. Hence a knife which if sharp is not defective if a consumer injures himself with it. Non-standard products will be unlikely ever to be protected in this fashion. Hi-tec products with warnings may fall within this category but, the warnings must affect the legitimate expectation of the public without reference to expert knowledge.

7.3 PRODUCT SAFETY

The General Product Safety Regulations 1994 ("GPS Regulations") implemented the General Product Safety Directive (92/59/EEC) into UK Law. The Revised GPSD was adopted by the Council of Ministers on 27 September 2001 and the European Parliament on 4 October 2001. It requires transposition into UK law within 2 years. In this paper I will refer to the current regulations save where the revised GPSD will make material changes as most of the material provision of the original GPSD have been carried over to the revised GPSD. The changes focus on clarifying the scope of the Directive and tackling product safety issues which come to light after products are placed on the market including recall obligations. The GPSD imposes a general safety requirement that Consumer Products placed on the market must be safe. It imposes a series of measures aimed at ensuring that Consumers have the information needed to assess risks and that safety problems will be properly dealt with when they emerge. For the first time, the "Precautionary Principle" is specifically to be taken into account.

The GPSD impose a regime of product safety law which is quite distinct from product liability law. Product liability encompasses the civil law governing questions of compensation to injured persons by manufacturers or suppliers of defective products, and is governed now mainly by the Consumer Protection Act 1987 Part I which derived from the Product Safety Directive (85/374/EEC) of July 1985. Compliance with product safety regulations is a question of enforcement by the State through the criminal courts whist product liability law is private law and a matter for civil courts.

Nevertheless the ultimate purpose of both product safety regulations and product liability law is the same, namely to achieve consumer protection from unsafe products. It is inappropriate to treat the two regimes in entirely separate compartments. Manufacturers, suppliers, and their insurers, who have regard to the necessary systems designed to ensure that their products are reasonably safe will achieve the dual purpose of compliance with product safety regulations and reduction of exposure to potential product liability claims.

Also, just as a prosecution for driving without due care and attention, or under the Health and Safety at Work Act often precede a civil claim, enforcement under safety regulations may proceed a civil claim where injuries have already resulted prior to enforcement. The viability of defending the product liability claim can be significantly influenced by the outcome of the enforcement action. It is for this reason that manufacturers, suppliers and insurers will need to pay close attention to enforcement action as well as to responding to complaints.

It is important to stress that a manufacturer or supplier that complies with standards laid down in the product safety regulations does not automatically provide a defence to a product liability claim, although it may be persuasive evidence that the product is not defective. This is somewhat different than in certain instances in the USA where the mere fact that a manufacturer has complied with certain public law provisions "pre-empts" any civil claim.

A producer has two general obligations under the GPS Regulations, namely:-

- to place only safe products on the market [Regulation 7];
- to provide safety information and to adopt appropriate measures to ensure that the products can be used safely [Regulation 8].

Safe products is defined in Regulation 2(1):

"Safe product" means any product which, under normal or reasonably foreseeable conditions of use, including duration, does not present any risk or only the minimum risks compatible with the product's use considered as acceptable and consistent with a high level of protection for the safety and health of persons, taking into account in particular:

- the characteristics of the product, including its composition, packaging, instructions for assembly and maintenance;
- the effect on other products, where it is reasonably foreseeable that it will be used with other products;
- the presentation of the product, the labelling, any instructions for its use and disposal and any other indication or information provided bay the producer; and
- the categories for consumers at serious risk when using the product, in particular children.

and the fact that higher levels of safety may be obtained or other products presenting a lesser degree of risk may be available shall not in itself cause the product to be considered other than a safe product".

The revised GPSD will require the assessment of whether a product is safe, where applicable, to take account of the putting into service or installation of a product and its maintenance needs.

Safe Products - Regulation 8

Regulation 8 provides obligations to gather information and it is here where there is a potential extension on existing law.

- within the limits of his activity, a producer shall:

- provide consumers with the relevant information to enable them to assess the risks inherent in a product throughout the normal or reasonably foreseeable period of its use, where such risks are not immediately obvious without adequate warnings, and to take precautions against those risks; and
- adopt measures commensurate with the characteristics of the products which he supplies to enable him to be informed of the risks which these products might present and to take appropriate action including, if necessary, withdrawing the product in question from the market to avoid those risks.

- The measures referred to in sub-paragraph (1) above may including, whenever appropriate:
  - marketing of the products or product batches in such a way that they can be identified;
  - sample testing or marketed products;
  - investigating complaints; and
  - keeping distributors informed of such monitoring.

A more interesting potential effect for civil liability cases may well be in the obligation on a producer to gather and provide information on the safe use of his product. The obligation, which is an on-going one, namely both at the time that the product is first marketed and, if necessary, through its normal and reasonably foreseeable period of use. The obligation is not only to provide warnings on risks which may be encountered with a product but to provide information on risks which are not immediately obvious without adequate warnings. Hence there is no obligation to give warnings about risks which are immediately obvious and the test is an objective one.

One has to grapple with how information is to be provided to the user of the product because publishing information that a risk exists is not, of itself, enough. The information must enable the user of the product to assess the risks so that precautions against the risks can be taken. The standard to which the information is provided must be geared to the anticipated user. An overly scientific or complex "warning" might prevent the user from assessing the risks. He would need to be able to understand the warning. This has been the subject of enormous debate in the context of pharmaceutical Patient Information Leaflets, another European initiative although medicines are excluded from the GPS Regulations. It is nevertheless illustrative.

Leaflets are now required to be inserted into packets, and have created liability implications for manufacturers which cut both ways. First, for many years the pressure from the public for more information about medicines exceeded the willingness of the doctors to provide it. The Courts generally supported the approach adopted by the medical professional that information should be on a "need to know basis". Manufacturers supplied the information to the doctors, the "learned intermediary", and they were then not responsible for what information ultimately was provided to the patient. Although there have been many failures to warn medical negligence cases, those have generally failed, the Court applying a standard of the professional norm, which is circular in its implications. Indeed the risks of burdening the patient with too much information had been a theme running through English case law in the past. It will be interesting to see if this is effected by the Hepatitis C Judgment.

A product can be "defective" if the information about is defective. It is clear that therefore the manufacturer has an onerous task in developing the Patient Information Leaflets. Complaint may be made that the explanation of the risk was too technical or the print was too small, or the risk was not highlighted. The GPS Regulations would seem to underline this with its emphasis on the risks which were not immediately obvious.

Such considerations apply not only to the original purchaser or user of the product, but also to others affected by it, for example if a dishwashing powder caused injury on skin contact or a washing powder damaged someone else's clothes, and it was not adequately warned against, it could be said to be an unsafe product not only in the hands of the person who purchased the powder, but anyone who came into contact with it or was affected by it.

It is worth pointing out that a breach of the information requirement is not a criminal offence. The UK chose not to introduce an offence for this, concluding that it was unnecessary, due to the alternative powers available including the powers under Part II of the Consumer Protection Act 1987 to remove unsafe products from circulation. Nevertheless the obligation is a statutory one. In this context, if breached, it can have product liability implications.

7.4 THE ROLE OF THE PRODUCT LIABILITY DIRECTIVE IN PRODUCT REGULATION

The Directive sets a standard of product safety. Penalties that arise through the civil system of course will only be relevant if an injured person sues the manufacturer. The Directive does have a significant role in the framework of product regulation because of the impact of potential civil claims. It serves as an enforcement mechanism. This is because regulatory regimes will not necessarily pursue and prosecute defaulting suppliers due to resource limitations. Injured Claimants and their lawyers will not be so hesitant and therefore may be the only effective enforcement intervention.

As indicated, the Product Liability Directive and the General Product Safety Directive do have potentially different safety standards. The Product Liability Directive imposes a liability if the product does not provide the safety which a person is entitled to expect, whilst the obligation under the General Product Safety Directive is to place only "safe products" on to the market. A safe product is one which under normal and reasonably foreseeable conditions does not prevent a risk or only the minimum risks compatible with the product's use considered acceptable and consistent with a high level of protection. Both definitions, by their nature, set imprecise standards of safety which create difficulties in determining whether compliance exists in borderline cases.

7.5 THE PRECAUTIONARY PRINCIPLE

The precautionary principle has its roots in environmental law rather than product liability law. It reflects an attempt to set out the approach to be taken to both risk assessment and risk management in situations where the state of scientific knowledge is not sufficiently advanced to allow a proper assessment of the nature or extent of the risk.

The Commission's communication on the precautionary principle (COM (2000)) states that it applies:-

"Where preliminary objective scientific evaluation indicates there are reasonable grounds for concern that the potentially dangerous effect on the environment, human, animal, or plant health, may be inconsistent with the high level of protection chosen for the community".

The difficultly with it is its lack of precision.

The 1992 Rio declaration defined it (in the context of environmental risk) as:

"Where there are threats of serious and irreversible damage, lack of full scientific certainty shall not be used as a reason for postponing cost effective measures to prevent environmental degradation".

Hence, in product regulation context, the principle would apply where there are serious risks to the health of consumers, appropriate action should be taken to protect such consumers without waiting until all necessary scientific knowledge is available.

Where action is deemed necessary, measures based on the precautionary principle should be, *inter alia:*

- proportional to the chosen level of protection;
- non-discriminatory in their application;
- consistent with similar measures already taken;
- based on an examination of the potential benefits and costs of action or lack of action (including, where appropriate and feasible, an economic cost/benefit analysis);
- subject to review, in the light of new scientific data, and
- capable of assigning responsibility for producing the scientific evidence necessary for a more comprehensive risk assessment.

Whilst at first consideration, the proposition appears perfectly reasonable, there are practical consequences if it is used to validate extreme regulatory action taken to restrict the distribution of products in response to ill-informed media attention or following pressure group interest. The danger exists because the precautionary principle mandates direct action in relation to a perceived risk to be taken at a political level even in the absence of scientific evidence suggesting that the risk exists.

Although the Commission's communication lists a number of restraints by way of guidelines to its application, including the need to act proportionally and consistently and in a non-discriminatory fashion, questions of who has the burden of proving that a product is safe once the precautionary principle has been envoked remain outstanding.

The Commission has indicated that the precautionary principle forms a part of its regulatory policy. It is the Commission's response to the perceived loss of consumer confidence in product safety regulation following BSE and similar crisis. It is certainly not referred to in the EC treaty except in the field of environmental protection and its introduction to product regulation is entirely an initiative of the Commission itself.

Article 8 (2) of the revised GPSD states that when the authorities take enforcement action, they shall act

"in such a way as to implement the measures in a manner proportional to the seriousness of the risk, and taking due account of the precautionary principle".

No where in the GPSD is there any attempt to define the principle further. The DTI in the Consultation paper on the Transposing of the revised GPSD are aware of the difficulties this presents. They quote from what they believe to be the best attempt of a definition by the European Policy Centre in April 2001.

"The Precautionary Principle is a principle of risk management under which decision-makers may take provisional, proportionate and cost-effective measures to reduce potential risks to human health or the environment, by applying an appropriate degree of prudence when weighing up alternatives in the light of the results of a scientific risk assessment".

The DTI's concern is that concepts such as "an appropriate degree of prudence" are inherently vague and pose difficulties in transposing the Article. It remains to be seen how they resolve these difficulties.

There is a fear that it can be:

- applied arbitrarily. Since any scientific assessment necessarily involves some degree of uncertainty, the application of the precautionary principle could be justified on almost any occasion. (France's refusal to permit UK beef despite scientific advisers' recommendations).
- The duplicate caution. The precautionary approach would duplicate the prudent approach already adopted by scientists during their assessment process.
- Politicians can second guess scientists. Politicians would be able to make decisions undermining the principle of that scientific assessment should be separate from political risk management.

In its extreme, the precautionary principle would be used politically to justify an action which has at its root the restriction of consumer choice and a barrier to world trade.

CLOSING REMARKS

Of Course it remains to be seen how the application of the precautionary principle in the field of EMF Regulation will impact on product regulation and product liability law.

The Government have already warned local authorities against imposing outright bans on the erection of mobile phone masts which seems to indicate that a degree of reasonableness and proportionality will be exercised but it is certainly not far fetched to envisage that in future lawyers pursuing product liability claims will envoke the precautionary principle as the standard for defining the level of safety that persons are generally entitled to expect rather than the high watermark that the Product Liability Directive case law has claimed via the Hepatitis C Judgment in the English Court. A sympathetic Court can then use the principle to justify a ruling which is favourable to Claimants by giving "hindsight" a respectable, factual and legal framework. It is the final nail in the coffin of the Development Risks Defence.

Questions & Discussion

Alistair Philips (UK Powerwatch, Sutton)

One of the key things Mr Pearl said was with respect to the effects of electromagnetic fields on health causation. What is his opinion on the recent High Court Judgement on the Fairchild Case about asbestos causing mesothelioma. The Judgment stated that not only is proof needed that asbestos fibre exposure causes mesothelioma, but also which company was responsible for that exposure. If the person was exposed to asbestos fibres in more than one place then, no one was responsible for paying compensation. We have a similar situation with regard to microwave radiation exposure from mobile phones. This Judgment seems to make it impossible to prove any cause affects health. What are your comments?

Simon Pearl (Davies Arnold Cooper, London)

It is not easy to answer this question until next week (late May 2002) because, as you probably know, the House of Lords just heard the appeal in the Fairchild Case. What happened in that case is the Court of Appeal, rather to everyone's surprise, decided that if someone had mesothelioma and had been exposed to asbestos dust during the course of his working career and worked for more than one employer it would be impossible to identify one fibre of asbestos dust which actually resulted in him developing his mesothelioma, and therefore on the balance of probability he could not prove which employer or any employer was liable and therefore he would not succeed against anyone. The House of Lords, from what I hear, will reverse that decision[1]. That is my prediction and in the contexts of asbestosis claims I am sure it is going to be a political decision. In the context of what we are dealing with here with cellphones I do not think necessarily that is going to assist enormously because the issues in asbestos are going to be "which fibre of asbestos". I do not think it necessarily applies here. I think what is going to be key is that if the scientific evidence is that its the long term effect that matters then it is no different than the person who smokes different brands of cigarettes or the person who takes different brands of aspirin. You would simply sue each manufacturer and if the evidence supports the position that the damage was caused over a period of time then the manufacturers will have to be liable on a time exposure basis which was the way these asbestos cases always used to be dealt with. I suspect the pragmatic approach will ultimately be the way forward.[1]

---

[1] **Editor's note:** *The House of Lords did reverse the decision.*

Dr Martin Bootman (The Babraham Institute, Cambridge)

Do the electromagnetic fields we experience really matter? By that I mean is there actually a mechanism by which cells can actually demodulate and interpret them? This morning we have only heard about a couple of possible mechanisms, one being the radical pair mechanism which has not been proved in a biological system. As far as I am aware the best evidence for the radical pair mechanism is from the work done by McLaughlin, Hore and Timmel in Oxford, who have shown that very large organic radicals can be influenced by strong static magnetic fields. For very small radicals such as nitric oxide or superoxide, however, this is very unlikely to be the case. A lot of the speakers today have suggested that as the number of cells in organisms increase there are an increasing number of nonlinear mechanisms that may be susceptible to external influences, including electromagnetic fields. One thing they have actually failed to recognise is that there are also a number of dampening systems inside cells and multi-cellular organisms that would actually abrogate a large number of these effects.

Professor Ross Adey (Loma Linda University School of Medicine, USA)

There is credible and growing evidence that exposures to environmental electromagnetic fields are of consequence at mechanistic levels in cells and tissues. I shall briefly summarise one example of experimental evidence linking physiological regulatory actions of these fields on nitric oxide (NO) in brain tissue, which is contrary to Dr. Bootman's position that appears to be dismissive about its possible role. As noted by Bawin et al., (Bioelectromagnetics 17:388-395, 1996), there is growing evidence that exogenous magnetic fields oscillating at extremely low frequencies (1-100 Hz) interact with cognitive processes and influence brain EEG activity in humans and laboratory animals at intensities of 70-1400 μT rms (Bell et al., Brain Res. **570**:307-315, 1992; Lyskov et al., Bioelectromagnetics **14**:87-95, 1993). NO-cGMP dependent mechanisms modulate bursts of EEG rhythmic slow activity (RSA) induced by cholinergic stimulation of isolated rat hippocampal tissue (Bawin et al., NeuroReport 5:1869-1872, 1994). Treatments with inhibitors of NO synthesis shortened and stabilized the interburst intervals. By contrast, applications of NO donors or cGMP analogues during the blockade of NO synthesis once more destabilised and lengthened the interburst intervals. In the same hippocampal tissue preparation, sinusoidal 1 Hz magnetic fields at intensities of 56 and 560 μT, but not at 5.6 μT, triggered irreversible destabilisation of the RSA intervals (Bawin et al., Bioelectromagnetics 17:388-395,1996). Fields had no effects on RSA when NO synthesis was pharmacologically inhibited, but chelation of extracellular NO diffusing from intracellular sites did not interfere with field actions.

These observations are consistent with Dr. Bootman's reference to conclusions drawn by McLaughlin at Oxford that free radical interactions can only involve ongoing chemical reactions, potentially altering their rate and amount of product, but not their essential nature. As options for further research, they are consistent with models proposed by McLaughlin, regarding possible spin-mixing sensitivities to magnetic fields approaching zero, to Lander's (FASEB J. 11:118-124, 1997) redox potential models for free radical transduction with sensitivities independent of lock-and-key mechanisms.

Finally, Dr. Bootman sees "dampening" tissue mechanisms as essentially limiting the role of intercellular communication in setting tissue, as opposed to cellular, sensitivities. My paper cites a series of examples in excitable and nonexcitable cellular tissues where nonlinear models are consistent with these experimental observations, and offer predictive models for future research into stimuli with energies below tissue thermal collision energies of kT.

Professor Denis Henshaw (University of Bristol, UK)

As a comment, I would point out that there are a number of papers describing a reduction in the nocturnal production of melatonin in human populations in the presence of power line magnetic fields down to 0.2 microtesla (μT). I am not a biologist and cannot describe the mechanisms involved, but it appears that there are non-visual cells in the retina that are part of the system communicating with the pineal gland, that may switch off melatonin secretion in the presence of magnetic fields. I was struck by Dr. Adey's comment this morning that the human eye responds to a single blue-green photon with energy of about 2.5 electron volts. The human body does indeed respond to very low levels of electromagnetic fields.

Professor Ross Adey

The detailed biology of interactions of electromagnetic fields with melatonin at cellular receptors is an important area at the cutting edge of molecular biology of cancer research. Liberdy (J. Pineal Res. 14:89-97, 1993) first described inhibition of growth of cultured human breast cancer cells by melatonin in physiological concentrations. This regulatory influence of melatonin was lost in the presence of low-level 60 Hz 1.2 μT magnetic fields. Ishido *et al.* (Carcinogenesis 22:1043-8, 2001) have recently elucidated a key step in the action of magnetic fields on the transductive coupling of the melatonin signal from cell surface receptors to intracellular metabolic enzymes. Using the same mammary carcinoma cell line as in the Liberdy experiments, Ishido et al. noted that only 1-a melatonin receptors appeared to be involved, and that exposures to either 1.2 or 100 μT 50 Hz magnetic fields for 3, 5 and 7 days both inhibited the intracellular accumulation of cAMP in a time-dependent manner. cAMP is produced at cell membranes from the key cellular fuel adenosine triphosphate (ATP), by action of the enzyme adenylyl cyclase. Neither melatonin receptor functions nor adenylyl cyclase activity were altered by field exposures. this provides the first evidence that magnetic fields may cause uncoupling of signal transduction from melatonin receptors to adenylyl cyclase.

Dr Bart Curvers (KBC Bank NV, Brussels)

I am the head of the medical department of our bank and insurance company. Since 1995 we have had 800 employees suffering from lipoatrophia semicircularis. That is semicircular atrophia of the fatty tissue on the front side of the thighs. Could there be an influence of the electromagnetic environment? Is there any influence of electromagnetic fields on fat cells?

Professor Ross Adey

I have no knowledge of the clinical condition that you describe. Fatty tissue has a significantly lower conductivity than most other body tissues. The expected behaviour of fatty tissue would be as an imperfect dielectric, rather than as an ionic conducting medium characteristic of most other body tissues. The typically sparse and scattered pattern of blood vessels in fatty tissue might offer preferred pathways for microwave fields. If the reported condition were associated with environmental microwave fields (clearly an unlikely possibility), histological examination of vascular health in biopsy specimens might be of interest, but for obvious cosmetic regions, unlikely to be acceptable.

Dr Danuta Sotnik (Applied Immunology, Poznan, Poland)

Beta-carotene (known to be a free radicals sweeper) tablets were used to prevent lung cancer amongst smokers in a double blind 5-year trial in 10 European countries, including Russia. After two years researchers found shocking results: in a beta-carotene taking group of smokers the lung cancer was increased by nearly 18%, other cancers by 24%. The trial was abandoned, but only one European government, the Danish, decided to put a warning for smokers on beta-carotene supplements. My question is: how to make the British Government and possibly governments in other countries acknowledge this fact and make the public aware of it?

Simon Pearl

Is there a product which has a product licence and is it been currently marketed?

Dr Danuta Sotnik

There are many beta-carotene tablets available in Britain, free of prescription.

Simon Pearl

The normal way that British Government evaluates pharmaceutical products and medicines is through advice from Committee for Safety of Medicine and then the Medicines Commission via the Medicines Control Agency and the European Bodies.

Professor Olle Johansson (Karolinska Institute, Stockholm)

Professor Marvin mentioned the power density of the sun and of radio transmissions. What is the power density from starlight at night time?

Professor Andrew Marvin (The University of York, UK)

I have absolutely no idea.

Professor Olle Johansson

My personal guess is, that measured in ordinary SAR-levels, it is actually zero watts per kilogram, and still you see the stars! Could you, please, explain that?

Professor Andrew Marvin (The University of York, UK)

If you can still see them, there must be some energy there in the first place. As you know, the interaction between the photon coming from the star and the molecules in the retina give rise to the physiological response.

Professor Olle Johansson

I know about the basic textbook knowledge, but my point is that you can have situations where the body may respond to energy levels that you practically cannot measure because the body may be much more sensitive to electromagnetic fields than we currently realise.

Dr Philip Chadwick (Microwave Consultants Lt, UK)

Remember that eyes are a very dedicated single photon detector system. The dark adapted eye can detect a single photon and it is very specifically tuned to detect single photons of light.

Part III
Emissions & Standards

What are We Exposed to?
The Most Significant Sources of Exposure
in the Built Environment

Philip Chadwick

8.1 INTRODUCTION

Electromagnetic fields in the built environment are truly ubiquitous. Every office contains VDUs, every large building has a substation and many have a mobile phone transmitter on the roof (or, maybe more significantly, the roof of the building opposite). In towns and cities there is a rash of public service and private mobile radio antennas, as well as local TV and radio relays. We are all aware of this and, as can gleaned from other contributions to this books, there is a range of points of view as to how significant it may be.

There are, though, other sources of electromagnetic field of which we are less conscious, but which may give rise to significant exposures. This paper will describe some of the obvious and less obvious sources of exposure in the built environment.

8.2 POWER-FREQUENCY FIELDS

The levels of power-frequency electromagnetic fields in the built environment have been well characterised. We know that magnetic flux densities very close to office equipment and domestic appliances can reach milliTesla levels, but that the fall-off with distance is quite rapid. Most *exposures*, as opposed to *emissions*, are at the microTesla level or less at distances of tens of centimetres from office equipment, and often the office environment does not contain any more significant sources of exposure to power-frequency fields than does the domestic environment (Allen *et al.*, 1994, European Commission 1996)

There are exceptions, though. One significant and often overlooked source of magnetic field is the bulk tape eraser. These are used in all TV and radio stations, police stations, prisons and anywhere else that requires audio, video or CCTV tapes to be erased before reuse. Bulk tape erasers use very large magnetic fields to erase tapes rapidly without having to rerecord: some models allow the tape to be erased by passing it once over the top of the eraser, either on a conveyor or held in the hand of the operator. Recently, manufacturers have become aware of the issues of human exposure to magnetic fields and modern erasers have an interlocked drawer arrangement in which the tape is erased after being inserted inside the device. These modern erasers have effective shielding and operator exposure is low. However, tape erasers are simple and reliable devices and there are many of the older, unshielded or open types still in use.

Tape eraser technology is very simple and many erasers use a single-frequency unmodulated and substantially undistorted 50 Hz magnetic field. But sometimes the waveform of the field can be distorted, usually by the addition of harmonics. Harmonics are additional frequency components, at multiples of the 50 Hz fundamental frequency, that often arise from magnetic saturation effects in the cores upon which electromagnets are wound. Harmonics can be important when considering the interaction of electromagnetic fields with people, when assessing compliance with exposure guidelines but also when considering the possibility of any health effects related to current flow in the body. The reason is that the effectiveness of magnetic fields in producing current flow in the body is dependent on their frequency, so for example the same field strength will give (more-or-less) twice the body current at 100 Hz than at 50 Hz. The harmonic content is quite variable, but Figure 8.1 shows a hypothetical example of a distorted magnetic field waveform which is not atypical of tape erasers. The 50 Hz fundamental has 50% of 100 Hz second harmonic and 25% of 200 Hz fourth harmonic added, in phase. Depending on how the exposure summation is done (a simple sum of all contributions or a sum-of-squares summation) the corresponding current in the body would be between 30% and 130% more than for a single-frequency 50 Hz field of the same strength.

This increase may be significant: magnetic flux densities close to tape erasers can be up to 1 mT, and whole-body exposures of hundreds of microTeslas are possible. These levels are comparable with maximum field strengths specified in human exposure guidelines.

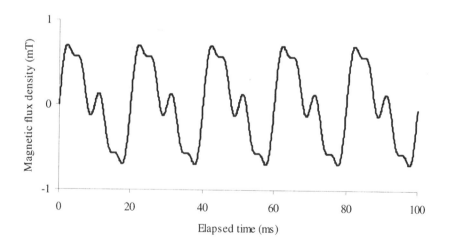

Figure 8.1 Magnetic field waveform typical of a tape eraser

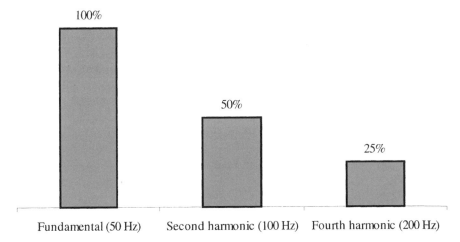

Figure 8.2 Spectrum of Figure 8.1 magnetic field waveform

8.3 VDUs

The emissions from VDUs have been well-characterised over many years, and there is little that can be added of any great originality. The technology of cathode ray tube (CRT) VDUs is summarised in Figure 8.3. The position at which the electron beam hits the inside of the screen is governed by two sets of coils in the neck of the tube. One set deflects the beam horizontally along each line, the other deflects it vertically down the screen. The horizontal scanning frequency is typically several tens of kHz, and the waveform is a sawtooth: the beam is scanned smoothly across the screen and then returned to the start position in around a tenth of the scan time. The beam is scanned vertically up and down the screen in the same way by the vertical deflection coils. The waveform of the magnetic field from these coils is also a sawtooth, with a typical frequency of 50 - 150 Hz. The reason that some VDUs can have higher magnetic field emissions than others is that the high voltage that is used to accelerate electrons toward the screen is derived from the horizontal deflection coil current. The transformer which steps up the voltage prior to rectification is the major field source in a VDU, and the exposure levels depend on the location of this with respect to the operator as well as the extent of any screening.

The magnetic fields from most modern VDUs are specified as being below the levels specified by the Swedish National Board for Measurement and Testing (MPR). The maximum MPR levels, to be measured at a distance of 50 cm from the VDU, are 250 nT for frequencies between 5 Hz and 2 kHz and 25 nT for frequencies between 2 kHz and 400 kHz.

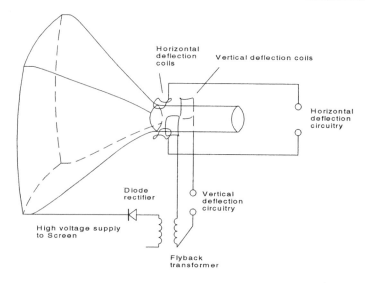

Figure 8.3 Schematic diagram of VDU circuitry. Reproduced courtesy of NRPB (NRPB 1994).

8.4 TRAINS AND TRAMS

The biggest exposures to electromagnetic fields on electrical transport systems come from the on-board traction components and electrical systems rather than the overhead lines or power rails.

There is very little published information about magnetic fields on UK trains and trams; the data reported here come from a 1998 paper (Chadwick and Lowes, 1998) which itself relied substantially on measurements made in the late 1980s and early 1990s. Nevertheless, it is interesting to observe the relationships between magnetic field exposures and the technologies used in electrical transport.

Intercity trains use 25 kV 50 Hz overhead lines to supply power to the engine. The traction components are located in the power cars, and exposures of passengers are substantially from on-train power supplies. Magnetic flux densities of several microTeslas are not untypical.

Local trains and tube trains usually have power supplied by a third "live" rail or by a lower voltage overhead line. The supply in both cases is "pseudo-DC", that is it has been rectified from a multiphase source but not smoothed. As a result there is a substantial AC ripple on the supply. On older trains, the supply would be smoothed and filtered on-board and motor power controlled by resistive current limitation. On newer trains, audio-frequency chopping is used to control power, and on the newest trains variable-frequency AC motors are used. Both systems may give rise to substantial magnetic field exposures because there will be line filters to stop the audiofrequency current propagating back to the power supply. More importantly, the power conditioning and traction components are generally under the floor of passenger cars. This means that the largest exposures can be to passengers rather than railway workers: it is possible to measure alternating magnetic fields of a milliTesla or more at floor level on a tube train.

8.5 ANTITHEFT SYSTEMS

Antitheft systems use magnetic fields to prevent theft of items from shops and libraries. The commonest technology uses pulsed or swept fields with a frequency of 8–13 MHz to detect self-adhesive labels attached to, for example, CDs or small consumer products. The tags are resonant at the operating frequency of the antitheft system and can be detected when the customer exits the store through the walk-through detection zone defined by the doorway arches. Similar technology is used to detect removable tags attached to clothing. The main magnetic field exposure from these devices is probably to customers, although some systems are installed close to check-out desks. The magnetic fields from these systems are quite localised (Figure 8.4, Bassen 2000) and although they can approach guideline levels closer to the transmit coils, the field strengths in the central volume are lower. Coupled with the short occupancy times, this means that exposures are generally well below guideline levels.

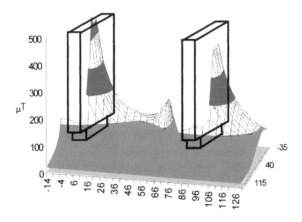

Figure 8.4 Magnetic field from antitheft system

The other common technology is that used to protect library books. The tags used to protect CD and consumer products can be removed or permanently deactivated at the point of sale, but library antitheft systems require that the tag can be repeatedly deactivated and activated when a book is borrowed and returned. Library antitheft systems often use a pulsed audiofrequency (1–2 kHz) field in conjunction with a high magnetic permeability tag whose magnetisation can be altered.

The instantaneous magnetic fields from library systems can be quite large, and exposure guidelines do not allow time-averaging at these frequencies. Figure 8.5 (Chadwick, 1998) shows the magnetic field from one such system; the instantaneous peak magnetic flux density in the pulse approaches 15 μT. It should be stressed that this is a maximum close to the transmit coil and true exposure of people will be lower than this.

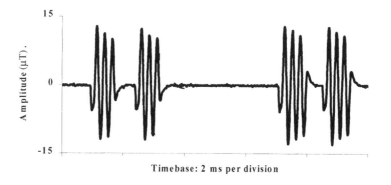

Figure 8.5 Library antitheft system pulsed magnetic field waveform

8.6 TELECOMMUNICATION SYSTEMS

Telecommunications systems are discussed in this paper because they are one of the more obvious sources of exposure to radiofrequency electromagnetic fields in the built environment. Very few office blocks or tower blocks have escaped the placement of a mobile phone base station on either on their roof or the roof of an adjacent building.

Mobile phone base station antennas are designed to direct radiofrequency signals away from the installation and generally the main beams of the antennas do not intercept the roof of the building upon which they are mounted. There may be side-lobes of the beam, which do intercept the roof, but the power of these is greatly reduced from that of the main beam. Roofing materials generally attenuate mobile phone signals quite strongly, and the results of measurements of field strength in offices directly below antenna installations indicate usually that exposure levels are not substantially higher than in buildings with no roof-top mobile phone installations. Indeed, the primary source of radiofrequency fields is often a base station on a nearby building. These exposure levels are invariably far below exposure guidelines, but it may be possible to measure electric field strengths approaching 1 V m^{-1} in some situations.

If access to the main beam of a base station antenna is possible, within a few metres of the installation, then it might be possible for exposures to approach guideline levels. This is likely to be possible only on rooftops where it is possible to approach the front of the antennas and where they are not mounted sufficiently far above the roof for the beam to be well above head height. In this situation, areas where exposure guidelines might be exceeded should be identified by mobile phone operators and access to these areas should be controlled carefully.

8.7 CONCLUSIONS

This paper has given an overview of the most significant sources of exposure to electromagnetic fields in the built environment. Of course, there are working environments which contain stronger sources of exposure than many of those listed here, but there are comparatively few people exposed in those environments. It is interesting to note that the most significant common sources of exposure to magnetic fields (tape erasers, trains, antitheft systems) are not those, which might immediately spring to mind. We often think of power station or railway workers as having the highest occupational exposures to magnetic fields; perhaps we should look a little harder at librarians who travel to work by train.

8.8 REFERENCES

Allen, S.G., Blackwell, R.P, Chadwick, P., Driscoll, C.M.H., Pearson, A.J., Unsworth, C. and Whillock, M.J., 1994, Review of occupational exposure to optical radiation and electric and magnetic fields with regard to the proposed CEC Physical Agents Directive. NRPB-R265. (Chilton: NRPB)

Bassen, H., 2000, Proceedings of WHO/ICNIRP meeting on antitheft systems, Helsinki, September 2000 (in press).

Chadwick, P., 1998, Occupational exposure to EMF: practical application of NRPB guidelines. NRPB-R3401. (Chilton: NRPB)

Chadwick, P. and Lowes, F., 1998, Magnetic fields on British trains. *Annals of Occupational Hygiene,* **42,** pp. 331-335.

European Commission D-GV 1996. Non-ionizing radiation: Sources, exposure and health effects. (Brussels: EC)

NRPB, 1994. Health effects related to the use of VDUs. Doc. NRPB 5,2. (Chilton: NRPB)

CHAPTER 9

Biological Effects of Electromagnetic Fields Leading to Standards

Zenon Sienkiewicz

9.1 INTRODUCTION

The biological effects of electromagnetic fields have been studied in laboratories for more than thirty years using a wide variety of exposure conditions, models and endpoints. At the same time, epidemiological studies have been carried out in homes and workplaces. This research has produced much valuable and useful information in helping to formulate public health policy although a number of uncertainties and possibilities remain to be resolved.

Numerous reviews, summaries and critiques of the biological effects literature are available. Older reviews include NRCP (1986), NRPB (1992), ORAU (1992), Saunders *et al.* (1991), Sienkiewicz *et al.* (1991; 1993), and WHO (1984, 1987; 1993). More recent reviews include Direction Général de la Santé, (2001), IARC (2001), IEGMP (2000), McKinlay *et al.* (1996; updated in 1999), NIEHS (1998), NRC (1997), NRPB (2001) and Royal Society of Canada (1999).

It is clear from an examination of this literature that many reported effects result from either the induction of surface electric charges, induced electric currents, or from increases in temperature (depending on the frequency of the field used). These particular effects and responses are used by national and international organisations, such the National Radiological Protection Board (NRPB) and the International Commission on Non-Ionizing Radiation Protection (ICNIRP) as scientific bases in setting guidelines and restrictions on exposure to electromagnetic fields for human populations (see NRPB, 1993, 1999; ICNIRP, 1998). Other responses have been reported, but they are far less clearly defined and any relevance to health may be unclear. At present none of these possibilities are considered sufficiently coherent to be used as a basis for setting standards.

The effects used as the biological bases for guidance in the UK are discussed here in two sections. The first section describes the effects of exposure to time-varying electric and magnetic fields with frequencies of less than 100 kHz, whereas the second section considers the effects of exposure to radiofrequency fields at frequencies between 100 kHz and 300 GHz.

In order to restrict the number of references necessary in this overview, please refer to NRPB (1993, 1999) for details of the citation to particular experimental studies.

9.2 LOW FREQUENCY ELECTRIC AND MAGNETIC FIELDS

For people, as with other living organisms, the major consequence of being exposed to low frequency fields of less than 100 kHz is the induction of electric fields and currents within their bodies. This principally occurs in either of two ways depending of the nature of the field.

Electric fields do not penetrate the body significantly, but they do build up an electrical charge on its surface. The polarity of this charge alternates at the frequency of the field, and as a result, electric currents flow through the body to and from the ground. The magnitude of these currents within the body is highly complex and depends on the electrical properties of the tissues involved.

In contrast, magnetic fields easily penetrate the body and cause circulating electric currents to flow within it. These currents do not necessarily flow to earth. The currents tend to be largest in peripheral tissues, and decrease in magnitude more towards the centre of the body.

Other interaction mechanisms between electric and magnetic fields with living tissues have been suggested, and although some enjoy limited experimental or theoretical support, none has yet gained universal acceptance.

9.2.1 Surface Electric Charge

Exposure to electric fields can result in perception effects due to the alternating electrical charge induced on the surface of the body. The perception results from the surface charge causing the hairs on the various parts of the body to vibrate. This effect may become stressful if exposure is prolonged. About ten percent of adults can perceive power frequency electric fields of 10 to 15 kV m^{-1}; the threshold for annoyance is around 15 to 20 kV m^{-1} (Figure 9.1). The factors that may affect sensitivity have not been well defined and it is possible that some individuals may show increased responsiveness.

Perception of an electric field by cutaneous stimulation is not considered hazardous. Nevertheless, it may affect ongoing activity and modify behaviour. In laboratory tests with rats and baboons, exposure to 60 Hz electric fields at up to 30 kV m^{-1} caused transitory changes in performance of learned tasks, and in arousal and social interactions. These effects were attributed to distraction and annoyance caused by field-induced vibration of the body hair.

9.2.2 Induced Current Effects

It is well established that electric currents applied directly to the body can stimulate peripheral nerves and muscles. These effects could prove fatal if breathing was inhibited or ventricular fibrillation was induced in the heart. Conceptually similar effects are considered possible with exposure to low frequency fields.

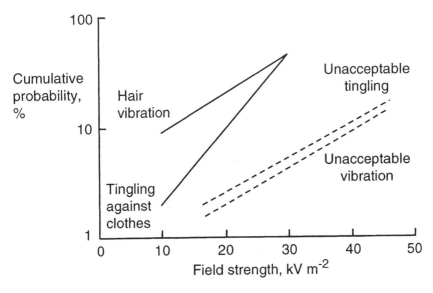

Figure 9.1 Thresholds for cutaneous perception effects in people exposed to electric field

Threshold current densities for direct nerve or muscle stimulation are about $1–10$ A m^{-2} at frequencies between 10-1000 Hz. The threshold rises progressively at frequencies above and below this range. Stimulation of peripheral nerves and muscles has been observed in volunteers exposed to rapidly alternating (1 kHz) gradient magnetic fields in experimental magnetic resonance imaging systems. In addition, the use of short, rapidly changing magnetic field pulses (generated using small search coils held close to the head) is an effective, non-invasive and painless method to stimulate nerves and muscles. Such techniques are used in clinical and experimental investigations.

Current densities at levels insufficient to stimulate nerve and muscles may affect ongoing electrical activity and influence neuronal excitability. The activity of nerve cells in various regions of the brain including the cerebral cortex are known to be sensitive to the endogenous electrical fields generated by the action of adjacent nerve cells at levels well below those required for direct stimulation. Neurophysiological studies using isolated nervous tissue preparations suggest that the threshold for these effects may be as low as 100 mA m^{-2}.

However, lower thresholds can be derived from experiments investigating the effects of weak fields on volunteers. This perhaps reflects the increased sensitivity of the intact organism compared to responses from isolated tissues. Faint, flickering, visual sensations, called magnetophosphenes, can be readily induced during exposure to magnetic fields above about 5 mT. This threshold depends on frequency, and at 50 Hz is about 10 mT (Figure 9.2). Magnetophosphenes can also be induced by the direct application of weak electric currents to the head with current densities of 10–20 mAm^{-2} and above.

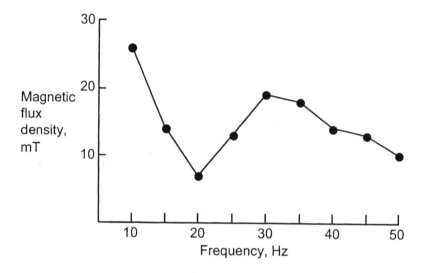

Figure 9.2 Thresholds for the induction of magnetophosphenes

One laboratory study investigated the effect of weak power frequency electric currents on human mental processes. Volunteers had electrodes attached to their head and shoulders through which weak electric currents could be passed and they were required to perform a number of reasoning tasks. During exposure, the current densities in the brain were estimated as around 10–40 mA m^{-2}. Compared to sham-exposure, the performance of most tasks was unaffected, but some changes were seen in alertness and in the performance of a complex reasoning task.

Overall these results suggest a conservative estimate of 10 mA m^{-2} for the current density threshold for the modulation of neuronal activity in the brain and central nervous system. (The possibility that some individuals may show increased responsiveness to induced electric currents is discussed in Chapter Twenty Six, this volume).

9.3 RADIO FREQUENCY FIELDS

The heating effects of radiofrequency (RF) and microwave fields with frequencies above 100 kHz are well established. It is important to recognise that the total whole body heat load experienced during exposure to RF fields is the sum of the RF energy absorbed (measured as the specific energy absorption rate or SAR in W kg^{-1}) and the endogenous rate of heat production (also measured in W kg^{-1}).

Endogenous heat production varies in normal individuals from about 1 W kg^{-1} at rest, to about 10 W kg^{-1} for short periods during hard physical exercise.

Power deposition within the body is never uniform, however. Differences in the electrical properties of tissues and the reflection and refraction of the fields at the interfaces of tissues of different electrical properties can result in localised regions of enhanced energy deposition. In addition, differences in local blood perfusion will affect heat dissipation characteristics, and some tissues are more sensitive to raised temperatures than others, or may be less able to effect repair to tissue damage.

9.3.1 Whole-body Responses

Most data on whole-body responses to RF fields are derived from animal studies. Animals, like humans, use various physiological and behavioural mechanisms in order to regulate body temperature. The responses of the thermoregulatory system to exposure to RF fields are well known and include altered rates of metabolic heat production, and the vasodilation of superficial blood vessels as well as changes in food intake, reductions in activity, and the behavioural selection of appropriate ambient temperatures. Thresholds for such responses have been reported in rodents and primates between about 0.3 W kg^{-1} and 5 W kg^{-1}.

The magnitudes of these responses appear to depend on many parameters, including the frequency of the fields used, and the orientation of the exposed animal with respect to the applied electric and magnetic field. These are caused by differences in power absorption by the deeper tissues of the body; and in particular, the less efficient stimulation of the temperature receptors in the skin results in less effective thermoregulation.

The performance of learned tasks in animals seems particularly sensitive to RF fields. These responses are robust and have been used as a biological basis for limiting exposure in some guidelines. Thresholds for decreased performance in both rats and primates have been reported as lying between 2.5 and 8 W kg^{-1}; concomitant rises in core body temperature were around 1 °C. Exposure to RF fields can also modify the action of drugs whose effectiveness can be altered by heat-induced changes in body physiology. Changes have been recorded in the duration of barbiturate-induced anaesthesia and in the permeability of the blood-brain barrier, but only at levels of exposure sufficient to raise body temperature.

Other easily demonstrable effects are generally consistent with responses to non-specific stressors such as heat. The acute exposure of primates to RF fields at SARs of 3 to 4 W kg^{-1}, sufficient to raise rectal temperature by 1 to 2 °C, resulted in increased stress hormone (plasma cortisol) levels; similar effects have been reported in rats. In addition, RF-induced changes have been reported in the levels of circulating white blood cells (increased levels of neutrophils and decreased lymphocyte levels) in rats and mice following thermal exposures which are similar to the changes induced by the injection of stress hormones, suggesting a common aetiology. Reported changes in white blood cell (natural killer cell and macrophage) activity have also been linked to heat-induced stress.

Similar changes would be expected in humans following RF-induced increases in body temperature. It is known that most healthy people can tolerate short-term rises in body temperature by up to about 1°C, although individuals vary widely in their ability to tolerate increased body temperatures. Some individuals cannot tolerate rectal temperatures of 38°C while others continue to perform well even at higher temperatures. However, prolonged exposure at body temperatures in excess of 38°C is known to increase the risk of heat exhaustion and reduce mental performance.

A few experiments have been carried out with volunteers investigating the relationship between whole body SAR, body temperature rise and the ensuing physiological responses. One study reported that following exposure to RF magnetic fields at a whole body SAR of 3 W kg^{-1} for 20 minutes, body temperature was increased by up to 0.7°C without having stabilised and the resting heart rate was elevated by up to 45%. The total heat load experienced by the volunteers, resulting from the sum of the SAR and metabolic heat production, can be estimated as about 5 W kg^{-1}, which represents a typical heavy workload for many industrial jobs.

Adverse environmental conditions and moderate physical exercise will reduce the tolerable level of RF or microwave energy absorption, whilst people under medication or with clinical conditions which compromise thermoregulation may be more sensitive to the effects of any RF-induced heating.

9.3.2 Localised Responses

In addition to any effects of whole-body heating, more localised heating of tissues can occur. Some tissues are considered to be more sensitive to any potential heating effects.

The lens of the eye is one tissue that is regarded as sensitive to RF fields. This is because of its lack of a blood supply and consequent limited cooling ability, and its tendency to accumulate damage and cellular debris. In anaesthetized rabbits, high local temperatures induced by acute exposure of the head to RF fields between about 1 and 10 GHz have been shown to induce cataracts. The SAR in the lens of the eye had to exceed a threshold of about 100-140 W kg^{-1}. Primate eyes were found to be less susceptible to cataract induction than rabbit eyes, possibly because of anatomical differences in their skulls, but also because primate eyes have thinner lenses which can dissipate heat more effectively. However, thresholds for chronic exposure have not been defined. In humans, cataracts have been historically associated with long-term occupational exposure to infrared radiation, indicating that some degree of caution on chronic RF effects should be exercised.

Testicular temperatures are normally several degrees below body temperature, and it has been known for some time that male germ cells are sensitive to elevated testicular temperatures. Long-term exposure to RF fields at about 6 W kg^{-1} may cause transient infertility in male rats. In one study, body temperature rose by about 1.5°C during exposure. In another study, testicular temperatures had to rise by 3.5°C or more to cause short-term infertility. In humans, it has been reported that repeated heating of the testis by 3–5°C will result in a decreased sperm count persisting for several weeks.

Heat has been shown to be teratogenic in various animal species including primates. Increased body temperature has been associated with central nervous system and facial defects in children whose mothers developed moderate to severe hyperthermia, especially during the first trimester of pregnancy. The embryo and fetus may be particularly sensitive to RF-induced heating, since heat loss across the placenta appears less effective than heat exchange with the environment. Acute exposure of rats at 11 W kg^{-1} was sufficient to induce embryo and fetal death and developmental abnormalities. Maternal body temperatures rose to 43°C in this study. Chronic exposure at 6 to 7 W kg^{-1}, usually raising maternal temperatures to between 39 and 41°C, was reported to induce growth retardation and subtle behavioural changes. In general, even prolonged exposures at less than 4 W kg^{-1} which do not raise maternal body temperature notably produce no significant teratological or developmental effects.

9.3.3 Surface Heating

The absorption of RF energy can be detected by temperature sensitive receptors in the skin. Power flux densities of around 300 W m^{-2} at 3 GHz have been detected experimentally during exposure for 10 s. Radiation of higher frequencies applied for similar lengths of time have been detected at lower power flux densities because of their greater absorption by the skin.

It is considered that the avoidance of the perception of skin warming for frequencies where the penetration depth is greater than the thickness of the skin (that is, frequencies below about 10 GHz) does not provide a reliable mechanism of protection against potentially harmful exposure. The avoidance of the perception of skin warming may give adequate protection at frequencies above 10 GHz, although its general effectiveness may be reduced if other factors in the environment compete for attention.

9.3.4 Pulsed Radiation

People with normal hearing may perceive pulse-modulated RF fields of frequencies between about 200 MHz and 6.5 GHz. They report they hear buzzing, clicking, hissing or popping noises. The nature of the effects are altered by the modulation characteristics of the field. Such auditory responses are not considered hazardous, although prolonged or repeated exposure might prove stressful.

It seems most likely that the effect results from the thermoelastic expansion of the brain tissue following a small but rapid increase in temperature on the absorption of the incident energy. This sets up a pressure wave within the brain which is interpreted as a noise by the auditory system. The perception threshold for pulses shorter than 30 µs depends on the energy density per pulse and has been estimated as about 400 mJ m^{-2} at 2.45 GHz, corresponding to an estimated specific energy absorption (SA) in the head of about 16 mJ kg^{-1}. However, a reduction in ambient noise has been reported to reduce this to about 280 mJ m^{-2}, an SA of 10 mJ kg^{-1} at 2.45 GHz.

9.4 POSSIBLE EFFECTS OF EXPOSURE

In addition to the effects described above, a large number of other biological effects have sometimes been reported. Most of these can be broadly grouped into effects on the brain and behaviour, on reproduction and development, and on cancer and carcinogenic processes, although other endpoints have been explored.

While interesting and thought-provoking, none of these possibilities at present provide a sufficiently coherent framework on which to base restrictions for human exposure. For example, many of the results using very low levels of exposure appear to defy conventional explanation, as the induced electric fields will be so minute that they will be completely masked by thermally generated electrical noise. Other studies have postulated the existence of specific frequency and amplitude "windows" for various effects, suggesting non-linear interactions. Also the results of some studies are questioned by many scientists due to their lack of successful replication by independent laboratories.

9.5 SUMMARY AND CONCLUSIONS

It is essential to base restrictions on human exposure to electromagnetic fields on well established, well defined effects, that reflect the totality of the published scientific evidence.

In addition these effects should reflect the consensus opinion of scientists working in this field, and also be acceptable to the majority of scientists and others working in related disciplines. The biological bases used by NRPB since 1993 are summarised in Table 9.1.

Exposure to electric and magnetic fields at frequencies below 100 kHz will result in the induction of electric fields and currents in living tissues. These may result some impaired or altered operation of the normal functions of the brain and nervous system. Such adverse effects can be avoided by restricting the current density induced in the head, neck and trunk.

Above about 100 kHz, the effects of acute exposure to RF fields result from induced heating. Adverse effects such as heat exhaustion and reduced mental performance resulting from whole body heating can be avoided by restricting the whole body SAR in order to limit any rise in temperature. In addition, adverse effects on heat-sensitive tissues can be avoided by restrictions on the localised SAR in these tissues.

Above 10 GHz, since absorption is largely confined to the skin and cornea, recommendations pertaining to long wavelength infrared radiation are applicable.

In addition, it is recommended that conditions under which the auditory effect can be invoked by exposure to pulsed RF fields should be avoided, as should the annoying effects of power frequency electric fields caused by the direct perception of surface charge. In many cases, these effects may be avoided by engineering or administrative controls.

Table 9.1 Biological bases for restrictions on human exposure[1]

Frequency	Dosimetric Quality	Adverse Effect	Restriction
Less than 100 kHz	Surface electric charge (50/60 Hz)	Stress of direct perception	Avoid[2]
	Induced current density (10 Hz – 1 kHz)	Visual or cognitive disturbance	10 mA m^{-2} head, neck, trunk
100 kHz – 10 GHz[3]	Specific energy Absorption Rate (SAR)	Body temperature $> 38°C$	0.4 W kg^{-1} whole body average
		Head and fetus temperature $> 38°C$	10 W kg^{-1} average over 10 g
		Trunk temperature $> 39°C$	10 W kg^{-1} average over 100 g
		Limb temperature $> 40°C$	20 W kg^{-1} average over 100 g
10 GHz – 300 GHz	Power density incident on body	Excess surface heating of the skin and cornea	100 W m^{-2}
Pulsed radiation 100 MHz – 10 GHz	Specific energy per pulse	Stress of perception	Avoid[2]

1   A full description of the basic restrictions is given in NRPB (1993).
2   These effects can also be avoided by administration or engineering controls.
3   In May 2000 the Board of NRPB accepted that the ICNIRP Guidelines for restricting exposures of the public should be adopted for mobile phone frequencies. This equates to a whole-body SAR of 0.08 W kg^{-1} and a local SAR in the head of 2 W kg^{-1}.

Finally, although the NRPB guidelines have remained essentially unchanged since 1993, NRPB is mindful of advances in the study of electromagnetic fields. New publications on biological and health effects are carefully examined and appraised, as are advances in dosimetry and field characterisation. The NRPB Advisory Group on Non-Ionising Radiation, under Sir Richard Doll, provides expert input in this respect. Similarly, guidelines and limits on exposure published by other organisations and authorities are scrutinised and considered. Hence the necessity for revision of the present guidelines is kept under review.

9.6 REFERENCES

Direction Général de la Santé, 2001, Les téléphones mobiles leurs stations de base et la Santé, Rapport du Directeur Général de la Santé, Denis Zmirou, président du groupe d'experts, www.sante.gouv.fr.

IARC, 2001, Non-ionising Radiation, part I: Static and Extremely Low Frequency Electric and Magnetic Fields, (19–26 June 2001), IARC Monographs on the Evaluation of Carcinogenic Risks to Humans Volume 80 (in preparation)

ICNIRP, 1998, Guidelines for limiting exposure to time-varying electric, magnetic and electromagnetic fields (up to 300 GHz). *Health Physics*, **74**, pp. 494–522.

IEGMP, 2000, *Mobile Phones and Health*, Independent Expert Group on Mobile Phones, Chairman Sir William Stewart, ISBN 0-85951-450-1.

McKinlay, A.F., *et al.*, 1996, *Possible Health Effects Related to the Use of Radiotelephones*, Proposal for a research programme by a European Commission Expert (Brussels, European Commission). With update, 1999.

NIEHS, 1999, Health *Effects from Exposure to Power-line Frequency Electric and Magnetic Fields*. NIH Publication No. 99-4493, www.niehs.nih.gov.

NRC, 1997, *Possible Health Effects of Exposure to Residential Electric and Magnetic Fields*. (Washington DC: National Academy Press).

NRCP, 1986, *Biological Effects and Exposure Criteria for Radiofrequency Electromagnetic Fields*, NCRP Report No 86, (Bethesda, MD: National Council on Radiation Protection and Measurements).

NRPB, 1992, Electromagnetic fields and the risk of cancer, A report of an Advisory group on Non-ionising Radiation, *Documents of the NRPB*, **3**(1), pp. 1-138.

NRPB, 1993, Board Statement on Restrictions on Human Exposure to Static and Time Varying Electromagnetic fields and Radiation, *Documents of the NRPB*, **4**(5), pp. 7-68.

NRPB, 1999, 1998 ICNIRP guidelines for limiting exposure to time-varying electric, magnetic and electromagnetic fields (up to 300 GHz): Advice on aspects of implementation in the UK, Documents *of the NRPB*, **10**(2), pp. 5–59.

NRPB, 2001, ELF electromagnetic fields and the risk of cancer, Report of an Advisory Group on Non-ionising Radiation, *Documents of the NRPB*, **12**(1), pp. 5-179.

ORAU, 1992, *Health Effects of Low-Frequency Electric and Magnetic Fields*, Prepared by an Oak Ridge Associated Universities Panel for the Committee on Interagency Radiation Research and Policy Coordination, ORAU 92/F8.

Royal Society of Canada, 1999, *A Review of the Potential Health Risks of Radiofrequency Fields from Wireless Telecommunication Devices*, An Expert Panel Report prepared at the request of the Royal Society of Canada for Health Canada , RSC.RPR 99-1, (Ottawa, Ontario: Royal Society of Canada).

Saunders, R.D., Kowalczuk, C.I. and Sienkiewicz, Z.J., 1991, The biological effects of exposure to non-ionising electromagnetic fields and radiations: III. Radiofrequency and microwave radiation. NRPB-R240, (London: HMSO).

Sienkiewicz, Z.J., Saunders, R.D. and Kowalczuk, C.I., 1991, The biological effects of exposure to non-ionising electromagnetic fields and radiations: II. Extremely low frequency electric and magnetic fields. NRPB-R239, (London: HMSO).

Sienkiewicz, Z.J., *et al.*, 1993, Biological effects of electromagnetic fields and radiation, In *The Review of Radio Science 1990–1992* edited by Stone, W.R. and Hyde, G., (New York: Oxford University Press), pp 737-770.

WHO, 1984, *Extremely Low Frequency (ELF) Fields*, Environmental Health Criteria 35, (Geneva: World Health Organisation).

WHO, 1987, *Magnetic Fields,* Environmental Health Criteria 69, (Geneva: World Health Organisation).

WHO, 1993, *Electromagnetic Fields (300 Hz – 300 GHz)*, Environmental Health Criteria 137, (Geneva: World Health Organisation).

CHAPTER 10

Biological Effects of Low Frequency Electromagnetic Fields[1]

Magda Havas

10.1 INTRODUCTION

The biological effects of low frequency electric and magnetic fields[2] (EMF) have become a topic of considerable scientific scrutiny during the past two decades. The flurry of research in this area has contributed greatly to our understanding of the complex electromagnetic environment to which we are exposed but it has not abated the controversy associated with the harmful effects of electromagnetic fields. If anything it has polarized scientists into two camps, those who think exposure to low frequency electromagnetic fields causes health effects and those who do not. Those who believe there is a causal association are trying to find the mechanism responsible and those who question the concept of causality think this research is a waste of time and money.

Controversy is the norm when complex environmental issues with substantial economic and health consequences are scientifically scrutinized. Asbestos, lead, acid rain, tobacco smoke, DDT, PCBs (and more recently oestrogen mimics) were all contentious issues and were debated for decades in scientific publications and in the popular press before their health effects and the mechanisms responsible were understood. In some cases the debate was scientifically legitimate, while in others interested parties deliberately confuse the issue to delay legislation (Havas *et al* 1984).

The public, uncomfortable with scientific controversy and unable to determine the legitimacy of a scientific debate, wants a clear answer to the question, "Are low frequency electric and magnetic fields harmful?"

As a direct response to public concern three major reports, with multiple contributors with diverse expertise, have been published recently on the health effects of low frequency electric and magnetic fields: one by the U.S. National Research Council (1997), another by the National Institute of Environmental Health Sciences (Portier and Wolfe, 1998), and the most recent, still in draft form,

---

[1] Note: An expanded version of this paper can be found in Environmental Reviews **8**: 173-253 (2000). Research published since 2000 has been incorporated in the present paper.

[2] Magnetic field and magnetic flux density are used interchangeably in this document although the correct term is the later.

by the California EMF Program (2001). These influential reports attempt to make sense of the many, and sometimes contradictory, documents from different fields of study, related to the health effects of power-line frequency fields.

The purpose of the present paper is three-fold:

(1) To characterize human exposure to low frequency electromagnetic fields;

(2) To identify key biological markers and possible mechanisms linked to EMF exposure;

(3) To comment on the concept of scientific consistency and bias.

The question "Are low frequency electric and magnetic fields harmful?" is valid and timely. The answer is likely to have far reaching consequences, considering our growing dependence on electric power, computer technology, and wireless communication, and it is likely to be of interest to a large population using, manufacturing, selling, and regulating this technology.

10.2 BACKGROUND INFORMATION

In the broadest sense, electromagnetic research involves three major sources of electromagnetic energy: those generated by the earth, sun and the rest of the cosmos (geofields); those generated by living organisms (biofields); and those generated by technology (technofields).

These fields interact and it is these interactions that most interest us. Solar flares sufficiently powerful to knock out satellites or to disrupt power distribution and the reflection of radio signals by the ionosphere are examples of geofield and technofield interactions. Photosynthesis, tanning, weather sensitivity are examples of geofield and biofield interactions.

The areas of current scientific debate are the interactions between living organisms (biofields) and technologically generated fields (technofields) at power line frequencies (60 Hz in North America and 50 Hz elsewhere) and at frequencies generated by computers and cell phones in the kilo (10^3), Mega (10^6) and Giga (10^9) Hertz range.

Until recently, frequencies below the microwave band were assumed to be "biologically safe". This began to change in the 1960s and early 1970s. Several months after the first 500 kV substations became operable in the Soviet Union high voltage switchyard workers began to complain of general ill health (Korobkova *et al.* 1971). The electric field, with maximum intensities between 15 and 25 kV/m, was assumed to be responsible for the health complaints. Personnel working with 500 and 750 kV lines were compared with workers at 110 and 220 kV substations. Biological effects were reported above 5 kV/m. The harmful effects of high voltage power lines on substation workers and their families have since been document elsewhere (Nordstrom *et al.* 1983, Nordenson *et al.* 1994).

Nancy Wertheimer and Ed Leeper were the first to report the potential harmful effects of power lines associated with residential rather than occupational exposure. An increased incidence of childhood leukaemia, lymphoma, and nervous system tumours was associated with residential exposure to power line-frequency fields in Denver Color ado (Wertheimer and Leeper 1979). Paul Brodeur did much to publicized this type of information in *The New York Times* and elsewhere (Brodeur 1993), alerting the public and enraging members of the scientific community who were unwilling to accept the Wertheimer and Leeper results.

The Wertheimer and Leeper study was repeated in various locations and by the early 1990s, more than a dozen studies were published on childhood cancer. While some studies found no effects others confirmed the Wertheimer and Leeper results.

Studies of childhood cancers were followed by studies of adult cancers in occupational as well as residential settings and by effects of electromagnetic fields on reproduction. Residential exposure was associated with miscarriages (Wertheimer and Leeper 1986, 1989) while occupational exposure was linked to various reproductive problems as well as adult cancers including primary brain tumours, leukaemia, and breast cancer. Similarities between childhood and adult cancers raised concern.

One problem with the early epidemiological studies was that information on exposure was scarce. Wire codes provided a surrogate metric for the magnetic field. In residential settings the magnetic field, which penetrates through walls, was assumed to be more important biologically than the electric field, which does not. Once portable gauss meters sensitive to power line frequencies became available, the spot measurement and 24-hour monitoring supplemented the wire codes. Of these three methods, the wire codes are highly associated (as measured by odds ratios or relative risk) and the spot measurements are poorly associated with magnetic field exposure and health effects in epidemiological studies (London *et al.* 1991, Feychting and Ahlbom 1993, Savitz *et al* 1988), although a reassessment of earlier studies points to a stronger association between wire codes and magnetic field measurements (Savitz and Poole 2001).

The odds ratio (OR) and relative risk (RR) are two metrics epidemiologists use to compare a test population (observed) with a control population (expected) for a specific endpoint (cancer for example). The higher the OR (ratio of observed to expected), the greater the association between an agent and an end point.

In the past decade appliances, rooms, and houses have been monitored and we have a much better understanding of the magnetic flux density to which we are exposed. Whether magnetic flux density is the only biologically important metric, or, indeed, the one we should be measuring remains to be determined.

Epidemiological studies were complemented by *in vivo* and *in vitro* studies attempting to explore the mechanisms underlying the EMF effect. Because of the novelty of this type of research there were (and still are) no standardized protocols for testing. Experimental intensities for magnetic flux density range from less than 0.1 µT to greater than 300,000 µT (300 mT); daily exposure varies from 30 minutes to 24 hours; and duration of exposure extends from days to years. Some of the tests involve continuous, homogeneous fields, others involved gradients, and still others used intermittent fields with on:off cycles ranging from seconds to hours. Interpreting such a wide array of exposure conditions is not an easy task and thus conflicting conclusions are to be expected depending on the scientific weight placed on individual studies.

10.3 EXPOSURE

10.3.1 Residential Exposure

In a residential setting there are three major sources of technologically generated magnetic fields: appliances, the indoor distribution system consisting of indoor wiring and grounding, and the outdoor distribution system consisting of either below or above ground wires and transformers. The early studies assumed that power lines provided the major source of magnetic field inside the home and both indoor wiring and appliances were ignored, although some studies attempted to minimize indoor sources by turning appliances and lights off. More recent studies recognize the importance of these additional sources and enable us to calculate cumulative and time-weighted average (TWA) magnetic flux densities for a given environment.

10.3.1.1 Outdoor Distribution System

Wire codes, used to estimate exposure to magnetic fields (based on distance and wire configuration) may provide a good relative surrogate for the magnetic flux density within a community; however, they become less reliable when different communities are compared. The magnetic flux density associated with outdoor wiring in a residential setting can range from less than 0.03 to greater than 8 μT, although the values are generally below 1 μT for most homes (Havas 2000, Table 7).

The electric field was not considered to be important in the residential epidemiological studies because they cannot penetrate building material. Electric fields immediately beneath overhead neighbourhood distribution lines are likely to be less than 30 V/m (Havas 2002, in press). However, there is a trend among electric utilities to increase the voltage of power distribution lines to minimize resistance and thus energy loss. As voltage increases so does the intensity of the electric field, and studies report that the harmful effects associated with magnetic field exposure may be worse in the presence of a strong electric field (Miller *et al.* 1996, see paper by Henshaw in these proceedings on particulate density near power lines).

10.3.1.2 Indoor Distribution System

Indoor wiring is another important source of magnetic fields in the home. Within a properly wired building far from a power line normal fields should not exceed 0.03 μT (Riley 1995). In a building with faulty wiring or with older knob-and-tube wiring, fields may be 0.2 to 3 µT, and even higher near walls, ceilings, and floors (Bennett 1994, Riley 1995).

EPRI (1993, as cited in NIEHS 1998) conducted a survey of 1000 homes and took both 24-h and spot measurements in different rooms. The median magnetic flux densities for 24-h measurements vary more than 10 fold with 50% of the homes exceeding 0.05 µT (and 1% of the homes exceeding 0.55 µT). The highest wire code category (VH, very high current configuration) in the Wertheimer and Leeper (1982) study was 0.25 µT and according to the EPRI study, 5% of the homes exceeded this value.

The spot measurements for magnetic flux density in the EPRI (1993) study differed in rooms and some were sufficiently high to suggest faulty wiring. Rooms with the highest average spot measurements ranged from 0.11 µT (50th percentile, 50% of homes exceeded this value) to 1.22 µT (99th percentile, 1% of homes exceeded this value).

Improperly installed indoor wiring can account for very high fields. In a survey of 150 buildings, Riley (1995) reported that the majority (66%) of the high fields above 3 mG (0.3 µT) were due to wiring and grounding problems, 18% were due to the proximity to power lines, and 3% were due to appliances. Of the wiring problems, 12% were due to knob-and-tube wiring used in older buildings, 22% were due to improper grounding to the plumbing system, and 65% were due to wiring violations. Knob-and-tube is a system of wiring used until the 1940s. The hot and neutral conductors are separated by several inches to several feet. The greater the separation the higher the magnetic field that is produced and the less it decreases with distance ($1*r^{-1}$ for a single line conductor rather than $1*r^{-2}$ for close parallel line conductors). Changes from knob-and-tube to twisted cables have reduced magnetic fields in modern homes.

Common wiring faults that lead to large magnetic fields include: neutral to ground connections, separation of conductors (as with knob-and-tube wiring), grounding to water pipes, and parallel neutrals (i.e. neutrals from different circuits connected together on the load side of the breaker box) (Riley 1995). Rerouting or adding ground return wires can produce background magnetic fields in the order of 1 µT in the home (Bennett 1994), a value that exceeds exposure in many occupations.

10.3.1.3 Appliances

EPA (1992) measured the magnetic fields produced by a variety of household and office appliances. According to this study, the magnetic fields generated by appliances differ enormously and drop off rapidly (generally $1*r^{-3}$) with distance. Magnetic flux densities, range from 150 µT for can openers to less than 0.1 µT for tape players. There are considerable model differences as well. For example, hair dryers can range from 70 µT to 0.1 µT depending on make and model.

The appliances of greatest concern are those with high magnetic flux densities and long exposure times. Electric blankets, for example, generate a field of 2 to 4 µT and are in contact with the body for several hours each night. New models, known as the positive temperature coefficient electric blankets, now generate magnetic fields that are one tenth or lower than those generated by the older models. Hair dryers and electric shavers generate a high magnetic field near the head. Power saws generate high magnetic fields and they may be of concern for the professional carpenter. Among household appliances electric can openers generate some of the highest fields recorded (50 to 150 µT at 15 cm).

10.3.1.4 Components of Residential Exposure

Maximum daily cumulative exposure can be attributed to appliances, indoor wiring, or outdoor power lines depending on the circumstances. Individuals living in the same building may be exposed to different magnetic fields based on the amount of time they spent in various rooms and the type of appliances they use. These differences, not considered in the early epidemiological studies, may account for some of the discrepancy in the results. Future epidemiological studies need to take them into consideration.

10.3.2 Occupational Exposure

Just as the early residential epidemiological studies used wire codes as surrogates for magnetic fields, the occupational studies initially based their result on job titles. As interest in occupational exposure increased, more measurements of magnetic fields in various occupational settings associated with individual exposure began to be documented. Because of the variability within and among occupations as well as between types of measurements (spot measurement vs. time weight averages), comparisons of occupations are difficult and can only be considered tentative at this time. Personal monitoring of workers provides the most information and, in the long term, may prove to be the most useful measurement.

Portier and Wolfe (1998) summarized a vast amount of data for time-weighted-average (TWA) magnetic field exposures according to industry type. The original data were ranked and classified into percentile groupings. The 95th percentile was at 0.66 µT and can be considered very high exposure with only 5% of the work force exposed to higher TWA magnetic fields. The 75th percentile was at 0.27 µT and is close to the values associated with very high current configuration (VH) for power lines (Wertheimer and Leeper 1982). The median (50th %) TWA magnetic flux density was at 0.17 µT and the 25th percentile was at 0.12 µT.

Despite the variability of occupational exposure, some general conclusions can be drawn. For instance, some of the highest exposures occur in the textile, utility, transportation and metallurgical industries. Among textile works, dressmakers and tailors who use industrial sewing machines are exposed to some of the highest fields (mean 3 µT, Havas 2000). In the utility industry, linemen, electricians, cable splicers, as well as power plant and substation operators are among those with the highest magnetic field exposure (mean 1.4 to 3.6 µT). In transportation, railway workers have high exposures (mean 4 µT). Among metal workers, welders and those who do electrogalvanizing or aluminium refining tend to have high magnetic field exposure (mean 2 µT).

Another industry with notable exposure is telecommunications, especially telephone linemen, technicians, and engineers (mean 0.35 to 0.43 µT). Individuals repairing electrical and elec8tronic equipment (0.16 to 0.25 µT) can also be exposed to above average magnetic fields, as can dental hygienists (mean 0.64 µT) and motion picture projectionists (mean 0.8 µT). Those involved in forestry and logging have a high average exposure of 2.48 µT (Havas 2000).

In an office environment, magnetic fields are generally at or below the 50th percentile (≤0.17 µT), except near computers, photocopiers, or other electronic equipment. People in sales, in computer services and in the construction industry are generally exposed to lower magnetic fields. Teachers have below average exposure with a TWA magnetic flux density of 0.15 µT.

Normally we think of high EMF exposure only or primarily in electrical occupations and perhaps in an office setting with computers and copy machines. However, a number of occupations not normally classified "electrical" can be exposed to high EMFs. These include airplane pilots, streetcar and trains conductors, hairdressers (hand-held hairdryers), carpenters (power tools), tailors and seamstresses (sewing machine), metal workers, loggers, and medical technicians.

10.3.3 Transportation

The few studies that document magnetic field exposure associated with transportation suggest that exposure can be quite high depending on the mode of travel.

Typical magnetic fields for commuter trains are much higher than for most occupational exposure. According to Bennett (1994), magnetic flux densities of 24 µT have been recorded 1 meter above the floor and 4 meters from the line of an electric commuter train. In the Amtrak train from Washington to New York, the average magnetic field at 25 Hz was 12.6 µT and the maximum field was 64 µT.

Passengers may not be on these commuter trains for long but workers are exposed to them all day. The MAGLEV (magnetic levitation) electric train generates varying frequencies and magnetic flux densities. Alternating currents in a set of magnets in the guide way change polarity to push/pull the train. The train is accelerated as the ac frequency is increased. Magnetic flux densities of 50,000 µT (50 mT) in the passenger compartment where people work have been reported (Bennett 1994).

Airplanes generate a 400 Hz electromagnetic field. The highest fields are in the cockpit with values greater than 10 μT near the conduits behind the pilot and co-pilot and near the windshield (heating element). In the passenger part of the airplane, values between 3 and 0.3 μT are more common (Havas, unpublished data). Since flights generally last several hours, cumulative exposure can be considerable. Employees and passengers are also exposed to higher than average cosmic radiation at these altitudes.

Extensive monitoring of automobiles has not been done, to my knowledge. Preliminary monitoring of a few vehicles suggests much lower magnetic fields than those associated with either commuter trains or airplanes (Havas, unpublished data). Drivers are exposed to higher magnetic fields in luxury vehicles with electronic equipment and in smaller than larger vehicles, presumably due to proximity to the alternator. The fan, air conditioning, heating, as well as the driving style (acceleration) all contribute to the ambient magnetic field. Motorbike riders are exposed to high magnetic fields in excess of 3 μT on the seat of the motorbike (Havas, unpublished data).

10.3.4 Complications with Exposure

Although we are beginning to get a sense of the magnetic environment we have created and can now estimate cumulative exposures, there is much we still do not know. It is not clear what attributes of the field are important biologically. Are values above a certain threshold critical, if so, what is that threshold? Are the rapid changes between high and low intensities biologically significant or should we focus on time-weighted cumulative exposure? We have yet to determine the metric of biological significance.

To complicate matters, the electromagnetic environment consists of an electric field as well as a magnetic field. Although the previous section and much of the literature have focused primarily on magnetic fields, conditions exist where both fields are present (a person standing directly under a power line or someone in contact with an electrical appliance). Also, external magnetic fields can generate internal electric fields, so a distinction between the two is not simple. The biological response is likely to be a function of the fields within our bodies rather than the external fields to which we are exposed and this is difficult to measure and equally difficult to calculate.

More than one frequency can be generated by the power distribution system. While the dominant frequency is 50/60 Hz, harmonics (multiples of the original frequency) and sub-harmonics (fractions of the original frequency) as well as transients (spikes generated by random on and off switching) are produced. Some of the studies suggest that biological effects are frequency and intensity specific (Blackman *et al.* 1979, Liboff 1985, Dutta *et al.* 1989). A slightly higher or lower frequency (or intensity) may not necessarily produce the same biological response. A good model for biological response may be one based on the radio tuned to a specific modulation (Frey 1994).

Biological response may also be influenced by the local magnetic field produced by the earth and this field may be spatially and temporally heterogeneous (Liboff 1985). What is becoming obvious is that this area of research, concerned with EMF exposure is complex and of utmost importance if we are to understand biological interactions with electromagnetic fields.

10.4 BIOLOGICAL RESPONSE TO EMFS

10.4.1 Cancer

Epidemiological studies of cancer have focused on two primary populations: children in residential settings and adults in occupational settings. The main cancers associated with EMF exposure are leukaemia, nervous system tumours and, to a lesser extent, lymphoma among children; and leukaemia, nervous system tumours, and breast cancer among adults.

10.4.1.1 Cancer in Children

Irrespective of which metric is used (wire codes, distance, measurements, or calculations of exposure), when viewed as a whole, many of the studies on childhood leukaemia suggest an odds ratio (OR) above 1. Critical distances appear to be approximately 50 m from a power line and critical magnetic flux densities are above 0.2 µT. Daytime spot measurements give the lowest ORs while median nighttime measurements give the highest.

Several studies suggest a dose/response relationship. Feychting *et al.* (1993, 1995) reported a significant OR of 2.7 above 0.2 µT and 3.8 above 0.3 µT. Schuz *et al.* (2001) reported a non-significant 1.33 OR between 0.1 and 0.2 µT, a significant 2.4 OR between 0.2 and 0.4 µT and 4.28 OR above 0.4 µT, based on nighttime exposure. These values are low compared with other known carcinogens like cigarettes and asbestos but are certainly well above background.

Two recent meta-analyses of childhood cancer conclude that exposure to magnetic flux densities in excess of 0.4 µT are associated with an increase risk of childhood leukaemia. The first of these meta-analyses (Ahlbom *et al.* 2000) includes data from 9 countries and is based on 3,203 cases and 10,338 controls. Above 0.4 µT the relative risk is 2.0, with a range of 1.27 to 3.13, which is statistically significant (P=0.002). This means there is a 2-fold increased risk for children developing leukaemia. Fortunately, a very small percentage (0.8%) of the children in this study were exposed to fields above 0.4 µT.

In the second meta-analysis based on 19 studies Wartenberg (2001) concludes that with widespread exposure to magnetic fields there may be a 15 to 25% increase in the rate of childhood leukaemia, which is "a large and important public health impact." In the United States as many as 175 to 240 cases of childhood leukaemia may be due to EMF exposure.

One point that must be kept in mind is that exposure to EMF is so "universal and unavoidable that even a very small proven adverse effect of exposure to electric and magnetic fields would need to be considered from a public health perspective: a very small adverse effect on virtually the entire population would mean that many people are affected." (NRC 1997).

10.4.1.2 Cancer in Adults

For adults, the link between EMF exposure and leukaemia, brain tumours, and breast cancer, is also convincing when viewed as a whole. Two forms of leukaemia seem to predominate: acute myeloid leukaemia (AML) and chronic lymphocytic leukaemia (CLL). As with childhood cancers there is some evidence for a dose/response relationship although it is very difficult to accurately estimate dose in an occupational setting. For this reason it is difficult to provide a threshold value, if indeed one exists, based on the information available. Studies suggest that cumulative exposure is important (Miller *et al.* 1996)

Among the cancers, the one with the highest odds ratio is breast cancer in men. Several studies indicate a relative risk above 4 for men (Demers *et al.* 1991, Tynes *et al.* 1992, Floderus *et al.* 1994), while the highest value for women is 2.17 (Loomis *et al.* 1994). This form of cancer is rare among men and the presence of one or two cases is likely to result in a high risk estimate. The lower OR of 2 for women should not be taken lightly since as many as 5000 women in Canada and as many as 44,000 in the United States die from breast cancer each year (WHO 1998).

Laboratory studies report an increase growth rate for estrogen-responsive breast cancer cells above 12 mG (1.2 µT) (Liburdy *et al.* 1993). These studies have been independently replicated by at least two other labs and show a causal relationship between magnetic fields and breast cancer growth.

Astrocytoma is the most common type of brain cancer associated with EMF exposure (Floederus *et al.* 1993, Theriault *et al.* 1994, Lin *et al.* 1985). Floederus *et al.* (1993) reported a dose-response relationship for astrocytoma with a non-significant increased OR of 1.3 below 0.19 µT; a statistically significant OR of 1.7 between 0.2 and 0.28 µT and a significant OR of 5.0 above 0.29 µT.

10.4.2 Reproduction

Adverse pregnancy outcomes, including miscarriages, still births, congenital deformities, and illness at birth have been associated with maternal occupational exposure to electromagnetic fields (Goldhaber *et al.* 1988) as well as residential use of electric blankets, heated waterbeds, conductive heating elements in bedroom ceilings (Wertheimer and Leeper 1986, 1989, Hatch *et al.* 1998). The development of childhood cancers (particularly brain tumours) and congenital deformities have been linked with paternal EMF exposure in occupational settings (Nordstrom *et al.* 1983, Wilkins and Koutras 1988, Johnson and Spitz 1989, Tornqvist 1998).

10.4.2.1 Residential Exposure

Two studies by Wertheimer and Leeper, one examining the use of electric blankets and heated waterbeds (1986) and the other examining ceiling cable electric heat (1989), showed that fetal loss increased when conception occurred during the months of increasing cold (October to January) for parents exposed to an EMF source during the night. Homes in which electric blankets and ceiling cables were not used did not show a seasonal pattern of fetal loss. Electric blankets can generate magnetic fields as high as 4 μT at a distance of 5 cm, and ceiling cable heating produces ambient magnetic fields of approximately 10 μT and electric fields of 10-50 V/m. Ambient fields in most homes, even those with baseboard heaters, tend to be less than 0.1 μT and 10 V/m (Wertheimer and Leeper 1989).

Timing of exposure may be of particular significance. Liburdy *et al.* (1993) reported that women sleeping under electric blankets had disrupted melatonin production. The threshold for effect was between 0.2 and 2 μT, well within the range of the Wertheimer and Leeper (1986, 1989) studies. Melatonin has many functions one of which is the regulation of sex hormones, oestrogen and progesterone, which are critical for full term pregnancies.

Li *et al.* (2002) reported an increased risk of miscarriage for women exposed for any length of time during a normal 24-hour period to a magnetic field above 16 mG (1.6 μT). The California EMF Program draft report (2001) calculates that as many as 40% of the miscarriages (24,000 miscarriages) each year in California may be attributed to magnetic field exposure.

10.4.2.2 Maternal VDT Use

Clusters of abnormal pregnancies associated with maternal use of video display terminals (VDT) during pregnancy have been reported in Canada, the United States, Britain, and Denmark (DeMatteo, 1986). A study of 803 pregnancies among data processors in the British Department of Employment indicated that abnormal pregnancies were 36% among VDT users but only 16% among non-VDT users (DeMatteo 1986).

Goldhaber *et al.* (1988) conducted a case-control study of 1583 pregnant women who attended one of three gynaecology clinics in Northern California during 1981 and 1982. They found a significantly elevated risk of miscarriages for the working-women who reported using VDTs for more than 20 hr each week during the first trimester of pregnancy compared to other working women who reported not using VDTs (OR 1.8, 95% CI: 1.2-2.8). The elevated risk could not be explained by age, education, smoking, or alcohol consumption. No significantly elevated risk for birth defects was found for moderate and high VDT exposure (OR 1.4, 95% CI: 0.7-2.7).

10.4.2.3 Paternal Exposure

Paternal occupational exposure to electromagnetic fields has also been linked to reduced fertility, lower male to female sex ratio in offspring, congenital malformations and teratogenic effects expressed in the form of childhood cancer (Nordstrom *et al.* 1983, Spitz and Johnson 1985, Wilkins and Koutras 1988, Tornqvist 1998, Feychting *et al.* 2000).

Nordstrom and colleagues (1983) did a retrospective study of pregnancy outcomes for 542 Swedish power plant employees working in high voltage (130 to 400 kV) substations. Employees who worked on lines no higher than 380/220 V served as the reference group. There was no significant difference in spontaneous abortions or perinatal deaths among the high voltage switchyard workers but there was an increase of congenital malformations for this group, especially for those with wives aged 30 plus, compared with the reference group (OR approximately 2.5). Two additional differences are worth noting. One is that the male to female sex ratio of offspring was slightly lower (0.92) for high-voltage switch yard workers compared with the reference group (1.16). The second is that couples experienced some difficulty conceiving when the husband worked in a high-voltage switch yard (OR approximately 2.5). *In vivo* studies with rats showed that exposure to high electric fields reduced plasma testosterone concentrations and reduced sperm viability (Andrienko *et al.* 1977; Free *et al.* 1981).

Feychting *et al.* (2000) reported a statistically significant association between paternal exposure to magnetic fields at or above 0.3 µT with a two-fold increase in childhood leukaemia but no risk with childhood brain tumours.

Wilkins and Koutras (1988) conducted a case-control study of Ohio-born children who had died of brain cancer during 1959 and 1978. Case fathers were more likely than control-fathers to be electrical assemblers, installers, and repairers (OR=2.7, 95% CI=1.2-6.1); welders and cutters (OR=2.7, 95% CI=0.9-8.1); or farmers (OR=2.0, 95% CI=1.0-4.1). Although chemicals cannot be ruled out as potential confounders, these industries (except perhaps farming) tend to have higher than average EMF exposure. A paternal occupational study that can differentiate between EMF and chemical exposure and the risk of childhood cancers is needed.

10.4.3 Depression

Several lines of evidence suggest that depression is associated with and may be induced by exposure to electromagnetic fields. Epidemiological studies have found higher ratios of depression-like symptoms (Poole *et al.* 1993) and higher rates of suicide (Reichmanis *et al.* 1979) among people living near transmission lines.

Poole *et al.* (1993) conducted a telephone survey of people living adjacent to a transmission line and a control population selected randomly from telephone directories. Questions related to depression were based on the Center for Epidemiological Studies-Depression scale. A higher percentage of depressive symptoms were recorded among people living near the line compared with the control population. The odds ratio was 2.1 (1.3-3.4, 95% confidence interval). Demographic characteristics, environmental attitudes, and reporting bias do not appear to influence the OR. The association between proximity to the transmission line and headaches (migraine and other) was much weaker (OR 1.2 and 1.4 respectively).

Depressive symptoms as well as fatigue, irritability, and headaches have also be reported for occupational exposures (DeMatteo 1986, Wilson 1988).

Another line of evidence comes from *in vivo* studies that report desynchronization in pineal melatonin synthesis in rats exposed to electromagnetic fields (Wilson 1988). The association between depression and disrupted melatonin secretion is well documented (see Breck-Friis *et al.* 1985, Lewy *et al.* 1982). Exposure to artificial light (a different part of the electromagnetic spectrum) in the evening also disturbs night-time melatonin synthesis (Lewy *et al.* 1987), which suggests that timing of EMF exposure may be critical and that night-time exposure may be more biologically critical than daytime exposure.

10.4.4 Alzheimer's Disease

In contrast to cancers, very few studies have examined the association between occupational EMF exposure and Alzheimer's disease. One case-control study by Sobel *et al.* (1995) included 3 independent clinical series of non-familial Alzheimer's disease in Finland (2 series) and California, USA (1 series). Non-familial Alzheimer's was selected to minimize the genetic influences in the aetiology of this disease. A total of 387 cases and 475 control were included in the combined series and were classified into two EMF categories (medium/high and low exposure in primary occupations). Significantly elevated odds ratios (OR 3.9, 1.7-8.9 95% CI) were observed for the combined data sets for females working primarily as seamstresses and dress makers. The OR for males was also above 1 (OR 1.9) but was not statistically significant.

Sewing machines generate very high magnetic fields, much higher than most electrical occupations. More studies focused on Alzheimer's disease and EMF exposure with a much broader occupation base are needed before any definitive statements can be made. The highly significant OR in this study is disturbing if the results can be generalized to a broader population.

10.4.5 Amyotrophic Lateral Sclerosis (ALS)

Several studies link EMF exposure to amyotrophic lateral sclerosis (ALS). Three studies have reported a statistically significant increase in ALS, with a relative risk from 1.3 to 3.8, for electric utility workers (Deapen and Hendersen 1986, Savitz *et al.* 1998a,b, Johansen and Olsen 1998). The California EMF Program classifies EMFs as possibly causal agents in ALS. Both Alzheimer's disease and ALS are neurodegenerative diseases.

10.4.6 Electromagnetic Sensitivity

One of the most detailed and carefully controlled experiments to determine the existence of electromagnetic field sensitivity was conducted by Rea and co-workers (1991). A four-phased approach was used that involved establishing a chemically and electromagnetically "clean" environment; screening 100 self-proclaimed EMF-sensitive patients for frequencies between 0 and 5 MHz; retesting positive cases (25 patients) and comparing them with controls; and finally retesting the most reactive patients (16 patients) with frequencies to which they were most sensitive during the previous challenge.

Sensitive individuals responded to several frequencies between 0.1 Hz and 5 MHz but not to blank challenges. The controls subjects did not respond to any of the frequencies tested.

Most of the reactions were neurological (such as tingling, sleepiness, headache, dizziness and in severe cases unconsciousness) although a variety of other symptoms were also observed including pain of various sorts, muscle tightness particularly in the chest, spasm, palpitation, flushing, tachycardia, oedema, nausea, belching, pressure in ears, burning and itching of eyes and skin.

In addition to the clinical symptoms, instrument recordings of pupil dilation, respiration and heart activity were also included in the study using a double-blind approach. Results indicate a 20% decrease in pulmonary function and a 40% increase in heart rate. Patients sometimes had delayed or prolonged responses. These objective instrumental recordings, in combination with the clinical symptoms, demonstrate that EMF sensitive individuals respond physiologically to certain frequencies.

People who claim to be electrically sensitive can't use computers and develop headaches and "brain fog", which they describe as an inability to think clearly, when they are exposed to fluorescent lighting for any length of time. The symptoms can be quite debilitating but often the medical profession's response is that the symptoms are probably psychosomatic. Hence the diagnosis creates more stress for the patient and does not correct the underlying cause of the problem.

10.4.7 The Elusive Mechanism

The effect of an environmental pollutant, such as DDT, lead, asbestos, is often observed long before the mechanism of action is understood. This delay does not negate the original observation. With respect to electric and magnetic fields, several promising mechanisms related to the biological responses are currently being considered. For low frequency, low intensity fields these include but are not limited to (1) melatonin production; (2) mitosis and DNA synthesis; and (3) ion fluxes particularly that of calcium.

10.4.7.1 Melatonin Production

Melatonin is a neurohormone that regulates sleep cycles, sex hormones, and reproduction. It is produced by the pineal gland, a light-sensitive pea-shaped gland located in the middle of the brain. In animals the pineal gland serves as a compass (it detects changes in the geomagnetic field), a clock (it sense changes in visible light, a part of the EMF spectrum, and induces sleep), and a calendar (it senses changes in photoperiod and induces hibernation as well as ovulation and thus controls reproductive cycles in seasonal breeding animals).

Melatonin follows several natural cycles. It is higher at night than during the day and is associated with restful sleep. It is higher in young people, particularly infants who spend a lot of time sleeping, as opposed to the elderly who have difficulty sleeping. It is higher in winter than in summer and has been linked with changes in serotonin levels and seasonal affective disorder (SAD), a form of depression that is accompanied by prolonged periods of fatigue. Melatonin has been used to treat sleep disturbances associated with jet lag.

The evidence linking changes in the melatonin cycle to EMF exposure is growing. We now know that the pineal gland can senses changes in electromagnetic frequencies other than those associated with visible light including static and power frequencies fields (Liburdy *et al.* 1993). Timing of exposure is critical for melatonin production. If EMF exposure occurs in the evening it can interfere with night-time concentrations of melatonin and affect sleep but if it occurs earlier in the day it has no effect on melatonin production (Reiter and Robinson 1995).

Melatonin also controls the concentrations of sex hormones. High levels of melatonin are associated with lower levels of oestrogen. Some types of breast cancer are oestrogen-responsive which means their growth is promoted by oestrogen. Post-menopausal women have an increased risk of developing breast cancer if they take oestrogen supplements. High levels of melatonin (which suppress oestrogen levels) may have a protective effect on this form of cancer. Conversely, if normal night-time peaks of melatonin are reduced and oestrogen levels remain high, this form of breast cancer is likely to be more aggressive.

Women sleeping under electric blankets have lower night-time melatonin levels (Wilson *et al.* 1990), which shows that melatonin regulation in influenced by power line frequency at intensities commonly found in the home.

Since melatonin controls reproductive cycles it may also explain some of the miscarriages experienced by women who either sleep in a high EMF environment (electric blankets, waterbeds, or ceiling-cable heating systems) or work with video display terminals that generate power frequency and higher frequency fields (Wertheimer and Leeper 1986, 1989; Goldhaber *et al.* 1988).

Melatonin has also been heralded as a natural anti-cancer chemical (Reiter and Robinson 1995). If endogenous melatonin concentrations are reduced, the natural ability of the body to fight cancerous cells may be compromised, resulting in a more aggressive spread of the cancer.

Melatonin is synthesized from serotonin, a neurotransmitter associated with depression (Reiter and Robinson 1995). Imbalances in the serotonin/melatonin cycle may account for depressive symptoms experienced by people living near power lines or working in high electromagnetic environments.

Melatonin is linked with some of the key responses to electromagnetic fields, namely breast cancer as well as other forms of cancer, miscarriages, and depression, and for this reason is one of the more likely candidates for explaining the mechanism responsible for some of the bioeffects of electromagnetic fields.

10.4.7.2 Mitosis and DNA Synthesis and Chromosomal Aberrations

The dynamics of cell proliferation is complex but changes in mitosis associated with fluctuations with the earth's magnetic field and with various ac frequencies has been reported. Liboff *et al.* (1984) examined the effect of electromagnetic fields on DNA synthesis in human fibroblasts. They exposed the cells to frequencies between 15 and 4 kHz and intensities from 2.3 to 560 μT and measured the incorporation of tritiated thymidine. DNA synthesis was enhanced during the 24-hour incubation. The threshold for this effect is estimated to be between 5 and 25 μT/sec (product of magnetic flux density (rms) and frequency) and is within the range associated with abnormal chick embryo development (10 μT/sec).

10.4.7.3 Ion Fluxes and Molecular Resonance

If resonance occurs in atoms or molecules (as has been suggested for some physiologically important monovalent and divalent ions, including lithium, potassium, sodium, and calcium) then these frequencies may very well have biological consequences (Blackman *et al.* 1994). The model that has received empirical support (but has also been criticized) is that of cyclotron resonance. The frequencies at which ions resonate depends on their mass, charge, and the strength of the static (geofield) magnetic field. Alternating current at the resonant frequency can transfer more energy to these ions and thus disturb their internal movement. The effects are location specific which may explain the discrepancy in some epidemiological and laboratory based studies.

Calcium has received the most attention in this regard. Brain tissue of newly hatched chicks released calcium ions when exposed to a radio frequency modulated at specific frequencies (15, 45, 75, 105 and 135 Hz, for example), which suggested that specific frequencies windows were important for biological effects (Adey 1980, Blackman 1985). Calcium is critical for many cell processes and changes in its flux could have significant and diverse effects on biota.

10.5 COMMENTS ON BIAS AND CONSISTENCY

10.5.1 Executive Summary of Three Major Reviews

Since 1997, three major reports have reviewed the literature on the biological effects of low frequency electromagnetic fields. Of interest is the shift in conclusions of these three reports during a 5-year period.

10.5.1.1 US National Research Council Expert Committee (1997)

The overall conclusions of the NRC Expert Committee, as stated in the Executive Summary, are as follows (NRC 1997, page 2):

" . . . *the current body of evidence does not show that exposure to these fields presents a human health hazard. Specifically, no conclusive and consistent evidence shows that exposures to residential electric and magnetic fields produce cancer, adverse neurobehavioral effects or reproductive and developmental effects.*"

"*An association between residential wiring configuration (called wire codes, defined below) and childhood leukemia persists in multiple studies, although the causative factor responsible for that statistical association has not been identified. No evidence links contemporary measurements of magnetic-field levels to childhood leukemia.*"

10.5.1.2 National Institute of Environmental Health Science Executive Summary (1998)

The evaluation of the majority of the Working Group is that extremely low frequency (ELF) EMF can be classified as "*possibly carcinogenic*" and that this "*is a conservative, public-health decision based on limited evidence of an increased risk for childhood leukemias with residential exposure and an increased occurrence of CLL (chronic lymphocytic leukemia) associated with occupational exposure. For these particular cancers, the results of in vivo, in vitro, and mechanistic studies do not confirm or refute the findings of the epidemiological studies.*" (Portier and Wolfe 1998, page 402).

They go on to state that "*Because of the complexity of the electromagnetic environment, the review of the epidemiological and other biological studies did not allow precise determination of the specific, critical conditions of exposure to ELF EMF associated with the disease endpoints studied.*" (Portier and Wolfe 1998, page 400).

10.5.1.3 California EMF Program, Executive Summary (Draft 3, 2001)

The California Department of Health Services initiated the California EMF Program on behalf of the California Public Utilities Commission. Three reviewers examined epidemiological studies linking EMFs to 13 health conditions to determine whether these links might be causal in nature. These assessments were based on previously developed Risk Evaluation Guidelines and criteria developed by the International Agency of Research on Cancer (IARC).

Based on IARC Guidelines, the reviewers state that electromagnetic fields are:

- *Possible Human Carcinogens to Human Carcinogen:* based on childhood and adult leukaemia

- *Possibly Causal*: based on adult brain cancer, miscarriage, and Lou Gehrig's disease, and that there is

- *Inadequate evidence* for male breast cancer, female breast cancer, childhood brain cancer, suicide, Alzheimer's disease, acute myocardial infarction, general cancer risk, birth defects, low birth weight or neonatal deaths, depression and electrical sensitivity.

The reviewers calculate that 1150 deaths per year with an additional 24,000 miscarriages annually may be attributed to EMFs. These estimates are much higher than the sum of annual non-fatal cancers associated with chloroform in chlorinated drinking water (49 cases), benzene in ambient air (100 cases); formaldehyde in indoor air (124 cases); or naturally occurring indoor radon (570 cases), all of which are currently regulated environmental agents. Over 1000 deaths with a much larger number of non-fatal cancers in California is a serious environmental hazard that requires serious regulatory attention.

During a relatively short period of 5 years we have moved from "no evidence links contemporary measurements of magnetic-field levels to childhood leukaemia" (NRC 1997); to electromagnetic fields being classified as a *possible carcinogen* based on childhood and adult leukaemia (Portier and Wolfe 1998); to electromagnetic fields classified as *possibly causal* for 5 health conditions, those identified by NIEHS as well as adult brain cancer, miscarriage, and Lou Gehrig's disease (California EMF Program 2001). If this trend continues, with better designed studies, more of the health conditions listed above are likely to be linked in a causal way with electromagnetic field exposure. The increasing connection between EMF exposure and Oestrogen-responsive breast cancer among younger woman rather than all forms of breast cancer among women of all ages is one case in point.

10.5.2 The Question of Bias

Prejudicial bias is something that scientists try to avoid since their credibility depends on an open unbiased approach to scientific hypothesis testing. By prejudicial bias I refer to someone with a firmly held opinion whose mind is not open to evidence that might contradict that opinion. Cultural bias, a type of bias associated with different scientific disciplines (and indeed different cultures), refers to the amount of proof needed before an opinion is considered valid. This type of bias, or level of acceptance, is considered the norm within a scientific subculture and is taught to young scientists as part of their training. Since variability among data sets and within scientific sub-disciplines differs, the standards for acceptance are culturally defined. Physical scientists are accustomed to precise measurements while biological scientists, particularly those who work in the field, are accustomed to considerable variability in their data sets and have developed techniques to detect low signal to noise ratios. For this reason, two scientists with different expertise will often interpret the same data differently. One sees the noise while the other sees the signal. Differentiating between prejudicial and cultural bias is difficult.

Two strong cultural biases are presented in the literature: One represented the views of epidemiologists and the other that of physiologists. These conflicting perspectives are both well presented in the NIEHS and California reviews.

The NRC (1997) document is culturally biased towards the physical sciences and is highly critical of positive associations between EMF exposure and effects to the point that it raises questions of prejudicial bias. Scientific studies that suggested detectable biological responses to electromagnetic fields in the section on cellular and molecular effects and in the section on animal and tissue effects were down played so frequently that I began to think, "Methinks, thou doth protest too much!" For a detailed assessment of this refer to Havas (2000). Positive studies (those finding an association between exposure and effects) were criticized, while negative studies (those finding no association) were accepted at face value.

Another example of bias is the absence of studies dealing with occupational exposure in the executive summary despite the fact that they were included in the body of the text. The following are quotes from this summary that indicate increased risk of cancer associated with occupational exposure to electromagnetic fields, none of which appears in the executive summary.

Across a wide range of geographic settings . . . and diverse study designs . . . workers engaged in electrical occupations have often been found to have slightly increased risks of leukaemia and brain cancer (Savitz and Ahlbom 1994). (pg. 179).

. . . a large well-designed study of utility workers in Canada and France provided evidence of a 2- to 3- fold increased risk of acute myeloid leukaemia among men with increased magnetic field exposure (Theriault et al. 1994). Brain cancer showed much more modest increases (relative risk of 1.5-2.8) with increased magnetic field exposure. (pg. 180).

A series of three studies reported an association between electrical occupations and male breast cancer (Tynes and Andersen 1990; Matanoski et al. 1991; Demers et al 1991) . . . (pg. 181).

Female breast cancer in relation to electrical occupations was evaluated by Loomis et al. 1994 . . . a modest increase in risk was found for women in electrical occupations, particularly telephone workers . . . (pg 181).

The relative risks in the upper categories of 2-3 reported in the high quality studies of Floderus et al. 1993 and Theriault et al 1994 cannot be ignored (pg 181). Yet this is exactly what the NRC report did, it ignored some vital pieces of information in its executive summary.

10.5.3 The Question of Consistency

The issue of "consistency" vs. "inconsistency" is an interesting one. For example, water boils at 100 °C but it can also boil at higher and lower temperatures depending on atmospheric pressure. Without understanding the importance atmospheric pressure we may claim that two studies, each of which report a different temperature for the boiling point of water, are inconsistent. It's not until we understand the role atmospheric pressure plays that we recognize the consistency.

Similarly in EMF research, we can state that a study showing the link between cancer and residential or occupational EMF exposure and that showing a link between bone healing and medical EMF exposure are inconsistent because one is linked with a harmful cancerous growth and the other with a beneficial bone growth. However, if the underlying mechanism is similar, namely that electromagnetic fields enhance the rate of cell division (and/or cell differentiation) then we again recognize the consistency.

Not all studies found an increased relative risk (odds ratio) between residential EMF exposure and one specific type of childhood cancer. Some found an increase in acute myeloid leukaemia, others in lymphomas, and still others in central nervous system tumours. Once again, this can be viewed as an inconsistency. Alternatively, if EMFs are involved in cancer promotion rather than cancer initiation (which is what the *in vitro* studies show), then the tumour type is not necessarily an inconsistency. The higher relative risk for different types of cancer may be viewed as a consistency if EMF promotes tumour growth that was initiated by a different agent. The type of tumour would be agent (or initiator) specific. Furthermore, an underlying mechanism that supports tumour promotion (of several types of tumours) is the melatonin hypothesis.

10.5.4 Classical Chemical Toxicology and EMF Exposure

Some of the apparently contradictory results may be due to the fact that the chemical toxicology model, with its emphasis on dose/response, may be the wrong model for electromagnetic bioeffects. We may be getting a distorted picture by viewing the results through this lens. Frey (1994) suggests that the radio with its frequency modulated carrier waves may provide a much better model for understanding electromagnetic bioeffects. The radio picks up a very weak electromagnetic signal and converts it into sound. The electromagnetic energies that interfere with the radio signal are not necessarily those that are the strongest but rather those that are tuned to the same frequencies or modulations. Similarly "if we impose a weak electromagnetic signal on a living being, it may interfere with normal function if it is properly tuned" (Frey 1994, page 4). This makes sense once we recognize that living organisms generate and use low frequency electromagnetic fields in everything from regeneration through cellular communication to nervous system function. Frey goes on to suggest that high frequency EM waves may carry low frequency EM signals to the cell.

10.6 CONCLUSIONS

After a decade of trying to make sense of data from diverse fields I have become increasingly convinced that electric and magnetic fields do affect living systems;

that these effects vary with individual sensitivities, with geography as influenced by the earth's magnetic field, and with daily and seasonal cycles; that they can occur at low frequencies and low intensities; and that we are very close to understanding several of the mechanisms involved.

If we wish to manage the risk of EMF we need to understand the parameters of exposure that are biologically important (this has yet to be done), and to identify biological end points and the mechanisms responsible for those endpoints. The scientific work is unfinished but this should not delay policy makers who are now in a position to introduce cost-effective, technologically feasible measures to limit EMF exposure.

The entire realm of EMF interactions is complex, but I am convinced that studies in this area will provide us with a novel view of how living systems work and, in the process, will open a new dimension into scientific exploration dealing with living energy systems. I am also convinced that this information will have many beneficial outcomes. We will better understand certain disorders and will learn to treat these and other ailments, for which we currently lack the tools.

10.7 REFERENCES

Adey, W.R. 1980. Frequency and power windowing in tissue interactions with weak electromagnetic fields. *IEEE* **68**:119-125.

Ahlbom, A. 2001. Neurodegenerative diseases, suicide and depressive symptoms in relation to EMF. *Bioelectromagnetics* **22**:S132-S143.

Andrieko, L.G. 1977. The effect of an electromagnetic of industrial frequency on the generative function in an experiment (in Russian). *Gig. Sanit.* **6**:22-25; Engl. Transl. *Gig. Sanit* **7**:27-31.

Beniashvili, D. S., V.G. Beniashvili, and M.Z. Menabde. 1991. Low-frequency electromagnetic radiation enhances the induction of rat mammary tumors by nitrosomethyl urea. *Cancer Letters* **61**:75-79.

Bennett, W.R. Jr. 1994. *Health and Low-Frequency Electromagnetic Fields,* (Yale University Press).

Blackman, C.F. 1985. Effects of ELF (1-120Hz) and modulated (50Hz) RF fields on the efflux of calcium ions from brain tissue in vitro. *Bioelectromagnetics* **6**:1-11.

Blackman, C.F., J.P. Blanchard, S.G. Benane, and D.E. House. 1994. Empirical test of an ion parametric resonance model for magnetic field interactions with PC-12 cells. *Bioelectromagnetics* **15**:239-260.

Blackman, C.F., J.A. Elder, C.M. Weil, S.G. Benane, D.C. Eichinger and D.E. House. 1979. Induction of calcium-ion efflux from brain tissue by radio-frequency radiation: Effects of modulation frequency and field strength. *Radio Science* **14**:93.

Breck-Friis, J., B.F. Kjellman, and L. Wetterberg. 1985. Serum melatonin in relation to clinical variables in patients with major depressive disorder and a hypothesis of low melatonin syndrome. *Acta Psychiatr. Scand.* **71**:319-330.

Brodeur, P. 1993. *The Great Power-Line Cover-Up.* (Little, Brown and Company (Canada) Limited).

California EMF Program. 2001. An evaluation of the possible risks from electric and magnetic fields (EMFs) from power lines, internal wiring, electrical occupations and appliances. Draft 3, April 2001. *California Department of Health Services*, Oakland, California.

Deapen, D.M., and B.E. Henderson. 1986. A case-control study of amyotrophic lateral sclerosis. *American Journal of Epidemiology* **123**:790-798.

DeMatteo, B. 1986. *Terminal Shock: The Health Hazards of Video Display Terminals.* (NC Press Ltd., Toronto).

Demers, P.A., D.B. Thomas, K.A Rosenblatt and L.M. Jimenez. 1991. Occupational exposure to electromagnetic fields and breast cancer in men. *American Journal of Epidemiology* **134**:340-347.

Dutta, S.K., B. Ghosh and C.F. Blackman. 1989. Radiofrequency radiation-induced calcium ion efflux enhancement from human and other neuroblastoma cells in culture. *Bioelectromagnetics* **10**:197-202.

EPA. 1992. EMF in your Environment: Magnetic Field Measurements of Everyday Electrical Devices. EPA/402/R-92/008. Office of Radiation and Indoor Air, U.S. Environmental Protection Agency, Washing D.C.

EPRI. 1993. Survey of Residential Magnetic Field Sources: Goals, Results and Conclusions, Vol. 1. Project RP3335-02. Rep. TR-102759-V1. Prepared by High voltage Transmission Research Center for Electric Power Research Institute, Palo Alto. Calif. (As cited in the NRC 1997 since this Volume 1 sells for $20,000 to academic institutions).

Feychting, M. and A. Ahlbom. 1993. Magnetic fields and cancer in children residing near Swedish high-voltage power lines. *American Journal of Epidemiology* **138**:467-481.

Feychting, M., G. Chulgen, J.H. Olsen and A. Ahlbom. 1995. magnetic fields and childhood cancer--a pooled analysis of two Scandinavian studies. *European Journal of Cancer* **31A**:2035-2039.

Feychting, M., B. Floderus and A. Ahlbom. 2000. Parental occupational exposure to magnetic fields and childhood cancer (Sweden). *Cancer Causes and Control* **11**:151-156.

Floederus, B., T. Persson, C. Stenlund, W. Wennberg, A. Ost, and B. Knave. 1993. Occupational exposure to electromagnetic fields in relation to leukemia and brain tumors: a case-control study in Sweden. *Cancer Causes Control* **4**:465-476.

Floderus, B., S. Tornqvist and C. Stenlund. 1994. Incidence of selected cancers in Swedish railway workers, 196179. *Cancer Causes and Control* **5**:189-194.

Free, M.J., W.T. Kaune, R.D. Phillips, and H.C. Cheng. 1981. Endocrinological effects of strong 60-Hz electric fields on rats. *Bioelectromagnetics* **2**:105-121.

Frey, A.H. 1994. *On the Nature of Electromagnetic Field Interactions with Biological Systems.* (R.G. Landes Co., Austin).

Goldhaber, M.K., M.R. Polen and R.A. Hiatt. 1988. The risk of miscarriage and birth defects among women who use visual display terminals during pregnancy. *American Journal of Industrial Medicine* **13**:695-706.

Hatch, E.E., M.S. Linet, R.A. Kleinerman, R.E. Tarone, R.K. Severson, C.T. Hartsoek, C. Haines, W.T. Kaune, D. Friedman, L.L. Robison and

S. Wacholder. 1998. Association between childhood acute lymphoblastic leukemia and use of electric appliances during pregnancy and childhood. *Epidemiology* **9**:234-245.

Havas, M. 2000. Biological effects of non-ionizing electromagnetic energy: A critical review of the reports by the US National Research Council and the US National Institute of Environmental Health Sciences as they relate to the broad realm of EMF bioeffects. *Environmental Reviews* **8**:173-253.

Havas, M. 2002. Intensity of Electric and Magnetic Fields from Power Lines within the Business District of Sixty Ontario Communities. *The Science of the Total Environment* (in press).

Havas, M., T.C. Hutchinson, and G.E. Likens. 1984. Red Herrings in Acid Rain Research. *Environmental Science and Technology* **18**:176A-186A

Johansen, C. and J. Olsen. 1998. Mortality from amyotrophic lateral sclerosis, other chronic disorders, and electric shocks among utility workers. *American Journal of Epidemiology* **148**:362-368.

Johnson, C.C. and M. Spitz. 1989. Childhood nervous system tumours: An assessment of risk associated with paternal occupations involving use, repair or manufacture of electrical and electronic equipment. *International Journal of Epidemiology* **18**:756-762.

Korobkova, V.P., Yu.A. Morozov, M.D. Stolarov, and Yu. A. Yakub. 1971. Influence of the electric field in 500 and 750 kV switchyards on maintenance staff and means for its protection. In: *International Conference on Large High Voltage Electric Systems*, Paris, August 1972, CIGRE, Paris 1977.

Lewy, A.J., H.A. Kern, N.E. Rosenthal, and T.A. Wehr. 1982. Bright artificial light treatment of a manic-depressive patient with a seasonal mood cycle. *American Journal of Psychiatry* **139**:1496-498.

Li, D.K., R. Odouli, S.Wi, T. Janevic, I. Golditch, T.D. Bracken, R. Senior, R. Rankin, and R. Iriye. 2002. A population-based prospective cohort study of personal exposure to magnetic fields during pregnancy and the risk of miscarriage. *Epidemiology* **13**:9-20.

Liboff, A.R. 1985. Geomagnetic cyclotron resonance in living cells. *Journal of Biological Physics* **13**:99-102.

Liboff, A.R., T. Williams, Jr., D.W. Strong, and R. Wistar Jr. 1984. Time-varying magnetic fields: Effect on DNA Synthesis. *Science* **223**:818-820.

Liburdy, R.P., T.R. Sloma, R. Sokolic, and P. Vaswen. 1993. ELF magnetic fields, breast cancer and melatonin: 60 Hz fields block melatonin's oncostatic action on ER positive breast cancer cell proliferation. *Journal of Pineal Research* **14**:89-97.

Lin, R.S., P.C. Dischinger, J. Conde and K.P. Farrell. 1985. Occupational exposure to electromagnetic fields and the occurrence of brain tumors. *Journal of Occupational Medicine* **27**:413-419.

London, S.J., D.C. Thomas, J.D. Bowman, E. Sobel, T.C Cheng and J.M. Peters. 1991. Exposure to residential electric and the risk of childhood leukemia. *American Journal of Epidemiology* **134**:923-937.

Loomis, D.P., D.A. Savitz and C.V. Ananth. 1994. Breast cancer mortality among female electrical workers in the United States. *Journal of National Cancer Institute* **86**:921-925.

Matanoski, G.M., P.N. Breysse and E.A. Elliott. 1991. Electromagnetic field exposure and male breast cancer. *Lancet* **337**:737.

Miller, A.B., T. To, D.A. Agnew, C. Wall and L.M. Green. 1996. Leukemia following occupational exposure to 60 Hz electric and magnetic fields among Ontario electricity utility workers. *American Journal of Epidemiology* **144**:150-160.

Nordenson, I., K. Hansson Mild, G. Anderson and M. Sandstrom. 1994. Chromosomal aberrations in human amniotic cells after intermittent exposure to fifty-hertz magnetic fields. *Bioelectromagnetics* **15**:293-301.

Nordenson, I., K. Hansoon Mild, S. Nordstrom, A. Sweins and E. Birke. 1984. Clastogenic effects in human lymphocytes of power frequency electric fields: *in vivo* and *in vitro* studies. *Radiation and Environmental Biophysics* **23**:191-201.

Nordstrom, S., E. Birke and L. Gustavsson. 1983. Reproductive Hazards among Workers at High Voltage Substations. *Bioelectromagnetics* **4**:91-101.

NRC. 1997. *Possible Health Effects of Exposure to Residential Electric and Magnetic Fields*. National Research Council (U.S.) Committee on the Possible Effects of Electromagnetic Fields on Biologic Systems. (National Academy Press, Washington D.C.).

Poole, C., R. Kavet, D.P. Funch, K. Donelan, J.M. Charry and N.A. Dreyer. 1993. Depressive symptoms and headaches in relation to proximity of residence to an alternating-current transmission line right-of-way. *American Journal of Epidemiology* **137**:318-330.

Portier, C.J. and M.S. Wolfe (Eds.). 1998. *Assessment of Health Effects from Exposure to Power-Line Frequency Electric and Magnetic Fields. National Institute of Environmental Health Sciences Working Group Report of the National Institutes of Health.* (NIH Publication No. 98-3981, Research Triangle Park, N.C.).

Rea, W.J., Y. Pan, E.J. Fenyves, I.Sunisawa, H. Suyama, N. Samadi and G.H. Ross. 1991. Electromagnetic field sensitivity. *Journal of Bioelectricity* **10**:241-256.

Reichmanis, M., F.S. Perry, A.A. Marino and R.O. Becker. 1979. Relation between suicide and the electromagnetic field of overhead power lines. *Physiology, Chemistry and Physic* **11**:395-403

Reiter, R.J. and J. Robinson. 1995. *Melatonin: Your Body's Natural Wonder Drug.* (Bantam Books, N.Y.).

Riley, K. 1995. T*racing EMFs in Building Wiring and Grounding.* (Magnetic Sciences International, Tucson Arizona).

Savitz, D.A. and A. Ahlbom. 1994. Epidemiologic evidence on cancer in relation to residential and occupational exposures. In: *Biological effects of electric and magnetic fields: Sources and mechanisms* Vol. 1. Edited by D. O. Carpenter and S. Ayrapetyan, (Academic Press, N.Y.).

Savitz, D.A., H. Checkoway, and D.P. Loomis. 1998a. Magnetic field exposure and neurodegenerative disease mortality among electric utility workers. *Epidemiology* **9**:398-404.

Savitz, D.A., D.P. Loomis, C-K.J. Tse. 1998b. Electrical occupations and neurodegenerative disease: Analysis of US mortality data. *Arch. Environmental Health* **53**:1-3

Savitz, D.A. and C. Poole. 2001. Do studies of wire code and childhood leukemia point towards or away from magnetic fields as the causal agent? *Bioelectromagnetics Supplement* **5**:S69-S85.

Savitz, D.A., H. Wachtel, F.A. Barnes, E.M. John and J.G. Tvrdik. 1988. Case-control study of childhood cancer and exposure to 60 Hz magnetic fields. *American Journal of Epidemiology* **128**:21-38.

Schuz, J., JP Grigat, K. Brinkmann, and J. Michaelis. 2001. Residential magnetic fields as a risk factor for childhood acute leukaemia: Results from a German population-based case-control study. *International Journal of Cancer* **91**:728-735.

Sobel, E., Z. Davanipour, R. Sulkava, T. Erkinjuntti, J. Wikstrom, V.W. Henderson, G. Buckwalter, J.D. Bowman and P-J Lee. 1995. Occupations with exposure to electromagnetic fields: A possible risk factor for Alzheimer's disease. *American Journal of Epidemiology* **142**:515-523

Spitz, M.R. and C.C. Johnson. 1985. Neuroblastoma and paternal occupation, a case-control analysis. *American Journal of Epidemiology* **121**:924-929.

Theriault, G., M. Goldber, A.B. Miller, B. Armstrong, P. Guenel, J. Deadman, E. Imbernon, T. To, A. Chevalier, D. Cyr and C. Wall. 1994. Cancer risks associated with occupational exposure to magnetic fields among electric utility workers in Ontario and Quebec, Canada, and France: 1970-1989. *American Journal of Epidemiology* **139**:550-572.

Tornqvist, S. 1998. Paternal work in the power industry: Effects on children at delivery. *Journal of Occupational and Environmental Medicine* **40**:111-117.

Tynes, T. and A. Andersen. 1990. Electromagnetic fields and male breast cancer. *Lancet* **336**:1596.

Tynes, T., A. Andersen and F. Langmark. 1992. Incidence of cancer in Norwegian workers potentially exposed to electromagnetic fields. *American Journal of Epidemiology* **136**:81-88.

Wartenberg, D. 2001. Residential EMF exposure and childhood leukemia: Metal-analysis and population attributable risk. *Bioelectromagnetics Supplement* **5**:S86-S104.

Wertheimer, N. and E. Leeper. 1979. Electrical wiring configuration and childhood cancer. *American Journal of Epidemiology* **109**:273-284.

Wertheimer, N. and E. Leeper. 1982. Adult cancer related to electrical wires near the home. *International Journal of Epidemiology* **11**:345-355.

Wertheimer, N. and E. Leeper. 1986. Possible effects of electric blankets and heated waterbeds on fetal development. *Bioelectromagnetics* **7**:13-22.

Wertheimer, N. and E. Leeper. 1989. Fetal loss associated with two seasonal sources of electromagnetic field exposure. *American Journal of Epidemiology* **129**:220-224.

WHO. 1998. *1996 World Health Statistics Annual.* (World Health Organization, Geneva, Switzerland).

Wilkins, J.R. III and R.A. Koutras. 1988. Paternal occupation and brain cancer in offspring: A mortality-based case-control study. *American Journal of Industrial Medicine* **14**:299-318.

Wilson, B.W. 1988. Chronic exposure to ELF fields may induce depression. *Bioelectromagnetics* **9**:195-205.

Wilson, B.W., C.W. Wright, J.E. Morris, R.. Buschbom, D.P. Brown, D.L. Miller, R. sommers-Flannigan, and L.E. Anderson. 1990. Evidence for an effect of ELF electromagnetic fields on human pineal gland function. *Journal of Pineal Research* **9**:259-269.

CHAPTER 11

Reduction and Shielding of RF and Microwaves

Dietrich Moldan and Peter Pauli

11.1 INTRODUCTION

In the age of telecommunication, engineers and the public users increasingly show the desire to perform data transfers with the highest possible complexity in any place. Just think of the extremely high growth rates in the sector of mobile communication, resulting in 15 million users of mobile cellular radio systems in Germany and about 100 million users in Europe in early 1999. By the end of the year 2001 there were over 60 million users in Germany alone. Today, analogue cordless telephones are offered only in very small numbers, because the digital standards DECT and GSM succeeded in Europe. In future phases, mobile phones will become internet compatible and modern office systems, such as laptops, printers, PC's, monitors, scanners and fax machines, are to communicate over distances of up to ca. 250 metres without cables – Bluetooth is one of the magic words.

For the operation of a good wireless connection, the lowest possible attenuation of the radiated power between transmitter and receiver is of great importance. In contrast, in medicine and particularly in intensive care units, the lowest possible irradiation on life-supporting systems from outside is demanded. In sensitive fields of electronic data processing, as well as in the armed forces, safety from interception is the first priority. Furthermore, high demands are made on the electromagnetic compatibility (EMC) of technical equipment. The industry has reacted with a variety of shielding possibilities for a variety of applications. In the field of biology, human and other living beings feel impaired and disturbed; some of them react with health disorders to the increasing pollution of the environment with electromagnetic waves. There are around 33,000 mobile radio stations and more than 81,000 amateur radio operators in Germany. Countless transmitters broadcast hundreds of radio programmes and dozens of television channels.

In the course of a large-scale examination at the University of the German Federal Armed Forces (Universität der Bundeswehr in München-Neubiberg) at Munich, more than 100 different building materials and combinations of these materials, as well as special shielding materials, were tested for their attenuation of technically produced electromagnetic waves. Numerous firms provided the materials. A few specimens were also purchased in the free trade. The results of 85 selected materials have been published in May 2000. They are intended to provide an overview for building owners, planners and architects, the users of wireless communication systems and the affected residents, as well as the medical profession.

11.2 GENERAL REMARKS ON RADIO-FREQUENCY AND MICROWAVE RADIATION

During radio operation, electromagnetic waves are transmitted in a wireless process from a transmitter to one or several receivers. The radio-frequency range (RF) extends from the frequency spectrum between ca. 10 and 100 kHz (1 kilohertz = one thousand oscillations per second), at the lower end, to the MHz range (1 megahertz = one million oscillations per second). Finally to the highest. technical used range from ca. 1 to 300 GHz (1 Gigahertz = one billion oscillations per second), called "microwaves".

Low and medium frequency radio stations broadcast in the kHz range. The MHz range is occupied by HF-, VHF- and UHF-frequencies in broadcasting, amateur radio, television, radio link systems, non-pulsed cordless telephones, radio-paging services, the German C-Net and D-Net mobile radio (D-Net: 900 MHz, GSM-Standard) and other systems. In addition, there is now the terrestrial digital audio broadcasting system T-DAB and the digital trunk radio TETRA and TETRAPOL. In the low GHz range, for example, the German E-Net mobile radio (1,800 MHz, GSM-Standard), digitally pulsed telephones in accordance with the DECT Standard, UMTS-frequencies (1,900 – 2,100 MHz), microwave ovens, satellite telephones and motion detectors can be found. In the range from several GHz to 150 GHz, there are applications of radio link systems, radar or modern traffic control technology.

Microwaves behave in a similar way as light, which can be reflected by materials (a phenomenon referred to as "reflection") or penetrate them (a phenomenon known as "transmission"). Both processes depend on the type and structure of the material, and also on the polarisation of the electromagnetic wave. In this respect, it is important to know that FM radio and VHF/UHF television stations usually transmit horizontally polarised waves, whereas long, medium and short wave broadcasting stations and mobile radio base stations normally transmit vertically polarised waves.

Until a few years ago, almost exclusively stations with a continuous transmission of analogue amplitude or frequency modulated waves were in operation. Only the development of a new technique made today's mobile phone boom possible: communication with pulsed frequencies. With the aid of this method, e.g. up to eight units and more can be served at the same time on one frequency, whereas eight frequencies and more were necessary in the past. In mobile radio according to GSM-standard, for instance, a call is newly set up 217 times per second between the mobile phone and the basic station antenna – a process referred to as "pulsing". This means that the mobile phone transmits for only 0.6 milliseconds followed by a silent period of 4 milliseconds. Cordless telephones in accordance with the DECT Standard establish a contact 100 times per second. There are numerous studies on the possible influence of pulsed radiation on the biological mechanisms of living beings, including plants. A negligible influence on the environment and living beings is attributed to continuously transmitted waves of low intensity. In contrast, studies performed by scientists of neutral research institutes throughout the world – sometimes also initiated by the telecommunications industry – again and again found effects of low-frequency pulsed high-RF-signals on biological processes and nervous systems, which can no longer be ignored. Numerous further examinations of this problem are currently being carried out. For this reason alone, preventive protection makes sense.

11.3 MEASURING ARRANGEMENT

The attenuation of the RF and microwave transmission was measured following MILSTD 285. In this procedure, the device under test is positioned between a transmitting and a receiving antenna, making sure that the waves find their way from the transmitter to the measuring receiver only through the specimen – if at all. The measurement examines the question: How much of the transmitted power can penetrate the test device and be recorded at the receiving antenna. Each antenna is located at a distance of about 30 cm from the surface of the test specimen. Significant diffraction effects occur at frequencies below 1 GHz and care was taken to reduce the transmission round the edges of the specimen. A 2.1 x 2.1-metre metal surface was mounted in front of the face side of a shielded room lined inside with RF-absorbers. In the middle of the metal surface there was a 60 x 80 cm opening. In front of this opening, the test pieces with dimensions of 1 x 1 m or 1 x 0.8 m were attached. The transmitting antenna was installed outside, in front of the specimen, and the receiving antenna inside the shielded room. In addition, appropriate contacting of the devices under test ensured that no RF-waves could penetrate laterally during the measuring procedure.

The degree of reduction of the transmitted power is given in decibels (dB). 3 dB means that the value of the penetrated power is halved or the shielding effect is doubled. Accordingly, 10 dB means a reduction to one-tenth or by 90%.

A linearly polarised wave was employed in the measurements; this means that the electric fieldstrength of the wave radiated by the transmitter was oscillating in a vertical direction in this case. In certain continuous manufacturing processes of shielding materials, which are similar to assembly line production, the high-frequency shielding additives or coatings may show an orientation in the direction of production. For this reason, measurements were performed with the polarisation vector parallel to the direction of production of the material, the longitudinal direction, and also across this direction. In most cases, no or only small differences were determined. If there are substantial differences, they are indicated in the published diagrams or in the text.

11.4 RESULTS

In the publication of May 2000 the results of the measurements are described in six chapters:

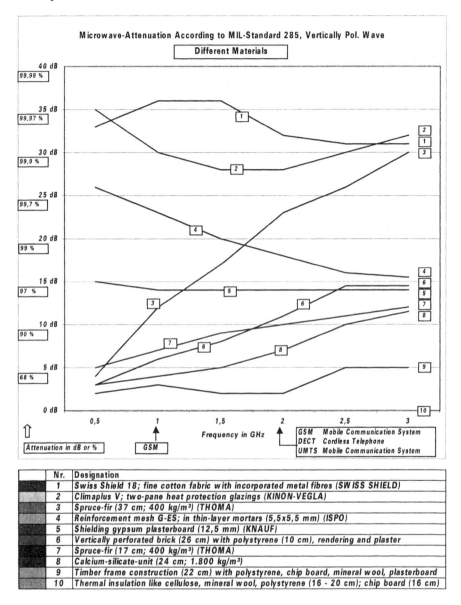

	Nr.	Designation
	1	Swiss Shield 18; fine cotton fabric with incorporated metal fibres (SWISS SHIELD)
	2	Climaplus V; two-pane heat protection glazings (KINON-VEGLA)
	3	Spruce-fir (37 cm; 400 kg/m³) (THOMA)
	4	Reinforcement mesh G-ES; in thin-layer mortars (5,5x5,5 mm) (ISPO)
	5	Shielding gypsum plasterboard (12,5 mm) (KNAUF)
	6	Vertically perforated brick (26 cm) with polystyrene (10 cm), rendering and plaster
	7	Spruce-fir (17 cm; 400 kg/m³) (THOMA)
	8	Calcium-silicate-unit (24 cm; 1.800 kg/m³)
	9	Timber frame construction (22 cm) with polystyrene, chip board, mineral wool, plasterboard
	10	Thermal insulation like cellulose, mineral wool, polystyrene (16 - 20 cm); chip board (16 cm)

Figure 11.1 Microwave attenuation according to MIL-Standard 285, vertically polarised wave; from 0.5 to 3 GHz.

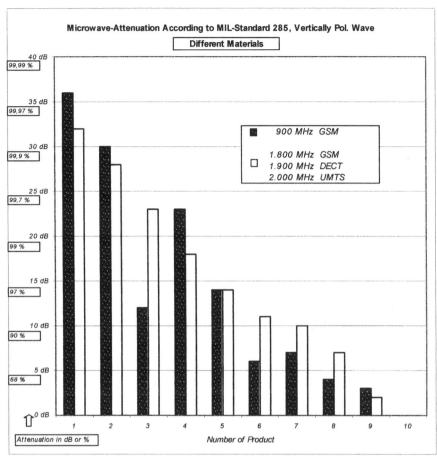

Nr.	Designation
1	Swiss Shield 18; fine cotton fabric with incorporated metal fibres (SWISS SHIELD)
2	Climaplus V; two-pane heat protection glazings (KINON-VEGLA)
3	Spruce-fir (37 cm; 400 kg/m³) (THOMA)
4	Reinforcement mesh G-ES; in thin-layer mortars (5,5x5,5 mm) (ISPO)
5	Shielding gypsum plasterboard (12,5 mm) (KNAUF)
6	Vertically perforated brick (26 cm) with polystyrene (10 cm), rendering and plaster
7	Spruce-fir (17 cm; 400 kg/m³) (THOMA)
8	Calcium-silicate-unit (24 cm; 1.800 kg/m³)
9	Timber frame construction (22 cm) with polystyrene, chip board, mineral wool, plasterboard
10	Thermal insulation like cellulose, mineral wool, polystyrene (16 - 20 cm); chip board (16 cm)

Figure 11.2 Microwave attenuation according to MIL-Standard 285, vertically polarised wave; columns at 0.9 GHz and 0.9 GHz.

- massive building materials, including bricks, concrete, calcium-silicate units, autoclaved aerated concrete, lightweight concrete
- wooden structures, including solid wood, timber-frame-constructions
- windows and accessories, including window glass, anti-sun foils, roller shutters
- wall coverings, including chip or gypsum boards, wallpaper, plaster

- roof-structures and insulation, including roof tiles, insulation materials
- textiles for RF-shielding, including fabrics, nets

In the diagrams, the attenuation values in the MHz and GHz ranges were depicted only up to a maximum value of 60 dB each.

In this article you can see interesting results of the 6 different groups within one diagram. In the first diagram (Figure 11.1) the microwave attenuation is shown from 0.5 to 3 GHz, while in Figure 11.2 only the results at 0.9 GHz (GSM mobile communication systems) and 1.9 GHz (1.8 GHz GSM mobile communication systems, 1.9 GHz DECT cordless telephone, 2.0 GHz UMTS mobile communication systems) are shown in columns.

11.4.1 Swiss Shield

Swiss Shield 18 is a fine cotton fabric with incorporated metal fibres and shielding values of more than 30 dB (99.9%). It can be used as a curtain, when you have an old window pane or your microwave shielding window is opened. Some people use it around their bed, if microwaves can not be reduced sufficiently.

11.4.2 Climaplus

Heat protection glazings consist of two panes. One pane is covered with a very thin layer of e.g. precious metal to reduce the transit of infrared waves. It effects the shielding of microwaves of more than 28 dB (> 99.8%). Nowadays these panes of glass are regularly used according to German legislation to increase thermal insulation.

11.4.3 Spruce-fir

THOMA, an Austrian company producing prefabricated houses without iron materials, invented this type of wall: several layers of wood, horizontal, vertical, diagonal – pressed together and connected with wooden plugs. The main scientific findings were that wood with a lot of resin like spruce (2 wt.-%) is much more shielding than wood with little resin like oak (0.4 wt.-%).

11.4.4 Reinforcement Mesh from ISPO

In thin-layer mortars you have to use a reinforcement mesh to avoid cracks in the rendering. This is especially necessary when using thermal wall insulation systems. When reinforcement mesh G-ES is used, attenuation values of more than 17 dB (98%) can be achieved for frequencies of up to 2 GHz. The mesh size leads to a decrease in attenuation with increasing frequencies, resulting in values of less than 10 dB (90%) from 6 GHz onwards.

11.4.5 Shielding Gypsum Wallboard from KNAUF

Carbon fibres in the outer layer of the backside cardboard lead to increasing attenuation values of 14 dB (96%) in comparison to normal gypsum wallboard without any shielding effects. When grounding the back side, electrical low frequency fields from electrical cables are shielded too. The installation of the special boards is – except the grounding – similar to normal boards.

11.4.6 Vertically Perforated Bricks

The wall thickness substantially influences the attenuation: the thicker the material, the greater the attenuation, but nevertheless in many cases it is not sufficient. Insulation itself does not cause any shielding effects.

11.4.7 Spruce-fir

The degree of shielding is anyway in accordance to the thickness. See Figure 14.3.

11.4.8 Calcium Silicate Units

The calcium silicate units show a lower attenuation behaviour in the MHz range in comparison to vertically perforated bricks, which is desirable in certain applications. Normally, the outsides of calcium silicate units are coated with thermal insulating composite systems. When using the reinforcement mesh G-ES, however, values of more than 20 dB can easily be exceeded.

11.4.9 Timber Frame Construction

Conventional timber frame construction with a density of ca. 170 kg/m³, employed in most ready-built houses, offers small to no reduction of electromagnetic waves in the MHz and GHz bands. The shielding values are sometimes in the range of the inaccuracy of the measuring system.

11.4.10 Thermal Insulation

Materials for thermal insulation such as cellulose, mineral wool, polystyrene and others do not effect the shielding of microwaves within independence of thickness. Similar results are given by gypsum wallboard, chip board and clay roof tiles.

11.4.11 Details about the Measurements

The measurements are limited by the type of experimental set-up and the employed frequencies and systems. In the lower MHz range, a "crosstalk" of the waves may occur – a small part of the waves reaches the receiving antenna through reflections in the room. This results in lower attenuation values than the test piece actually exhibits. In the higher GHz range, starting from about 7 GHz, the attenuation is reduced to about 60 dB due to the limited measuring dynamics.

Generally, it has to be pointed out that the materials were tested under laboratory conditions. In practice, deviating results occur when rooms are not uniformly made of one material or have different coverings. Openings, such as windows, window frames, doors and transitions from one building material to another may be ideal gaps where high-frequency radiation can "slip through". The measurement of the exact irradiance values on site and advice from knowledgeable measuring engineers helps to find appropriate shielding materials, if necessary. The most expensive material is not always the best one. The right choice largely depends on the place of application, the desired effect and the possibilities that can be realised. Less is often more.

In the case of metallic, metal-coated or electrically conductive materials, it has to be taken into account that they may adopt the electric potential of the environment, e.g. of current-carrying power supply cables. For this reason, it is important that these materials are capable of contacting and can be connected with the grounding points of the building to achieve a potential equalisation.

The diagrams are intended to provide a general outline, rather than to compare all materials, because this is impossible due to the differences between the raw materials alone, such as clay, wood, sand, glass and textiles, and the specific properties of every type of building material.

11.5 SUMMARY

The attenuation of high frequency waves by both everyday building materials and the use of special shielding materials was tested for 100 specimens. Depending on the material, there were small or great reductions, caused by the type of material on the one hand, but also influenced by the thickness on the other hand. The perfect shielding of a room, technically required for certain measuring rooms, is normally not necessary in general construction. The guidelines concerning the transmissibility or attenuation of a structure and the knowledge of material-specific properties are meant to allow planners, designers, building owners and measuring engineers in the field of baubiology to select the appropriate material or combination, taking individual aspects into account.

CHAPTER 12

Are We Measuring the Right Things?
Windows, Viewpoints and Sensitivity

Alasdair Philips

12.1 ARE WE MEASURING THE RIGHT THINGS?

This is a crucial question if we are ever to understand the complex way that electric and magnetic fields interact with living beings. I believe that the metrics being used at present are not only inadequate, but largely inappropriate, and suggest some of the extra ones that are needed.

"Good dosimetry" is only of any use if you measure the right things in the right ways, and thenanalyse the resulting data appropriately. I would have to classify much of the published electromagnetic field (EMF) science to date as "poor science" and of little real practical use at this stage of this important debate. It is based on simplistic reductionist understandings from classical physics and equilibrium thermodynamics. It is disappointing that the first-round funding by the UK Mobile Telephone Health Research (MTHR) committee [MTHR, 2002], mainly funds yet more research ideas that seem well past their "sell by" date. Thermally related effects are now generally well understood and should not need further funding.

We urgently need to adopt a more holistic approach that addresses issues at the very core of the biological organisation of life processes. These include homoeostasis, ontogenesis and phylogenesis, as set out by Rosen [1967]. Presman [1970] summarised leading Soviet bioelectromagnetic insights up to that time. His work contains an outline of a holistic electromagnetic field theory of living organisms and their relationship to their environment. There is now plenty of evidence for endogenous EMFs and that significant bioeffects can result from external EMFs. One western scientist who has regularly explored these areas since the 1970s is Ross Adey [e.g. Adey, 1990]. It is now established that living organisms can react sensitively to weak EMFs. We know that weak endogenous EMFs are involved in the regeneration and growth of new tissue. EMFs (including biophotons [Brugemann, 1993]) are emitted from living beings, and communication using EMF signals is established for some fishes and insects, and is strongly suspected as being utilised at some level by all living organisms.

Scientists need to direct their attention to 'wholeness' and ask radically new questions. Not only is our universe electromagnetic, but we are also electromagnetic beings. When the electricity is no longer within our being, our physical body ceases to function at the level we describe as living [Ashworth,

2001]. For all our clever molecular biological genetic "fiddling" with the matter of life, we are no nearer being able to give life to a dead mass of cells.

12.1.1 Windows

1. A 'window' is defined by boundaries (intensities, frequencies, etc) between which an external stimulus will have a biological effect on a living being.
2. We are dealing with living beings constantly seeking homoeostasis, and a stronger signal does not necessarily mean that it will have a larger or more serious effect. Incoming information with virtually zero energy can have a dramatic effect on a person's state of wellbeing (e.g. a doctor telling a patient that they have cancer). "Understandable information" (such as sound level, language, visual information, etc) goes in through definable windows.
3. Likewise, pulsing at a regular (coherent) rate can have a dramatic effect if the repetition rate finds a natural resonance in the system upon which it impinges. That is why troops were ordered to break step when crossing a bridge. Resonant signals go in through various, definable, windows.
4. There are other windows. Some are influenced by our personal genetic history and sensitivities developed during our life.

12.1.2 Viewpoints

1. The public want benefits and also the best protection (against every hazard, all of the time).
2. The Government want a quiet life, a thriving economy, cheapness and popularity.
3. Big business wants a thriving economy with maximal profits and no onerous duties or liabilities.
4. Scientists want interesting problems and continued research funding.
5. Pragmatists* and the insurance industry want a "fair balance". *(* e.g. the Electromagnetic Biocompatibility Association, EMBA)*

12.1.3 Sensitivity

Is affected by:
1. Age, gender, psychosocial load and other stresses.
2. Physical wellness, including skin condition and conductivity.
3. The biocompatibility of the incoming signals (both in energy and informational content).
4. Exposure to other insults (e.g. chemicals).
5. Stability of the point of optimum homoeostasis.
6. Response latencies and relaxation times.
7. Genetic and life-history factors.

12.2 BACKGROUND

Over thousands of generations, life has evolved in an electromagnetic environment that ranges from the Earth's geomagnetic field, through radio frequency and light to X-ray and gamma waves. This paper concentrates on the use of frequencies up to about 100 GHz.

We are all part of a great electromagnetic experiment. People living one hundred years ago would not have been bathed in the many unnatural forms of electromagnetic energy that we now live in. Marconi had just managed to send the first radio signal across the Atlantic and man-made EMF pollution was almost non-existent. In 1920 the Marconi Company began the first public speech transmissions from their Chelmsford (UK) factory, amplitude modulated on the long wavelength of 2750 metres (109 kHz). Most of Great Britain still did not have electricity and some areas that did were supplied with Direct Current that did not vibrate 50 or 60 times every second. The large Cambridgeshire village where I live did not receive mains electricity until the autumn of 1939, just over 62 years ago. Amazingly, my grandmother chose not to have mains electricity in her London house until the mid-1970s.

The real growth of commercial radio broadcasting started in the early 1930s. In December 1932 the Wireless Constructor magazine was reporting: *"Every week one reads of some station planning to radiate enormous power, some fiddling little continental (station) will suddenly develop into an overpowering giant"*. It warned *"you may find yourself in the position of a paralysed man watching the rising of a tide which will ultimately drown him."* Prophetic words?

The world's first public television service was started in November 1936 from Alexandra Palace in London. Regular TV broadcasts in Sweden did not start until 1957 [Andersson and Westlund,1991]. It is important to keep these short timescales in mind.

An otherwise unexplained (by staffing or treatment protocols) downturn in the survival time of patients treated for Chronic Myeloid Leukaemia (CML) coincided with the start of high-powered television broadcasting in Western Australia. From 1950 to 1963 the survival time was fairly static, with 50% of patients surviving 55 months from diagnosis. From 1964-1967 this fell to only 21 months [Woodliffe and Dougan, 1980].

Leukaemia, breast cancer and some other cancers and neurodegenerative diseases such as Alzheimer's Disease, Amytropic Lateral Syndrome (ALS, a form of motor neurone disease) and miscarriage are among the adverse health problems that have been found to be associated with EMFs. The incidences of all these are increasing, despite better (and often expensive) "cures". Pharmaceutical company thinking has a considerable influence on cancer research and drug-curative research dominates funded projects, with very little money going into environmentally related preventative research.

12.3 WINDOWS

Childhood (mainly acute lymphoblastic) leukaemia first appeared in the 1920s and is recognised as a "modern industrialised society" disease. The incidence is continuing to rise. Dr Sam Milham published a paper tracing the rise of childhood leukaemia with electrification across the U.S.A. [Milham and Ossiander, 2001].

This suggests that either the electric field component is the more important, or that there is a very low threshold effect with magnetic field exposure, which does not follow a conventional, relatively simple, (e.g. linear, supralinear, logarithmic, etc) dose-response relationship. It has already been proven that incidence is not associated with the total electrical power used by society (with the higher current producing higher magnetic fields).

I propose that it is likely that adverse health effects caused by EMFs have a biphasic response curve causing a low level dosage window response.

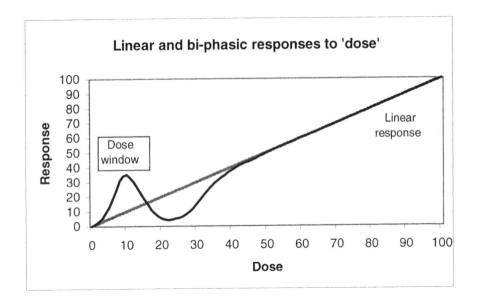

Figure 12.1 Biphasic and linear dose-response curves

This may be deduced both from the regular anecdotal reports of electrical hypersensitivity (discussed later), and from a substantial amount of published peer-reviewed research into the biological responses of animals and humans to very low doses of many pharmacological substances [Biphasic, 2002].

It is quite possible that cancers caused by exposure to low-levels of ionising radiation also follow this type of response curve. There are real low-level exposure effects that may initiate cancer and other adverse health effects before the living system starts to detect and repair the damage.

The sides of the early response peak define the dangerous exposure window. Damage starts to occur at very low levels of long-term chronic exposure in ways that are not detected by the immune system. Then, a level is reached where cellular repair mechanisms start to operate. These provide protection until the exposure reaches high levels when there can be too much damage to be repaired. The response then follows a more typical dose-response curve.

I suggest that this is the case with childhood leukaemia, and the 0.3 or 0.4 microTesla power frequency exposure level that is now internationally agreed as a point where the incidence doubles [Ahlbom, *et al,* 2000] [Greenland *et al,* 2000], is actually the main curve threshold. There are also, however, numbers of leukaemia cases caused by the *(probably co-)* carcinogenic EMF exposure that occurs at very much lower levels. These have not been identified, as they are lost in normal statistical analyses that assume a single increasing dose-response relationship.

Dr Alice Stewart proposes that the ionising radiation limits, as set by the Hiroshima and Nagasaki atomic bomb data, were based on a 'radiation hardened population' who had survived the first few months when susceptible people died of many 'problems'. [Stewart, 1998, 2000] Military electronics are purposely 'radiation hardened' to minimise the effect of ionising radiation exposure.

12.3.1 Resonance Windows

A number of possible frequency bands have been suggested. There is some research-based evidence of types of "ion cyclotron" and "Larmor" resonance effects that occur at various low frequency magnetic fields. These are caused by molecular resonances in the Earth's geomagnetic field. They include, amongst other effects, important cellular calcium efflux changes that have been reported by many laboratory studies. These were recently discussed in considerable detail by a leading EMF-bio-effects expert, Professor Ross Adey [Adey, 1999].

12.3.2 Endogenous, Entrainment & Interference Windows

We also know, mainly from (originally classified) military work [e.g. DIA, 1976], that certain frequency bands are more psychoactive than others. Signals pulsing in the range of normal brainwave and other bio-system signals have more impact, especially if they are amplitude modulated on to RF carrier frequencies.

Primarily these are in the range from one to a few hundreds of hertz, though brainwave activity extends to at least several kHz.

12.3.3 Natural EMF Noise Windows

Examples of different kinds of window are the naturally "electromagnetically quiet" regions in the ambient EMF spectrum. The main cosmic quiet "radio window" happens to be just in the place where we locate the mobile phone bands (900 and 1800 MHz) and the microwave oven frequency (2450 MHz). This can be seen in Figure 12.2. [derived from NASA 1994, and also Kraus and Fleisch, 1999]

Using basic physics we can calculate that the ICNIRP exposure guidelines allow radio signals in the mobile phone bands some 10^{15} or more times higher than the natural background levels that we were exposed to only 50 years ago.

Non-modulated (i.e. CW) and FM (e.g. VHF radio) signals are merely likely to mask the subtle signals in which Mankind has evolved. The pulsing amplitudes of many modern data signals (e.g. mobile phone GSM TDMA signals) vibrate strongly at ELF/VLF frequencies similar to those of our own bodies' endogenous signals. These are very likely to have biological consequences.

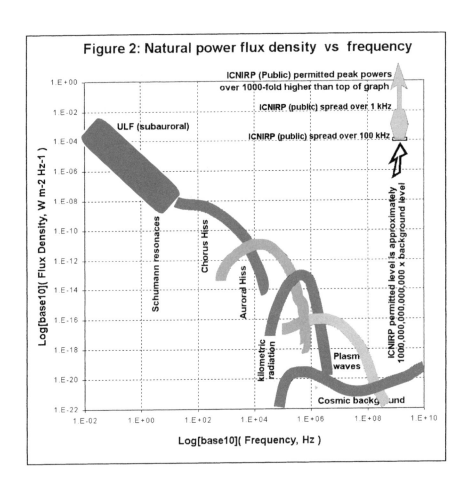

Figure 12.2 Biphasic and linear dose-response curves

12.4 VIEWPOINTS

"The time has come," the Walrus said, *"to talk of many things: of shoes – and ships – and sealing wax – of cabbages and kings – why the sea is boiling hot – and whether pigs have wings."* [Carroll, 1832-98] Some of the EMF-Health debate has been just as strange and disjointed as that quotation.

THE PUBLIC want.the benefits of the latest technologies, but want them to be risk-free. Unfortunately, life is a risky process. The only thing we can be sure about is that we will eventually die. Before that happens, most of us wish for a long and healthy life. People want to be told "safe" or "not safe". They do not want to make personal risk-benefit choices. They want to rely on the government and legislation to protect them.

GOVERNMENTS want a quiet life, a thriving economy, cheap solutions to problems, and popularity. Governments backed industry in denying that smoking caused lung cancer; tax income from tobacco greatly exceeds smoking-related NHS costs and smokers live shorter lives (saving billions in extra pension payments).

The time scale for disease caused by chronic environmental exposure to insults is often very long. In the UK, mesothelioma deaths from the inhalation of asbestos fibres, again long-denied cause of lung disease are not expected to peak until about the year 2023. This is despite UK human exposure to these fibres being strictly controlled since the 1970s.

A UK Government Minister force-fed his young daughter on TV with a beef-burger roll to "prove" that (n)vCJD was not related to eating beef products. In fact, many thinking scientists and lay people had already decided that a real link was likely. As a result of inadequate official action, we now have an unknown number (1000s to 100,000s) of cases of vCJD likely to develop over the next 30 years. Such examples leave the public with little faith in advice from official scientists and politicians.

BUSINESS wants a thriving economy with maximum profits and no liabilities. The dangers of cigarette smoking and asbestos exposure were long denied, with government support.

When tetra-ethyl lead (an anti-knock agent and also a recognised brain poison) was removed from petrol by law, the petrochemical industry lobbied hard to replace it with benzene as *"by far the cheapest and best option"*. It just so happens that benzene is a toxic waste product from petrochemical refining and the industry was paying to dispose of large quantities of it. It is one of the few known causes of myeloid leukaemia. A brain poison was removed and replaced with a known carcinogen purely due to industry lobby pressure.

Now, despite good evidence [e.g. Hansson Mild, *et al*, 1998] that some people are experiencing adverse health effects from cell phone use, the industry denies that any problems exist other than alarm caused by activists.

SCIENTISTS want interesting problems and continued research funding. There has always been a gulf between 'biologists & clinicians' and 'physicists & engineers'. The majority of members of both groups generally give up the other's area of science before university. Chemistry is between the two, but biochemistry has a much larger following than biophysics. This has greatly slowed the development of a holistic science and has led to life-processes being mainly seen in terms of biochemical reactions.

At the deepest level, however, it is electronic forces that control the shape of molecules and how they interact with each other. Electric and magnetic field interactions are the known physical fundamental manifestations of curved space-time that shapes our universe. The presence of man-made electromagnetic fields can change the outcomes of naturally occurring biophysical interactions, especially in living beings. It is not easy to persuade eminent scientists that they may have to change their views built up over many years of mechanistic science.

Science that provides technological advances or finds ways of saving money is usually popular. Science that points out problems that will cost extra money to solve is not. This has affected science for at least the last two centuries, and the effects of funding changes in the last 20 years means that scientists' work is more and more being decided by the commercial interests of multi-national companies.

Epidemiology looks for health effects in the community. In the 1840s, the doctor John Snow, credited with starting modern epidemiology, identified the cause of a cholera outbreak in London as a particular water pump – citing the results of his field investigations. Despite being proven correct, during his life he was attacked and outcast by the medical establishment and it was not until 15 years after his death that they started to accept his methods.

When Dr Alice Stewart first tried to publish her findings in the late 1950s that X-rays in early pregnancy caused abnormalities and childhood leukaemia, it was vigorously denied. She was ostracised by the UK medical establishment, as she was being critical of standard medical radiography practice; by the mid-1970s great care was being taken to avoid X-ray exposure of pregnant women [Green, 1999].

In 1998, the Doll-Hill [Doll and Hill, 1956] smoking risk figures were re-examined by Sam Milham [Milham, 1998]. The Relative Risk (RR) for heavy smokers with respect to non-smokers is 23.7; compared with light smokers it falls to 3.5 and with medium smokers to a mere 1.9. These are typical RRs we see in many epidemiological studies into possible EMF related adverse health outcomes. We are all exposed to light or medium levels of EMF pollution, so even if EMFs cause a lot of chronic health problems, they would be unlikely to stand out from the background noise in whole (or random) population studies.

The gulf between conventional mechanistic-world-view scientists (and regulatory authorities) and most leading edge EMF-bio-effects scientists is a large one. I believe that we need a change of scientific paradigm regarding what "life" is all about. It is not just a matter of tinkering at the edges and trying to decide at what level an effect occurs, but accepting that living beings interact with the universe in ways far more complex than was thought possible. Quantum mechanics have established the primacy of the inseparable whole.

Being alive changes things. Being conscious can change them even more.

PRAGMATISTS. The insurance industry often has to pay the final bill for inaction. Swiss Re, one of the largest re-insurance companies, published a report in 1996 called Electrosmog – **a phantom risk** [Brauner, 1996]. This is a landmark publication and contains much wisdom that is missing from official documents on the possible risks of EMFs.

They asked the question: *"Do electromagnetic fields (EMFs) Impair Health?"* and came to the conclusion that the only reliable answer is *"Perhaps"*.

A living organism can amplify coherent incoming signals to levels where information patterns that they contain triggers a biological response. Swiss Re acknowledges this: *"It is necessary to distinguish between energy effects and signal effects as two different dangers posed by electromagnetic phenomena."*

The relationships between EMF exposure and disease are not merely complicated but are so complex that we cannot yet identify them even using modern tools and methods.

Bioscience has moved on from *"Yes"* or *"No"* answers to deep questions, to the realisation that all causal laws are merely statistical observations, and there is a fundamental and qualitative difference between *"certain"* and *"highly probable"*, between *"must"* and *"can"*, and between *"yes/no"* and *"perhaps"*. It is the difference between *"knowledge"* and *"conjecture"*. Swiss Re write: *"Because all scientific knowledge is based on statistical observations, the knowledge of science is mere presumptive knowledge. While classical science considered a cause to be only that which must necessarily bring about an effect as a result of the causal principle, today a cause is also considered to be that which may bring about an effect. The possibility that electromagnetic exposure might favour the incidence of certain diseases cannot be excluded. According to our present understanding, electromagnetic fields would then be a cause of disease just like a flu virus which may, but need not necessarily, result in influenza."* [Brauner, 1996]

12.5 ELECTROMAGNETIC BIOCOMPATIBILITY

All living beings detect and use information in order to survive. This essential fact is not taken into account in most of the EMF research that has assumed "averaged energy" is the active factor. Our direct senses of sight, sound, etc., are only of use because we extract information from the physical responses of our sensors. Language, music, art, science and other human endeavours only exist because we interpret and use informational input.

If we could send a modern computer data Compact Disk, with an encyclopaedia on it, back one hundred years, and ask the best scientific minds of the time to try to work out what it was, even with unlimited financial resources they would not have been able to succeed in this task.

We could take these analogies further. Imagine attending a performance of the Swan Lake ballet. Conventional physical and medical science could record the movements and sounds and analyse them into data sets and look for patterns. It could also analyse the clothes of the dancer, and the ballet shoes, their materials and method of construction. But it would completely miss the whole point of the ballet, and would be able to say nothing about the human (invoked) response to the ballet.

12.6 SENSITIVITY

The auditory vibration sensitivity of a normal human ear is quite amazing at around 10^{11} m, about the diameter of a hydrogen atom. This quantum limit to detection is achieved despite large amounts of thermal noise. To achieve this the inner ear must possess amplifiers whose noise performance could only be achieved by traditional electronics circuitry working at near $0°K$. The only way that this performance could be achieved at normal body temperature is if large numbers of cells are working in a highly co-operative and coherent way. This sensitivity cannot be described by any mechanistic chemical kinetic model, and may be representative of a more general 'living tissue' property. [Adey, 1998]

We are now surrounded by unnatural pulsing electromagnetic signals millions of times stronger than were present only 50 years ago. We are "broad-band receivers" whose cells and tissue can act in non-linear ways [e.g. Wessel, 1999] to "detect" incoming RF signals; we are not frequency selective, though resonances do occur (e.g. body size resonances at VHF frequencies, and under-wired bras can resonate at cell-phone frequencies) providing windows where effects will be enhanced.

Electronics is used in almost everything now. Current trends are to make everything work faster so that we will soon even be able watch the latest movies in colour on our multipurpose phone handsets.

A potentially much more bioactive change has also taken place. The "digital revolution" has caused signals to become "lumpy", with bursts of full amplitude data pulses often emitted at human endogenous bio-signalling rates. The form of our exposure has changed dramatically over the last 15 years. GSM phone signals are *very* different in character from analogue TV and FM radio transmissions.

Are these changes relevant? There is good evidence that they are. As higher speeds were introduced in the 1980, so were reports of cases of electrical hypersensitivity. [Katajainen and Knave, 1995; Smith and Choy, 1986; Choy, Monro and Smith, 1987]

People, animals and even plants, can be amazingly sensitive to environmental fields. There is much we still do not know, and most main-line scientists do not seem to be even looking in the right directions.

Professor Eric Laithwaite (Electrical Engineer) gave a Friday evening lecture at the Royal Institution in 1970. Stimulated by a theory of Dr Callahan, a professor of entomology at the University of Florida who had published an article suggesting that moths and butterflies communicated using far-infra-red (terahertz) microwaves. [Callahan, 1965] Conventional wisdom stated that butterflies locate their mates only using their sense of smell.

Laithwaite repeatedly placed caged breeding-ready females in the middle of a field and waited to see if males would arrive. They did when the females were in the air-sealed plastic box, but did not when they were in the open sided but electromagnetically screened 'Faraday Cage', showing that it was most likely that radio-wave communication was involved, certainly with the initial long-range mate detection process.

THE EARTH's ambient geomagnetic field varies around the world and ranges over about 20 to 70 microTesla (20,000 - 70,000 nanoTesla or gamma) range. The level in any one location varies slightly in diurnal, lunar and sidereal time frames.

At super-low frequencies we have magnetic noise from changes in the Earth's magnetic core current flows. These come to the surface of the Earth in various ways depending on the magnetic and electrical properties of the underlying strata. Old "country wisdom" has long recognised that some places are not good to live and sleep in. People used to pen cattle into fields in areas where they wanted to build a house and watch to see which parts they would choose to settle in *(=good)* and which parts they would avoid *(=bad)*. Cancers and other serious illnesses were thought likely to result when people lived and slept in geopathically active areas. In Germany it is common practice for oncologists to work with dowsers to check the houses and bed places of cancer victims for geopathically active zones.

As we approach the extremely low frequencies (ELF) we have low level, but fairly coherent, waves generated by lightning strikes powering natural Earth-Ionosphere cavity resonances. These Schumann resonances are in the range 8 to 40 Hz, i.e. the frequency range of most endogenous human and animal body "vital signs" signals and are claimed to be important to life and health.

Figure 12.3 shows average geomagnetic variations, and the main Schumann resonances, against frequency.

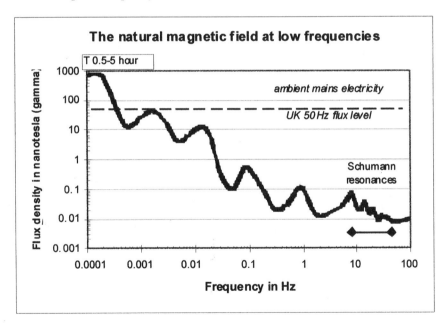

Figure 12.3 Geomagnetic fields[*derived from Campbell, 1997; Spaceweb, 2002*]

The current ambient power-frequency 50 Hz magnetic field in most UK homes and offices is around 30 - 50 nanoTesla, whereas the natural Schumann signals are a factor of 1000 lower, around 50 picoTesla. If bodily awareness of the Schumann rhythm is important and necessary for wellbeing, then swamping this rhythm with power-frequency or pulsing microwave fields may be unwise. Because GSM mobile phone base stations pulse at ELF and VLF (217 Hz, etc) and the ICNIRP safety guidance allows 138 nT at 900 MHz and 195 nT and 1800 MHz, these 'pollutant' pulsing signals can effectively y be detected by biological tissue as over 10,000 times higher than natural ELF/VLF fields.

The human pineal gland synthesises melatonin and may be one of our main magnetic field sensors. Though some laboratories have found no effects, at least six have published the results of studies that show low-intensity ELF EMFs can suppress night time melatonin levels. We also have evidence that low levels of melatonin are associated with a number of cancers, including breast cancer.

Dr Cyril Smith calculates [Smith, 1985; Smith and Best, 1989] that a quantum of magnetic flux through a typical human pineal gland would result in a flux of 75 picoTesla (pT), and that the minimum detectable magnetic flux needed to overcome the random thermal energy in the pineal would be 240 pT. This suggests that we would only be aware of the 50 pT Schumann waves if several parts of our body were co-operatively involved with the detection process.

We do know that some birds and insects are very sensitive to the background magnetic field while flying, being able to detect changes in the order of 1 nanoTesla [Keeton, 1979].

NOT EVERYONE IS AFFECTED. As initial biological effects as well as any associated adverse health effect depend on aliveness, they depend on the state of the person when being exposed to the radiation. This can be seen in the mobile phone study cited earlier [Hansson Mild, 1998]. The factors include the person's already prevailing level of stress, the robustness of their immune system, and the stability of their brain rhythms in the presence of external interference. Unlike the case of electronic measuring instruments, identical exposure to exactly the same radiation will result in different responses in different people. This variance is not specific to EMF exposure as, even with smoking, not everyone develops smoking related health problems. Exposure to an electromagnetic field may simply supply the final contribution that raises a particular person's level of stress above some critical value, thereby triggering the manifestation of a particular pathology [Hyland, 2001; Rea, 1991].

At the 2001 Bradford-Hill Memorial Lecture [Strachan, 2001], Dr David Strachan proposed that the modern task of epidemiology is to help provide "safety for the susceptible". In Sweden, Professor Kjell Mild has estimated that about 2% of the population are hypersensitive (that is they get idiopathic or allergic stress reactions) to environmental pollution (electrical and chemical sensitivities) [Hansson Mild, 2001]. This group shows a high Relative Risk (RR) when exposed to such pollution, but when data is averaged over the whole population the rise in RR becomes very small and is usually statistically insignificant.

From my contact with sufferers over the last 25 years, I believe that up to about 5% of the general population are highly sensitive to EMFs, with maybe a third, or more, of the population experiencing undiagnosed symptoms. These include headaches, poor sleep quality, general lassitude and asthenias, and probably a compromised immune response. I suggest that the problem is a hidden cost for industrialised countries of billions of pounds per year.

This cost is rising. A Swedish Trade Union (SIF) has found that the number of people reporting symptoms associated with hypersensitivity is rising rapidly [SIF, 1998]. The number of members reporting serious symptoms rose from 11% in 1993 to over 20% in 1996, when 8% reported that their symptoms were extremely serious.

12.7 WHAT METRICS SHOULD WE USE?

Most EMF regulatory guidance around the world today is only intended to protect against electric shock, radio-frequency heating, and the induction of currents that directly affect neurological processes in gross ways.

When I first started digital electronics design in the 1970s, clock rates (the timing of binary bits) were around 100 kHz. Early computer designs clocked at 1 or 2 MHz. The latest PC computer chips clock at over 2 GHz - that is a higher frequency than the microwaves used to carry mobile phone conversations (900 and 1800 MHz or 1.8 GHz). These fast clock rates cause electromagnetic noise that can interfere with both equipment function and people's health. In addition, many systems now use 'bursts' of data that cause amplitude modulation of these microwave signals by low frequency (ELF/VLF) components.

Both electric and magnetic fields induce signals that are proportional to the rate of change of the field (i.e. to dV/dt and dB/dt). Yet almost all published studies have used magnetic flux levels corrected for frequency, as if 1 microTesla at 5000 Hz will have the same effects as one at 50 Hz. This is most surprising as both the ICNIRP and the older NRPB exposure guidance levels (for gross effects) **do** have frequency dependent terms. Also, the few studies that have looked at transients (fast, often short lasting, changes) have shown increasing levels of ill health with increasing transient activity.

12.7.1 As the traditional metrics of electric and magnetic field strengths (or flux density) we need to use ones of electric and magnetic induction potential based on some measure of the rates of change of these fields. (Proportional to dV/dt and dB/dt).

12.7.2 We should also investigate the amplitude modulation patterns of different types of radio-frequency signal. In particular we should identify coherent (i.e. regular) pulsing frequency components that are allowed, by ICNIRP Guidance, to be over 10,000 times higher than naturally occurring ELF/VLF magnetic flux signals.

It is noteworthy that the Salzburg Resolution on Mobile Telecommunications Base Stations, dated 8th June 2000, highlights concern over low frequency pulsing, stating: *"For preventative public health protection a preliminary guideline level*

*for the sum total of exposures from all ELF pulse modulated high-frequency
facilities such as GSM base stations of 0.1 $\mu W/cm^2$ (0.6 V/m) is recommended."*
This is 100 times lower power than they recommend for the total of all RF
radiation. [Salzburg, 2000]

12.7.3 We should look for low-level window effects when analysing our data.

To identify these we will need to identify susceptible sub-groups of the population,
otherwise they will get lost in the noise floor of data from the whole (or random)
population under study. I believe that it will be necessary to properly investigate
anecdotal reports of adverse health problems, and robustly defend the accusation
of "Texas sharp shooting" by conventional epidemiologists.

12.7.4 We should look for living biological system sensors.

It is clear that living beings respond very differently to environmental insults than
does dead tissue. It is also clear that some people are far more sensitive than
others. I suggest that we need to consider real long-term health effects from,
presently generally unrecognised, (sub/supra)consciousness and other subtle
'living being' factors. The way to assess these factors will be to include actual
people in the sensory loop – combining selective epidemiology and bio-physics.

12.8 REFERENCES

Adey, R. 1990, Electromagnetic Fields and the Essence of Living Systems, in
    Modern Radio Science, pp 1-37, J.B. Andersen (ed), Oxford U.Press, Adey, R.
    1999, Cell and Molecular Biology associated with Radiation Fields of Mobile
    Telephones, pp 845-872, in *Review of Radio Science, 1996-1999*, Stone &
    Ueno, Editors, Oxford University Press.
Ahlbom, A. *et al* 2000, "A pooled analysis of magnetic fields and childhood
    leukaemia", Br.J.Cancer, **83**, 692-698
Andersson, M, and Westlund, L. 1991, *Hypersensitivity to electricity – can it be
    prevented?* ISBN 91-630-0661-8
Ashworth, D. 2001, *Dancing with the Devil*, Crucible. ISBN 10 902 733037
Biphasic, 2002. There is too much research to list. Use an internet search for
    "biphasic dose response".
Brauner, C. 1996, *Electrosmog - a phantom risk*, see www.swissre.com
Brugemann, H. 1993 (Ed), *Bioresonance and mulitresonance therapy* Haug,
    ISBN2 8043 4010 4
Callahan, P. 1965, It seems difficult to track down published papers, but books
    include: *Nature's Silent Music* and *Paramagnetism – Rediscovering Nature's
    Secret Force of Growth*, see the publisher's web site: www.acresusa.com

Campbell, W. 1997, *Introduction to Geomagnetic Fields,* Fig. 3.45, Cambridge University Press, ISBN 0521 57193 6

Carroll, L. (1832-98), *Through the Looking Glass,* Chapter 4.

Choy, R., Monro, J., Smith, C. 1987, Electrical Sensitivities in Allergy Patients, *Clinical Ecology,* Vol.4, No.3, 119-128

Defense Intelligence Agency (USA). 1976, DST-1810S-074-76

Doll, R. and Hill, A.B. 1956, Lung cancer and other causes of death in relation to smoking: a second report on the mortality of British doctors, *British Medical Journal,* **2,** 1071-1081.

Green, G. 1999, The Woman Who Knew too Much; Alice Stewart and the Secrets of Radiation, Univ. Michigan Press, ISBN 0472 111 078.

Greenland, S. *et al.* 2000, A pooled analysis of magnetic fields, wire codes, and childhood leukaemia, *Epidemiology,* **11,** 624-634

Hansson Mild, K. *et al.* 1998, *Comparison of symptoms experienced by users of analogue and digital mobile phones,* Arbetslivsrapport, NIWL, Sweden. ISBN 1401-2928

Hansson Mild, K. 2001, at *Mobile Telephones and Health,* London, 6th June. also see http://www.feb.se/news/abstracts010927.pdf

Hyland, G. 2001, *The Physiological and Environmental Effects of Non-ionising Electromagnetic Radiation,* European Parliament report STOA/2000/07/03.

Katajainen, J. and Knave, B. (Eds), 1995, Electromagnetic Hypersensitivity, Proceedings of the Second Copenhagen Conference, Denmark, May 1995. ISBN 87-981270-2-0

Keeton, W.T. 1979 Avian orientation and navigation, *British Birds,* Vol.72.

Kraus, J. and Fleisch, D. 1999, *Electromagnetics with Applications,* ISBN 007 116429 4, p.336.

Milham, S. 1998, Carcinogenicity of Electromagnetic Fields, *European Journal of Oncology,* **3,** 93-100.

Milham,S. and Ossiander. 2001, Historical evidence that residential electrification caused the emergence of the childhood leukaemia peak, *Medical Hypotheses,* **56 (3),** 290-295

MTHR, 2002, see http://www.mthr.org.uk

NASA, 1994, "Natural Orbital Environment Guidelines for Use in Aerospace Vehicle Development", B.J.Anderson, editor and R.E. Smith, compiler; NASA TM 4527, chapters 6 and 9, June 1994

Presman, A. 1970, Electromagnetic Fields and Lufe, New York, Plenum Press.

Rea, W. 1991, Electromagnetic Field Sensitivity, Journal of Bioelectricity, 10(1&2), 241-256

Rosen, R. 1967, Optimality Principles in Biology, New York, Plenum Press.

Salzburg, 2000. In Proceeding of International Conference on Cell Tower Siting Univeristy of Vienna and Public Health Dept. Federal State of Salzburg, Austria. See www.land-sbg.gv.at/umweltmedizin

SIF, 1998, *Hypersensitive in IT environments,* Swedish Union of Clerical and Technical Employees in Industry). See http://www.sif.se

Smith, C.W., Choy, R. 1986, Electrical Sensitivities in Allergy Patients, *Clinical Ecology,* Vol.4, No.3. 93-102

Smith, C.W. 1985 High sensitivity biosensors and weak environmental stimuli, Colloquium on Bioelectronics and Biosensors, University College of North Wales, Bangor, April 1985, pp17-19

Smith, C.W. and Best, S. 1989, *Electromagnetic Man*, DENT, p44-46.

Spaceweb, 2002, see http://www.oulu.fi/~spaceweb/

Strachan, D. 2001, Bradford Hill memorial lecture, London School of Hygiene and Tropical Medicine.

Stewart, A. 1998, European Parliament STOA Workshop 5th February 1998.

Stewart, A. 2000, A-Bomb survivors, *Int.J.Epi.*, vol.29, No.4, 4th Aug 2000.

Wessel, R. 1999, Supralinear Summantion of Synaptic Inputs by an Invertebrate Neuron: Dendritic Gain is Mediated by an "Inward Rectifier" K+ Current, Journal of Neuroscience, July 15th 1999, **19(14):**5875-5888

Woodliffe, H.J. and Dougan,L. 1980, Survival in chronic granulocytic leukaemia in Western Australia; reprinted in letter in *The Medical Journal of Australia*, Vol.1(67) No.12, 14th June 1980.

Questions & Discussion

Dr Jane Cuthbert (Private Individual, UK)

I understand that the NRPB focuses on animal research only. Could Dr Sienkiewicz please tell us when the NRPB intends to start dealing with the problem of people hypersensitive to electromagnetic fields? The genetic differences between the various animals make them an unsuitable model for understanding effects on humans.

Dr Zenon Sienkiewicz (National Radiological Protection Board, UK)

Specifically, regarding the possibility of NRPB studying hypersensitivity to EMFs, you are asking me to make policy decisions, which I am not qualified to make. However I would add that although I study animal behaviour, NRPB has successfully collaborated on human epidemiological studies investigating diseases associated with exposure to electromagnetic fields, perhaps most notably as part of the recent UK Childhood Cancer Study.

Eva Marsalek (PMI Austria)

Dr. Sienkiewicz, do you see some problems in the existing threshold values and do you agree that there is a possibility that they are not well established? I feel like a guinea pig when you maintain these values rather than apply the precautionary principle. I do not understand why you protect the industry instead of people.

My other question is concerning shielding: I really do not understand why people should have to pay for protection by shielding. I do not understand or agree that shielding should be paid for by the government using taxpayers' money. In my opinion shielding is not the long-term solution. Prevention is better than cure. How can we shield plants and/or animals?

Dr Zenon Sienkiewicz

NRPB formulates guidelines on exposure to protect the health of people that are based on the scientific evidence. The costs of implementing our recommendations are not part of this process. Our advice on the science is formulated taking into account the opinions of groups of national and international experts, such as the Advisory Group on Non-ionising Radiation under the chair of Sir Richard Doll. There is very much a consensus among many of these various groups regarding effects of electromagnetic fields. If consequences of exposure are established we can use them to formulate guidance but it is very difficult to offer guidance if the evidence is largely anecdotal. Regarding the Precautionary Principle, as I understand it, it is useful for Governments and those making decisions on the management of risks where the science is suggestive but is still subject to some uncertainty. But NRPB is a scientific advisory body and it has to make recommendations based on the science.

I do not believe that there is any scientific need to advocate the need of shielding or other protective measures provided exposure values are below those recommended in national or international guidelines. However, if individuals wish to have what they believe to be a greater measure of protection, then it is a matter of their personal choice.

Dr Dietrich Moldan (Baubiologie-Umwelltanalytik-Molden, Germany)

The problem is mobile antennae are set up because they are not prohibited based on uncertain knowledge. They affect some sensitive people and that is certain. These people have then to pay for shielding solutions. This is unfair.

The best way would be to reduce the electric fields. Shielding is a problem. People have to pay for the shielding; they did not set up the mobile mast antennae.

Eva Marsalek (PMI, Austria)

We need better regulations using the precautionary principle as a basis. Shielding at best is an interim solution only.

Dr Dietrich Moldan

We need to reduce the electromagnetic fields. Birds avoid those fields which may harm them.

Dr David Dowson (Avon Complementary, UK)

I am particularly interested in health and magnetic fields. I believe that this question is actually for all three speakers and I expect I might get three different answers! It has been suggested to me that the voltages down the national grid lines are not constant and there may be spikes reaching several millions of volts when different parts of the grid are switched on and off. This may well lead to exposure, albeit for very short periods of time, to levels of electromagnetic fields over and above those recommended as the maximum by the National Radiological Protection Board. Should I be concerned about this in the same way as a convict who is going to the electric chair is not concerned with the average voltage over the next few minutes, but the voltage in the first milliseconds?

Professor Denis Henshaw (University of Bristol, UK)

I do not have any detailed knowledge about such transients and indeed these may have hitherto unrecognised health effects. Right now I am more concerned about the point that microTesla magnetic field levels can double the leukaemia risk but NRPB guidance does not address this.

Professor Denis Henshaw

You mention anecdotal data concerning the doubling of child leukaemia risk as if this is the result of media circulation. The brain cancer studies showing increased risk is based on a number of studies. The evidence is reviewed in the California Health Department EMF Report which has concluded that it is 50 - 80% likely that exposure to power frequency magnetic fields increases the risk of brain cancer.

There is also the issue of the link between magnetic fields and depression and suicide, which has not been reviewed by NRPB. There is a degree of consistency in the literature suggesting an increased risk of depression and suicide for magnetic field exposures above 0.2 µT (see Medical Hypotheses 2002, Vol. 59: 39-51). Interestingly, this is also the level (0.2 µT) where human studies have indicated a depression in the nocturnal production of melatonin by the pineal gland. The suppression of the growth of breast cancer cells *in vitro* occurs around 1.2 µT. I personally wish to avoid exposures in the range 0.2 to 1.2 µT. It does not seem sensible to have recommended limits which are 4000 times higher at the current NRPB limit of 1,600 µT, or 250 times higher at current ICNIRP limit of 100 µT.

I would also say that some of the observations are firmly established experimentally and I give the example of the paper published by Ishido *et al* 2002 (Carcinogenesis 2001, Vol. 22, No. 7, 1043-1048). The authors studied the effect of melatonin in suppressing the growth of breast cancer itself in vitro and how this suppression was reduced in the presence of 1.2 µT magnetic fields. This study was first carried out by Liburdy in 1993. At the time no one believed his findings, but Ishido *et al* are now the fifth laboratory worldwide to report independently this interference by magnetic fields of the oncostatic action of melatonin. That to me is an experimentally established result. It is not anecdotal, nor is it simply an association.

Dr Zenon Sienkiewicz

You present an interesting hypothesis that does seem to suggest that magnetic fields are capable of causing a very wide range of health effects, but is one which seems contrary to the opinions of many other experts in this field. In addition to the draft report from California, many other expert groups and committees have reviewed the data relating to the risk of childhood leukaemia, including NIEHS in the USA and AGNIR in the UK. None of these conclude that there is sufficient evidence to suggest magnetic fields are a cause of cancer in children. They accept there may be an association between exposure to fields above 0.4 µT and increased risk, but none believe that the evidence is sufficient to conclude that such exposure causes cancer. Similarly, regarding the possibility that circulating levels of melatonin in humans are affected by night-time exposure to magnetic fields, I believe the data are more equivocal that perhaps you suggest. Several studies performed with care have failed to find any significant field-dependent effects. This inconsistency also applies to the results of studies using rats, mice and other animals. However I do agree this possibility is most interesting and worth studying more.

The data related to suppression of melatonin in MCF-7 cells is also most interesting, but it is a considerable leap comparing effects in cancer cells in culture to normal cells in living humans. And, if I remember correctly, there are only certain sub-clones of this type of cell that show this particular effect. So this may be a true biological effect but I am very hesitant about applying these results more generally.

Blake Levitt (Journalist, USA)

Mr Chadwick has mentioned that computer manufacturers follow the Swedish MPR II standards for EMF emissions applicable to cathode ray tube computer models but my understanding is that laptop models do not adhere to these standards. Am I correct?

Dr Philip Chadwick (Microwave Consultants Ltd, UK)

That is correct because MPR II is primarily referring to the emissions from cathode ray tube displays as I described in my paper. Laptops will have a different range of emissions. I do not think they emit very substantial fields and the frequencies and wave forms will be different from CRTs.

Blake Levitt

But my understanding was that the emission for laptops are actually higher in some of the models for some frequency ranges and that the manufacturers of laptops are no longer voluntarily adhering to the MPR II standards. I would also think that with a laptop there is less shielding because the cases are smaller and a person uses it closer to the body.

Dr Philip Chadwick

If you take a particular narrow frequency range I imagine a laptop has less shielding and you are closer to so within that range the emissions might be higher than a CRT, but the fields are very very low. The predominant frequencies are 30 to 80 kHz electric fields.

Blake Levitt

My next question has to do with the proximity of laptops to the genital area. It is known that the eyes and genital areas are more sensitive to emf exposures. Is there any additional concern regarding this and hence for recommending lower emissions for laptop models?

Dr Zenon Sienkiewicz

This is a difficult question to answer because I know of no specific research into the possibility of changes in fertility from the use of laptops. As you say, rapidly dividing cells are by definition more sensitive to many toxic insults, but generally low frequency fields do not affect fertility in animals. However, sperm production in mammals is sensitive to the effects of heat. But the heating must be sufficient to induce a rise in temperature of several degrees. Many years ago, NRPB investigated sperm production in animals exposed to either microwaves or hot water, and in both cases the same reductions in fertility were seen that depended on the induced temperature. So in the absence of heating, effects from laptops seem unlikely.

Dr Philip Chadwick

I think what will be interesting will be when we start putting things like GPRS cards and UMTS cards into the back of laptops. Then you really will have something which does represent quite a high source of emissions, as you say and we are using them in a rather odd position.

Andrew Hunt (Ministry of Defence, UK)

We hear a lot about the national and international standards for compliance eg NRPB, ICNIRP. Are there any plans for British/ International Standards for how to measure the field strength/power density from emitters in particular those emitters with complex pulse modulation waveforms?

Dr Philip Chadwick

Currently, we have CENELEC standards for measuring emissions from a range of emitters. We have already a standard for mobile phones; the base stations standard is going through right now. Then there are generic standards for the measurement of all the things that do not have their own product standard. Technically they are directed towards putting things on the market. The argument is that if the manufacturer makes a product they have a liability under either the Radio Telecommunications Terminal Equipment or the Low Voltage Directive, to show that it is safe. The way this can be done is to use one of the CELENEC so called "harmonised" standards. You could use those standards for existing systems including those with pulsed waveforms. That is within Europe. Internationally the same job is being undertaken by the International Electrotechnical Commission.

Malcolm Charsley (ETS Lindgren, UK)

On products such as computers and mobile phones the cases are often made of plastic rather than metal. Lightweight is a major selling point in these types of products. Unfortunately, this approach goes against good shielding practice. Industry will not produce better shielded products in these areas unless the market place is influenced by standards/legislation that override the current market pressures for " smaller is better, lighter is better."

Dr Brian Wearing (EA Technology, UK)

I heard the comment that the birds do not sit on mobile phone masts. However, birds sit on transmission towers as well as medium voltage phase conductors so they must not be affected by EMF at these very high levels. Has anyone studied the effects on these birds? Are there cancers appearing in starlings or other birds that perennially use towers and conductors as perches?

Alasdair Philips (UK Powerwatch, Sutton)

I think there is a big difference between people and birds because of the size of the heads. One of the biggest effects that have been reported anecdotally is sleep disturbance from mobile phone masts. People hear a buzzing noise, which disturbs their sleep.

Dr Philip Chadwick

This reminds me of an observation I made about 10 years ago. It does not relate to communications, it relates to radar. I visited a radar test range which used to be on the Essex coast. They can park the radar pointing out to sea, stationary, while they do measurements on the beam. The day I was there, there was a very definite concentration of seagulls in the water in the area of the beam. I was told that all the seagulls moved around and stayed sitting in the beam as it was moved. The reason was that it is damned cold in the North Sea in December!

Anne Silk (Vice Chairman of Electromagnetic Biocompatibility Association)

I did some work for the Forestry Commission on very rare conditions of oak trees: " Oak Die Back". The tree dies from the top down, not from the roots upwards as is usual. I was given over 100 sites in the UK using OS map references and was able to pinpoint the common distance between the sick trees and high multi- mast user. Perhaps we can learn from trees as well as birds.

Part IV

Health Effects of Electromagnetic Environments

CHAPTER 13

Biological Effects of Neutralising Vaccines: the Effects of Weak Electromagnetic Fields and the Concordance between the Two

Jean A. Monro

13.1 BACKGROUND

In the course of treating patients with migraine at the National Hospital for Nervous Diseases at Queen Square, London, it was clear that many of these seriously ill patients presenting at a tertiary referral centre had problems with food. A number of sensitivities were identified using the RAST (radioallergosorbent test) technique for IgE mediated food allergies. The interpretation of the test was as documented (Monro, *et al.*, 1980) in a previous paper. The confirmation of their sensitivities was through undertaking an elimination diet followed by challenges. The elimination diet was for 5 days followed by sequential single food challenges with observation of symptoms and pulse changes of 10 bpm at 20 or 40 minutes. They also had intradermal skin testing using low-dose neutralising vaccines for foods. The neutralising testing was undertaken according to the principles that had been set down by Lee and Miller. This technique, refined in 1960 by Dr J B Miller, is a safe, effective treatment for sensitivities of all kinds, food, chemical or inhalant.

13.2 METHOD

Lee (1961) found that, by limiting skin testing to single foods, using different dilutions of the food extract, reactions were induced at some strengths and relieved at others. He subsequently used the technique for inhalant allergens, insect bites, drug reactions, and fungus and yeast infections. Theron Randolph later used this technique for food and Binkley used it for gases, air pollutants and plant terpenes; Miller (1979) modified the method for treatment of Herpes and other active viral infections and used hormonal neutralising doses for the relief of dysmenorrhoea and premenstrual tension.

Miller (1977) described "Food Allergy Provocation Testing and Injection Therapy." Serial dilutions of antigens are prepared, starting with a stock extract that is diluted in a 1:5 ratio in a diluent.

Final concentrations in Vials

Vial number	Degree of dilution	Final concentration
Concentrate	0	1:20
1	5 x	1:100
2	25 x	1:500
3	125 x	1:2,500
4	625 x	1:12,500
5	3,125 x	1:62,500
6	15,625 x	1:312,500
7	78,125	1:1,562,500
8	390,625 x	1:7,812,500
9	1,953,125 x	1:39,062,500

Included by permission *Ann Allergy Asthma Immunol* 1977;38:185-191. Copyright © 1977

Patients are injected intradermally with a starting dose of antigen - usually 0.01 ml of a 1:500 strength. The size and characteristics of the wheal are recorded initially and after 10 minutes. Those wheals that give "positive" reactions are described:

1. Growth of at least 2 mm in average diameter
2. Blanching
3. Hardness
4. Raised
5. Discoid (thick, circular, with cliff-like, sharply demarcated edges).

During the 10-minute interval symptoms may be induced, such as headache, sleepiness, lethargy, depression, elation and weakness of limbs. Pulse changes may occur with arrhythmia and chest pain, and peripheral vascular changes, abdominal bloating, abdominal pain, oedema, skin itching and urticaria are common.

Further intradermal injections are given with sequential decreasing strengths of vaccines. The neutralising dilution is usually that dose which gives the first negative wheal that does not grow more than 2 mm in each direction and no symptoms.

The following were identified:

GRAIN

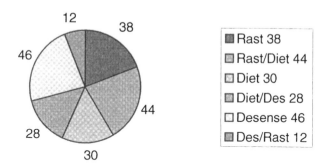

No. of patients

Any indications **219** All indications **21** Sample size **286**

DAIRY PRODUCTS

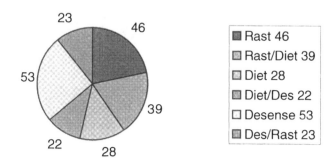

No. of patients

Any indications **224** All indications **13** Sample size **286**

MILK

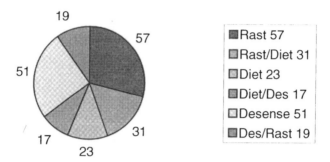

No. of patients

Any indications **208** All indications **10** Sample size **286**

WHEAT

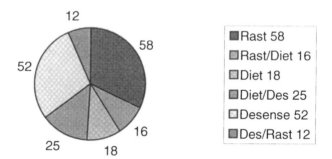

No. of patients

Any indications **199** All indications **18** Sample size **286**

EGG

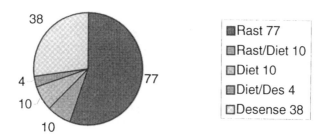

38

4

10

10

77

■	Rast 77
■	Rast/Diet 10
▫	Diet 10
■	Diet/Des 4
▫	Desense 38

No. of patients

Any indications **162** All indications **8** Sample size **286**

CHEESE

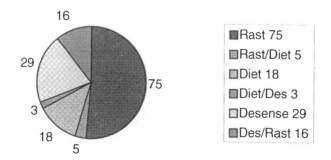

16

29

3

18

5

75

■	Rast 75
■	Rast/Diet 5
▫	Diet 18
■	Diet/Des 3
▫	Desense 29
■	Des/Rast 16

No. of patients

Any indications **161** All indications **5** Sample size **286**

TOMATO

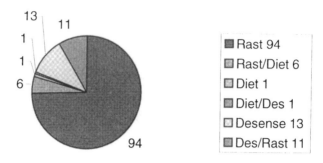

No. of patients

Any indications **126** All indications **0** Sample size **286**

RICE

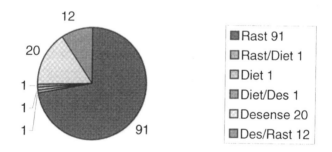

No. of patients

Any indications **126** All indications **0** Sample size **286**

FISH

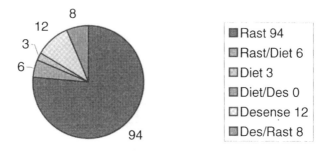

No. of patients

Any indications **123** All indications **0** Sample size **286**

TEA

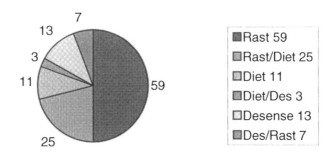

No. of patients

Any indications **122** All indications **4** Sample size **286**

APPLE & ORANGE

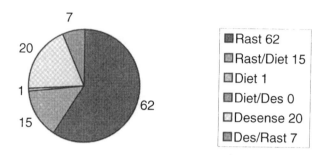

No. of patients

Any indications **110** All indications **5** Sample size **286**

MAIZE & OATS

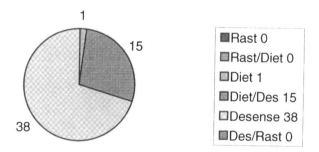

No. of patients

Any indications **54** All indications **0** Sample size 286

This indicates that the reactions provoked and neutralised depend not on antigen/antibody reactions at a particular site, but on reactions, which can be induced and negated in periodic fashion by a total body response. In a similar way an individual may react in an addictive way to food - one slice of bread may be an "underdose" in an individual who may compulsively require three for satiation. The equivalent situation in drug, tobacco or alcohol addicts is that they may require a particular "dose" before craving is satisfied.

These observations imply that the antigen/antibody response, though it may be one of the pathways for the body's recognition and handling of foreign material, is not the only mechanism. One must distinguish between correlation and cause. Because of the known involvement of the endorphin system in addiction, it is likely that this, too, is involved in the reactions that have been described in provocation/neutralisation.

The dilutions were initially prepared with a preservative of 0.4% phenol, and it was clear that the phenol preservative that was present in every dilution was causing problems in that many patients had persistent symptoms. The standard method of managing this would have been to desensitise to the phenol using a phenol solution in saline in a series of sequence of dilutions so that a neutralising end-point could be achieved. However, because of the severity of symptoms in these migraineurs it was thought more suitable to prepare the dilutions in saline with no preservatives and to freeze the vaccines between clinics when they would be used. At the clinics a row of different strengths of vaccines was provided, each in a glass phial (vial) – all were frozen initially.

Obviously the solution had to be thawed before use and at this point serendipity played a part. Because the patients were eager to have their treatment, they began to hold vaccine tubes with frozen material in them. They began to evince symptoms similar to the symptoms they had when the material was injected. This was a puzzling phenomenon and it was thought that perhaps there had been a contaminant on the outside of the vials which were then washed and the patient given the phial to hold again. However, the same symptoms occurred, whether the material was frozen or thawed. It was then thought it was possible that they were reacting to cold, as it is known that cold can induce immunological responses, but the vaccines, when thawed and at room temperature, could have the same effect, even though contained in a phial.

The glass containers were sent to the National Physical Laboratory with the enquiry as to what could be transmitted through the glass of the phials. The response was that there were frequencies that could penetrate the glass of the vials within the range of radio wave frequencies. It was clear that "antigens" within the phials were having an interactive effect with the patient through the glass phial.

Patients were then exposed to vials that were screened from them either by solid plates or by meshes. Where the mesh pore size was large enough the vaccine could have an effect. Where the screening was with a solid plate there was no effect.

Where an intermediate mesh pore size was used, on some occasions there was a screening effect with no symptoms being induced, and on others symptoms were produced. Hence it was clear that the interactive effect was an electromagnetic one penetrating through meshes but screened by solid metal plates.

It was then thought appropriate to apply some of the liquid vaccine to the skin of the patients. Patients could react to these if they were very sensitive. The more acutely sensitive the patients, the more reactive they were. As the patients attending the facility were migraineurs it was thought that this could apply particularly to patients with migraine.

It has been postulated by Sicuteri (1961) that migraine is an endorphin withdrawal syndrome. Indeed, most of the effects of injected antigens are similar to those induced by endorphins, i.e. drowsiness, altered mood, autonomic nervous system changes, such as alteration in colour of hands or face – either flushing or pallor, widening of the pupils and gastrointestinal symptoms. These effects are often seen when the provocation/neutralisation technique is used.

These observations can be explained by postulating that the basic mechanism for the reaction to antigen is mediated electromagnetically. It is known, for example, that endorphin production can be stimulated by electrical means.

Biological systems use the same atoms and molecules as physical systems, and life has evolved in an atmosphere flooded with electromagnetic radiation. Simply described, the earth is an electromagnet with North and South poles. For over a decade, Dr Cyril Smith has studied the interactions of electromagnetic fields with biological materials (Smith and Best, 1989). He observed that the growth rate of bacterial cultures of E. coli varied in different magnetic field strengths – low field strengths of the order of a few milliTesla were effective in altering growth rate. As the field strength was increased the effect did not increase in magnitude. These Mean Generation Time studies were followed by enzyme studies, selected for more refined examination of the enzymatic lac operon system in E. coli.

The transcription of the B-galactosidase gene is controlled by a repressor protein that binds very strongly to a specific site on the DNA located just outside the structural gene. In this position the repressor physically prevents the polymerase from moving into the structural gene, since the repressor interposes itself between them. Repressor protein molecules are continually being synthesized and degraded, so that the system is in a state of dynamic equilibrium. There is, therefore, always a finite, though small, probability that the repressor site will be unoccupied and the B-galactosidase synthesis can occur.

The response of this system to chemical disturbance is very rapid and reversible. It is clear that such a responsive system is likely to be sensitive to external factors, such as the low-strength magnetic fields that have already been shown to have significant effects on the growth of E. coli.

Three hundred cultures of E. coli were grown in controlled experimental conditions of a fixed magnetic field and B-galactosidase synthesis assayed. It was found that certain critical field strengths could control B-galactosidase synthesis through the presence or absence of the repressor protein on the DNA chain. The enzyme lysozyme was also found to be influenced by weak E.L.F. fields and the mean generation time of yeast cultures *Saccharomyces cerevisiae* grown in E.L.F. fields varied.

These studies formed the basis of an understanding that biological systems could be affected by weak fields. Interestingly, the effects were not magnified by increased strength of field.

Dr Smith had also demonstrated that reactivity to weak electromagnetic fields could be discerned by other living systems. Collaboration with Dr Smith then followed. Patients were taken to Salford University and put into a Faraday cage that was completely screened from external electromagnetic frequencies. Within the Faraday cage the patients were exposed blind to frequencies which were created from a frequency generator through the electromagnetic spectrum from 1 Hz to 2 GHz. They were positioned well away from the equipment to which they were not directly exposed (3 metres away) and the symptoms of the very weak electromagnetic fields produced by the frequency generators were cyclically induced. They were similar to the symptoms that the patients experienced with the low-dose neutralising vaccines until the points where the neutralising dose had been achieved. The concordance between the effects of electromagnetic frequencies and the effects of low-dose neutralising vaccines was exact for any individual patient's symptoms.

In view of this, to continue investigations with consent from our own patients, preparations of antigens in saline dilution were made in series of 1:5 dilution starting with 0.5% dilutions well below Avogadro's Number. The cyclical changes occurring in patients sometimes at different dilutions were recorded. Antigen dilution was assumed therefore to produce the effects electromagnetically.

In view of these observations, it was decided to investigate homoeopathic dilutions and their effects on patients. Fifteen patients were selected. Each of these patients had been previously diagnosed as being allergic to wheat, milk and egg, both by elimination diet followed by challenge which induced symptoms or observable physiological changes, and by previous skin testing using the provocation/neutralization method, with vaccines on the scale shown at the start of this paper. The vaccines were prepared by Ainsworth Homoeopathic Pharmacy in dilutions of 1x, 6x, 10x, 30x.

Patients were exposed to each of these strengths within their vials, and also injected intradermally, with a 0.05 ml wheal being raised. Symptoms were noted and charted. Patients then held the vial and symptoms were noted. Where symptoms occurred, intermediary preparations of vaccines were obtained and charted.

13.3 RESULTS

Symptoms to milk at different dilutions

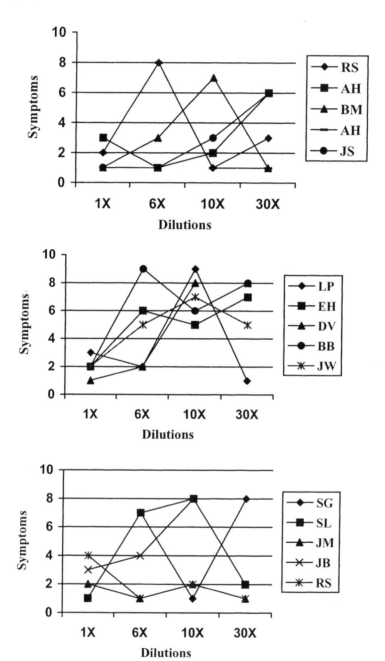

13.4 DISCUSSION

Dilutions of antigens below Avogadro's number, viz homoeopathic remedies, behaved in a manner similar to antigens injected sequentially, as in the Miller provocation/neutralization technique, and homoeopathic remedies have a similar pattern of provoking and neutralising symptoms.

The secondary effect of exposure to antigens is alteration of antigen/antibody response. It may be that the antigen exposure in weak dilutions triggers the endorphin system, particularly in view of the fact that responses are almost immediate on exposure to antigens. That dramatic response was similar in timescale to anaphylaxis, which can occur in fractions of a second. Unless this were an electrically-mediated phenomenon, it would be impossible for a protracted chemical antigen/antibody response to provoke states of collapse such as anaphylaxis in the very short time following exposure to antigen. It has been clearly shown that electromagnetic fields can alter endorphin production.

Because antigen effects can be transmitted through the glass vial of vaccine containers, they must be exercising an electromagnetic effect and because these effects are so similar to those seen in morphine withdrawal states, the endogenous opioid system must have been triggered and an opioid withdrawal effect provoked. Homoeopathic remedies act electromagnetically and can trigger antigen/antibody responses in a cyclical fashion as well as opioid effects, and opioid withdrawal effects in a similar pattern of response.

13.5 CONCLUSION

There is an absolute concordance between neutralising vaccines, electromagnetic fields and homoeopathy. Each impinges on recognition systems in the individual which have a final common pathway and can produce identical symptoms or nullify these symptoms. The response of these influences cannot be a cumbersome immunological action as recognised by antibody responses, as the responses are very swift. It must therefore lie in the chemical sphere with such delicate mechanisms as the endorphin system or intercellular memory such as cytokines (Rook and Zimla, 1997) and a recognition system for electromagnetic frequencies as yet unidentified.

13.6 REFERENCES

Lee, C., 1961, A new test for diagnosis and treatment of food allergies. *Buchanan County Medical Bulletin,* **25**, p. 9.

Miller, J. B., 1977, A double-blind study of food extract injection therapy: a preliminary report. *Annals of Allergy,* **38**, pp. 185-191.

Miller, J. B., 1979, Treatment of active Herpes virus infections with influenza virus vaccine. *Annals of Allergy,* **42**, pp. 295-305.

Monro, J. A., Brostoff, J., Carini, C. and Zilkha, K., 1980, Food allergy in migraine: study of dietary exclusion and RAST. *Lancet,* **2**, pp. 1-4.

Rook, G. A. W. and Zimla, A., 1997, Gulf War syndrome: is it due to a systemic shift in cytokine balance towards a TH2 profile? *Lancet,* **349,** pp. 1831-1833.

Sicuteri F., Testi A. and Anselmi, B., 1961, Biochemical investigations in headache: increase in hydroxyindoleacetic acid excretion during migraine attacks. *International Archives of Allergy and Applied Immunology,* **19,** pp. 55-58.

Smith, C. W. and Best, S., 1989, *Electromagnetic man,* (London: J. M. Dent).

13.7 ACKNOWLEDGEMENTS

My thanks are offered to Dr Kevin Zilkha in whose department at the National Hospital for Nervous Diseases in Queen Square I had the opportunity to see the patients and undertake these studies, and to Dr Jonathan Brostoff and Dr Càrini who undertook the RAST tests for the patients. Dr Smith's collaboration has been invaluable. Mr Cliff Edwards kindly performed the analyses of the records to produce the pie charts.

CHAPTER 14

Electroclinical Syndromes – Live Wires in Your Office?

Anne C. Silk

14.1 TERRA INCOGNITA

Terra Incognita, in the 21st C describes the uncontrolled environment where individuals have neither knowledge nor control of their exposure to Radio Frequency (RF) and Microwave (MW) fields. The Department of Health in seeking proposals for their recent RF Research Programme in 2000 stated in 6.2. "These priorities reflect areas of research where there are unanswered health questions, as well as public concerns." The DoH sought research into health effects of public exposure to pulsed and transient RF fields and also intermediate frequencies from security systems broadcast transmitters and electrified transport.

They also asked whether there are sub-populations with undue susceptibility to radiation and whether there are critical groups, stages or times for exposure. Also "are there sub-groups of the population exposed to higher levels of radioactivity or electromagnetic fields?" Healthcare professionals recognise that it is seldom that a single event it the cause of a problem. It is much more likely to be a series of factors which are cumulative over time. The paper will address from the clinical viewpoint this cumulative summation of specific factors which predispose an individual to adverse effects from RF and MW exposure to which the acronym CASTLE applies.

Certain parts of the brain are particularly sensitive to external magnetic fields, in particular the temporal lobes and their subcortical components, the amygdala and the hippocampus. Both are prone to kindling, the process whereby the tissues display electrical impulses over time as the brain is repeatedly, and environmentally, exposed to exogenous EMFs. Coupling efficiency of the body to Extremely Low Frequency EMFs is relatively low, however a time-varying (or pulsed) EMF will induce currents in a conducting medium like body tissue, with its varying dielectrics, with a magnitude which is proportional to the rate of change of the magnetic field, rather than its absolute value. The cell membrane is normally a formidable barrier to the transport of ions and charged molecules, however electroporation (as from an electric shock) which may be described as an electrostatically driven structural re-arrangement of the cell membrane (an athermal, not thermal process) results in a large increase in cell membrane conductance, caused by ion transport through temporary membrane openings. The blood-brain barrier is a further portal which can be breached by EMFs.

The WHO (1993) describes the four regions in the human body for RF absorption. Up to 30 MHz is the subresonance region; 30 - 300 MHz is the resonance region where the whole body and parts can resonate: 300 MHz - 3 GHz is the 'hotspot' region where significant heating of areas from one to several cms can occur; above 3 GHz is the surface absorption range. Eddy currents can be generated in the brain, buried metals within the skull can act as re-radiating antennas, and shape of skull, sinuses and ventricles are also important factors.

There are many compounding factors when addressing RF and MW exposure, ground reflections, albedo, vegetation, local buildings and local sources as well as domestic sources. Some case studies will be presented and the hypersensitivity question addressed.

The WHO on January 23rd 2002 published a statement in which they referred to Fact sheet 193 of June 2000. This stated in respect of mobile phones and base stations. "However there are gaps in knowledge that have been identified ... to better assess health risks. It will take about 3-4 years for the required RF research to be completed".

The public are participants in both the benefits and the costs of modern technology. In the first they are eager participants, but in the second, for the present, the jury is still out on the question of human effects.

14.2 ELECTROCLINICAL SYNDROMES

This is the term used by Professor H Luders of Berlin to describe the newly recognised syndromes described as electromagnetic origin. Many of the patients studied exhibited signs that he felt in 1987 "should no longer be considered rare" whereas in 1975 such signs of electrical activity in the brain were described as indicating "a rare condition" (Luders 1987). Luders goes on to state these syndromes "may constitute two ends of a spectrum of clinical behaviour". Looking at the temporal lobe where electrical spikes were recorded, symptoms ranged from "strange indefinable feelings, experiential hallucinations, interpretive illusions, motionless stare", with short terms memory loss (amnesia) and 4-6 Hz sharp waves in the EEG. When the amygdala was stimulated, Luders found rising epigastric discomfort, nausea, pallor, facial flushing, pupil dilation, fear, panic and confusion. Here he found 16-28 Hz rhythmic spikes. When the frontal lobe received electrical stimulation, postural effects are seen, there is some speech distortion, partial motor tonic effects, olfactory hallucinations and automatism may ensue.

The pineal gland is particularly sensitive to environmental change in electromagnetic fields and it is affected by light, sound, temperature and olfactory stimulus as well as steroids and adrenaline. Unusually this tiny gland has been found to have no blood-brain barrier around it and is thus more vulnerable to changes in its milieu (Reiter 1981).

There are three distinguishing risks:

- Electromagnetic Field health risks, the possibility that fields and radiation have a negative effect on our health.
- The classical development risk, the possibility that future scientific findings will demonstrate EMF health risk greater than now assumed.
- The socio-political risk, the possibility that society changes its values and laws.

14.3 MIGRAINE VARIANTS – A NEW NOSOLOGICAL ENTITY

Is migraine, with all its attendant rubrics, traits and variants Nature's warning signal when some factor in the environments is perceived by the body as a threat to biostasis? Migraine Variants (MVs) are less recognized, less understood, and far less common than migraines, consequently, as Birbeck writes (Birbeck 2002) "little population based data is available describing the incidence or prevalence of MVs." There are currently in neurology several puzzling syndromes and symptoms – these include Exploding Head Syndrome, the SUNCT Syndrome (Pareja 1997), Thunderclap headaches (Dodick 2001) some Drop Attacks, the Electroclinical Syndromes (Luders 1987) and Epistaxis of unknown cause.

Are we looking at a new nosological entity triggered by exposure to electromagnetic fields?

Certainly the rubrics of migraine need not involve the way most people think of this condition. Those who experience a wide range of symptoms, some very strange, may never have told the doctor. Those who over time have found that a particular set of circumstances, or even a particular place, triggers their symptoms may have suffered in silence of years, trying this or that remedy. Certainly this applies to the many men the writer has interviewed. It is thought that such people cannot habituate incoming signals – it is as if a continuous tape loop is running in the brain for the period of the attack. Thunderclap headache may be a unique and recurrent, but benign, primary headache disorder. Harling (Harling 1989) asks, "Is it migraine?" There are other causes for this very severe pain, but in many cases these have been excluded, and still the patient experiences these headaches and other symptoms. Hughes of the US Dept of Labor, Washington DC (Hughes 1998) has stated that Electromagnetic hypersensitivity is "a progressively disabling disease associated with exposure to Electromagnetic fields created by electronic equipment" and "the importance of ES should be recognised and appreciated by the medical community". A report by the US National Institute for Environmental Health Sciences stated "Over time, the symptoms may become more pronounced and persistent and begin to interfere with the ability to work or stay in the general work environment".

14.4 WHAT ARE THE SYMPTOMS OF ELECTROSENSITIVITY?

The wise words of Sir Austin Bradford Hill in 1965 are especially relevant today – he wrote "what is biologically plausible depends on the biological knowledge of the day". We know that bioeffects encompass many variables including genetic predisposition, the state of the immune system, size, shape orientation of the body to EMFs, distance from sources of EMF's time duration of exposure, amplitude wavelength and wave form of exogenous signals (the "window" effect).

Sir John Krebs, the Government Chief Scientist, writing in the Times on June 3rd stated "the fact that no evidence of adverse effects has yet to be reported ..... cannot be interpreted as meaning that such effects do not exist, as no data have yet been collected". Whilst Sir John was referring to the effects of folic acid supplements, the same logic must be applied to electrosenstivity. It is very doubtful whether the NRPB are collecting such data, referring instead to the HSE. But apart from clinicians in many other countries, there is now a considerable corpus of evidence of anecdotal cases in the UK. The NRPB claimed "there is little scientific evidence to support the idea of electrosensitivity!" But they are quite wrong when they also state ".... Nor are there any accepted mechanisms to explain hypersensitivity".

	"hum" or "buzz"	Phonophobia
Vision difficulties	Disorientation	Nausea
(focus, colour loss)		
Dry eyes –	Irritability	Urge to void (gut)
Sudden, no obvious cause		
Benign lid fasciculation (tic)	Short term memory loss	Extreme thirst
Vertigo	Sleep disturbance	Dry Mouth
Pulsing headaches	"bubbling" in ears	Profuse yawning
"Icepick Syndrome"	Skin rashes	Cardiac effects
(stabbing head pain)		
Resonance and	Rosacea (reddening)	Photophobia
Variant Migraines		

Pulsed fields, whether photic (as flickering images) acoustic (as drumming, dripping taps etc), tactile (as Chinese water torture) or modulated signals in the endogenous human brainwave regions will kindle, or drive, the system into highly exaggerated behaviour. It is now time to assess ES symptoms, with environmental factors, to see whether they fit into any known diagnostic template. Migraines divide into several classifications, with symptoms in the premonitory area, the Aura, and Sensory Sensitivity. But by no means all have headaches, and by no means all experience an aura. But all are paroxysmal disorders due to electrical overload in the brain. Current research indicates involvement of the brainstem in the pathophysiology of migraine (Weiller 1995) Increased blood flow has been found in the human brain during spontaneous attacks in patients with migraine without aura. The neural hypothesis involving the Brainstem replaces the traditional vascular hypothesis. The route is : Activation of the trigeminal system, deceased serotonin leading to dilation of the cranial blood vessels. Calcium channel activity is also involved, with genetic links to the gene responsible. 15% of women and 6% of men are reported to develop migraine but the prevalence has increased by 40% over 10 years, according to "Update in Neurology" from the Royal College of General Practitioners (2000)

The Editor of the British Medical Journal, Keith Petrie (Petrie 2002), wrote in March 2002 "we believe that these concerns about technological change which have *been largely unrecognised by researchers*, have important implications" (my italics). It is here proposed that the many strange symptoms increasingly reported by those who are electrically sensitive are in fact Janus-like manifestations of the nosological entity known as Migraine Variants/Resonance Migraines/Migraine Traits. A further pointer is that of the dozens of people questioned by the writer, most admit to common or classical or basilar migraines when young but all state they *ceased to experience these* before they became electrically sensitive. Why should this be so? Are neural circuits overloaded? Why are GPs failing to recognise ES? What is the effect of hyperstimulation over time to the brainstem? It is due time for the medical professions to address this problem – just because doctors were not taught about electrosensitivity at Medical School does not mean it is not there. It is there, the numbers are growing and effects on productivity and health are affecting the quality of life of very many people.

RF covers a wide spectrum and increasingly exposure is global. The evidence for health effects is diverse and in 2001 not only cancers, but also neurological diseases are under scrutiny. In particular the concept of "nidal" as used by medical geographers has come to the fore. It is worth here noting a curious phrase in the Dept. of Health RF Research Programme (1). It states under Priority questions 6.2 EMF's "These priorities reflect areas of research where there are unanswered health questions, as well as public concerns, (viii) Health effects of public exposure to pulsed and transient RF fields excluding telecomms and also (ix) Health effects of intermediate frequencies from security systems, broadcast transmitters and electrified transport."

The concept of cumulative summation from multiple RF sources does not appear to be covered.

The use of mean power data has predominated in studies for decades; however increasingly physicists are opining that Peak powers are of far greater significance than mean power. Further, the wave form is now known to be important. Is the wave sine, square, sawtooth? This brings in rise and fall periods. In a US Department of Defense (sic) memorandum of 1965, the physicist who wrote the report stated, after studying the many Russian reports he had been given that "there should be a detailed study to determine the precise nature of Microwave fields, including continuous versus pulsed emissions, repetition frequency of pulsed operations, peak and average power densities, *standing wave* plots, *multipath summations* including topographic reflections from intervening buildings and ground return from internal corner reflectors".

14.5 RF HOTSPOT

The WHO have described a RF Hotspot as "a highly localised area in which the local values of Electric or Magnetic fields strengths are significantly elevated, Such a Hotspot may be produced by the intersection of narrow beams of RF energy from directional antennas, by the reflection of fields from conductive surfaces (standing waves) or by induced current flowing in conductive objects exposed to ambient RF fields (5). RF Hotspots are characterized by very rapid spatial variation of the fields and typically result in partial body exposures of individuals at the Hotspots". The WHO makes clear that such Hotspots are not to be confused with an actual thermal hotspot within the body.

The purpose of an antenna is to receive or transmit energy signals. Those natural signals from the Earth are in the Radio Frequency band and are known to seismologists as Whistlers and to Radio amateurs as Sferics. In the region 40 - 100 MHz are many sources of manmade signals (which cover Police, Fire, Communications, amateur FM, VHF TV, diathermy, emergency medical radio links, air traffic controls, MRI Units, dielectric heating, plastic welding machines, wood drying and glueing machines, certain food processing machines, and plasma heating).

In this 40 - 100 MHz region the human body will act as a tuned antenna and the amount of radiated energy absorbed increases dramatically. In the Extremely Low Frequency range (0 - 300 Hz) electrical currents can arise within the body and may produce biological effects.

From 300 MHz to 3 GHz curved surfaces will tend to focus microwaves. This will include the bones of the skull and will lead to local hotspots within the head, the depth depending on the frequency and amplitude of the signal.

EDDY CURRENTS are generated in the brain area from exogenous signal. The area affected will vary due to several factors: shape of skull, metallic implants (aneurysm clips, amalgam fillings etc.) One example is 'seeing stars' after a blow to the head when the cortex receives a high energy input due to the blow to the head. A more subtle example is that of microwave hearing; a click, hum or buzz is "heard" when no external source is present. This is due to a thermo-elastic wave of acoustic pressure that travels via the bone conduction to the inner ear where it activates the cochlear receptors via the same mechanism for normal hearing. There has been no systematic research as to the long term effects of such supra-threshold RF pulses on humans.

When a biological system is exposed to Radio frequency or Microwave radiation, electrical and magnetic fields are induced within it. These internal fields then give rise to ionic currents and molecular excitations which result in heating - the thermal effect.

The interior of the human brain and body is electrically complex due to the presence of insulating membranes and tissues of various impedances. The capacitance of the body is dependent on size, body fat, shape and orientation of the body. Internal currents will differ between fat and thin persons, whether standing, sitting or reclining, whether walking barefoot, wearing shoes with leather or synthetic soles, those standing on a non-conductive platform etc. Thus we see that electrical and magnetic effects will differ in different people standing or working at the same place.

RADIO FREQUENCY RANGES FOR ABSORPTION IN THE HUMAN BODY

There are four regions for absorption in the human body, varying with frequency

- Up to 30 MHz - sub-resonance region. The trunk surface neck and legs show significant absorption
- 30-300 MHz - resonance region - the whole body and parts can resonate
- 300 MHz - 3 GHz - 'hot spot' region - significant heating of areas from 1 cm to several cms.
  This region covers mobile phone frequencies in UK
- Above 3 GHz - surface absorption range - energy is absorbed by, and heating limited to,
  the surface of the body
- (WHO 'Electromagnetic Fields 300 Hz - 300 GHz' Geneva 1993)

14.6 REFRACTION OF RF ENERGY IN THE BRAIN

Within any material the wavelength at the velocity of propagation slows within that material. For example, 1 GHz in a dielectric of 9 relative permittivity will shorten to 3 GHz. Thus a lower frequency field impinging on the skull will become higher at deeper levels within the brain. And if there is a metallic boundary within the head (e.g. amalgam fillings, implants, root canals, grommets) there will be a phase change at the interface.

Stimulation of the brain, whether by inserted electrodes or exogenous EM fields or by eddy currents, can produce biochemical change, free radical production (unpaired electrons) and emotional reactions.

14.7 WHAT ARE THE PERCEIVED PROBLEMS?

These cover both clinical and subclinical effects and include DNA strand breaks in brain cells, blood-brain barrier permeability changes, electroporation, effects on memory processing, brain wave potentials, promotion of Heat Shock proteins, production of free radicals (unpaired electrons), effects on the Autonomic Nervous System, asthenopia, ocular problems, the MW auditory problem, cancer induction and promotion, dermatological problems.

Amongst the sub-clinical effects are Migraine Variants (no headaches), behavioural anomalies, panic, nausea. Also noted is Electrosensitivity; a newly recognized problem when people are affected by ambient RF fields of varying frequency.

14.8 BRAIN BIOCHEMISTRY AND EXOGENOUS ELECTRICAL SIGNALS

The brain is a vast chemical production unit with billions of cells and neurons connected by cross-links of enormous complexity. Endogenous (internal) electrical signals are vital to biostasis, controlling neural transmissions, receptors, enzyme production etc. However just like a phone, TV or radio, the Central and Autonomic nervous systems are subject to interference from exogenous (external) electrical signals. Thus if pain receptors in the skin or brain receive an anomalous electrical signal of the frequency to 'fire' the 'on' signal, a sharp pain will be felt in the eye, ear, leg, spine etc. (WHO, 1987). If the signal triggers the visual cortex, a virtual visual effect will be generated, which the mind will then project into actual space ahead of the individual. The interpretation of such an effect will depend on the mindset of that individual. On the molecular scale all living processes must be understood in terms of electromagnetic fields and forces. Without these, life is no longer possible. Excessive levels, relative to each cell, each neuron, each synapse, can lead to problems, the seriousness of these depending on the genetic makeup, immunocompetence and to some extent, lifestyle, of the individual.

The skull, brain and ventricles consist of many differing layers, types of tissue, types of fluid (e.g. blood, grey matter, white matter, cerebrospinal fluid etc.) The ventricles, hollow air filled spaces within the brain, vary in size and shape. Thus Lenz's law, which states that if an electromagnetic field cuts through conducting surfaces, currents will be induced, applies to the brain over and above the EMFs generated by the body itself.

14.9 HOTSPOTS IN THE BRAIN

There is some controversy regarding thermal and athermal effects in brain tissue. Athermal or non-thermogenic effects will occur when the energy input does not increase temperature by more than 1 degree C. However there are a number of potential pitfalls in this idea. For example, some exposure set-ups are capable of reflecting of focussing RF radiation in such a way that subjects received a higher SAR than the experimenters may realise, leading to erroneous conclusions.

Secondly as different tissues have differing dielectric constants. Internal refraction and reflection at tissue interfaces has the potential for causing micro 'hotspots'. But we must be careful. If the small hotspot is so tiny as to heat only a few neighbouring molecules the debate becomes semantic. But if an area like the hypothalamus or the brain stem heats up, what are the consequences?

Studies at Lund University in Sweden by Prof Salford in 1999 found that the blood-brain barrier could be breached after very short exposures of radiation at the same levels as Mobile Phones. Molecules such as proteins and toxins can pass out of the blood while the phone is switched on and pass into the brain. (10)

COUPLING EFFICIENCY of the body to electric and magnetic fields at ELF is relatively low, however a time-varying field will induce electric currents in a conducting medium like body tissue with a magnitude which is proportional to the time rate of change of the magnetic field rather than its absolute value. Ordinarily the cell membrane is a formidable barrier to the transport of ions and charged molecules. However electroporation, and electrostatically driven structural re-arrangements of the cell membrane (an athermal, not thermal process) results in a large increase in cell membrane conductance, now believed to be caused by ion transport through temporary membrane openings (pores). The driving force for electroporation is the physical interaction of the electric field with two deformable materials of differing dielectric constants, lips and aqueous electrolytes. As Weaver (1995) states the ability of a physical stimulus like a strong electrical field pulse to compromise this barrier is of considerable significance. E field pulses that are too large or too long are found to cause plasma membrane destruction and are associated with cell killing.

14.10 ELECTROSENSITIVITY AND SIGNAL TRANSDUCTION PATHWAYS

Are pulsed signals the "on" switch for electrosensitivity? It is well known that flickering images at ELF rates will, in many subjects, trigger migraines, with and without headaches, and generate nausea, disorientation, dizziness, flushing, "spaced out" sensation, even microsleep. This pulsing effect also applies to tactile pulsed signals - Chinese Water torture of old being a prime example. In the same way, the beat or pulsation of music opens up a motorway into our emotions and rhythmic dancing is known to lead to altered states in some people. In fact low frequency pulsed EMFs, whether as modulated. field effects from one signal pulsing against another, or as acoustic or visual signals, or seismic as ELF Earthquake precursors, can act as sensory inputs to living organisms, with the Autonomic Nervous System (ANS) being particularly vulnerable.

Knave (1994) has described Electrosensitivity (ES) as multifactorial in origin, with some studies indicating that flickering lights for hours each day may be one cause. Both Kjell-Hans Mild and Cyril Smith implicate malfunction of the ANS. Professor Hughes of the US Department of Labor in Washington went further, describing ES as "a progressively disabling disease associated with exposure to electromagnetic fields created by electronic equipment". He also stated "the importance of electromagnetic sensitivity should be recognized and appreciated by the medical community". Yet despite increasing recognition of ES in Scandinavia, Russia, Europe and the US, there is as yet very little awareness in the UK, with a very few exceptions.

There are also strong indications that a little known form of Migraine is involved in ES. Known in neurological literature as Resonance Migraines, Migraine Traits or Migraine Variants, sufferers do not have the classical headaches, photophobia and other symptoms. But a wide spectrum of anomalous behaviour is reported in the literature and depending on the environment or locale of the individual, can affect driving, productivity, coherence, and lifestyle. Electrical overload in the brain, and especially brain stem, are currently considered strong pointers to this little reported area.

Spikes of ambient energy, surges and transients in the brain will affect all sensory modalities, the Somatory (sudden tingling in limbs, digits, nose), the Visual (sudden flashes of light, coloured balls of light etc.), the Auditory (sudden ringing sound, as of bells, sudden buzzing, as of bees) hearing one's own name called; the Olfactory (intense smells which may be lovely (as lilies) or ghastly (as decay, rotting odours); Positional (vertigo - dizziness without rational cause). In the Autonomic nervous system, sudden anxiety or panic, profuse sweating in normal temperatures and hairs standing on end (piloerection).

Activity in deep brain structures may generate or modulate hallucinations, the particular detail 'seen' due to the areas of the neocortex which have been stimulated. If the temporal lobe is subject to exogenous stimulation a variety of effects can be produced; intense fear, terror, utter isolation, 'alone in the Universe', very strong feelings of disgust, revulsion, for no reason; very strong depression without cause; feelings of familiarity (deja vu), a revelation that one can solve all the problems in the world. This stimulation need not be directly applied, external sources will suffice to trigger strange perceptions. How such perceptions are interpreted depends on the mindset and situation of the one experiencing them. The neurons of the hypothalamus organise aggression and the manifestation of rage. Experiments in the US and Holland have shown that specific frequencies will cause both man and animals to attack violently the next living thing seen after the stimulus is applied, the amygdala being especially vulnerable.

14.11 BIOLOGICAL RESPONSES

Radio frequency energy does not behave like a drug. The rate of absorption and distribution of energy within a body depends on many factors. These include:-

- Dielectrics of the irradiated tissue (bones absorb less energy than muscles)
- Size of object (body) to radiating source
- Shape, geometry and orientation of object, or body
- Proximity of object (body) to radiating source
- Re-radiating artefacts (e.g. metallic implants) in body and environment
- Polarization of RF source
- Immunocompetence of the irradiated body
- Penetration of RF/MW energy in body, the lower the frequency, the deeper the penetration

(NB The human body will maximally absorb RF in the MHz bands acting as a half-wave dipole)

14.12 INDIVIDUAL RESPONSE TO 'HOTSPOTS'

On the perceptual level there are three major and significant factors for any individual

- The response of the system to electrical current depends on the state of the system itself.
- Moderate magnitudes of electricity do not produce permanent changes in the system when it is in a globally homeostatic state. But when the system is, or has been stressed, whether due to trauma,   shock, pathological conditions, subclinical condition in the continuum of human effects or chemical (drugs), electric shock or lightning strike or 'splash', it is very much more vulnerable.
- Extreme magnitudes of electrical current lead to the detriment of the system regardless of its state at time of application.
- In a recent (draft) document under 'Prudent Avoidance' is the statement 'However, it is often the case that exposures from distant TV and radio transmitters may be higher than those from nearby mobile phone masts.' Also 'touching one's head with a mobile phone antenna while engaged in conversation, increases the amount of RF energy that is absorbed by the head and reduces the amount that reaches the base station antenna. A result, the base station instructs the phone to increase its power output.'

There is a vast area in the continuum between day-to-day usual perceptions and sufficiently frequent and unusual sensations which are labelled by the medical profession as syndromes. Human sensitivity varies very considerably between those who barely notice anomalous sensations and those for whom such awareness makes life intolerable. It must be noted that those living and working in RF fields of high amplitude are very frequently found to be photophobic (light sensitive) phonophobic (sound-sensitive) and osmophobic (smell sensitive). They are also very sensitive to touch, however gentle on certain parts, e.g. upper arm, forehead, legs. Short term memory loss is encountered (missing time) and some report hearing their own name called in an empty house - a pointer to high Alpha wave activity in the brain. Response to fields depends on many factors, height, thickness and shape of skull, immunocompetence, pre-sensitisation, position, whether horizontal or vertical, polarisation of ambient field, skin resistivity, coupling to ground, etc. Metallic dental fillings can act as re-radiating antennae, magnetic implants act the same way.

Before an individual can respond to any stimulus, there must be a change whereby the input of stimulus energy becomes converted into a code the neurons will understand and act on. Such responses depend on the source of the stimulus, light and the eye, acoustics and the ear, tactile sensation and the skin, radiofrequency pulses and the brain.

14.13 ARE THERE CONFOUNDING FACTORS?

Yes, there are many. These include inaccurate diagnosis, bias, too short exposure time, data smoothing, mean power rates used rather than Peak Envelope Power, spikes and surges. Much research in field of RF effects has been funded by Manufacturers of RF emitting equipment and an inbuilt bias is apparent in the structure of many such studies.

CONFOUNDING FACTORS (CFs) Epidemiologists are still investigating the relevant precursor factors to use in their study of RF and MW field effects. Working from the specific of human 'real world' individuals, this paper hopes to draw attention to the specific variables and CFS which are common to ES and the upper continuum of RF effects. Amongst these are: Blood groups: Chemotyping - slow oxidisers, poor sulphoxidisers, poor acetylators; Electric shock even years prior to present, Lightning strike, even years prior to present; closed head injury even years prior to present.

Questionnaires are frequently used and often omit factors which, to the writer, appear of prime importance. For example, it is rare to find Electric Shock as a life event mentioned, yet it is highly relevant to Electroporation (a weakening of the cell membrane).

14.14 CANARIES IN THE COAL MINE? - POINTERS TO HYPERSENSITIVITY

Coal miners of old used to take caged canaries down the mines with them - the presence of the dreaded methane gas would be signified when the canaries stopped singing and dropped to the base of the cage. The miners knew that was the time to get back to the surface as speedily as possible, before they too dropped senseless to the ground. In Pozzuoli, in Italy, the Lazaretto (Guardian) of a cave had a similar trick to show those on the Grand Tour in the 19th C. He would take a dog just inside the cave, and in a few minutes it would collapse and be totally unresponsive. Then he would haul the unconscious dog out, where, in the fresh air it would recover. But the dog did not always recover and on occasions the Lazaretto would haul out a canine corpse.

It is feasible that the growing numbers of people who respond adversely to electromagnetic fields of varying frequencies may be our 21st C canaries or Pozzuoli pets. Just like them, in dozens of cases interviewed, when the individuals are away from the place, office, or house where they are affected, their symptoms disappear. What are the symptoms? These include fatigue, nausea, sleeplessness (at home), pains without cause, acoustic effects (the Hum), visual anomalies (magnetophosphenes, flickering white, rarely blue, images at edge of the visual field which vanish when the head turns to examine the image), benign lid fasciculation (the twitch which can barely be seen, but feels very obvious), fasciculation (visible twitchings below the skin on arms, hands). Sudden sense of fear, terror, sense of presence when no-one else is there, a sensation of being pushed.

Health problems arise from a variety of sources, including genetic predisposition, biochemical imbalance and shocks of many types. It is very probable that the very signs and symptoms now being reported in a growing percentage of people including heavy users of mobile phones, may act as a useful tool in electro-diagnosis of pre-existing chemical imbalances.

In the 21st Century it is now timely for the medical and allied professions to address "E 3" – as the engineers already describe Environmental Effects of Electromagnetism. Great concerns are involved in ensuring Electromagnetic Compatibility (EMC) for equipment from planes to computers to elevators (to list but three). So surely the human body, as an electrochemical construct, is just as important? And possibly even more liable to suffer Interference effects?

14.15 REFERENCES

Birbeck G, 2002, Migraine Variants – Emedicine Journal **3**:2 1-12

Dodick DW, 2002, Thunderclap Headache J Neurol Neurosurg Psychiatry **72**: 6-11

Harling DW *et al*, 1989, Thunderclap Headache – Is it Migraine? Cephalalgia 9:87-90

Hughes MM, 1998, A review of Studies reporting Human sensitivity to Electromagnetic Fields Abstracts Bioelectromagnetics Society Conf, June. 1998

Knave B, 1994, Electrosensitivity Scan. J of Work, Environment & Health **20**:84

Lee RC, 1992, Electrical Trauma CUP

Luders H, 1987, Electroclinical Syndromes - Springer Verlag – Berlin

Pareja J *et al*, 1997, SUNCT Syndrome Headache **37**:4 195-202

Petrie KJ, 2002, Modern worries – new Technology and Medicine Brit. Med. J 324-690

Reiter R, 1981, The mammalian Pineal Gland – AM. J Anatomy **162(4)** 287-313

Salford LG, 1993, Permeability of the Blood-brain barrier induced by 915 MHz Electromagnetic radiation Bioelectrochem. Bioenergy **30** :293-301

Silk A.C., 1999, Mobile Phones-hot heals-hot eyes? IEE Conference

Silk A.C., 2002, Human Efects of Radiofrequency Fields. National Phisical Lab conference, March

Weaver J., 1995, Electroporation in Cells and tissues Radio Science US **30**:1 205-221

Weiller C, May A *et al*, 1995, Brain stem activation in spontaneous Human Migraine attacks Nature Medicine **1**:7 658-664

WHO, 1993, Electromagnetic Fields 300Hz – 300GHz – Geneva

14.16 FURTHER REFERENCES FOR THE CHAPTER

14.16.1 Books

Radiation Ed. D. Brune (2001) Scandinavian Science Publishers 550pp
Electromagnetic Compatibility (1997) J Scott & C van Zyl Newnes
Radar Ed P S Hall (1997) Brassey Military Publications
Mobile Phones and Health (2000) Sir William Steward UK Government
   Independent Expert Group
Electromagnetic Fields 300 Hz - 300 GHz (1993) WHO Geneva

14.16.2 Conference Abstracts

Microwave and Radiofrequency Fields - Safety issues (1999) Institute of Physics
   in Medicine Conference
Bioelectromagnetics Society conference Abstracts - (2001) 23rd meeting - St. Paul -
   USA
Bioelectromagnetics Society Conference Abstracts - (2001) 22nd meeting -
   Munich - Germany
Mobile Phones - Is there a Health Risk? - (2001) International Conference
   abstracts - London
STOA EU Committee Meeting - Brussels - (2001)
Possible Health effects related to use of Mobile Phones - COST Action 244 bis
   (1999) - International Congress on Occupational Health - 26th - August 2999 -
   Singapore

CHAPTER 15

Health Effects of High Voltage Powerlines

Denis L. Henshaw and A. Peter Fews

15.1 INTRODUCTION

High voltage powerlines can ionise the air, emitting a stream of so-called corona ions into the atmosphere. These ions attach to aerosol-sized particles of air pollution, such as those containing potentially carcinogenic polycyclic aromatic hydrocarbons, PAHs, increasing the electric charge state on these aerosols. The resulting cloud of charged aerosols is carried by the wind up to several hundred metres from powerlines, occasionally extending to several kilometres away. When inhaled, aerosol particles with electric charge have a higher probability of lung deposition compared with uncharged aerosols. Directly under powerlines, 50 Hz oscillation of pollutant aerosols results in increased deposition on the body. One consequence of this phenomenon is an unusually high radiation dose rate to the skin from the naturally occurring radioactive radon decay product aerosols in air. Increased exposure by these mechanisms implies increased risk. A risk analysis suggests that between 2,000 and 3,000 cases of pollution-related ill health may occur annually among the 2.7 million of the population living within 400 m of 132, 275 and 400 kV powerlines in the UK.

There is much current interest in the adverse health effects of exposure to power frequency electric and magnetic fields. Evidence for the latter has been significantly strengthened in recent months with the release for public discussion of the California Health Department EMF Report (2001). The Report suggests that an added risk of miscarriage, childhood and adult leukaemia, brain cancer and greater incidence of suicide are some of the health risks associated with exposure to magnetic fields such as those that radiate from powerlines. Evidence of the depression in nocturnal production of melatonin in women exposed to magnetic fields of 0.2 µT or lower (Davis *et al*, 2001; Levallois *et al*, 2001) and of the inhibition of the oncostatic action of melatonin by magnetic fields of 1.2 µT (Ishido *et al*, 2001), adds weight to our growing understanding of the mechanisms by which exposure to power frequency magnetic fields may result in adverse health effects.

In addition to these developments, we have been studying the way in which electric field effects, specifically associated with high voltage powerlines, interact with aerosol-sized particles of air pollution in a manner which increases exposure to such pollution. Increased exposure implies increased risk of those illnesses already known to be associated with air pollution. A risk analysis suggests that across the UK as a whole, the number of excess cases of illness near powerlines is likely to be at a level of public health significance.

Here we summarise the two most important mechanisms by which electric field effects associated with high voltage powerlines mediate the increased exposure to ambient air pollution, together with an assessment of the magnitude of the effect in terms of public health.

15.2 CORONA ION EMISSION FROM HIGH VOLTAGE POWERLINES

The most important mechanism by which powerlines mediate increased exposure to air pollution is via corona ions. The basic processes of corona ion production and behaviour in the atmosphere are shown in Figures 15.1 and 15.2. High voltage powerlines can ionise the air producing free electrons and molecular (positive) ions of oxygen and nitrogen. The electrical mobilities of electrons and positive ions are quite different and as a result, separate clouds of positive and negative ions are emitted into the atmosphere. Within 100 ns, these ions attract polar molecules, mainly water, forming molecular clusters around 0.4 to 1.2 nm in size. Such molecular clusters constitute what is known as small ions or nuclei mode aerosols. Apart from their size, they differ from larger aerosols in that they are bound only by the electric charge of the ion. If this charge was to disappear, for example by neutralisation, then the cluster would fall apart.

Power companies assess corona in terms of the power lost along cables. Losses of up to 0.1 mA per metre of cable are estimated (Abdel-Salam and Abdel-Aziz, 1994). This corresponds to 6.25×10^{14} ions per metre per second. Even if most of these ions are subsequently re-absorbed by the line, the potential number of ions emitted into the atmosphere is large.

In the atmosphere these small ions disperse in two ways. They may disperse in their own right or they may attach themselves to particles of air pollution at a rate governed by their ion/aerosol attachment coefficient (Hoppel and Frick, 1986), thereby increasing the electric charge on such particles over and above any natural charge they may hold. Thus the cloud of charge, consisting of a combination of both small ions and electrically charged aerosols, constitutes a space charge, which only becomes neutralised when dispersed well away from powerlines. This is a non-equilibrium steady state situation in which space charge is maintained in the air near powerlines by the continuous production of corona ions, balanced by the eventual neutralisation some distance away.

Corona ion space charge can be carried considerable distances from powerlines by the wind. In the 1950s such space charge was detected up to 7 km away (Chalmers, 1952; Mülheissen, 1953). Corona ion emission has been extensively researched both theoretically (Morrow, 1997; Abdel-Salam and Abdel-Aziz, 1994) and experimentally using test lines (Carter and Johnson, 1988). We have measured corona ion emission from 132, 275 and 400 kV AC powerlines in South-west England (Fews *et al*, 1999a). Significant effects were found up to 400 metres downwind. In ongoing work, significant effects have since been detected between 2.7 and 7 km downwind of a 400 kV powerline in Somerset.

Our measurements and those of both previous authors have used changes in the observed ground-level DC field of the earth, as an indicator of the presence of corona ion space charge near powerlines.

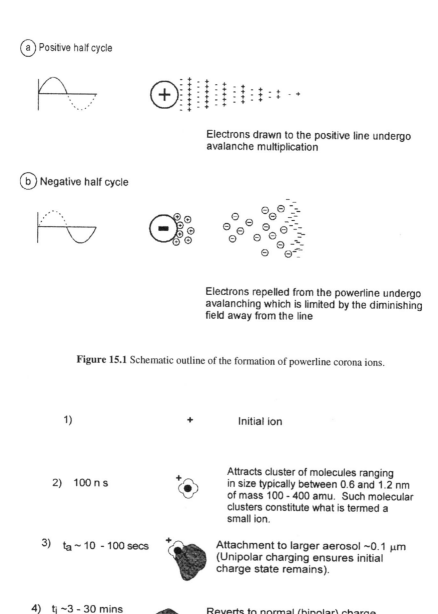

(a) Positive half cycle

Electrons drawn to the positive line undergo
avalanche multiplication

(b) Negative half cycle

Electrons repelled from the powerline undergo
avalanching which is limited by the diminishing
field away from the line

Figure 15.1 Schematic outline of the formation of powerline corona ions.

1)   +   Initial ion

2)   100 n s   Attracts cluster of molecules ranging
                      in size typically between 0.6 and 1.2 nm
                      of mass 100 - 400 amu. Such molecular
                      clusters constitute what is termed a
                      small ion.

3) t_a ~ 10 - 100 secs   Attachment to larger aerosol ~0.1 μm
                      (Unipolar charging ensures initial
                      charge state remains).

4) t_i ~3 - 30 mins   Reverts to normal (bipolar) charge
                      distribution

Figure 15.2 Schematic diagram showing the fate of corona ions in the atmosphere.

In Figure 15.3, simultaneous measurements are shown of the variation with time of the DC electric field 50 m upwind and 50 m downwind of a 132 kV powerline in South Gloucestershire. Measurements were made using a calibrated Chubb JCI 140C DC field mill meter mounted one metre above the ground. Upwind, the field is relatively stable at ~160 V m^{-1} over the 15-minute period of the measurement. Downwind, the field is much higher, ~700–800 V m^{-1} and is erratic on a time scale of seconds. Modelling of airborne space charge shows that this is consistent with a flux of positively charged small ions and charged aerosols flowing past the meter (Wilding, 2002). The observations suggest that the concentration of excess unipolar charges in air is of the order of a few thousand per cm^3. This should be compared with the pollutant aerosol particle concentration, which is measured alongside the DC field measurements using a TSI 3010 condensation particle counter. The aerosol concentration varies widely, but a typical average value is around 15,000 per cm^3, suggesting that in the steady state situation, between 15 and 20% of pollutant aerosols may become charged by corona ions. Work in progress is using charged aerosol spectrometers to measure the size spectrum of both small ions and charged aerosols near powerlines.

It is known that between 50 and 90% of outdoor pollutant aerosols penetrate indoors in normal ventilation (Hussein *et al*, 2001). It may therefore be assumed that near powerlines a proportion of pollutant aerosols electrically charged by corona ions will be inhaled in the indoor environment.

15.3 HEALTH IMPLICATIONS OF CORONA IONS EMITTED FROM HIGH VOLTAGE POWERLINES

When inhaled, electrically charged aerosol particles have a higher probability of being deposited in the lung compared with uncharged aerosols. This occurs by mirror charge effects. Basically, when an electric charge approaches a conducting surface, it sees a reflection of itself in that surface in the form of an aerosol of opposite charge to its own. It then becomes attracted to that surface by this self-attraction mechanism. The phenomenon is short range, operating only a few tens of microns from conducting surfaces. It is therefore effective only in confined spaces of dimensions similar to the range of mirror charge forces. Such a situation occurs in the small airways and alveoli of the human lung. No external electric field is involved in this process, which means that the effect can occur well away from the direct effect of the powerline AC electric field. It should be further noted that deposition on surfaces by mirror charge outside the body is not a predominant mechanism owing to the short-range nature of the effect.

In considering the increased lung deposition of inhaled electrically charged aerosol-sized particles of air pollution, we are particularly interested in those falling in the peak of the number density distribution, the so-called ultrafine aerosols which lie in the approximate size range 20–100 nm. These possess a number of important features compared with larger aerosols.

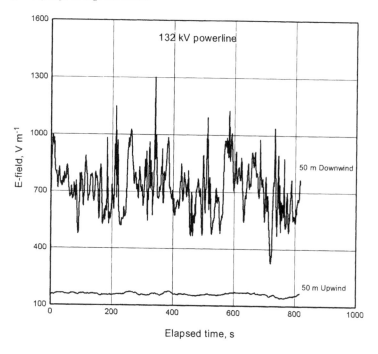

Figure 15.3 Time profile of the DC electric field measured 1 m above the ground, 50 m upwind and downwind of a 132 kV powerline in South Gloucestershire.

They are able to penetrate deeply into the lung, depositing mainly in the tracheobronchial and alveolar regions. For this reason they are postulated to be more toxic than larger aerosols, such as the PM10s, which are aerosol particles of around 10 µm in size (Seaton *et al*, 1995). The ultrafine aerosols are also important because they contain a significant proportion of the potentially carcinogenic polycyclic aromatic hydrocarbons (PAHs), including the important lung carcinogen benzo[α]pyrene (BaP), (Allen *et al*, 1996).

On inhalation in their natural state, the ICRP 66 lung model (1994) estimates that aerosols in the size range 100–200 nm have only a 30–40% probability of being deposited in the lung (Figure 15.4). Direct measurements on human volunteers suggest that for 100 nm aerosols the lung deposition is lower, in the range 24–27% (Kim and Jaques, 2000). The total deposition fraction estimated for the inhalation of PAH aerosols has been estimated at around 30% (Venkataraman and Raymond, 1998). Cohen *et al* (1998) has measured the deposition probability of ultrafine aerosols in metal alloy casts of the human tracheobronchial tree. For 20 and 125 nm aerosol particles, the authors found a respective increase in deposition of 3.4 ± 0.3 and 2.3 ± 0.3 when the aerosols were singly charged, compared with their natural (mainly uncharged) state. The effect of electric charge in significantly increasing the probability of deposition is considered important because a high proportion of lung cancers originate in the tracheobronchial region.

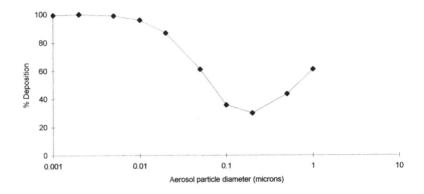

Figure 15.4 Total lung deposition of inhaled particles as a function of size (AMTD) in the absence of electric charge effects, according to the ICRP 66 lung model (1994).

15.4 RISK OF AIR POLLUTION RELATED ILLNESS NEAR POWERLINES

We therefore postulate that corona ions mediate increased exposure to air pollution up to 400 m from high voltage powerlines, the predominant effect occurring downwind of the prevailing south-westerly wind across the UK. As stated above, this includes the effects of indoor exposure from the flux of outdoor aerosols which penetrate indoors. A risk analysis can be developed to estimate the increase in illness that might occur annually near powerlines in the UK.

There is evidence that apart from smoking, lung cancer risk is associated with air pollution. Katsouyanni and Pershagen (1997) reviewed 12 case-control studies and 9 cohort studies carried out between 1955 and 1995. More recently Nyberg *et al* (2000) reported the results of a study in Stockholm. Together these showed an excess relative risk of lung cancer in relation to urban air pollution in the range 1.3–2.5.

For singly charged ultrafine aerosols, Cohen *et al* (1998) found increased tracheobronchial deposition factors of 3.4 and 2.3 for 20 and 125 nm aerosols. This is an average increased deposition factor of 2.85, or an absolute increase in deposition of 1.85 or 185%. Following the analysis by Venkataraman and Raymond (1998), this suggests that 100% ultrafine aerosol charging may almost double the tracheobronchial deposition of inhaled PAH ultrafine aerosols.

For the risk assessment, let us make the conservative assumption that corona ion exposure results in a 30% increase in deposition in the tracheobronchial region of the lung. Take the affected population as living within 400 m of high voltage powerlines, in the direction downwind of the prevailing south-westerly wind across the UK. In 1996 there were 25,700 cases of lung cancer in men and 15,200 cases in women in the UK, a rate of 69 per 100,000 of the population (UK statistics on cancer, 2002). From Table 15.1, the estimated number of people living within 400 m of powerlines $\geq$ 132 kV is 4.6%, or 2.7 million people. If a 30% increase in lung cancer risk downwind compared with upwind is assumed, this yields 285 cases annually. The range quoted in Table 15.2 of 200–400 cases annually takes into account two possibilities: (i) that on average corona ion effects may not extend to 400 m from powerlines or (ii) that a contribution to risk in those living upwind of the prevailing south-westerly wind should be included.

A similar estimate may be made for other illnesses associated with air pollution such as cardiovascular and respiratory illnesses, aggravated asthma and allergies. In Table 15.2 we estimate that between 2,000 and 3,000 excess cases of such illnesses occur annually. Some studies have linked childhood leukaemia risk to air pollution, notably from vehicle exhausts (Savitz and Feingold, 1989; Knox and Gilman, 1997; Nordlinder and Järvholm, 1997; Feychting *et al*, 1998; Harrison *et al*, 1999; Pearson *et al*, 2000; Reynolds *et al*, 2002). Raaschou-Nielson *et al* (2001) found an association between traffic pollution and Hodgkin's disease but not with childhood leukaemia. There is also evidence relating parental exposure to PAHs to an increased leukaemia risk in their offspring (Savitz *et al*, 1990; Shu *et al*, 1999). In Table 15.2, we estimate that 2–6 excess cases of childhood leukaemia annually might occur in those living within 400 m of high voltage powerlines.

Table 15.1 Approximate number of people living near high voltage powerlines in the UK.

Distance (m)	$\geq$ 275 kV powerline Number / (% UK population)[*]	$\geq$ 132 kV powerline Number / (% UK population)[§]
25	35,400 (0.06)	94,400 (0.16)
50	70,800 (0.12)	177,000 (0.30)
100	142,00 (0.24)	354,000 (0.60)
150	248,000 (0.42)	620,000 (1.05)
200	372,000 (0.63)	932,000 (1.58)
250	519,000 (0.88)	1,300,000 (2.20)
300	690,000 (1.17)	1,730,000 (2.93)
350	897,000 (1.52)	2,240,000 (3.80)
400	1,100,000 (1.86)	2,740,000 (4.65)

*From UKCCS (2000)
[§]Estimated here

Table 15.2 Possible excess number of cases annually of ill health in persons living near
high voltage powerlines in the UK.

Condition		Possible excess cases annually in the UK near high voltage powerlines
Corona Ions		
(i)	Lung cancer mortality	200–400
(ii)	Other illnesses associated with air pollution	2,000–3,000
(iii)	Childhood leukaemia incidence	2–6
50 Hz Electric Fields		
(i)	Skin cancer incidence (non-melanoma)	17

15.5 OSCILLATION OF AEROSOLS IN A 50 HZ POWERLINE ELECTRIC FIELD

A second mode of interaction between powerline electric fields and air pollution is that of the 50 Hz oscillation of aerosols in the powerline electric field. This has been investigated using natural radon decay product aerosols as a marker for general aerosol behaviour. Electric fields are distorted around conducting objects, leading to an enhanced field gradient near that object. The human body is conducting and as a result there is an 18-fold increase in electric field around the human head near high voltage powerlines. For a typical 5 kV m^{-1} field strength directly under a powerline, the field around the human head would be 90 kV m^{-1}.

Fews (1999b) reported a 1.4 to 2.9-fold increase in deposition of the alpha-particle emitting ^{218}Po and ^{214}Po radon decay product aerosols on models of the human head placed under high voltage powerlines outdoors, compared with control heads placed some distance away. The observations have consequences for the radiation dose rate to the sensitive cells at the basal layer of the skin from radon decay products which is up to two orders of magnitude higher outdoors compared with indoors. It is possible that people spending only a small proportion of their time under high voltage powerlines outdoors would exceed the International Commission on Radiological Protection (ICRP) recommended skin dose limit to the general public of 50 mSv y^{-1} (NRPB, 1997). Such a situation would be relevant to people whose homes are directly under high voltage powerlines (within ±25 m of the centre line), about 94,000 people in the UK. NRPB (1997) estimates that the excess relative risk of non-melanoma skin cancer is 60% per Sv of radiation. Taking account of the increased deposition of ^{218}Po and ^{214}Po aerosols on exposed areas of the body such as the head under powerlines, this suggests an increased risk of skin cancer.

We have carried out a risk analysis resulting from the increased radiation dose to the basal layer of the skin within alpha-particle range of deposited ^{218}Po and ^{214}Po aerosols (Fews *et al*, 2000) in people living under powerlines in the UK. In Table 2, we estimate such exposure results in 17 excess cases of non-melanoma skin cancer annually. As indicated above, we are in this case concerned with the time that people spend outdoors under high voltage powerlines.

15.6 DISCUSSION

All overhead powerlines of 132 kV and above appear to be in a state of low-level corona ion emission most of the time. Powerlines are designed to be corona free, but for the voltage carried, the onset field for corona is close to that on the cable surface (Wilding, 2002). Once installed, powerline cables progressively attract pollutants, in part by the oscillation and deposition of small aerosol particles, leading to the build-up of small edges and point corona discharge along their length. Thus the level of corona emission from lines is variable and is also dependent on weather conditions. On some lines the effects are severe with corona ion space charge detectable several kilometres downwind.

In an attempt to look for evidence of adverse health effects resulting from powerline corona ion emission, Preece *et al* (2001) have compared cancer incidence within 400 m upwind and downwind of powerlines in South-west England. The authors found a statistically significant higher incidence of both lung and mouth cancer on the downwind compared with the upwind side of powerlines. This work is being extended to other areas of the UK.

Separately, the UK National Radiological Protection Board has recently set up an *ad hoc* committee to investigate the health implications of powerline corona ion emission. Future work should focus both on more detailed measurements of charged aerosol particles near high voltage powerlines, as well as epidemiological studies of illnesses linked to air pollution in populations living near powerlines. Such work should include the effects of increased deposition on the body of pollutant aerosols under powerlines as a result of their 50 Hz oscillation.

Even an approximate realisation of the risk estimates outlined in this paper would be of public health significance. This would suggest that the building of overhead powerlines near populated areas should be avoided, in favour of under-grounding, for which the technology is well established.

15.7 ACKNOWLEDGEMENTS

This work is supported by Children with Leukaemia, Registered Charity No. 298405, Medical Research Council Programme Grant No. G8319972 and the Department of Health.

15.8 REFERENCES

Allen, J.O., Dookeran, N.M., Smith, K.A. and Sarofim, A.F., 1996, Measurement of polycyclic aromatic hydrocarbons associated with size-segregated atmospheric aerosols in Massachusetts. *Environmental Science & Technology*, **30**, pp. 1023-1031.

Abdel-Salam, M. and Abdel-Aziz, E.Z., 1994, A charge simulation based method for calculating corona loss on AC power transmission lines. *Journal of Physics D: Applied Physics*, **27**, pp. 2570-2579.

Carter, P.J. and Johnson, G.B., 1988, Space charge measurements downwind from a monopolar 500 kV HVDC test line. *IEEE Transactions on Power Delivery*, **3**, pp. 2056-2063.

California Health Department, 2001, An evaluation of the possible risks from electric and magnetic fields (EMFs) from power lines, internal wiring, electrical occupations and appliances. (California EMF Program, 1515 Clay Street, 17th Floor, Oakland, CA 94612) http://www.dhs.ca.gov/ehib/emf/RiskEvaluation/riskeval.html

Chalmers J.A., 1952, Negative electric fields in mist and fog. *Journal of Atmospheric and. Terrestrial Physics*. **2,** pp. 155-9.

Cohen, B.S., Xiong, J.Q., Ching-Ping, F. and Li, W., 1998, Deposition of charged particles on lung airways. *Health Physics*, **74**(5), pp. 554-560.

Davis, S., Kaune, W.T., Mirick, D.K., Chen, C. and Stevens, R.G., 2001, Residential magnetic fields, light-at-night, and nocturnal urinary 6-sulfatoxymelatonin concentration in women. *American Journal of Epidemiology*, **154**, pp. 591-600.

Fews, A.P., Henshaw, D.L., Wilding, R.J. and Keitch, P.A., 1999a, Corona ions from powerlines and increased exposure to pollutant aerosols. *International Journal of Radiation Biology*, **75**(12), pp. 1523-1531.

Fews, A.P., Henshaw, D.L., Keitch, P.A., Close, J.J. and Wilding, R.J., 1999b, Increased exposure to pollutant aerosols under high voltage powerlines. *International Journal of Radiation Biology*, **75**(12), pp. 1505-1521.

Fews, A.P., Henshaw, D.L., Wilding, R. J. and Keitch, P.A., 2000, Dose to the skin basal layer from increased plateout of radon decay product aerosols under high voltage powerlines. *22nd Annual Meeting of The Bioelectromagnetics Society* (In Cooperation with The European Bioelectromagnetics Association), Munich, 11–16 June, 2000.

Feychting, M., Svensson, D. and Ahlbom, A., 1998, Exposure to motor vehicle exhaust and childhood cancer. *Scandinavian Journal of Work Environmental Health,* **24**, pp. 8-11.

Harrison, R.M., Leung, P.L., Sommervaille, L., Smith, R. and Gilman, E., 1999, Analysis of incidence of childhood cancer in the West Midlands of the United Kingdom in relation to proximity to main roads and petrol stations. *Occupational and Environmental Medicine,* **56**, pp. 774-780.

Hoppel, W.A. and Frick, G.M., 1986, Ion-aerosol attachment coefficients and the steady-state charge distribution on aerosols in a bipolar ion environment. *Aerosol Science and Technology*, **5**, pp. 1 – 21.

Hussein, T.F., Kakko, L, Aalto, P., Hämeri, K. and Kulmala, M., 2001, Indoor-outdoor aerosols: Particle size characterisation in a suburban area. *Journal of Aerosol Science*, **32**, Suppl. 1, pp. S149-150.

ICRP Publication 66, 1994, Human respiratory tract model for radiological protection. *Ann ICRP (UK)* **24**, Nos. 1-3, (Pergamon Press).

Ishido, M., Nitta, H. and Kabuto, M., 2001. Magnetic fields (MF) of 50 Hz at 1.2 μT as well as 100 μT cause uncoupling of inhibitory pathways of adenylyl cyclase mediated by melatonin 1a receptor in MF-sensitive MCF-7 cells. *Carcinogenesis*, **22** (7), 1043-1048.

Katsouyanni, K. and Pershagen, G., 1997, Ambient air pollution exposure and cancer. *Cancer Causes and Control*, **8**, pp. 284-291.

Kim, C.S. and Jaques, P.A., 2000. Respiratory dose of inhaled ultrafine particles in healthy adults. *Ultrafine particles in the atmosphere. Philosophical Transactions of The Royal Society*, pp. **358**, 2693.

Knox, E.G. and Gilman, E.A., 1997, Hazard proximities of childhood cancers in Great Britain from 1953-80. *Journal of Epidemiology Community Health*, **51**, pp. 151-159.

Levallois, P., Dumont, M., Touitou, Y., Gingras, S., Mâsse, B., Gauvin, D., Kröger, E., Bourdages, M. and Douville, P., 2001, Effects of electric and magnetic fields from high-power lines on female urinary excretion of 6-sulfatoxymelatonin. *American Journal of Epidemiology*, **154**, pp. 601-609.

Morrow, R., 1997, The theory of positive glow corona. *Journal of Physics D: Applied. Physics*, **30**, pp. 3099-3114.

Mühleisen, R., 1953, Die luftelektrischen Elemente im Großstadtbereich. *Z. Geophysics*. **29**, pp. 142-160.

National Radiological Protection Board (NRPB), 1997, Assessment of skin doses. *Documents of the NRPB*, **3**, No. 3, (NRPB, Chilton, UK).

Nordlinder, R. and Järvholm, B., 1997, Environmental exposure to gasoline and leukaemia in children and young adults – an ecological study. *International Archives of Occupational and Environmental Health*, **70**, pp. 57-60.

Nyberg, F., Gustavsson. P., Järup. L., Bellander. T., Berglind. N., Jakobsson. R. and Pershagen. G., 2000, Urban air pollution and lung cancer in Stockholm. *Epidemiology*, **11**, pp. 487-495.

Pearson, R.L., Wachtel, H. and Ebi, K.L., 2000, Distance-weighted traffic density in proximity to a home is a risk factor for leukaemia and other childhood cancers. *Journal of Air Waste Management Association*, **50**, pp. 175-180.

Preece, A.W., Wright, M.G., Iwi, G.R., Dunn, E. and Etherington, D.J., 2001, Cancer and high voltage powerlines with respect to wind direction. *23rd Annual Meeting of the Bioelectromagnetic Society, St Paul, Minnesota, June 2001*.

Raaschou-Nielson, O., Hertel, O., Thomsen, B.L. and Olsen J.H., 2001, Pollution from traffic at the residence of children with cancer. *American Journal of Epidemiology*, **153**(5), pp. 433-443.

Reynolds, P., Elkin, E., Scalf, R., Von Behren, J. and Neutra, R.R., 2001, A case-control pilot study of traffic exposures and early childhood leukaemia using a geographic information system. *Bioelectromagnetics*, Supplement 5, pp. S58-S68.

Savitz, D.A. and Feingold, L., 1989, Association of childhood cancer with residential traffic density. *Scandinavian Journal of Work Environmental Health,* **15**, pp. 360-363.

Savitz, D.A. and Chen, J., 1990, Parental occupation and childhood cancer: Review of epidemiologic studies. *Environmental Health Perspectives,* **88**, pp. 325-337.

Seaton, A., MacNee, W., Donaldson, K. and Godden, D., 1995, Particulate air pollution and acute health effects. *The Lancet,* **345**, pp. 176-78.

Shu, X.O., Stewart, P., Wen, W. et al., 1999, Parental occupational exposure to hydrocarbons and risk of acute lymphocytic leukaemia in offspring. *Cancer Epidemiology, Biomarkers and Prevention,* **8**, pp. 783-791.

UK statistics on cancer, suicide and other illnesses, 2002, http://www.statistics.gov.uk

UK Childhood Cancer Study Investigators, 2000, Childhood cancer and residential proximity to power lines. *British Journal of Cancer,* **83**(11), pp. 1573-1580.

Venkataraman, C. and Raymond, J., 1998, Estimating the lung deposition of particulate polycyclic aromatic hydrocarbons associated with multimodal urban aerosols. *Inhalation Toxicology,* **10**, pp. 183-204.

Wilding, R., 2002, Corona ions and implications for human health. Ph.D. thesis, University of Bristol.

CHAPTER 16

Effects of 50 Hz Magnetic Field Exposure on Mammalian Cells in Culture

Barry D. Michael, Kevin M. Prise, Melvyn Folkard,
Stephen Mitchell and Stuart Gilchrist

16.1 INTRODUCTION

16.1.1 Background

Human exposure to electromagnetic fields (EMFs) continues to rise with the general increase in applications of electrotechnical devices. Man-made fields differ very greatly in strength and temporal distribution from those that occur in nature, such as the earth's magnetic field and the electric fields and disturbances that arise in the atmosphere. It is therefore inevitable that society should need to examine whether there is any detriment to human health that results from exposure to these artificial fields. The range of man-made frequencies extends from static fields through power, audio and radio frequencies up to microwaves. From a simple physical point of view, there is little to suggest that the EMFs to which people are generally exposed could have any effect on their health. This is quite different from the situation with ionising radiations (x-rays, alpha-particles etc.) where their intrinsic energy is sufficient to change molecular structure (e.g., by breaking DNA) and where the connection between the induction of genotoxic damage and the causation of cancer and heritable defects seems fairly clear. However, some epidemiological studies have indicated a possible correlation between EMF exposure and the incidences of certain types of cancer, although other studies have failed to confirm an association. At the present time, research continues to examine whether there are any real risks and, if so, what are their magnitudes. The aim is to provide the facts needed for regulation, risk management and public information. Current research is on three main fronts: epidemiology, animal studies and cellular studies. This combination of approaches is required to establish whether there is evidence of a health effect in exposed populations, whether the magnitude or incidence of any effect relates to the level and/or duration of exposure and whether there is a candidate mechanism for any effect. To establish a cause-and-effect relationship, clear evidence of all three of these aspects is required. The study reported in this chapter investigates previous work published by others which has indicated that the mutagenic effect of ionising radiation on cells in culture can be enhanced by exposing them to power-frequency EMFs. Thus this work had further suggested that the level of genotoxic damage present in the population due to natural and man-made ionising radiation exposures, as well as from other processes known to cause damage to DNA, might be amplified by EMF exposure and lead to an increase in cancer. For reasons set out below, the uncertainties associated with studies of the type reported previously are considerable. The purpose of the present study was to repeat independently some aspects of the earlier work and so help to verify whether or not the reported amplifying effect of EMF was reproducible and significant.

16.1.2 Earlier Work

Studies on the role of EMF in the interactions with biological systems have predominantly concentrated on a search for any direct effects. Despite a range of studies, few consistent, reproducible or verifiable effects have been reported. Some studies have suggested that rather than a direct genotoxic response, EMF exposure may modulate, for example, a promotional step in the pathway to carcinogenesis. A study by Walleczek *et al.,* (1999) reported interactions between EMF exposure and the effects of ionising radiation. An increased frequency of mutations at the hypoxanthine-guanine phosphoribosyl transferase (HPRT) gene was found if Chinese hamster ovary (CHO) cells in culture were exposed to an EMF field after irradiation with 2 Gy of γ-rays. A 1.8-fold increase in the radiation-induced mutation frequency was observed by subsequent exposure for 12 hours to a 0.7 mT, 60 Hz field. No evidence was found for changes in background mutation frequency when the cells were exposed to EMF only. A similar study using higher field strengths over longer exposure times also found an influence on the yield of X-ray induced mutations (Miyakoshi *et al.,* 1999). A number of other studies have pointed to influences of EMF exposure on the effects of DNA-damaging agents (LaGroye and Poncy, 1997, Maes *et al.,* 2000, Yaguchi *et al.,* 2000, and Miyakoshi *et al.,* 2000) while the results of other studies have indicated that there is no such interaction (Ansari and Hei, 2000, Suri *et al.,* 1996, Mittler, 1973).

16.2 METHODS

The study reported here was designed to replicate the experiments of Walleczek *et al.* (1999), following the same experimental protocol which consisted of a series of repeat experiments alternating between one exposed coil and one sham coil (alternating the exposed coil with each exposure) and double sham exposures. CHO cells were passaged twice in HAT (hypoxanthine, aminopterin and thymidine) medium and then seeded into T75 flasks. They were grown to a level below confluence (~2 x 10^5 cells/flask) and the flasks were exposed to 2 Gy of cobalt-60 γ-rays. The flasks were then placed in an EMF irradiator developed in-house which consisted of two separate solenoid coils individually shielded with mu-metal within a cell culture incubator. Careful attention was paid in the design to provide good magnetic isolation (including from the earth's and other fields), magnetic field uniformity (<1% variation), freedom from vibration and temperature control and uniformity (<<0.1° C). The flasks were placed within the EMF irradiator within 5 minutes of the end of γ-irradiation and were maintained there for 12 hours' EMF exposure. The two coils were energised under computer control such that they were randomly and blindly selected to deliver 0.7 or 0.0 mT exposures, so that in a series of experiments both actual and sham EMF exposures took place. The sequence of actual and sham exposures was kept blind until after the mutations were scored to prevent any possibility of expectation and bias in their evaluation. After EMF exposure, the cells were grown for three days, diluted and subcultured for a further two days to ensure that functional HPRT protein would be diluted out of the progeny of cells in which mutations in the gene had taken place. This was necessary to ensure that in the HPRT assay (in which the cells were challenged with 6-thioguanine) the cells that had a mutation survived to form colonies and those without a mutation died. An aliquot of the same subculture was not challenged with 6-thioguanine and the colony count from this gave the number of surviving non-mutant cells present so that the results could be calculated in the conventional way as mutants per survivor. These procedures were essentially the same as those used by Walleczek *et al.*

16.3 RESULTS

Our studies so far have shown significantly lower mutation frequencies than those reported by Walleczek *et al.* The values measured by us, however, are more consistent with other published studies in the literature using the HPRT system. As with the work of Walleczek *et al.*, there was wide variation from experiment to experiment. Our initial data from 16 experiments are summarised in Table 16.1.

Table 16.1 Mean mutation frequencies in CHO cells exposed to 2 Gy of γ-rays and to 0.7 mT or 0.0 mT ("sham") with standard errors of mean

	Mean Mutation Frequency $(x10^5)$	S.E.M. $(x10^5)$
Right coil exposed	2.31	0.71
Left coil exposed	1.96	0.42
Either coil exposed	2.15	0.42
Sham exposure	1.89	0.47

The SEMs shown in Table 16.1 were calculated assuming normal distributions and this is discussed further below. There is no significant difference between the mutation frequencies observed in the right and left coils and their average, 2.15 x 10^5 was greater by a factor of 1.14 than the value of 1.89 x 10^5 for the sham exposures. The uncertainties on the ratio 1.14 are clearly substantial and caution therefore is needed in comparing it with the corresponding value of 1.8 determined by Walleczek *et al.* for the mutation-enhancing effect of exposure to a 0.7 mT field.

We found that the wide variation in our data was largely attributable to the low mutation frequencies in our experiments, compared those found by Walleczek *et al.* The average number of viable mutants initially induced in a flask of cells was about 4. Because of the need to grow the cells on for a number of generations as described above, at each dilution and subculture, (or passage) the average number of mutants present again reduced to about 4. Thus the statistical uncertainties had been increased because of three successive Poisson processes, each with a mean of about 4. The effect of this on the statistics is illustrated in Figure 16.1. This is a Monte Carlo simulation of three successive Poisson processes each with a mean of 4. It shows that the distribution at passage 2 is skewed and is broadened compared with the simple Poisson distribution for the induced mutants.

A similar Monte Carlo approach was used to estimate the statistical uncertainty in the ratio of 1.14 derived above for the enhancing effect of EMF exposure and this is illustrated in Figure 16.2. Here the curve shows the distribution of probabilities that a true mean ratio of 1.14 might be observed at other values when the average number of mutants present at each stage was 4, as in our experiments. It illustrates the wide error limits and, while our preliminary result does not differ significantly from the ratio reported by Walleczek *et al.*, a ratio nearer unity is more probable.

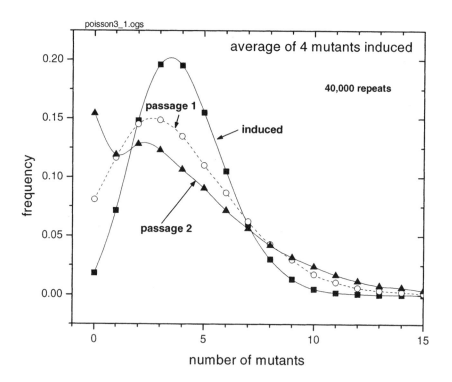

Figure 16.1 Monte Carlo simulation of the frequency distributions of mutants present initially and at each stage of subculture, based on a mean of 4 mutants present at each stage.

16.4 CONCLUSION

In our initial studies, no significant effect of an exposure to EMF after irradiation has been observed although a high level of variation from experiment to experiment was found due to the low numbers of mutants being processed through the selection system. To reduce the uncertainties, we are currently modifying the experimental protocol to allow more mutants to be generated, initially by increasing the number of cells exposed and also by increasing the radiation dose. There was no obvious reason for the 5- to 10-fold lower yield of mutants in our experiments than in those of Walleczek *et al.*, but this will have decreased the numbers of mutants present in the system and increased the statistical uncertainty.

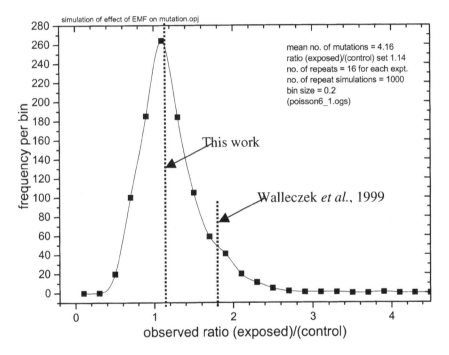

Figure 16.2 Comparison of the enhancing effect of 0.7 mT EMF exposure on γ-ray-induced mutations in CHO cells in this preliminary study with the value reported by Walleczek *et al.*, illustrating the statistical uncertainty in our data.

16.5 ACKNOWLEDGMENTS

The authors wish to acknowledge that this work was supported by a grant from the EMF Biological Research Trust.

16.6 REFERENCES

Ansari, R.M., and Hei, T.K., 2000, Effects of 60 Hz extremely low frequency magnetic fields (EMF) on radiation- and chemical-induced mutagenesis in mammalian cells. *Carcinogenesis*, **21**, pp. 1221-1226.

Lagroye, I., and Poncy, J.L., 1997, The effect of 50 Hz electromagnetic fields on the formation of micronuclei in rodent cell lines exposed to gamma radiation. *International Journal of Radiation Biology*, **72**, pp. 249-254.

Maes, A,, Collier, M., Vandoninck, S., Scarpa, P., and Verschaeve, L., 2000, Cytogenetic effects of 50 Hz magnetic fields of different magnetic flux densities. *Bioelectromagnetics*, **21**, pp. 589-596.

Mittler, S., 1973, Magnetism and x-ray induced sex-linked recessive lethals in Drosophila. *Journal of Heredity*, **64**, p. 233.

Miyakoshi, J., Koji, Y., Wakasa, T., and Takebe H., 1999, Long-term exposure to a magnetic field (5 mT at 60 Hz) increases X-ray-induced Mutations. *Journal of Radiation Research (Tokyo)*, **40**, pp. 13-21.

Miyakoshi, J., Yoshida, M., Yaguchi, H., and Ding, G.R., 2000, Exposure to extremely low frequency magnetic fields suppresses x-ray-induced transformation in mouse C3H10T1/2 cells. *Biochemical and Biophysical Research Communications,***271**, pp. 323-327.

Suri, A., deBoer, J., Kusser, W., and Glickman, B.W., 1996, A 3 milliTesla 60 Hz magnetic field is neither mutagenic nor co-mutagenic in the presence of menadione and MNU in a transgenic rat cell line. *Mutation Research*, **372**, pp. 23-31.

Walleczek, J., Shiu, E. C., and Hahn, G. M., 1999, Increase in radiation-induced HPRT gene mutation frequency after nonthermal exposure to nonionizing 60 Hz electromagnetic fields. *Radiation Research*, **151**, pp. 489-497.

Yaguchi, H., Yoshida, M., Ding, G.R., Shingu, K., and Miyakoshi, J., 2000, Increased chromatid-type chromosomal aberrations in mouse m5S cells exposed to power-line frequency magnetic fields. *International Journal of Radiation Biology*, **76**, pp. 1677-1684.

CHAPTER 17

Electromagnetic Fields - Interactions with the Human Body

Jeffrey W. Hand

17.1 INTRODUCTION

Electromagnetic (E-M) fields are ubiquitous in today's environment, being inherent to communications, power and other needs of modern society. The proliferation in the use of E-M fields has been accompanied by an increased concern regarding their safety.

Non-ionising E-M radiations differ essentially from ionising radiations in that the photon energy, given by the product of the frequency and Planck's constant (= 6.626 x 10^{-34} J s), is insufficient to cause ionisation. Traditionally, the consensus of scientific opinion is that interactions between E-M radiations and the human body are thermal, although there have been claims for other mechanisms of interaction.

This chapter discusses aspects of interactions between the body and non-ionising E-M fields such as microwaves (MW), radiofrequency (RF) fields and extremely low frequency (ELF) electric and magnetic fields. Dosimetry, exposure guidelines and some specific examples of E-M fields encountered within buildings are addressed briefly. The terms 'ELF' and 'radiofrequency' are often used in the biological effects and occupational health literature to cover the ranges from above static fields (> 0 Hz) to 3 kHz and from 3 kHz to 300 GHz, respectively. The 300 MHz to 300 GHz frequency range is also referred to as the 'microwave' range.

17.2 ELECTRIC AND MAGNETIC FIELDS

Electric and magnetic fields are produced by electric charges and their motion. Their behaviour and relationship are described by the set of equations known as Maxwell's equations. A time-varying electric field produces a magnetic field and vice versa. This is strictly true for E-M fields of all frequencies but in the case of low frequency fields, for example ELF fields, the electric and magnetic field components can be considered individually. Such fields are said to be 'quasi-static'.

An electric field may be described by its magnitude, E (V m^{-1}), and the electric flux density, D (coulomb m^{-2}). These quantities are related through the electrical properties of the medium described by the permittivity, ε:

$$D = \varepsilon E \qquad (17.1)$$

A magnetic field is described by its magnitude, H (A m^{-1}), and the magnetic flux density B (Tesla (T)). These are related through the magnetic properties of the medium described by its permeability, μ:

$$B = \mu H \qquad\qquad\qquad (17.2)$$

The permeability of biological materials μ is taken to be equal to μ_0, the permeability of free-space ($= 4\pi \times 10^{-7}$ H m^{-1}).

17.2.1 ELF Fields

Electric and magnetic fields associated with the generation, transmission, or use of electrical power in Europe generally have a frequency of 50 Hz and a wavelength of 6000 km. In the USA the frequency is 60 Hz and the wavelength is 5000 km. Fields produced by sources with dimensions that are much smaller than the wavelength are not associated with radiation of energy. Instead, the electric and magnetic fields, which under these circumstances can be considered independently, can produce fields within nearby objects, including the human body. This type of exposure is often referred to as being in the "near-field".

The energy absorbed is directly related to E-M fields inside the body and not those incident upon the body. The internal and incident E-M fields can be quite different, depending on the size and shape of the body, its electrical properties, its orientation with respect to the incident E-M fields, and the frequency of the incident fields. However, since direct measurement of the incident fields is easier and more practical than measurement of the internal fields, especially in people, dosimetry is used to relate the internal fields to the incident fields.

Interactions between a low frequency electric field and the body differ from the magnetic field case in that the body perturbs the electric field but not the magnetic field. Whilst exposure to either field results in induction of internal electric fields and currents, there are differences in the spatial distribution and magnitudes of these induced fields. For example, Figures 17.1 and 17.2 show the predicted electric field and current density when a 1.77 m tall man model is exposed to a 60 Hz electric field of 1 kV m^{-1} directed from foot to head, and a 60 Hz uniform magnetic flux density of 1 μT directed from front to back, respectively. These data indicate that a relatively high electric field or current density may be induced in regions such as the ankles, knees and neck.

External electric and magnetic fields required to induce average and maximum electric fields of 1 mV m^{-1} in individual organs within a similar man model (grounded man, foot to head electric field, front to back magnetic field) are discussed in detail by Stuchly and Dawson (2000), and the ranges of these external fields are listed in Table 17.1. This induced field level is achieved when the external field is at the low end of the ranges shown in Table 17.1 in the cases of bone (for $E_{\text{organ-avg}}$ and $E_{\text{organ-max}}$ for electric field exposure and $E_{\text{organ-max}}$ for magnetic field exposure) and liver (for $E_{\text{organ-avg}}$ and magnetic field exposure). In contrast, external fields at the high end of the ranges indicated in Table 17.1 are required in the cases of CSF (for $E_{\text{organ-avg}}$ for electric field exposure, and $E_{\text{organ-avg}}$

and $E_{organ-max}$ for magnetic field exposure) and thyroid (for $E_{organ-max}$ for electric field exposure).

The natural electric field consists of a static component which is in the order of 100 V m^{-1} (Lide, 2000) together with a smaller time-dependent component. The latter is about 10^{-4} V m^{-1} at 50 Hz (Bernhardt, 1992). The natural magnetic field is typically 40 µT (static component) with a small time-varying component; at 50 Hz the latter is about 10^{-6} µT (Bernhardt, 1992).

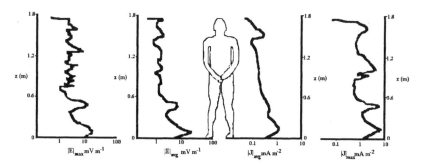

Figure 17.1 Electric field and current density predicted within a man model exposed to a uniform 60 Hz, 1 kV m^{-1} vertical (foot to head) electric field (adapted from Dawson *et al* 1998). From left to right: maximum electric field, layer averaged electric field, layer averaged current density, and maximum current density. The 1.77 m tall man is assumed to be separated from a horizontal ground plane by 1.44 cm of air below the soles of the feet.

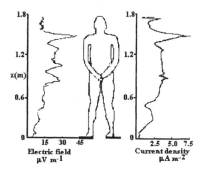

Figure 17.2 Layer-averaged electric field and current density predicted within a 1.77 m tall man model exposed to a uniform 60 Hz magnetic flux density of 1 µT (directed from front to back). (adapted from Stuchly and Dawson 2000).

Table 17.1 Ranges of external electric and magnetic fields (60 Hz) required to induce an organ-average ($E_{organ-avg}$) or organ-maximum ($E_{organ-max}$) electric field of 1 mV m^{-1} (from Stuchly and Dawson 2000)

Exposure	$E_{organ-avg}$ = 1mV m^{-1}	$E_{organ-max}$ = 1mV m^{-1}
Electric field	310 - 3250 V m^{-1}	22 - 1040 V m^{-1}
Magnetic field	37.9 - 233 µT	6.0 - 51.8 µT

Background levels in homes are generated mostly by wiring, particularly where there is a net current carried on sets of wiring. The UK 'ring main' system is particularly difficult to balance and commonly accounts for the high field sometimes measured in houses. As a result both electric and magnetic fields in homes can vary enormously. Average values are 10 V m^{-1} and 0.05 µT but magnitudes can vary from zero to 70 V m^{-1} and 0.68 µT, respectively (Preece *et al.*, 1996). Similarly, Stuchly and Dawson (2000) reported that typical magnetic background levels in homes are within the range 0.05-0.13 µT, with levels in most homes below 0.1 µT, and that typical electric fields in the home and office are in the range 1-20 V m^{-1}, although local fields close to appliances can be higher (up to about 300 V m^{-1}). Office electrical and electronic equipment and domestic appliances can produce strong local fields (typically 0.5-50 µT at 0.15 m) but these decrease rapidly with increasing distance from the device. A survey of fields in buildings (mostly commercial offices of less than 20 stories) in Ontario and Quebec reported that the mean ± standard deviation of E-field and H-field magnitudes were 1.19 ± 0.9 V m^{-1} and 0.16 ± 0.11 µT, respectively (Lau and Cohen 1990). The maximum values recorded in this study were 8.75 V m^{-1} and 1.19 µT.

Generally exposure of most people is similar in work place or home, but some occupations (for example in the electricity utility industry) are associated with much higher exposures.

There are several mechanisms through which ELF fields could produce biological effects. Electric fields can exert forces on molecules or cellular structures that may lead to movement of charged particles, orientation of dipolar molecules, geometrical changes in cellular structures, or induce voltages across cell membranes. Forces exerted by magnetic fields directly on cellular structures are very small, since tissues are essentially nonmagnetic. However, magnetic fields may cause biological effects by inducing electric fields within the body, although currents induced in the body by fields of less than about 1000 V m^{-1} or 50 µT are weaker than those that occur naturally in the body (Adair, 1991; Valberg *et al.*, 1997). Modelling (Xi *et al.*, 1994) also suggests that induced currents associated with most environmental exposure conditions are weak compared to endogenous currents, but points to the fact that their spatial distributions are different. Any forces arising from interaction with the fields must be compared with those due to random thermal motion of molecules and cellular structures. In general external fields significantly greater than those typically encountered in the home or workplace environment would appear to be required to result in effects that exceed natural thermal ones.

The reaction rates of chemical reactions that involve free radical pairs may be affected by the presence of magnetic fields and both theoretical and experimental data suggest that biochemical effects could occur in the presence of fields above about 1 mT or more (Eichwald and Walleczek, 1998; Eveson *et al.*, 2000).

Guidelines for limiting exposure to time varying electric and magnetic fields have been issued by several agencies. For example, basic restrictions (based on established health effects, particularly changes to nervous system function) and reference levels (provided for comparison with measured values of physical quantities) for exposure of the general public to fields within the range <1 Hz to 100 kHz recommended by ICNIRP (1998) are shown in Table 17.2. Although based on the same scientific evidence, guidelines issued by other agencies and bodies can differ in their detailed interpretation of the basic data (Erdreich and Klauenberg, 2001).

Table 17.2 Basic restrictions and Reference levels for general public exposure to low frequency electric and magnetic fields (ICNIRP 1998)

Basic restrictions (general public exposure)	
Frequency Range (Hz)	**Current density for head and trunk (mA m^{-2}) (RMS)**
< 1	8
1-4	8/f
4-10^3	2
10^3-10^5	f/500

f in Hz

Reference levels (general public exposure)			
Frequency Range	**Electric field strength (V m^{-1})**	**Magnetic field strength (A m^{-1})**	**Magnetic flux density (μT)**
<1 Hz	-	3.2 x 10^4	4 x 10^4
1-8 Hz	10^4	3.2 x 10^4/f^2	4 x 10^4/f^2
8-25 Hz	10^4	4 x 10^3/f	5 x 10^3/f
0.025-0.8 kHz	250/f	4/f	5/f
0.8-3 kHz	250/f	5	6.25

f as indicated in the frequency range

ICNIRP (1998), in summarising biological effects and epidemiological studies (frequencies up to 100 kHz), concluded:

'... there is currently no convincing evidence for carcinogenic effects and these (experimental) data cannot be used as a basis for developing exposure guidelines ...'

'... In the absence of support from laboratory studies, the epidemiological data are insufficient to allow an exposure guideline to be established...'

A recent comprehensive review of experimental and epidemiological studies relevant to an assessment of the possible risk of cancer resulting from exposures to power-frequency E-M fields (NRPB, 2001) concluded that:

'Laboratory experiments have provided no good evidence that extremely low frequency electromagnetic fields are capable of producing cancer, nor do human epidemiological studies suggest that they cause cancer in general. There is, however, some epidemiological evidence that prolonged exposure to higher levels of power frequency magnetic fields is associated with a small risk of leukaemia in children. In practice, such levels of exposure are seldom encountered by the general public in the UK. In the absence of clear evidence of a carcinogenic effect in adults, or of a plausible explanation from experiments on animals or isolated cells, the epidemiological evidence is currently not strong enough to justify a firm conclusion that such fields cause leukaemia in children. Unless, however, further research indicates that the finding is due to chance or some currently unrecognised artefact, the possibility remains that intense and prolonged exposures to magnetic fields can increase the risk of leukaemia in children'.

17.2.2 RF and MW E-M Fields

When characterising the electric and magnetic field from sources comparable with or larger than the wavelength the concepts of both near- and far-fields must be considered. The transition between these regions occurs at a distance r_{n-f}, given approximately by

$$r_{n-f} = \frac{D^2}{\lambda} \tag{17.3}$$

where D is the largest dimension of the source and λ is the wavelength. This boundary is not sharp since near fields become less important as the distance from the source increases. The power flux density S (W m^{-2}) is determined through the vector product

$$S = E \times H \tag{17.4}$$

S is also known as the Poynting vector. At distances $r > r_{n-f}$, S decreases as $1/r^2$. Furthermore, in this far-field region there is a fixed relationship between E and H, namely

$$\frac{E}{H} = \eta \tag{17.5}$$

where η is the wave impedance. In free space is η is approximately 377 Ω. The power flux density may be determined from knowledge of either E or H using

$$S = \frac{E^2}{\eta} = \eta H^2 \tag{17.6}$$

The region in which r is very much smaller than λ is known as the reactive part of the near field. The region in which r is comparable with λ is often called the intermediate near field. There is no constant relationship between E and H in the near field and both E and H, together with their relative phases, must be known to fully characterise the E-M field. The time averaged Poynting vector associated with the near field is zero.

The field structure close to a RF or MW antenna or source may be highly inhomogeneous with considerable variation in the wave impedance (a few $\Omega <$ $E/H <$ 1000s of Ωs). In regions out of the reactive part of the near field but still within the intermediate near field, there can be considerable spatial variations of E and H. Although the near field components do not contribute to the radiated energy, they may interact strongly with, and lead to energy deposition in, bodies located in those regions.

When considering exposure of the human body to such fields, 3 frequency ranges may be considered, namely sub-resonance (< 30 MHz), resonance (30 - 400 MHz), and supra resonance. At frequencies below about 30 MHz, energy absorption by the body decreases rapidly with decreasing frequency. Membrane effects are predominant at frequencies below approximately 100 kHz. At frequencies in the approximate range 30-400 MHz, the wavelength is comparable to the dimensions of the human body or body parts. Relatively high absorption can occur under these conditions. At higher frequencies the wavelength becomes smaller than body dimensions and the electrical conductivity of many tissues begins to increase more rapidly with increasing frequency. The result is increasingly rapid attenuation and decreased penetration of the E-M field into the body. At frequencies between approximately 200 MHz and 3 GHz, tissue curvature and refraction can give rise to local hot spots within the body.

Figure 17.3 shows data from Dimbylow (1997) that illustrate the frequency dependence of the whole-body averaged SAR for 4 body sizes (adult to 1 year-old child). In each case it is assumed that the feet are in contact with an electrical ground. Resonance occurs when the height of the body is approximately one quarter of the free-space wavelength. In these examples the resonant frequencies are 35, 55, 70, and 100 MHz, respectively. When the body is not grounded, the resonant frequency is doubled. Figure 17.4 shows the (layer-averaged) power absorption predicted in the case of the man model at two frequencies - one close to resonance (30 MHz) and one above resonance (120 MHz). In each case the whole body averaged SAR is 0.4 W kg^{-1}. Close to resonance the SAR peaks locally within the ankle and knee regions. Above and away from resonance such local peaks are not so pronounced and occur in the abdomen and neck. Such resonant behaviour is accounted for in exposure guidelines. For example, Table 17.3 lists basic restrictions and reference levels for exposure of the general public recommended by ICNIRP (1998). It is well established that RF radiation at high power levels can be harmful due to its ability to heat biological tissue. Areas of the body that are poorly perfused (e.g. the eyes and the testes) are susceptible on account of their inability to dissipate an abnormal thermal load. For example, short-term exposure to RF radiation at a power density of 100 - 200 mW cm^{-2} can cause cataracts in experimental animals (Cutz, 1989, Kramar *et al.*, 1987) whilst temporary sterility, caused by such effects as

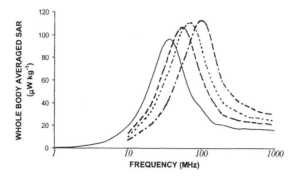

Figure 17.3 Whole body SAR for grounded bodies of various size. Curves, from left to right, correspond to adult (1.76 m tall), 10 year-old child (1.38 m), 5 year-old child (1.10 m), and 1 year-old child (0.75 m). The Incident electric field in all cases is 1 V m^{-1} (RMS) and is vertically polarised and incident upon the front of the body. (adapted from Dimbylow, 1997).

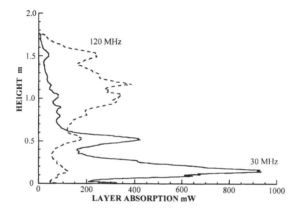

Figure 17.4 Power absorption by layer for a 1.76 m tall man model exposed to plane wave irradiation from the front for frequencies of 30 MHz (near resonance) and 120 MHz such that whole body averaged SAR is 0.4 W kg^{-1} (adapted from Dimbylow, 1997).

changes in sperm count and in sperm motility, is possible after exposure of the testes to high level RF radiation (Lebovitz *et al.*, 1989).

There are other effects that, although not resulting in an overall change in tissue temperature, nevertheless have a thermal basis. One example is the 'microwave auditory effect' in which, under certain specific conditions of frequency, signal modulation and intensity, animals and humans can perceive the RF/MW signal as a buzzing or clicking sound (NCRP 1986; Puranen and Jokela 1996). The most widely accepted explanation is that absorption of energy from the MW field leads to a thermoelastic interaction in the auditory cortex of the brain.

The evidence for harmful biological effects following exposure to RF radiation at field intensities lower than those that would produce significant and measurable heating is less clear. 'Non-thermal' effects reported have included changes in the immune system, neurological effects, behavioural effects, evidence for a link between MW exposure and the action of certain drugs and compounds, and a 'calcium efflux' effect in brain tissue.

There are experimental results that suggest ELF and MW fields might be involved in cancer promotion under certain conditions. However, contradictory experimental results have also been reported in many of these cases, and further experiments are needed to determine the generality of these effects and whether 'non-thermal' mechanisms exist that could cause harmful biological effects in animals and humans exposed to E-M radiation.

Table 17.3 Basic restrictions and Reference levels for general public exposure to electric and magnetic fields (1 kHz and above) (ICNIRP, 1998)

Basic restrictions (general public exposure)			
Frequency Range (Hz)	Current density for head & trunk (mA m⁻²) (RMS)	Whole-body average SAR (W kg⁻¹)	Localised SAR (W kg⁻¹)
10^3-10^5	f/500		-
10^5-10^7	f/500	0.08	2 (head & trunk) 4 (limbs)
10^7-10^{10}	-	0.08	2 (head & trunk) 4 (limbs)

f in Hz

There is also a basic restriction to the general public exposure of 10 W m⁻² for power density for frequencies between 10 and 300 GHz.

Reference levels (general public exposure)				
Frequency Range	Electric field strength (V m⁻¹)	Magnetic field strength (A m⁻¹)	Magnetic flux density (µT)	Equivalent plane wave power density (W m⁻²)
3-150 kHz	87	5	6.25	-
0.15-1 MHz	87	0.73/f	0.92/f	-
1-10 MHz	87/ f ½	0.73/f	0.92/f	-
10-400 MHz	28	0.073	0.092	2
400-2000 MHz	1.375 f ½	0.0037 f ½	0.0046 f ½	f/200
2-300 GHz	61	0.16	0.20	10

f as indicated in the frequency range

Several scientific bodies have reviewed the present state of knowledge regarding untoward effects of low level E-M fields and radiation. ICNIRP (1998), in summarising biological effects and epidemiological studies involving frequencies in the range 100 kHz to 300 GHz, concluded:

'Epidemiological studies on exposed workers and the general public have shown no major health effects associated with typical exposure environments. Although there are deficiencies in the epidemiological work, such as poor exposure assessment, the studies have yielded no convincing evidence that typical exposure levels lead to adverse reproductive outcomes or an increased cancer risk in exposed individuals' and '… In general, the effects of exposure of biological systems to athermal levels of amplitude-modulated electromagnetic fields are small and are very difficult to relate to potential health effects'.

17.3 EXAMPLES OF E-M EXPOSURE IN BUILDINGS

17.3.1 Electronic Article Surveillance Systems

Electronic article surveillance (EAS) systems used to detect theft of items from shops, libraries, and hospitals are common in today's environment. Most systems employ one of four technologies - audio-frequency (AF) magnetic, acousto-magnetic, swept RF, and MW. Common features are a detection system situated at the exit to the premises under surveillance and detectable tags that are attached to good or other items being protected.

AF magnetic systems usually operate at a frequency between 10 Hz and 20 kHz and the detection system utilises the near field produced by a current-carrying loop. Typically, the maximum magnetic flux density ranges from about 190 μT to 25 μT depending upon operating frequency, although the field falls off rapidly with increasing distance from the loop. The tag contains a component made from high permeability material and this modulates the field with a specific signature when activated, detectable by the receiver coil.

Acousto-magnetic systems operate within a frequency range of approximately 30-135 kHz and also use the near field produced by a current carrying loop in which the maximum flux density is typically 13-4 μT depending upon operating frequency. In these systems tags are made from a magnetostrictive material. When activated they are driven resonantly by the pulsed excitation field and continue to transmit for a short time after the pulse has ended, enabling their presence to be detected by the receiver coil.

Swept RF systems operate within the frequency range 1-10 MHz and also use the near field produced by a current-carrying loop in the detection process. In these systems the magnetic flux density is typically 0.25 μT. A pulsed (50-100 Hz repetition frequency), frequency swept (typically over 2 MHz) inductive field interacts with the tag which consists of a disposable antenna and resonant circuit. The receiver coil detects the tag's response to this excitation.

MW systems operate typically with the 800 MHz to 2.5 GHz range. A propagating field is produced by a directional antenna and emission levels are typically less than 1 µW cm^{-2}. Detection of an activated tag is achieved through the non-linear response of a disposable antenna and diode.

Fields associated with EAS systems often exceed ICNIRP (1998) reference level values but nevertheless comply with limits determined by basic restrictions (Davies, 1999). However, implantable electronic medical devices may be affected by the E-M fields produced (McIvor *et al.*, 1999, Harris *et al.*, 2000). Devices potentially at risk to interference include pacemakers, neurological stimulators, and implantable defibrillators. It is therefore advisable not to remain close to an EAS system for an unnecessarily long time.

17.3.2 Base Stations for Mobile Communications Systems

Base station antennas are usually mounted on purpose-built free-standing masts, often 15 m or more in height, or on the roofs or sides of buildings. Typically, a group of 3 antennas, each covering a $2\pi/3$ sector, is employed and the main beam is tilted downwards so that it reaches ground level 50-200 m away. The divergence of the main beam vertically is usually a few degrees.

Estimating the field levels within an urban environment, in contrast to free space, is complicated by the presence of reflected fields from nearby structures and by the attenuating properties of those structures. Bernardi *et al.* (1999) showed that significant increases in exposure could occur in the presence of reflecting walls forming a partially closed environment, highlighting the need for more detailed studies. Even so, exposure levels are likely to be compliant with international guidelines (Bernadi *et al.*, 2000).

According to Mann *et al.* (2000) the radiated power from antennas used with macrocellular base stations in the UK ranges from a few to a few 10s of watts, with typical maximum powers around 80 W. Close to the antenna, an exclusion zone within human exposure could exceed guideline values, is defined. This can extend up to approximately 8-10 m in front of an antenna to be compliant with ICNIRP guidelines for exposure of the general public (FEI 1999, Mann *et al.,* 2000). Since microcellular base stations use powers of up to a few watts, compliance distances in these cases are expected not to exceed a few 10s of cms.

Petersen *et al.* (1997) describe measurements made in the top floor apartments of a building with high-gain sector antennas mounted outside just above the apartments, and found a maximum power density of 0.4 µW cm^{-2}. Measurements made in a corridor in the floor directly below a roof-top base station in which the antennas were located 3 m above the main roof showed a maximum power density of 8 µW cm^{-2}. In both cases the base stations were operating at their maximum capacity.

Mann *et al.* (2000) measured power density at 188 locations around 17 base station sites. For all locations and for distances up to 250 m from the base stations, power density at the measurement positions did not show any trend to decrease with increasing distance. Spectral measurements were obtained over the 30 MHz to

2.9 GHz range at 73 of the locations. The data varied over several decades. The geometric mean total exposure arising from all radio signals at the locations considered was 0.0018% of the ICNIRP (1998) public reference level. The maximum exposure at any location was 0.18% of the ICNIRP reference public level. Most of the measurements were for towers used by one operator; levels are expected to be higher near to towers used by more than one operator.

Line *et al.* (2000) reported measurement from 14 locations throughout Australia. Antennas were mounted on towers with a typical height of 25 m (range from 20 to 40 m). The highest average level (averaged over 24 hours) recorded was 0.052 μW/cm^2, and the maximum value recorded was 0.082 μW cm^{-2}. The mean value averaged for all survey locations over a 24 hour period was 0.0016 μW/cm^2. The worst case figure (assuming that 4 transmitters operated at full power at one of the sites) obtained was 0.178 μW cm^{-2}. RF E-M emissions from other sources were also monitored in this study and it was found that the most significant contributor was AM radio. AM and FM radio were responsible for >91% and 4.7%, respectively, of the measured power flux density, compared with 2% from base stations. When frequency weighting was applied, reflecting the frequency dependence of exposure limits between 1 and 10 MHz, these contributions became approximately 51%, 26%, and 11%, respectively. Another survey of RF fields in the general and work environments (Mantiply *et al.*, 1997) found that fields in the general urban environment are principally associated with radio and TV broadcast services.

Thansandote *et al.* (1999) measured RF levels in several schools in Vancouver that had base stations on them or near to them. The maximum RF level recorded was 2.6 μW cm^{-2}. Aniolczyk (1999) found that measurements of power density made around 20 GSM base stations and inside buildings with antennas in Poland did not exceed the appropriate recommended limit (0.1 W m^{-2}).

17.3.3 E-M Fields in the Hospital Environment

An indirect effect of E-M fields on human health is the potential hazard posed by E-M interference with medical devices (Boyd *et al.*, 1999). The incidence of reports of adverse effects of this type has increased over recent years. Problems with devices such as apnea monitors, electrically powered wheelchairs and pacemakers have been encountered and consequences have ranged from inconvenience to death. Witters *et al.* (2001) draw attention to the increasing number of incident reports that involve electronic security systems. Standards (eg IEC 2000, 2001a) specify a minimum immunity level and a recommended separation distance (between the device and the RF source) over specified frequency ranges and general guidance in evaluating E-M immunity of medical devices to RF fields is available (IEC 2001b, AAMI 1997, IEEE 1997).

Factors that determine the level of interference include the level of coupling between device and the source and their separation, the frequency of the carrier signal and the nature of any modulation. Care must be taken that any modification to the design or housing of a medical device does not result in an increased susceptibility to E-M interference; similar checks should be made after servicing or repair.

Boyd *et al.* (1997) found that broadband RF electric field strength measurements made in a range of hospital areas approached 30 V m^{-1} at varying distances from source devices. The strongest sources of RF fields included electrosurgical units, hand-held radios, and video display terminals. Power frequency magnetic field strength measurements approached 5 A m^{-1} at varying distances from source devices. The strongest sources of ELF fields included power lines and supplies, patient monitoring equipment, video display terminals, and electrosurgical units. Field strength measurements were highly dependent on environmental conditions. Subsequently this group (Boyd *et al.*, 1999) reported that 63% of all electric field strength measurements and 7% of all magnetic field strength measurements collected in the hospital exceeded the proposed IEC immunity requirements for medical devices.

Davis *et al.* (2000) studied propagation of 850 MHz and 1.9 GHz fields within hospital corridors and subsequently (Davis *et al.*, 2001) determined the spatial variation of 1.9 GHz fields in this environment. They found that the field level near to the central axis of the corridor tended to be greatest and also measured a reduced path loss near the floor compared with that at greater heights from the floor, although the field strength at floor level was lower (by about 10 dB) close to the source. Based on these findings, the authors consider that adoption of a minimum separation distance is acceptable for low power RF sources but suggest that caution is called for in the case of higher power sources or simultaneous usage of several low power sources, when an additional safety factor may need to be considered.

Effects on pacemakers arising from electrocautery, RF ablation, radar, welding systems, leaking MW ovens, and E-M surveillance systems that may cause asynchronous pacing have been identified. Asynchronous pacing may be arrhythmogenic and may provoke ventricular fibrillation. Pacemakers can be affected by both electric and magnetic power-frequency fields but their sensitivity and the severity of effects are very dependent on design and model.

It has been recommended that for patients with implanted pacemakers, hand-held telephones should be worn at least 15-20 cm from the pacemaker and preferably used with the ear contra-lateral to the location of the pacemaker.

The effects of a wide range of radio handsets on 178 different medical devices have been reported by MDA (1997). Handsets were grouped as emergency radios (as used by emergency services personnel and operating between 28 and 470 MHz with power as high as 10 W), security radios (2-way radios and VHF/UHF radio handsets used by security, maintenance and portering staff), cell phones (analogue and digital mobile phones) and cordless phones (pagers, radio computer local area networks). Overall, the medical devices suffered EMI from handsets in 23% of tests and of these cases, 43% were considered serious since they would have had a direct impact on patient care. The likelihood of interference was strongly dependent upon the type of handset involved. At 1 m distance, the percentages of the medical devices that suffered EMI from emergency radios, security radios and cell phones were 41%, 35%, and 4%, respectively. No significant effects were recorded due to the cordless phone group. Physiological monitors, defibrillators, and external pacemakers were the most severely affected.

An informative summary of the problem of EMI in the hospital environment is given by Lyznicki *et al.* (2001).

17.4 CONCLUSION

Current guidelines for limiting exposure of the public to E-M fields over a wide range of frequencies are based on an extensive set of theoretical and experimental data relating how these fields interact with the human body, primarily in terms of energy deposition. Although understanding of the phenomena involved is in general extensive, questions regarding certain health aspects of such exposure remain to be fully answered. In general, public exposure to ELF and RF E-M fields is compliant with such guidelines within a wide margin. However, when exposure takes place close to the E-M source, dosimetry suggests such cases can present a greater challenge to the recommended limits. Another potential hazard to health can arise through interference with medical equipment.

17.5 REFERENCES

AAMI (Association for the Advancement of Medical Instrumentation), 1997, Guidance on electromagnetic compatibility of medical devices for clinical/biomedical engineers -Part 1: Radiated radio-frequency electromagnetic energy. (TIR 18) AAMI, Arlington, VA.

Adair, R.K., 1991, Constraints on biological effects of weak extremely-low-frequency electromagnetic fields. *Physics Review A*, **43**, pp. 1039-1048.

Aniolczyk, H., 1999, Electromagnetic field pattern in the environment of GSM base stations. *International Journal of Occupational Medicine and Environmental Health,* **12**, pp. 47-58.

Bernardi, P., Cavagnaro, M., Pisa, S., and Piuzzi, E., 1999, Evaluation of the power absorbed in subjects exposed to EM fields in partially closed environments by using a combined analytical-FDTD method. Proceedings of the IEEE MTT-S International Microwave Symposium Digest Vol. 2, (Piscataway, NJ: IEEE), pp. 599-602.

Bernardi, P., Cavagnaro, M., Pisa, S., and Piuzzi, E., 2000, Human exposure to radio base-station antennas in urban environment, *IEEE Transactions on Microwave Theory and Techniques*, **48**, pp. 1996-2002.

Bernhardt, J.H., 1992, Non-ionizing radiation safety: radiofreqeuncy radiation, electric and magnetic fields. *Physics in Medicine and Biology*, **37**, pp. 807-844.

Boyd, S., Boivin, W.S., Coletta, J.N., Harris, C.D., and Neunaber, L.M., 1997, Characterization of the electromagnetic (EM) fields in critical medical device environments. Proceedings of 1997 FDA Forum on Regulatory Sciences, 8-9 December 1997, NIH, Bethesda, MD. Abstract C01.

Boyd, S.M., Boivin, W.S., Coletta, J.N., Harris, C.D., and Neunaber, L.M., 1999, Documenting radiated electromagnetic field strength in the hospital environment *Journal of Clinical Engineering*, **24**, pp.124-132.

Cutz, A., 1989, Effect of microwave radiation on the eye: the occupational health perspective. *Lens and Eye Toxicity Research*, **6**, pp. 379-86.

Davies, G., 1999, Electronic article surveillance (EAS) and radio frequency identification: current and future trends in relation to exposure to EM fields. Proceedings of COST244bis Action on Biomedical Effects of Electromagnetic Fields: Workshop on Emerging Technologies, 6-7 November 1999, Southampton, pp. 1/1-1/9.

Davis, D., Segal, B., Trueman, C.W., Calzadilla, R., and Pavlasek, T., 2000, Measurement of indoor propagation at 850 MHz and 1.9 GHz in hospital corridors. Proceeding of 2000 IEEE-APS Conference on Antennas and Propagation for Wireless Communications, (Piscataway, NJ: IEEE), pp. 77-80

Davis, D., Segal, B., Martucci, D.M., and Pavlasek, T.J.F., 2001, Volumetric 1.9 GHz fields in a hospital corridor: electromagnetic compatibility implications. Proceeding of 2001 IEEE International Symposium on Electromagnetic Compatibility Vol 2, (Piscataway, NJ: IEEE), pp.1131-1134

Dawson, T.W., Caputa, K. and Stuchly, M.A., 1998. High resolution organ dosimetry for human exposure to low frequency electric fields. *IEEE Transactions on Power Delivery*, **13**(2), pp.366-373.

Dimbylow, P.J., 1997, FDTD calculations of the whole-body averaged SAR in an anatomically realistic voxel model of the human body from 1 MHz to 1 GHz. *Physics in Medicine and Biology*, **42**, pp. 479-490.

Erdreich, L.S., and Klauenberg, B.J., 2001, Radio frequency radiation exposure standards: Considerations for harmonization. *Health Physics*, **80**, pp. 430-439.

Eichwald, C., and Walleczek, J., 1998, Magnetic field perturbations as a tool for controlling enzyme-regulated and oscillatory biochemical reactions. *Biophysical Chemistry*, **74**, pp. 209-224.

Eveson, R.W., Timmel, C.R., Brocklehurst, B., Hore, P.J., and McLauchlan, K.A., 2000, The effects of weak magnetic fields on radical recombination reactions in micelles. *International Journal of Radiation Biology*, **76**, pp. 1509-1522.

FEI (Federation of the Electronics Industry), 1999, Submission to the Independent Expert Group on mobile phones. November 1999. (London:FEI).

Harris, C., Boivin, W., Boyd, S., Coletta, J., Kerr, L., Kempa, K., and Aronow, S., 2000, Electromagnetic field strength levels surrounding electronic article surveillence (EAS) systems. *Health Physics*, **78**, pp.21-27.

ICNIRP (International Commission on Non-Ionizing Radiation Protection), 1998, Guidelines for limiting exposure to time-varying electric, magnetic, and electromagnetic fields (up to 300 GHz), *Health Physics*, **7**, pp. 494-522.

IEC (International Electrotechnical Committee), 2000, Medical electrical equipment - part 1: General requirements for safety 1: Collateral standard: Safety requirements for medical electrical systems (IEC 60601-1-1), (Geneva: IEC).

IEC (International Electrotechnical Committee), 2001a, Medical electrical equipment - part 1: General requirements for safety 2. Collateral standard:

electromagnetic compatibility - requirements and tests (IEC 60601-1-2), (Geneva: IEC).

IEC (International Electrotechnical Committee), 2001b, Electromagnetic compatibility (EMC)- Part 4-3: Testing and measurement techniques - Radiated, radio-frequency, electromagnetic field immunity test (IEC 61000-4-3), (Geneva: IEC).

IEEE (Institute of Electrical and Electronics Engineers), 1997, Recommended practice for an on-site ad hoc test method for estimating radiated electromagnetic immunity of medical devices to specific radio-frequency transmitters (C63.18), (New York: IEEE).

Kramar, P., Harris, C. and Guy, A.W., 1987, Thermal cataract formation in rabbits. *Bioelectromagnetics*, **8**, pp. 397-406.

Lau, P.-K. and Cohen, M.M., 1990, Characteristation of electromagnetic fields in buildings. *Electronics Letters*, **26**, pp.1778-1779.

Lebovitz, R.M., Johnson, L., and Samson, W.K., 1987, Effects of pulse-modulated microwave radiation and conventional heating on sperm production. *Journal of Applied Physiology*, **62**, pp. 245-52.

Lide, D.R., 2000. *Handbook of Chemistry and Physics*, 81st edition, edited by Lide, D.R. (Boca Rotan, FL: CRC Press), section 14, p. 31

Line, P., Cornelius, W.A., Bangay, M.J., and Grollo, M., 2000, *Levels of radiofrequency radiation from GSM telephone base stations.* Technical Report 129 (Yallambie:Australian Radiation Protection and Nuclear Safety Agency).

Lyznicki, J.M., Altman, R.D., and Williams, M.A., 2001, Report of the American Medical Association (AMA) Council on Scientific Affairs and AMA recommendations to medical professional staff on the use of wireless radio-frequency equipment in hospitals. *Biomedical Instrumentation & Technology*. **35**, pp.189-195.

Mann, S.M., Cooper, T.G., Allen, S.G., Blackwell, R.P., and Lowe, A.J., 2000, *Exposure to radio waves near mobile phone base stations.* NRPB Report R321 (Chilton: NRPB).

Mantiply, E.D., Pohl, K.R., Poppell, S.W., and Murphy, J.A., 1997, Summary of measured radiofrequency electric and magnetic fields (10kHz to 30 GHz) in the general and work environment. *Bioelectromagnetics,* **18**, pp. 563-77.

McIvor, M.E., Reddinger, J., Floden, E., and Sheppard, R.C., 1998, Study of pacemaker and implantable cardioverter defibrillator triggering by electronic article surveillance devices (SPICED TEAS). *Pacing and Clinical Electrophysiology*, **21**, pp.1847-1861.

MDA (Medical Devices Agency), 1997, Electromagnetic compatibility of medical devices with mobile communications. (MDA Device Bulletin DB9702), (Wetherby: MDA, Department of Health).

NCRP (National Council on Radiation Protection and Measurements), 1986, *Biological effects and exposure criteria for radiofrequency electromagnetic fields NCRP Report No 86.* (Bethesda, MD:NCRP).

NRPB (National Radiological Protection Board), 2001, ELF electromagnetic fields and the risk of cancer, *Documents of the NRPB*, 12(1).

Petersen, R.C., Fahy-Elwood, A.K., Testagrossa, P.A., and Zeman, G.H., 1997, Wireless telecommunications Part A: Technology and RF safety issues. In *Nonionizing radiation: An overview of the physics and biology*, edited by Meltz, M.L., Glickman, R.D., and Hardy, K., (Madison, WI: Medical Physics Publishing), pp. 197-226.

Preece, A.W., Grainger, P., Golding, J. and Kaune, W.T., 1996, Domestic magnetic field exposures in Avon. *Physics in Medicine and Biology,* **41**, pp. 71-81.

Puranen, L., and Jokela, K., 1996, Radiation hazard assessment of pulsed microwave radars. *Journal of Microwave Power and Electromagnetic Energy*, **31**, pp. 165-177.

Stuchly, M.A. and Dawson, T.W., 2000, Interaction of low frequency electric and magnetic fields with the human body. *Proceedings of IEEE*, **88**, pp. 643-664.

Thansandote, A., Gajda, G.B., and Lecuyer, D.W., 1999, Radiofrequency radiation in five Vancouver schools: exposure standards not exceeded. *Canadian Medical Association Journal,* **160**, pp. 1311-1312.

Valberg, P.A., Kavet, R., Rafferty, C.N., 1997, Can low-level 50/60 Hz electric and magnetic fields cause biological effects? *Radiation Research* **148**, pp.2-21 (Erratum: 1997, *Radiation Research*, **148**, p528)

Witters, D., Portnoy, S., Casamento, J., Ruggera, P., and Bassen, H., 2001, Medical device EMI: FDA analysis of incident reports, and recent concerns for security systems and wireless medical telemetry. Proceedings of IEEE International Symposium on Electromagnetic Compatibility Vol. 2, (Piscataway, NJ.: IEEE), pp.1289-1291.

Xi, W., Stuchly, M.A. and Gandhi, O.P., 1994, Induced electric currents in models of man and rodents from 60 Hz magnetic fields. *IEEE Transactions on Biomedical Engineering*, **41**, pp.1018-1023.

CHAPTER 18

Evidence to Support the Hypothesis that Electromagnetic Fields and Radiation are a Ubiquitous Universal Genotoxic Carcinogen

Neil Cherry*

18.1 INTRODUCTION

The whole world is exposed to shortwave and satellite broadcast and telecommunication electromagnetic radiation (EMR). Most areas, especially developed urban areas are exposed to many TV, radio, mobile phone radiation and ELF fields (EMF) in all buildings, including homes, and along most streets. Therefore the exposures are ubiquitous and the whole body is exposed making the cancer effects universal (over the whole body). All of the radiation and ELF fields induce electric currents in the body that produce oscillating electromagnetic fields throughout the body but especially higher in strong conduction organs, including the brain, the blood circulation system and bone marrow. These signals are genotoxic, damaging and altering the DNA in exposed cells, causing cancer and enhancing cell death rates.

DNA damage is common from daily activity, including from oxygen free radicals from breathing. Hence DNA repair is a vital part of the body's protection system. DNA repair can lead to a mistake with the survival of an altered and mutated cell. Over time the cumulative cell damage can lead to the development of cancer and other illnesses. Exposure to genotoxic substances in the environment enhances the mutated cell population and accelerates the development of detectable cancer. In children this can happen only months after birth if the sperm or egg contain mutated DNA or the fetus was exposed to a toxin. A common early childhood Leukaemia (ALL) peaks at 2-3 years. The first residential EMF cancer studies, Wertheimer and Leeper (1979, 1982), found an adult cancer latency peak at 7 years.

It was well established in the 1970s that the higher the carrier frequency, the lower the human tissue dielectric constant becomes. As a consequence, the higher the carrier frequency, the higher the induced electric current becomes in the tissue for a given unit external field strength. Johnson and Guy (1972), Schwan and Foster (1980), Adey (1988), Vignati and Giuliani (1997). This is now called the EMR Spectrum Principle. The higher the carrier frequency, the higher the expected

---

* Dr Neil Cherry died on 24 May 2003, aged 56, from motor neurone disease.

biological and epidemiological impact is. This is also supported by Brain Cancer RRs from Electric Utility Workers under high mean exposures, Theriault *et al.* (1994), OR = 28.48 (1.76-461.3). Zaret (1977) reports that a group of 18 radar exposed workers, had 2 with Astrocytomas. Compared with the SEER Astrocytoma rate in the average young male population of the US in the 1970-74, RR = 1634 (385-6939), p<0.000001.

There is now a very large body of epidemiological research showing that EMF and EMR enhances cancer rates in exposed groups from residential and occupational exposures. A large body of laboratory research shows that across the spectrum these fields and signals damage and alter DNA through Chromosome Aberrations, Micronuclei Formation, DNA strand breakage, altered gene activity, neoplastic transformation, and altered cell proliferation in animal and human cells. Modern assay methods, especially the comet assay, confirm direct DNA strand breakage at non-thermal, isothermal and low induced current situations, proving that it is a direct genotoxic effect. As early as 1959 a published study showed that pulsed RF exposure for 5 minutes increased significant chromosome aberrations in an iso-thermal situation, Heller and Teixeira-Pinto (1959). The authors concluded that the RF exposure is "a powerful and controlled mutagenic agent" and "The effects noted mimicked those produced by ionizing radiation and c-mitotic substances".

The only scientific hypothesis that makes sense of this massive published research, including many of the apparently conflicting results, is that EMF/EMR is a Ubiquitous Universal Genotoxic Carcinogen (UUGC).

Many of the weak and apparently inconsistent epidemiological and laboratory results are contributed by the absence of a No-Exposure reference group. This leads to extend the Healthy Worker Effect to include an adjustment factor to deal with the No Unexposed Reference Group Effect. Some studies conclusions are related to the failure to recognise the Universal nature of whole-body exposure. Elwood (1999) concludes that the RF exposures relationship to cancer is weak and inconsistent, largely because like the carcinogen effects of chemicals, they are related to a single type of cancer. However, Table 3 of Elwood shows elevated cancer RR in many organs in multiple independent studies. This supports the UUGC hypothesis. Explaining this in a Planning Court appeal in Australia led to a win for the council I was appearing for as an expert witness in an Appeal against a cell site base station near an office block.

18.2 HISTORICAL CANCER TRENDS

We now have reliable proof that EMF fields have significantly contributed to the many dose-response increasing trends of cancer and many other health effects. Over 20 studies show dose-response increased childhood cancer in residential situations. This proof involves direct connection with the introduction of household wiring, backed by many studies showing elevated and dose-response elevated cancer rates in residential and occupational situations. The significant rising long-term cancer trend is masking and reducing the Relative Risk rates in more recent studies, producing weaker and weaker conclusions. Actually the evidence, when considered in total and in an open-minded context, is getting stronger and stronger.

Cancer rates have risen progressively and significantly over the past century in developed countries, Figure 18.1. Current research shows that a major proportion of these trends is caused by the EMF and EMR. Strategies, standards and low emission technology will significantly reduce cancer rates and many other health effects.

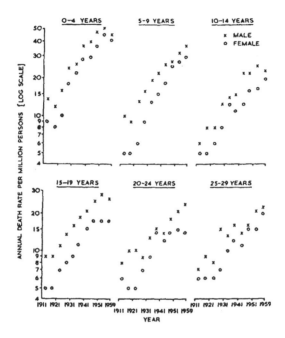

Figure 18.1 Trend in leukaemia mortality with time for England and Wales for 5-year age-groups by sex, from 1911-1959, Court-Brown and Doll (1961).

Fraumeni and Miller (1966) and Burnet (1958) show similar rising leukaemia trends in the first half of the 20th Century in England, Wales, Ireland, Denmark, Australia and the United States. Court-Brown and Doll identified a new early childhood cancer aged peak for Lymphatic Leukaemia peaking between 2 and 4 years that had not existed before 1920 in England, Figure 18.2. They state that the data "may suggest that a new leukaemogenic agent was introduced". We now know that that was the electromagnetic fields produced by household wiring, Milham and Ossiander (2001). Milham and Ossiander found that the only factor that related to the location and time when this early childhood cancer peak occurred, in situations all around the world, including in U.S. states, was the household electric wiring. They concluded "the childhood leukaemia peak of cALL is attributed to residential electrification. 75% of childhood cALL and 60% of all childhood leukaemia may be avoidable." Their results are independently confirmed by Kraut *et al.* (1995), Hatch *et al.* (1998), Green *et al.* (1999), and over 20 other studies that show dose-response increased trends of childhood cancer from resident exposures to electromagnetic fields.

Milham and Ossiander point out that the originally new electromagnetic fields are very small because the power was only used for lights, radio and irons. Therefore the chronic mean personal exposure would have been much lower than 1mG. This relationship is independently confirmed by and this is far more than classically required for a causal link (Hill, 1965).

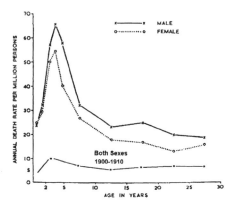

Figure 18.2 Age-specific death rates from leukaemia under the age of 30 years, by sex in England and Wales, 1945-1959, Court-Brown and Doll (1961).

18.3 REVIEWS OF EMF/EMR CANCER RELATIONSHIPS

A totally independent team of Swedish medical scientists, reviewed almost 100 epidemiological papers published up to July 1994, (Hardell *et al.* 1995). They concluded that there are possible associations between:

- an increased risk of leukaemia in children and the existence of, or distance to, power lines in the vicinity of their residence,
- an increased risk of chronic lymphatic leukaemia and occupational exposure to low frequency electromagnetic fields and,
- an increased risk of breast cancer, malignant melanoma of the skin, nervous system tumours, non-Hodgkin lymphoma, acute lymphatic leukaemia or acute myeloid leukaemia and certain occupations.

Many more studies have now been published to strengthen these conclusions.

Milham (1998) reviewed the ELF published epidemiological studies and shows that "In the 20 years since publication of the original Wertheimer and Leeper study associating magnetic fields with childhood cancers, about 40 residential and 100 occupational epidemiological studies have been published with nearly 500 separate Risk Ratios. For every lowered risk there are about six that are elevated. A number of these studies show significant dose-response between magnetic field and cancer incidence. The low risks in these studies may be due to the fact that there are no unexposed control groups available. In some cases the magnetic meters used failed to detect the actual field strength.

A review of the epidemiology of Brain Cancer has been specifically carried out for this chapter. Close to 100 published studies now exist. They start with the studies of Wertheimer and Leeper (1979, 1982) and Lester and Moore (1982) which show significant dose-responses in All Cancer for children and adults, from residential ELF and radar exposures. All Cancer includes Brain Cancer. In children about 26% of Cancers are Brain Cancer according to the SEER registry. The majority of the studies involve "Electrical occupations" and many involve residential situations. Also included are Pilots and Aircrew, computer and cell phone users and military radio and radar exposed personnel. There are over 400 exposed groups showing elevated Brain Cancer, over 140 are significant and over 55 show dose-response trends.

A conservative and reasonable No-Exposure Factor (NEF=4) is based on Milham and Ossiander's 60% of the trends in the 20th Century. If this is applied then almost all of the 400 groups RRs would be significantly elevated and most would be highly significantly raised.

The Dorlands Medical Dictionary, states that a genotoxic substance causes cancer. The following section summarizes the large body of evidence showing that ELF and RF/MW exposure causes DNA damage.

18.4 EVIDENCE OF ELF GENOTOXICITY

18.4.1 Chromosome Damage from ELF Exposure

The well established biological mechanism of ELF cancer causation is the genotoxicity of the ELF fields. That is, they directly damage the DNA as is shown by Chromosome Aberrations (CA), Micronuclei Formation and DNA strand breakage. El Nahas and Oraby (1989) observed significant dose-response dependent micronuclei increase in 50 Hz exposed mice somatic cells. Elevated CAs have been recorded in a number of workers in electrical occupations. In Sweden Nordenson *et al.* (1988) found significant CA in 400 kV-substation workers and with 50 Hz exposures to peripheral human lymphocytes, Nordenson *et al.* (1984) and exposures human amniotic cells, Nordenson *et al.* (1994). Significant CA in human lymphocytes exposed to 50 Hz fields are also reported by Rosenthal and Obe (1989), Khalil and Qassem (1991), Garcia-Sagredo and Monteagudo (1991), Valjus *et al.* (1993) and Skyberg *et al.* (1993). Skyberg *et al.* collected their samples from high-voltage laboratory cable splicers and Valjus *et al.* from power linesmen. Other studies showing ELF associated CAs include Cook and Morris (1981), Cohen *et al.* (1986 a,b), Lisiewicz (1993), and Timchenko and Ianchevskaia (1995). Skyberg *et al.* (2001) found employees in a Norwegian factory exposed to strong ELF fields and mineral oil had significantly increased Chromosome Aberrations. This currently involves 15 studies showing that ELF exposures cause Chromosome Aberrations.

18.4.2 The Comet Assay Method

A very advanced assay of DNA strand breakage has been developed by Dr N.P. Singh at the University of Washington. This is called the microgel electrophoresis or Comet Assay, Singh *et al.* (1994). The Comet Assay involves isolating single cells and combining them with agarose to form a gel, removing the bound protein using proteinase K, because DNA is negatively charged and the bound protein is positively charged, applying an intense fluorescent dye (YOYO-1) and using electrophoresis to place the cell in an electric field gradient so that any DNA broken segments, that are negatively charged, flowing down towards the positive pole forming a comet-like tail, Figure 18.3. The fluorescent cell and segments are examined in a vertical fluorescent microscope. For double-strand breaks the RNA is removed using ribonuclease A.

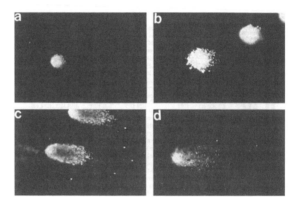

Figure 18.3 Photographs of double-strand break DNA migration pattern of individual brain cells from rats exposed to (a) bucking condition (0.1 mT), (b) magnetic fields of 0.1 mT, (c) 0.25 mT and (d) 0.5 mT, Lai and Singh (1997a). The "bucking mode" is the condition to reverse the field to cancel the magnetic fields with all else remaining constant.

The Comet Assay shows that electromagnetic fields cause DNA-strand breaks (Figure 18.3).

18.4.3 DNA Strand Breakage from ELF Exposure

Six independent laboratories have also published data on ELF induced DNA strand breaks confirming that ELF EMR damages DNA strands; Lai and Singh (1997a), Svedenstal *et al.* (1999a,b), Ahuja *et al.* (1997, 1999), Miyakoshi *et al.* (2000) and Zmyslony *et al.* (2000). Lai and Singh (1997a) also demonstrate the involvement of free radicals and the protective effect of melatonin. The evidence above suggests that EMR reduces melatonin confirming that reduced melatonin causes higher concentrations of free radicals which produce more DNA strand breaks from EMR exposure (ELF and to RF/MW frequencies). Increased DNA strand breaks will result in increased chromosome aberrations.

Svedenstal *et al.* (1999a) installed a cage of mice under some high voltage powerlines at a field strength of 8µT. Their DNA-strand breakage was compared with a group of sham exposed mice. After 32 days the exposed mice had 1.4 mT and sham exposed mice had 0.8mT, p<0.001. An example of a dose-response is given by Ahuja *et al.* (1999), (Figure 18.4). Early papers raised the question as to whether sparks and high currents from shocks were the cause of the Chromosome Aberrations. It is now shown that this is not necessary. The ELF oscillating magnetic fields cause direct DNA strand breakage and this is reflected in Chromosome Aberrations, Micronuclei Formation and altered genetic activity.

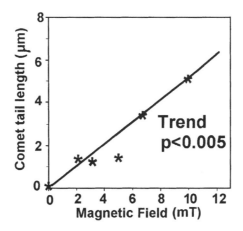

Figure 18.4 Comet tail length difference in before and after blood samples of peripheral blood leukocytes from human male subjects exposed to ELF magnetic fields of varying intensities for 1 hr, Ahuja *et al.* (1999).

A dose-response is supportive of a causal effect, Hill (1965). When backed by other studies it is confirmed to be causal. Miyakoshi *et al.* (2000) and Zmyslony *et al.* (2000) both show synergistic effects of 50Hz magnetic fields with X-rays and iron cations, respectively, and enhanced DNA-strand breakage. Li and Chow (2001) showed that 50 Hz, 1.2 mT magnetic fields and found that regarding E Coli. cells:

"That without protection of heat shock response, magnetic field exposure indeed induced DNA degradation and this deleterious effect could be diminished by the presence of an antioxidant, Trolox C. In our in vitro test, we also showed that the magnetic field could potentiate the activity of oxidant radicals."

Thus acute high exposure and chronic low exposure to 50/60 Hz magnetic fields causes significant Chromosome Aberrations, Micronuclei Formation and DNA-strand breakage in multiple, independent laboratories. This is classically sufficient to conclude that there is a causal effect. This robustly supports the hypothesis that ELF electromagnetic fields are a Universal Genotoxic Carcinogen.

The EMR Spectrum Principle predicts that if genotoxic effects are found for ELF exposures, then there should be stronger evidence for RF/MW exposures being genotoxic, even at much lower mean field strengths.

18.5 EVIDENCE OF EMR GENOTOXICITY

18.5.1 Introduction

Substances that produce Chromosome Aberrations cause cancer. Smerhovsky *et al.* (2001) studied a large group of miners who had been exposed to radon. They found that a 1% increase in chromosome aberrations resulted in a 64 % increase in the incidence of cancer, p<0.0001.

18.5.2 Earlier Published Statement

When there is direct evidence that EMR induces significant increases in chromosome damage, with significant dose response relationships, this gives evidence of a causal effect when replicated or extended by independent laboratories. Baranski and Czerski (1976) is a book on the biological effects of microwaves based on studies published up to that time. In their section on Chromosome and possible genetic effects, they open with the statement:

"Chromosome aberrations and mitotic abnormalities may be induced, at least under certain conditions and in certain cell types, by exposure to microwave or radiofrequency fields. This is a well-established fact, as several reports from at least five independent laboratories exist".

Baranski and Czerski note some uncertainties about exposure conditions. Also that in some experiments no temperature increases were observed. The evidence available in 2002 is much stronger and confirms that the genetic damage effects occur at non-thermal levels of RF/MW exposure and they do not require heat to be involved for DNA damage to occur. The following evidence on RF/MW and ELF genotoxicity was first presented in my paper that I was invited to present to the conference at the European Parliament in Brussels on June 29th, 2000. This paper showed that there was strong evidence that RF/MW were genotoxic and it was supported by epidemiological studies showing dose-response relationships for Cancer, Cardiac, Reproductive and Neurological (CCRN) health and mortality effects, with trends pointing to zero exposure for the Level of No Adverse Effects.

18.5.3 Chromosome Damage from RF/MW Exposure

The first identified study that showed that pulsed RF radiation could cause significant chromosome aberrations was that of Heller and Teixeira-Pinto (1959). Garlic roots were exposed to 27 MHz pulsed at 80 to 180 Hz for 5 min and then they were examined 24 hrs later. The concluded that this RF signal mimicked the chromosomal aberration produced by ionizing radiation and c-mitotic substances. No increased temperature was observed.

Blood samples were taken from the staff of the U.S. Embassy in Moscow. They had been chronically exposed to a very low intensity radar signal. Significant increases in chromosome damage were reported, Jacobson (1966) cited in

Goldsmith (1997). The above average compared with average chromosome aberrations yields RR = 2.1 (1.22-3.58), p=0.004.

Yao (1982) exposed rat kangaroo RH5 and RH16 cells to 2.45 GHz microwaves, maintaining the temperature at 37°C in the incubator. After 50 passages with microwave exposure there were 30 passages without. Significant chromosome aberrations were measured after 20 microwave exposed passages. Yao (1978) also found elevated chromosome damage in microwave exposed eyes of Chinese Hamsters.

Garaj-Vrhovac *et al.* (1990) noted the differences and similarities between the mutagenicity of microwaves and VCM (vinyl chloride monomer). They studied a group of workers who were exposed to 10 to 50 $\mu W/cm^2$ of radar produced microwaves. Some were also exposed to about 5ppm of VCM, a known carcinogen. Exposure to each of these substances (microwaves and VCM) produced highly significant (p<0.01 to p<0.001) increases in Chromatid breaks, Chromosome breaks, acentric and dicentric breaks in human lymphocytes from blood taken from exposed workers. The results were consistent across two assays, a micronucleus test and chromosome aberration assay. Ranking the exposures into low, middle and high showed that chromosome aberrations and micronuclei are significantly higher than the controls, (p<0.05, p<0.001, p<0.0001), for each of the exposure intensities.

Garaj-Vrhovac *et al.* (1991) exposed Chinese Hamster cells to 7.7 GHz microwave radiation to determine cell survival and chromosome damage. They assayed chromosome aberrations and micronuclei and found that microwaves increased these in a dose response manner, Figure 18.5, to levels that were highly significantly elevated (p<0.02 to p<0.01).

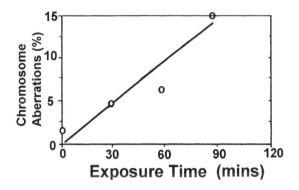

Figure 18.5 Chromosome aberrations in V79 Chinese Hamster cells exposed to 7.7 GHz microwaves at 30 mW/cm², Garaj-Vrhovac *et al.* (1991).

An exposure level of 30 mW/cm² is usually able to slightly raise the temperature over an hour. This experiment was undertaken under isothermal conditions, with samples being kept within 0.4°C of 22°C. The consistency of the time exposure and the survival assay at non-thermal exposure levels, confirms that this is a non-thermal effect.

This is very strong evidence of genotoxic effects from RF/MW exposures. When chromosomes are damaged one of the primary protective measures is for the immune system natural killer cells to eliminate the damaged cells. Alternatively the cells can enter programmed cell suicide, apoptosis. Garaj-Vrhovac *et al.* (1991) measured the cell survival rates. They found that cell survival reduced and the cell death increased in a time dependent and exposure dose response manner, Figure 18.6.

Figure 18.6 shows that cell death varies with time and intensity of exposure, down to very low exposure levels. An apparent 'saturation' at high levels is also evident. This is probably an indication of the lethal effect of high intensity microwaves for time periods as little as 30 mins. Since this is an isothermal experiment it raised important questions about the reasons for the cell death as acute genetic damage which is continuously related to microwave exposure down to non-thermal levels.

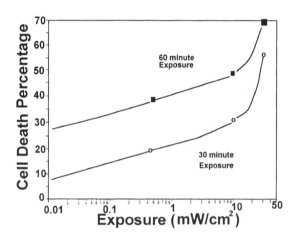

Figure 18.6 Cell death percentage of Chinese hamster cells exposed to 7.7 GHz microwaves (CW) for 30 minutes and 60 minutes in an isothermal exposure system, Garaj-Vrhovac *et al.* (1991).

Note that the general public ICNIRP guideline for microwaves above 2GHz is $1mW/cm^2$, and for workers is $5mW/cm^2$. Even at 100 times below the public exposure guideline a 60 minute exposure kills 28% of the cells and 30 minutes kills 8% of the cells. Garaj-Vrhovac *et al.* (1992) exposed human lymphocytes and showed that microwave radiation produced a dose-response increase in chromosome aberrations, Figure 18.7.

Having established that microwave exposure damaged chromosomes, this research team was asked to analyse blood samples from workers who had been exposed to pulsed microwaves generated by air traffic control radars while they were repairing them. Garaj-Vrhovac and Fucic (1993) assessed the chromosome aberrations (CAs) in 6 of these men and found elevated CAs in the range 3% to 33%, all being significantly higher than unexposed people. The repair rate over time was monitored. Figure 18.8 shows the repair rate over 30 weeks for the 33% case.

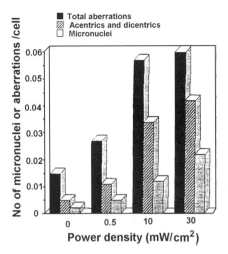

Figure 18.7 The relation of total chromosome aberrations. micronuclei and specific chromosome aberrations for each cell in human lymphocyte cultures in the dose of microwave radiation in itro, Garaj-Vrhovac *et al.* (1992).

Figure 18.8 shows the one male subject who was monitored over 33 % CA which was followed over 30 weeks following this exposure. The repair rate follows a significant linear rate (r=0.98), dropping from 33% to 3% over 30 weeks, 1%/week. Two different rates are evident. Two other subjects had repair rates at 0.6 to 1.1%/week and two at 0.25 to 0.3%/week. The authors note that Sagripanti and Swicord (1986) showed that microwave radiation damaged single-strand DNA and the Szmigielski (1991) showed that out of 29 epidemiological studies in the previous decade, 22 suggested a relationship between various neoplasms and exposure to electromagnetic fields.

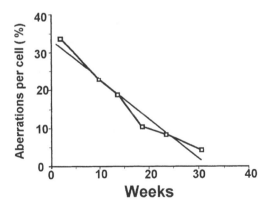

Figure 18.8 The time-dependent decrease in the number of chromosome aberrations for subjects with high numbers of chromosomal impairments, y = 0.318 - 0.010x, r = 0.98, Garaj-Vrhovac and Fucic (1993).

One of the difficulties of assaying genetic damage is the size of cells, chromosomes and DNA molecules. Chromosomes consist of twisted and folded DNA to form them. Figure 18.9 shows the actual microscopic images of chromosomes in human blood taken from a man exposed to radar.

There is no doubt that the radar exposure damaged the chromosomes. The damage is highly visible. Figure 18.9 shows that after microwave exposure there were many acentric, dicentric, polycentric, fragments, chromatid, ring chromosomes, chromosome breaks and chromatid interchange.

Figure 18.9 Chromosomes from the highly exposed subject, (a) before exposure and (B) after accidental exposure to a microwave radar signal, Garaj-Vrhovac and Fucic (1993).

Independently, Maes *et al.* (1993) found highly significant (p<0.001) increases in the frequency of chromosome aberrations (including dicentric and acentric fragments) and micronuclei in human blood exposed to 2.45 GHz microwaves to 30 to 120 minutes *in vitro*. The micronuclei assay showed a dose response with time, Figure 18.10.

Maes *et al.* (1997) observed elevated CAs from microwave exposure. Koveshnikova and Antipenko (1991a,b), Haider *et al.* (1994) Timchenko and Ianchevskaia (1995), Balode (1996), Mailhas *et al.* (1997) and Vijayalaxmi *et al.* (1997), and Pavel *et al.* (1998) have reported significant chromosome aberrations from RF/MW exposures.

Vijayalaxmi *et al.* (1997) chronically exposed cancer prone mice to 2.45 GHz CW microwaves at an SAR of 1 W/kg for 20 hr/day, 7 days/week for 18 months. Their aim was to determine whether microwaves were genotoxic through determining if there was significant micronuclei formation. They found highly significant increases in micronuclei in peripheral blood, from 8 per 2000 cells in sham exposed mice to 9 per 2000 cells microwave exposed mice, an increase of 12.5%, p<0.01. There was a significant increase of 6.6%, p<0.025, of micronuclei in the bone marrow. They also observed a significant 41% increase in tumours in the exposed mice compared to the sham exposed mice.

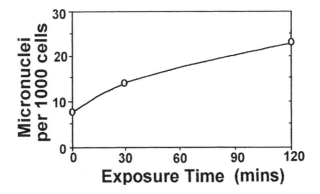

Figure 18.10 Micronuclei in microwave exposed human lymphocytes, the average of 4 donors,
Maes *et al.* (1993). Exposure was to 75 W/kg, 2.45 GHz microwaves pulsed at 50 Hz,
under controlled isothermal conditions.

This was a totally unexpected result from this group. A great deal of
effort was put into playing down the implications. They describe the increase in
peripheral blood as a 0.05%, by dividing the increase of 1 by 2000. This is not a
significant increase and this is not the right comparison. It is a deliberate attempt to
disguise their true result that shows that microwaves are genotoxic.

Garaj-Vrhovac (1999) used a micronuclei assay and lymphocyte mitotic
activity to assess genetic damage of a group of 12 men occupationally exposed to
microwave radiation. Exposures ranged from 10 μW/cm^2 to 20 mW/cm^2, in the
frequency range 1250-1350 MHz. It was found that there was significantly
enhanced micronuclei formation and significantly altered mitotic activity. The
exposed group had more than 3 times more micronuclei, 174 vs 50, $p<0.0001$.

A relatively high exposure from a cell phone antenna resulted in significant
chromosome aberrations, including gaps, acentric and dicentric aberrations, in
human lymphocytes Maes *et al.* (1996).

Tice, *et al.* (2002) showed chromosome damage from all cell phones
tested, all being statistically significant and all but one highly significant with
dose-response relationships up to a factor of three increase in chromosome
aberrations. They repeated the experiment and confirmed that the results were
robust and not an artifact.

18.5.4 Chromosome Aberrations Conclusions

Multiple independent studies, over 25 papers, show increases and most show
significant increases in chromosome aberrations from RF/MW exposure. Four
studies show dose-response relationships. This is more than adequate to classify
RF/MW radiation as genotoxic.

Many studies, from independent laboratories, have shown that ELF, RF/MW
and cell phone radiation, significantly increases chromosome aberrations in
exposed cells and animals, and including cells taken from human beings who have

been exposed to EMR in occupational situations. Even at very low intensity radar exposures that were experienced at the U.S. Embassy in Moscow, significant increases in chromosome damage were measured from human blood samples. This evidence shows conclusively that across the EMR spectrum, EMR is genotoxic. Hence electromagnetic radiation is carcinogenic.

18.5.5 Neoplastic Transformation

Direct evidence that microwave genotoxic damage leads to neoplastic transformation of cells into cancer cells was first provided in 1967, Stodolnik-Babanska (1967). He also reported non-thermal microwave induced chromosome aberrations. Balcer-Kubiczek and Harrison (1989) observed a significant dose-response increase of neoplastic transformation in a standard cell set (C3H/10T1/2) from a 24 hr exposure to 2.45 GHz microwaves with a co-carcinogen TPA.

The transformation was assayed after 8 weeks of exposure to a known cancer promoter chemical TPA, Figure 18.11. The method was confirmed with a positive control using X-rays. This also showed that 60 Hz magnetic fields also significantly increased neoplastic transformation.

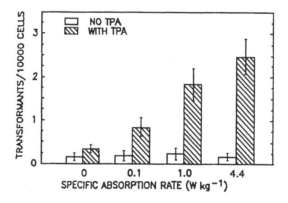

Figure 18.11 Dose-response relationship for induction of neoplastic transformation in C3H/10T1/2 cells by a 24h exposure to 2.45 GHz microwaves at the specific absorption rate (SAR) with and without TPA post-treatment for 8 weeks, Balcer-Kubiczek and Harrison (1991).

18.5.6 Direct Evidence of DNA-strand Alteration and Breakage

Sarkar *et al.* (1994) used Southern Blots of mouse DNA to study sections of the genome in cells with and without microwave exposure. The intensity used was that accepted as safe by the Non-ionizing Radiation Committee of IRPA (International Radiation Protection Association - the predecessor of ICNIRP), 1 mW/cm^2 (SAR = 1.18 W/kg). After a 2-hr exposure they observed a significant change of the DNA in the range 7-8 kb showing a mutagenic effect of microwaves.

Typically 100 or more cells are assayed and the comet tail length or moment is calculated from the frequency distribution. Observed tail lengths in this assay were as long as 250 µm, Figure 18.12. Lai and Singh (1995) showed, using the Comet Assay, that continuous and pulsed 245 GHz microwaves significantly enhanced DNA single strand breakage (p<0.001 after 4 hours), in a dose-response manner from 0, 0.6 and 1.2 W/kg SARs in living rat brains. Lai and Singh (1996) showed that microwaves at non-thermal SAR levels cause single and double DNA strand breaks, Figure 18.12. Pulsed microwaves caused more significant damage (p<0.01) than continuous microwaves (p<0.05).

Because Phelan *et al.* (1992) had shown that microwave exposure enhanced free radical activity, Lai and Singh (1997) used Melatonin, the body's highly potent antioxidant, and a free radical scavenging chemical (PBN) to see what happened to the demonstrated microwave induced DNA strand breaks. Both Melatonin (Figure 18.13) and PBN eliminated the induced DNA strand breaks, confirming the free radical mechanism for microwave induced DNA strand breaks

Three other laboratories have shown, using a range of Comet Assays, that microwaves, including cell phone radiation, directly damage DNA, enhancing strand breakage. Verschaeve and Maes (1998), who used a GSM cell phone signal to expose human and rat peripheral blood lymphocytes, found significantly increased strand breaks at high, but non-thermal exposure levels.

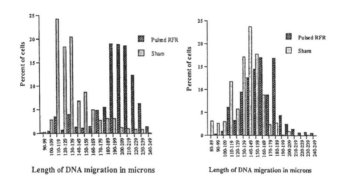

Figure 18.12 Single- (left) and double-strand (right) DNA breaks frequency distribution for percentage of cells of a given tail length from pulsed RFR and sham exposed brain cells, from 8 animals and 100 cells per animal, Lai and Singh (1996).

Phillips *et al.* (1998) exposed Molt-4 T-Lymphoblastoid cells to a range of cell phone radiation in the SAR range 0.0024 W/kg to 0.026 W/kg for both iDEN and TDMA signals. Using the basic equations, these SARs at the 813-836 MHz range [SAR = $\sigma E^2/2\rho$, σ=1 S/m, ρ=950 kg/m^3, and S = $E^2/3.77$ µW/cm^2, E: the electric field gradient in V/m and S the exposure in µW/cm^2] result in 1.3 to 13.0µW/cm^2. A 2 hr exposure to these very low levels of cell phone radiation significantly increased (p<0.0001) or decreased (p<0.0001) the DNA damage. Decreased DNA damage is evidence of increased repair that is evidence of damage, Meltz (1995). Significance at these levels is often taken as causal.

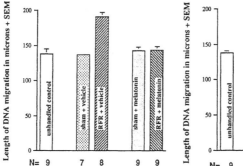
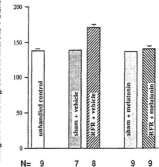

Figure 18.13 Effect of treatment with melatonin for RFR-induced increase in DNA single-strand (left) and double-strand (right) breaks in rats brain cells. Data was analysed using the one-way ANOVA, which showed a significant treatment effect (p<0.001) for both cases. "vehicle" involves injecting with the physiological saline without the active substance. Lai and Singh (1997)

Flight crews that are working in ELF and RF/MW fields have elevated DNA strand breaks, Cavallo *et al.* (2002), with reduction seen from eating antioxidants, confirming free radical involvement.

18.5.7 Motorola Funded Counter Research on DNA breakage

Motorola funded Dr Joseph Roti Roti's group at Washington University, St Louis, to replicate the Lai/Singh DNA damage research and to extend it to cell phone frequencies. "Replication" requires the work to be very closely following the method and conditions of the earlier study, usually carried out by independent researchers who are well qualified. Both groups used 2.45 GHz microwaves for exposure. However, the follow-up study used a cell-line (C3H/10T1/2) compared to Lai/Singh's living rats. The St Louis group also used a very different DNA damage assay based on Olive *et al.* (1992) not Singh *et al.* (1994).

This follow-up study also used a much weaker fluorescent stain, an overall weaker electrophoresis field (0.6 V/cm for 25 mins c.f. 0.4 V/cm for 60 mins). Most importantly, they did not use proteinase K to separate the bound protein from the DNA strands. It is therefore understandably evident that they used a much less sensitive method. The sensitivity of this method compared with the Singh Comet Assay is shown by the comet tail lengths of about 30microns in this assay compared with 250microns in Singh's assay.

However, despite this and their claims to find no DNA breakage from 2.45 GHz nor cell phone radiation, the data shows that they actually did. The first example was from Malyapa *et al.* (1997a), Figure 18.5, shown in Figure 18.14. It is visually evident that the 2hour exposure comet tail distribution is different from the sham exposure situation, with a number of tail lengths longer than 30 microns.

By splitting the distribution and counting the total number of tail lengths less than or equal to, and greater than 28microns, a 2x2 statistical analysis. Table 18.1, shows even with this weak assay technique, that 2 hours of microwave exposure of human brain cells (U87MG) significantly increases the DNA strand breaks (p=0.0067). Longer exposure periods induce enhanced DNA repair rates that are vital for brain cells because central neurons do not regenerate. However, it is the DNA damage that leads to cell death and mutations and mutations also occur from mistakes during the DNA repair process. Hence enhanced DNA repair rates, normally induced by DNA damage, and also enhances the cell death and cancer rates.

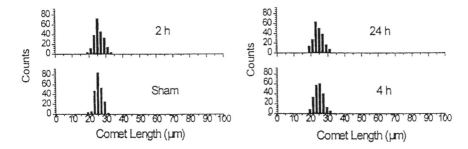

Figure 18.14 Frequency distribution of comet tail lengths for 2.45GHz exposed U87MG cells, Malyapa *et al.* (1997a).

Table 18.1 The 2x2 table of results for DNA strand breakage after exposure of U87MG cells to 2.45 GHz microwaves, from Figure 18.14

	Comet Length Class					
Time	**≤28μm**	**>28μm**	**RR**	**95%CI**	χ^2	**p-value**
Sham	196	29	1.00			
2hr	174	51	1.75	1.16 -2.76	7.34	0.0067
4 hr	206	20	0.06	0.40 -1.18	1.90	0.169
24 hr	197	25	0.87	0.53 -1.44	0.28	0.60

The time sequence of variations reveals a significant increase in DNA strand breakage after 2 hours and then the repair process kicks in and over compensates, Figure 18.15.

This confirms the Lai and Singh results rather than contradicting them. This shows significant DNA strand breakage after 2 hours.

Two other Malyapa *et al* (1997a and b) comet tail distributions were analysed using this analysis method, the first for rodent cells exposed to 2.45GHz microwaves, Figure 18.16, and the second human brain cells exposed to cell phone radiation, Figure 18.17.

Figure 18.16 shows highly significant DNA strand breakage, p<0.0000014, after 4 hours of 2.45 GHz exposure of rodent cells, and a much slower repair rate than the human brain cells demonstrate. Figure 18.17 shows the same pattern as Figure 18.15, demonstrating that CDMA cell phone radiation induces DNA damage (p = 0.000005) and repair after 24 hours (p = 0.01) in human brain cells.

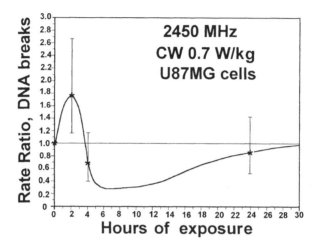

Figure 18.15 The time sequence of DNA damage and enhanced repair for Figure 5 in Malyapa *et al.* (1997).

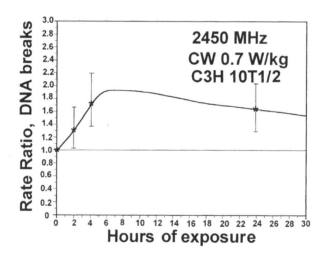

Figure 18.16 DNA strand breakage Risk Ratio and 95% confidence intervals for the frequency distribution of the Normalised Comet Moment of Malyapa *et al.* (1997a), Figure 6. The time line is a fitted estimate.

These results confirm the Lai and Singh results and confirm that microwave radiation and cell phone radiation significantly damages DNA stands and induces repair and significant repair after 4 hours in some cases. The C3H 10T1/2 cells show much slower DNA repair rates then the U87MG cells, indication a cell-specific characteristic. It is also well known that the damage and repair rates are strongly dependent on the position in the cell cycle, Durante *et al.* (1994).

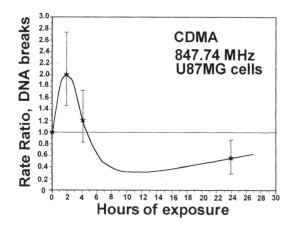

Figure 18.17 DNA strand breakage Risk Ratio and 95% confidence intervals for the frequency distribution in Figure 2 Normalised Comet Moment of Malyapa *et al.* (1997b), for 0.6W/kg SAR exposure of human brain cells to CDMA cell phone radiation.

The results of Malyapa *et al.* (1997a,b) puts the results of Phillips *et al.* (1998) into context. Phillips *et al.* (1998) found highly significant (p<0.0001) DNA-strand breakage at 0.0024 W/kg exposure to cell phone radiation. They also found significant DNA-strand repair (p<0.0001) with other exposure regimes at a similar SAR level. Significant DNA-strand repair is initiated by DNA-strand breakage. This is why earlier assays were based on looking for induced DNA repair as an indicator of DNA damage, Meltz (1995).

The lower sensitivity of the assay used by Malyapa et al. is directly demonstrated by the comet tail lengths. The longest comet tail lengths were 32 microns for Malyapa *et al.* and 250 microns for Lai/Singh. Despite the lower sensitivity of the Malyapa *et al.* assay, the results confirm the initial results of Lai and Singh (1995), that pulsed microwaves, including cell phone radiation, significantly enhances DNA-strand breakage.

18.5.8 Genotoxicity Conclusions

Hence RF/MW radiation has been confirmed to enhance DNA damage under RF/MW exposure from radar-like and cell phone exposures in four independent laboratories. One study included an exposure level which is 0.22% of the ICNIRP guideline, ICNIRP (1998).

A large body of scientific evidence is cited above that shows that electromagnetic radiation, across the EMR spectrum, causes comprehensive damage to the genetic material in cells. It also impairs the body's ability to repair and eliminate damaged cells through reducing melatonin, altering calcium ion signalling and impairing the performance of the immune system.

There is strong and robust evidence of chromosome aberrations, micronuclei formation, DNA-strand breakage altered oncogene activity and neoplastic transformation of cells, to conclude that EMR across the spectrum from ELF to RF/MW is genotoxic. This is independently confirmed by the established biological mechanisms of calcium ion efflux and melatonin reduction.

This is also totally independently confirmed by over a hundred occupational groups showing elevated cancer from EMR exposure, scores showing significantly to very highly significantly elevated cancer incidence and mortality, and several with dose-response relationships.

18.6 OTHER BIOLOGICAL EMF/EMR MECHANISMS

Carcinogens are important toxic substances that act in at least one of several ways to enhance genetic damage of cells or to reduce the repair of cells by altering the cell-to-cell communication, Li *et al.* (1999), altering the cellular calcium ion homeostasis, Fanelli *et al.* (1999), reducing melatonin, Reiter (1994).

The cell-to-cell communication, for example through gap-junctions, is a vital part of the checking process to detect genetically damaged cells and to initiate apoptosis (programmed cell death). Enhanced cell death is associated with a total process called premature ageing. Increased cancer rates also occur with age, particularly over 65 years.

The cellular ion concentration plays a central role in the cell's decision as to whether to survive to become a neoplastic cancer cell or to initiate apoptosis. Elevated cellular calcium ions favour cellular survival and cancer formation. EMR induced alteration of cellular calcium ion homeostasis is a well-established effect, Blackman (1990).

Melatonin is a highly potent free radical scavenger. Hence any substance that reduces the melatonin allows naturally occurring free radicals to produce more DNA damage, leading to greater cell death and cancer cell formation. Melatonin reduction from EMR/EMF exposure has been observed in multiple independent studies, for example Arnetz and Berg (1996), Burch *et al.* (1997, 1998, 1999a,b, 2000), Davis (1997), Graham *et al.* (1994, 2000), Karasek *et al.* (1998), Juutilainen *et al.* (2000), Pfluger and Minder (1996) and Wilson *et al.* (1990). Melatonin is also reduced by ELF fields at residential levels, Levallois *et al.* (2001) and in a dose-response manner, Davis *et al.* (2001). This is sufficient evidence for an established biological effect.

18.7 EPIDEMIOLOGICAL EVIDENCE

It is shown above that there is robust evidence that EMF and EMR are genotoxic. Hence they cause enhanced rates of cell death and cancer in exposed populations with no safe threshold level because cellular DNA is damaged cell-by-cell. Therefore, since electromagnetic fields and radiation expose the whole body, cancer rates are increased across many body organs, but most frequently in the brain, CNS, blood, bone marrow and lymph systems, leading to Brain Cancer, Leukaemia and Lymphoma. These have been shown by over 150 studies for ELF exposures, including many with dose-response trends.

Table 18.2 A summary of epidemiological studies involving adult leukaemia mortality or incidence, ranked by probable RF/MW exposure category

Study	Reference	Exposure Category	Leukaemia Type	Risk Ratio	95% Confidence Interval
Polish Military (Mortality)	Szmigielski (1996)	High	ALL	5.75	1.22-18.16
			CML	13.90	6.72-22.12
			CLL	3.68	1.45-5.18
			AML	8.62	3.54-13.67
			All Leuk.	6.31	3.12-14.32
Korean War Radar Exposure	Robinette *et al.* (1980) (Mortality)	High AT/ET	Leuk/ Lymp	2.22	1.02-4.81
Radio and TV Repairmen	Milham (1985)	Moderate	Acute Leuk.	3.44	
			Leuk.	1.76	
Amateur Radio (Mortality)	Milham (1988)	Moderate	AML	1.79	1.03-2.85
UK Sutton Coldfield <=2k m	Dolk *et al.* (1997a)	Moderate	Leuk	1.83	1.22-2.74
North Sydney TV/FM Towers (Mortality)	Hocking *et al.* (1996)	Low	All Leuk.	1.17	0.96-1.43
			ALL+CLL	1.39	1.00-1.92
			AML+CML	1.01	0.82-1.24
			Other Leuk	1.57	1.01-2.46
UK TV/FM (Incidence)	Dolk *et al.* (1997b)	Low	Adult Leuk.	1.03	1.00-1.07

Notes: ALL Acute Lymphatic Leukemia; CLL: Chronic Lymphatic Leukaemia; AML Acute Myeloid Leukaemia; CML: Chronic Myeloid Leukaemia; and All Leuk.: All Adult Leukaemia.

The RF/MW signals have stronger evidence of being genotoxic, as predicted by the EMR Spectrum Principle. However, the number of epidemiological studies involving RF/MW exposures is somewhat smaller. Despite the small number, they do confirm the UUGC hypothesis and show a global dose-response increase in Leukaemia, Table 18.2, and elevated cancer rates across many body organs in multiple independent studies, Table 18.3.

Table 18.3 Summary of all site cancers from Robinette *et al.* (1980), using AT/ET except for Brain cancer (FT/ET), Milham (1988), Szmigielski (1996) and for Dolk (1997a,b) using the maximum and/or significant result in the radial patterns

Exposure Regime	**Robinette** RF/MW High RR	**Milham** RF/MW Mod. PMR	**Szmigielski** Mixed High RR	**Dolk(a)** RF/MW VeryLow O/E	**Dolk(b)** RF/MW Very Low O/E
Relationship					
Sample Size(N)	202	2649	55,500	17409	13372
Symptoms					
All Cancer	1.66*	106**	2.07*	1.20*	
Throat/Stomach			3.24**		
Respiratory/ Lung	1.75	114**	1.06		
Colorectal/ bladder (1)			3.19**	1.36/1.76	1.10
Liver, pancreas		117*	1.47		
Skin, Melanoma	2.66		1.67*	2.39*	1.11
Thyroid			1.54		
Brain, CNS (2)	2.39	143**	1.91*	1.31	1.06
Leukaemia	2.22*	136*	6.31***	1.74*	1.15
NHL		164**	5.82***	1.30*	
ALL		162**	5.75*	3.57	1.04
AML.			8.62***	1.02	1.17
CML.			13.90***	1.23	
CLL			3.68**	2.56*	1.20

p-values: * <0.05; ** <0.01; *** <0.001

Note (1): Colorectal for Szmigielski and the left Dolk(a) and bladder for the right Dolk(a) and Dolk(b).
Note (2): In Milham 16 of the unspecified neoplasms were brain tumours that have been added to this group.

Occupational studies show highly significantly elevated cancer rates and residential studies provide strong and robust evidence that there is no safe threshold because at a distance of 5 to 10 km from regional broadcast towers, cancer rates are elevated in multiple studies in dose-response trends, Dolk *et al.*

It is important to recognise that in residential situations, the direct far-field RF/MW exposures from Radio and TV towers are much higher than the levels in gardens and lawns around homes because of the scattering and shielding effects of trees, buildings and hills. Measurements from Australia, McKenzie *et al.* (1998), show that these are about 2% of the direct roof level field strengths. Inside it is even lower because of the shielding effects of walls and roofs. Measurements place these at less than 0.6% of the direct field strength. Therefore a conservative adjustment factor, allowing for dominantly inside activity and going away for work, school, shopping, recreation, etc, gives a chronic personal mean exposure of about 1% of the direct roof-level exposure.

For the Sutton Coldfield TV Tower study, Dolk *et al.* (1997a) the All Cancer rate was still significantly but lowly elevated O/E =1.04 at 10 km. For Non-Hodgkin's Lymphoma O/E = 1.30 at 10 km. The peak combined VHF and UHF direct signals were reported at 7 $\mu W/cm^2$. This would have occurred inside 3 km. At 10 km the direct signals would be about $1/10^{th}$ of the inside 3 km levels. Adjusting for the foliage and building shielding effects gives the chronic mean personal exposure at 10 km of 0.007 $\mu W/cm^2$. This, and the dose-response relationships point to a near zero level of no effects, which is consistent with the RF/MW signals being genotoxic.

18.8 CONCLUSIONS

The genotoxic and epidemiological evidence strongly support and reinforce each other as confirmation of the hypothesis that:

Electromagnetic Radiation across the spectrum is a Ubiquitous Universal Genotoxic Carcinogen.

This means that the safe level of chronic exposure is zero. We all live in these fields, from the power supplies, appliances, computers, play stations, radio, TV, radar, satellite and mobile phone signals. Modern technology is heavily reliant on RF/MW signals for many of its activities and services. By ignoring or dismissing the robust evidence that EMF and EMR signals are genotoxic and cause cancer, cardiac, reproductive and neurological effects, Cherry (2000), exposure guidelines and standards are set to chronically expose workers and the public to signals that are very hazardous and are causing massive elevation of rates of adverse health effects across the whole exposed population – all people.

Cherry (2002) found that there was very strong evidence that the health and biological effects correlated with Solar and Geomagnetic Activity (S-GMA) are caused by the biophysical mechanism of the Schumann Resonance (SR) signal being detected and responded by the brain. The S-GMA modulates the SR signal through the Solar Wind and the Ionosphere. The brain has a natural ELF ion signal production and detection system for seeing, thinking, intelligence and emotion. It almost exactly matches the frequency range of Nature's SR signal, resulting in resonant absorption.

The altered brain waves changes the melatonin production by the pineal gland, and thereby causing modulation in a homeostasis mode to cause variation of cancer, cardiac, reproductive and neurological health and mortality rates in people around the world. The evidence also shows that the SR signal synchronizes the brain rhythm patterns on a daily and ELF scale, to keep them stable so that intelligence had been developed in a biological soft, adaptive ELF electromagnetic organ. Because the brain detects and responds to a 0.1 pW/cm^2 signal, and because of varying melatonin levels which can cause modulation of cancer rates, it is plausible that human invented signals over 1000 to a billion times higher will also cause CCRN health effects through a genotoxic mechanism.

This also puts the Schwarzenberg results of the study of sleep disturbance around a shortwave radio tower in Switzerland in a biophysical context. In an experiment the tower was secretly turned off for three days. Sleep quality improved significantly in all surveyed groups, including the far distant reference Group C, $p<0.0001$. Their mean exposure was 0.4 nW/cm^2. In modern technological physics it looks very small. In Nature's intelligent biophysical environment this is plausible as it is 4000 times stronger than the Schumann Resonance signal. These insights also give a plausible scientific basis to the widespread and growing syndrome called Electrosensitivity, shown even by the Director of the WHO. Dr Gro Harlem Brundtland. reacts to her own and other people's cell phones and she promotes sensible caution, backed up by sound science shown here, especially for children.

18.9 RECOMMENDATIONS

- Based on this robust and extensive evidence, or, if there is any doubt, the Precautionary Principle, then public health protection standards should have residential and occupational levels lower than:

$$0.1 \ \mu W/cm^2 \ (0.6 \ V/m)$$

- Because serious health effects occur at and below this level, the areas where signals are currently lower than 10nW/cm^2 should be maintained that way and other areas above 10nW/cm^2 should be adjusted to achieve mean exposures less than 10nW/cm^2 by 2010.
- Cell phone radiation is genotoxic, therefore it causes cancer. Cellular telephones can be made more than 99% safer. Exposing the user's head and body to less than 1% of the present SAR. This involves a hand-set shield, a directional aerial and a fibre optic hands-free kit. Therefore authorities should require the development and provision of the safer cell phone technologies.
- Cell phone base stations should generally be sited at last 200m away from residences and workplaces, with side-lobe protection, to significantly reduce the near vicinity exposure levels that have been shown to cause neurological, cardiac and cancer health problems. A Spanish Judge ordered that 49 antennas be removed from a site near a school where 4 children developed cancer over 18 months following their installation.
- Legislative measures to enforce these provisions are vital for protecting public health.

18.10 REFERENCES

Aden, W.R., 1988: Cell membranes: The electromagnetic environment and cancer promotion., *Petrochemical Research*, **13** (7): 671-677.

Ahuja, Y.R., Bhargava, A., Sircar, S., Rizwani, W., Lima, S., Devadas, A.H. and Bhargava S.C., 1997: Comet assay to evaluate DNA damage caused by magnetic fields., In: *Proceedings of the International Conference on Electromagnetic Interference and Compatibility, India Hyderabad*, December 1997: 272-276.

Ahuja, Y.R., Vijayashree, B., Saran, R., Jayashri, E.L., Manoranjani, J.K. and Bhargava S.C., 1999: In Vitro effects of low-level, low-frequency electromagnetic fields on DNA damage in human leucocytes by comet assay *Indian J Biochem and Biophys* **36**: 318-322.

Antipenko, E.N. and Koveshnikova, I.V., 1987: Cytogenetic effects of microwaves of non-thermal intensity in mammals. *Dok. Akad. Nauk USSR*, **296(3): 724-726** [In Russian].

Archimbaud, E., Charrin, C., Guyotat, D., and Viala, J-J, 1989: Acute myelogenous leukaemia following exposure to microwaves. *British Journal of Haematology*, **73(2): 272-273.**

Arnetz, B.B. and Berg, M., 1996: Melatonin and Andrenocorticotropic Hormone levels in video display unit workers during work and leisure. J Occup Med **38(11):** 1108-1110.

Balcer-Kubiczek, E.K. and Harrison, G.H., 1989: Induction of neoplastic transformation in C3H/10T1/2 cells by 2.45 GHz microwaves and phorbol ester. *Radiation Research* **117**: 531-537.

Balode, Z., 1996: Assessment of radio-frequency electromagnetic radiation by the micronucleus test in Bovine peripheral erythrocytes. *The Science of the Total Environment*, **180**: 81-86.

Baranski, S. and Czerski, P., 1976: *Biological effects of microwaves.* Publ: Dowden, Hutchinson and Ross Inc, Stroudsburg, Pennsylvania, U.S.A.

Beaglehole, R., Bonita, R. and Kjellstrom, T., 1993: *Basic Epidemiology.* World Health Organization, Geneva, Switzerland.

Blackman, C.E., 1990: ELF effects on calcium homeostasis. In *Extremely low freguency electromagnetic fields: The guestion of cancer*, BW Wilson, RG Stevens, LE Anderson Eds, Publ. Battelle Press Columbus: 1990; 187-208.

Burch, J.B., Reif, J.S., Pittrat, C.A., Keefe, T.J. and Yost, M.G., 1997: Cellular telephone use and excretion of a urinary melatonin metabolite. In: *Annual review of Research in Biological Effects of electric and magnetic fields from the generation, delivery and use of electricity, San Diego, CA*, Nov. 9-13, P-52.

Burch, J.B., Reif, J.S., Yost, MG., Keefe, T.J. and Pittrat, C.A., 1998: Nocturnal excretion of urinary melatonin metabolite among utility workers. *Scand J Work Environ Health* **24(3): 183-189.**

Burch, J.B., Reif J.S. and Yost, MG., 1999a: Geomagnetic disturbances are associated with reduced nocturnal excretion of melatonin metabolite in humans. *Neurosci Lett* **266(3)**: 209-212.

Burch, J.B., Reif, J.S., Yost, MG., Keefe, T.J. and Pittrat, C.A., 1999b: Reduced excretion of a melatonin metabolite among workers exposed to 60 Hz magnetic fields. *Am J Epidemiology* **150(1)**: 27-36.

Burch, J.B., Reif, J.S., Noonan, C.W. and Yost, M.G., 2000: Melatonin metabolite levels in workers exposed to 60-Hz magnetic fields: work in substations and with 3-phase conductors. *J of Occupational and Environmental Medicine,* **42(2):** 136-142.

Burnet, M., 1958: Leukemia as a problem in preventive medicine. *New Engl J Med;* **259**: 423-431.

Cavallo, D., Tomao, P., Marinaccio, A., Perniconi, B., Setini, A., Palmi, S. and Iavicoli, S., 2002: Evaluation of DNA damage in flight personnel by Comet Assay. *Mutation Research* **516(1-2)**: 148-152.

Cherry, N.J., 2001: Re: Cancer incidence near radio and television transmitters in Great Britain, II All high power transmitters, Dolk et al. 1997 a,b in American J. of Epidemiology, **145(1)**:1-9 and 10-17. *Comment in American J of Epidemiology* **153(2)**: 204-205.

Cherry, N.J., 2002: Schumann Resonances, a plausible biophysical mechanism for the human health effects of Solar/Geomagnetic Activity. *Natural Hazards* **26**: 279-331.

Cohen, M.M., Kunska, A., Astemborski, J.A., McCulloch, D., and Packewitz, D.A., 1986: The effect of low-level 60-Hz electromagnetic fields on human lymphoid cells: I. Mitotoc rate and chromosome breakage in human peripheral lymphocytes. *Bioelectromagnetics* **7(4)**: 415-523.

Cohen, M.M., Kunska, A., Astemborski, J.A., and McCulloch, D., 1986: The effect of low-level 60-Hz electromagnetic fields on human lymphoid cells: II. Sister-chromatid exchanges in peripheral lymphocytes and lymphoblastoid cell lines. *Mutat Res* **172(2)**: 177-184.

Court-Brown, W. M., Doll R., 1961: Leukaemia in childhood and young adult life: Trends in mortality in relation to aetiology. *BMJ* **26**:981-988.

Davis, S., 1997: Weak residential Magnetic Fields affect Melatonin in Humans, *Microwave News,* Nov/Dec 1997.

Davis, S., Mirick, D.K. and Stevens, R.G., 2002: Residential magnetic fields and the risk of breast cancer. *Am J Epidemiology* **155(5)**: 446-454.

Dolk H, Shaddick G, Walls P, Grundy C, Thakrar B, Kleinschmidt I, Elliott P., 1997a: Cancer incidence near radio and television transmitters in Great Britain, I Sutton Coldfield transmitter. *Am J Epidemiol* **145(1)**: 1-9.

Dolk H, Elliott P. Shaddick G, Walls P Grundy C, Thakrar B., 1997b: Cancer incidence near radio and television transmitters in Great Britain, II All high power transmitters. *Am J Epidemiol* **145(1)**: 10-17.

Dorland, 1994: *Dorland's illustrated Medical Dictionary, 28th Edition.* Publ. W.B. Saunders, Philadelphia, 1940pp.

El Nahas, SM. and Oraby, HA., 1989: Micronuclei formation in somatic cells of mice exposed to 50 Hz electric fields. *Environ Mol Mutagen* **13(2)**: 107-111.

Elwood, J.M., 1999: A critical review of epidemiological studies of radio requency exposure and human cancers. *Environ Health Perspective* Feb (Suppli 1): 155-168.

Fanelli, C., Coppola, S., Barone, R., Colussi, C., Gualandi, G., Volpe, P. and Ghibelli, L., 1999: Magnetic fields increase cell survival by inhibiting apoptosis via modulation of Ca^{2+} influx. *FASEB Journal* **13(1)**: 95-102.

Garaj-Vrhovac, V., Fucic, A, and Horvat, D., 1990a: Comparison of chromosome aberration and micronucleus induction in human lymphocytes after occupational exposure to vinyl chloride monomer and microwave radiation., *Periodicum Biologorum*, Vol 92, No.4, pp 411-416.

Garaj-Vrhovac, V., Horvat, D. and Koren, Z., 1990b: The effect of microwave radiation on the cell genome. *Mutat Res* **243**: 87-93.

Garaj-Vrhovac, V., Horvat, D. and Koren, Z., 1991: The relationship between colony-forming ability, chromosome aberrations and incidence of micronuclei in V79 Chinese Hamster cells exposed to microwave radiation. *Mutat Res* **263**: 143-149.

Garaj-Vrhovac, V. Fucic, A, and Horvat, D., 1992: The correlation between the frequency of micronuclei and specific aberrations in human lymphocytes exposed to microwave radiation in vitro. *Mutation Research* **261**: 161-166.

Garaj-Vrhovac, V., Fucic, A. and Pevalek-Kozlina, B. 1993: The rate of elimination of chromosomal aberrations after accidental exposure to microwave radiation. *Bioelectrochemistry and Bioenergetics*, **30**: 319-325.

Garaj-Vrhovac, V. 1999: Micronucleus assay and lymphocyte mitotic activity in risk assessment of occupational exposure to microwave radiation. *Chemosphere* **39(13)**: 2301-2312.

Garcia-Sagredo, J.M. and Monteagudo, J.L., 1991: Effect of low-level pulsed electromagnetic fields on human chromosomes in vitro: analysis of chromosome aberrations. *Hereditas* **115(1)**: 9-11.

Goldsmith, J.R., 1997: Epidemiologic evidence relevant to radar (microwave) effects. *Environmental Health Perspectives*, **105** (Suppl 6): 1579-1587.

Graham, C., Cook, M.R., Cohen, H.D. and Gerkovich, M.M., 1994: A dose response study of human exposure to 60Hz electric and magnetic fields. *Bioelectromagnetics* **15**: 447-463.

Graham, C., Cook, M.R., Sastre, A., Riffle, D.W. and Gerkovich, M.M., 2000: Multi-night exposure to 60 Hz magnetic fields: effects on melatonin and its enzymatic metabolite. *J Pineal Res* **28(1)**: 1-8.

Green, L.M., Miller, A.B., Agnew, D.A., Greenberg, M.L., Li, J., Villeneuve, P.J. and Tibshirani, R., 1999: Childhood leukaemia and personal monitoring of residential exposures to electric and magnetic fields in Ontario, Canada. *Cancer Cause and Control* **10**: 233-243.

Haider, T., Knasmueller, S., Kundi, M, and Haider, M., 1994: Clastogenic effects of radiofrequency radiation on chromosomes of Tradescantia. *Mutation Research*, **324**: 65-68.

Hardell, L., Holmberg, B., Malker, H., and Paulsson, L.E., 1995: Exposure to extremely low frequency electromagnetic fields and the risk of malignant diseases--an evaluation of epidemiological and experimental findings. *Eur. J. Cancer Prevention*, 1995 Sep;4 Suppl 1:3-107.

Hardell, L, Nasman, A, Pahlson, A, Hallquist, A, Hansson Mild, K, 1999: Use of cellular telephones and the risk for brain tumours: A case-control study. *Int J Oncol* **15(1)**:113-116.

Hardell, L, Nasman, A, Haliquist, A, 2000: Case-control study of radiology work, medical X-ray investigations and use of cellular telephones as risk factors. *J of General Medicine* www.medscape.com/Medscape/GeneralMedicine/journal/2000/v02.n03/

Hardell, L, Hansson Mild, K, Hallquist, A., Carlberg, M., Pahlson, A. and Larssen, I., 2001: Swedish study on use of cellular and cordless telephones and the risk for brain tumours. In: *Proceeding London Conference on Mobile phones and Health*, June 6-7, 2001.

Hatch, E.E., Linet, M.S., Kleinerman, R.A., Tarone, R.E., Severson, R.K., Hartsock, C.T., Haines, C., Kaune, W.T., Friedman, D., Robison, L.L. and Wacholder, S., 1998: Association between childhood acute lymphoblastic leukemia and use of electrical appliances during pregnancy and childhood. *Epidemiology* **9(3)**: 234-245.

Heller JH, Teixeira-Pinto *AA.,* 1959: A new physical method of creating chromosome aberrations. Nature **183 (4665)**: 905-906.

Hill, A. B., 1965: The Environment and Disease: Association or Causation? *Proc. Royal Society of Medicine (U.K.).* 295-300.

Hocking, B., Gordon, I.R., Grain, H.L., and Hatfield, G.E., 1996: Cancer incidence and mortality and proximity to TV towers. *Medical Journal of Australia*, Vol 165, 2/16 December, pp 601-605.

Hocking, B., Gordon, I.R. and Hatfield, G.E.,1999: Childhood leukaemia an TV towers revisited. Letters to the editor, *Australian and New Zealand J Public Health* **23(1)**: 94-95.

ICNIRP, 1998: Guidelines for limiting exposure to time-varying electric, magnetic and electromagnetic fields (up to 300 GHz). *Health Physics* **74(4)**: 494-522.

IEGMP, 2000: Mobile phones and health. *Independent expert group on mobile phones, Chairman Sir William Stewart, United Kingdom Government requested inquiry* www.iegmp.org.uk/

Jacobson, C.B., 1969: Progress report on SCC 31732, (Cytogenic analysis of blood from the staff at the U.S. Embassy in Moscow), *George Washington University, Reproductive Genetics Unit, Dept. of Obstertics and Genocolgy*, February 4, 1969.

Johnson, C.C. and Guy, A.W., 1972: Non-ionizing electromagnetic wave effects in biological materials and systems. *Proc IEEE* **60(6)**: 692-718.

Juutilainen, J., Stevens, R.G., Anderson, L.E., Hansen, N.H., Kilpelainen, M., Laitinen, J.T., Sobel, E. and Wilson, B.W., 2000: Nocturnal 6-hydroxymelatonin sulphate excretion in female workers exposed to magnetic fields. *J Pineal Research* **28(2)**: 97-104.

Karasek, M., Woldanska-Okonska, M., Czernicki, J., Zylinska, K. and Swietoslawski, J., 1998: Chronic exposure to 2.9 mT, 40 Hz magnetic field reduces melatonin concentrations in humans. *J Pineal Research* **25(4)**: 240-244.

Khaili, AM. and Qassem, W., 1991: Cytogenetic effects of pulsing electromagnetic field on human lymphocytes in vitro: chromosome aberrations sister-chromatid exchanges and cell kinetics. *Mutat Res* **247**: 141-146.

Kraut A, Tate R, Tran NM., 1994: Residential electric consumption and childhood cancer in Canada (1971-1986). *Arch Environ Health* **49(3)**: 156-159.

Koveshnikov, IV. and Antipenko, EN., 1991a: Quantitative patterns in the cytogenetic action of microwaves. *Radiobiologiia* **31(1)**: 149-151.

Koveshnikova, IV. and Antipenko, E.N., 1991b: The participation of thyroid hormones in modifying the mutagenic effect of microwaves. *Radiobiologiia* **31(1)**: 147-149.

Lai, H. and Singh, NP., 1995: Acute low-intensity microwave exposure increases DNA single-strand breaks in rat brain cells. *Bioelectromagnetics* **16**: 207-210.

Lai, H. and Singh, NP., 1996: Single- and double-strand DNA breaks in rat brain cells after acute exposure to radiofrequency electromagnetic radiation. Int. J. *Radiation Biology* **69(4)**: 513-521.

Lai, H., and Singh, NP., 1997a: Melatonin and N-tert-butyl-a-phenylnitrone Block 60 Hz magnetic field-induced DNA single- and double-strands Breaks in Rat Brain Cells. *Journal of Pineal Research* **22**:152-162.

Lai, H., and Singh, NP., 1997b: Melatonin and Spin-Trap compound Block Radiofrequency Electromagnetic Radiation-induced DNA Strands Breaks in Rat Brain Cells. *Bioelectromagnetics* **18**: 446-454.

Lester, J.R., and Moore, D.F., 1982a: Cancer incidence and electromagnetic radiation. *Journal of Bioelectricity*, **1(1)**: 59-76.

Lester, J.R., and Moore, D.F., 1982b: Cancer mortality and air force bases. *Journal of Bioelectricity*, **1(1)**: 77-82.

Lester, J.R., 1985: Reply to: Cancer mortality and air force bases, a reevaluation. *Journal of Bioelectricity*, **4(1)**: 129-131.

Levallois, P., Dumont, M., Touitou, Y., Gingras, S., Masse, B., Gauvinm, D., Kroger, E., Bourdages, M. and Douville, P., 2002: Effects of electric and magnetic fields from high-power lines on female urinary excretion of 6-sulfatoxymelatonin. *Am J Epidemiol.* **154(7)**: 601-609.

Li, S.H. and Chow, K-C., 2001: Magnetic field exposure induces DNA degradation. *Biochemical and Biophysical Research Communication* **280**: 1385-1388.

Li, C.M., Chiang, H., Fu, Y.D., Shao, B.J., Shi, J.R. and Yao, G.D., 1999: Effects of 50Hz magnetic fields on gap junction intercellular communication. *Bioelectromagnetics* **20(5)**:290-294.

Lilienfeld, A.M., Tonascia, J., Tonascia, S., Libauer, C.A. and Cauthen, G.M., 1978: Foreign Service health status study - evaluation of health status of foreign service and other employees from selected eastern European posts. *Final Report (Contract number 6025-619073) to the U.S. Department of State* July 31, 1978, 429pp.

Lisiewicz, J., 1993: Immunotoxic and hematoxic effects of occupational exposures. *Folia Med Cracov* **34(1-4)**: 29-47.

Maes, A., Verschaeve, L., Arroyo, A. De Wagter, C. and Vercruyssen, L., 1993: In vitro effects of 2454 MHz waves on human peripheral blood lymphocytes. *Bioelectromagnetics* **14**: 495-501.

Maes, A. Collier, M., Slaets, D., and Verschaeve, L., 1996: 954 MHz Microwaves enhance the mutagenic properties of Mitomycin C. *Environmental and Molecular Mutagenesis*, **28**: 26-30.

Maes, A., Collier M., Van Gorp, U., Vandoninck, S. and Verschaeve, L., 1997: Cytogenetic effects of 935.2-MHz (GSM) microwaves alone and in combination with mitomycin C. *Mutat Res* **393(1-2)**: 151-156.

Malyapa, R.S., Ahern, E.W., Straube, W.L., Moros, E.G., Pickard, WE. and Roti Roti, J.L., 1997a: Measurement of DNA damage after exposure to 2450 MHz electromagnetic radiation. *Radiation Research* **148**: 608-617.

Malyapa, R.S., Ahern, EW., Bi, C. Straube, W.L., Moros, E.G., Pickard, W.F. and Roti Roti, J.L., 1997b: Measurement of DNA damage after exposure to electromagnetic radiation in the cellular phone communication frequency band (835.62 and 847.74 MHz). *Radiation Research* **145**: 618-627.

Mailhas, J.B., Young, D., Marino, A.A. and London, SN., 1997: Electromagnetic fields enhance chemically-induced hyperploidy in mammalian oocytes. *Mutagenesis* **12(5)**: 347-351.

Manikowska-Czerska E, Czerski P Leach WM., 1985: Effects of 2.45 GHz microwaves on meiotic chromosomes of male CBA/CAY mice. J Hered **76(1)**: 71-73.

Maskarinec, G., and Cooper, J., 1993: Investigation of a childhood leukemia cluster near low-frequency radio towers in Hawaii. SER Meeting, Keystone, Colorado, June 16-18, 1993. *Am. J. Epidemiology*, **138**: 666, 1993.

McKenzie, D.R., Yin, Y. and Morrell, S., 1998: Childhood incidence of acute lymhoblastic leukaemia and exposure to broadcast radiation in Sydney - a second look. *Aust NZ J Pub Health* **22 (3)**: 360-367.

Meltz, M.L., 1995: Biological effects versus health effects: an investigation of the genotoxicity of microwave radiation. In: *Radiofrequency Radiation Standards, NATO ASI Series* (B.J. Klauebberg, Grandolfo and Erwin Eds). New York, Plenum Press, 1995: 235-241.

Michelozzi, P., Capon, A., Kirchmayer, U., Forastiere, F., Biggeri, A., Baraca, A. and Perucci, C.A., 2002: Adult and childhood Leukaemia near a high-power radio station in Rome, Italy. *Am J. Epidemiology* **155(12)**: 1096-1103.

Milham, S., 1982: Mortality from leukemia in workers exposed to electric and magnetic fields. *New England J. of Med.*, **307**: 249-250.

Milham S., 1985: Mortality in workers exposed to electromagnetic fields. Environ Health Perspectives **62**: 297-300.

Milham S., 1988: Increased mortality in amateur radio operators due to lymphatic and hematopoietic malignancies. *Am J Epidemiol* **127(1)**: 50-54.

Milham S Jr. 1996: Increased incidence of cancer in a cohort of office workers exposed to strong magnetic fields. *Am J Ind Med* **30(5)**: 702-704.

Milham, S., 1998: Carcinogenicity of electromagnetic fields. *Eur J Oncol* **3 (2)**: 93-100.

Milham, S and Ossiander, E.M., 2001: Historical evidence that residential electrification caused the emergence of the childhood leukemia peak. *Medical Hypotheses* **56(3)**: 1-6.

Miyakoshi, J., Yoshida, M., Shibuya, K. and Hiraoka, M., 2000: Exposure to strong magnetic fields at power frequency potentiates X-ray-induced DNA strand breaks. *J Radiation Research (Tokyo)* **41(3)**: 293-302.

Nordenson, I., Mild, K.H., Nordstrom, S., Sweins, A. and Birke, E., 1984: Clastogenic effects in human lymphocytes of power frequency electric fields. *Radiat Environ Biophys* **23(3)**: 191-201.

Nordenson, I., Mild, K.H., Ostinan, U. and Ljungberg, H., 1988: Chromosome effects in lymphocytes of 400 kV-substation workers. Radiat Environ Biophys **27(1)**: 39-47.

Nordenson, I, Mild, K.H., Andersson, G., and Sanderson, M,.1994: Chromosomal aberrations in human amniotic cells after intermittent exposure to fifty Hertz magnetic fields. *Bioelectromagnetics* **15(4)**: 293-301.

Olive, P.L., Wlodek, D., Durand, R.E. and Banath, J.P., 1992: Factors influencing DNA migration from individual cells subject to gel electrophoresis. *Exp Cell Res.* **198**: 259-267.

Pavel A, Ungureanu CE Bara II, Gassner F, Creanga DE., 1998: Cytogenetic changes induced by low-intensity microwaves in the species Triticum aestivum. *Rev Med Chir Soc Med Nat Iasi* **102(3-4)**: 89-92.

Pfluger, D.M. and Minder, C.E., 1996: Effects of 16.7 Hz magnetic fields on urinary 6-hydroxymelatonin sulfate excretion of Swiss railway workers. J Pineal Research **21(2)**: 91-100.

Phillips J.L., lvaschuk, 0., Ishida-Jones, T., Jones, RA., Campbell-Beachler, M. and Haggnen, W., 1998: DNA damage in molt-4 T-lymphoblastoid cells exposed to cellular telephone radiofrequency fields in vitro. *Bioelectrochem Bioenerg* **45**:103-110.

Reiter, R.J., 1994: Melatonin suppression by static and extremely low frequency electromagnetic fields: relationship to the reported increased incidence of cancer. *Reviews on Environmental Health.* **10(3-4)**:171-86, 1994.

Repacholi MH, Basten A, Gebski V, Noonan D, Finnie JH, Harris AW., 1997: Lymphomas in *Ep.-Piml* Transgenic Mice Exposed to Pulsed 900 MHz Electromagnetic Fields. *Radiation Research* **147**: 631-640.

Robinette, C.D., Silverman, C. and Jablon, S, 1980: Effects upon health of occupational exposure to microwave radiation (radar). *Am J Epidemiol* **112(1)**: 39-53.

Rosenthal, M. and Obe, G., 1989: Effects of 50 Hz electromagnetic fields on proliferation and on chromosomal alterations in human peripheral lymphocytes untreated and pretreated with chemical mutagens. *Mutation Research* **210(2)**: 329-335.

Sagripanti, J. and Swicord, M.L., 1976: DNA structural changes caused by microwave radiation. *Int. J. of Rad. Bio.*, **50(1)**, pp 47-50, 1986.

Sarkar, S., Sher, A. and Behari, J., 1994: Effect of low power microwave on the mouse genome: A direct DNA analysis. *Mutation Research*, **320**: 141-147.

Scarfi, M.R., Prisco, F., Lioi, M.B., Zeni, O., Della Noce, M., Di Pietro, R., Franceschi, C., Iafusco, D., Motta, M. and Bersani, F., 1997: Cytogenetic effects induced by extremely low frequency pulsed magnetic fields in lymphocytes from Turner's syndrome subjects. *Bioelectrochemistry and Bioenergetics* **43**: 221-226.

Schwan, H.P. and Foster, K.R., 1980: RF-Field interactions with biological systems: electrical properties and biophysical mechanisms. *Proc IEEE* **68(1)**: 104-113.

Selvin, S., Schulman, J. and Merrill, D.W., 1992: Distance and risk measures for the analysis of spatial data: a study of childhood cancers. *Soc. Sci. Med.,* **34(7)**:769-777.

Simko, M., Kriehuber, R., Weiss, D.G. and Luben, R.A., 1998: Effects of 50Hz EMF exposure on micronucleus formation and apoptosis in transformed and nontransformed human cell lines. *Bioelectromagnetics* **19L** 85-91.

Singh, N.P., Stephens, R.E. and Schneider, E.L., 1994: Modification of alkaline microgel electrophoresis for sensitive detection of DNA damage. *Int J Radia Biol* **66**: 23-28.

Skyberg, K., Hansteen, IL., and Vistnes, Al., 1993: Chromosome aberrations in lymphocytes of high-voltage laboratory cable splicers exposed to electromagnetic fields. *Scandinavian Journal of Work Environment & Health* **19(1)**:29-34.

Skyberg, K., Hansteen, IL., and Vistnes, Al., 2001: Chromosome aberrations in lymphocytes of employees in transformer and generator production exposed to electromagnetic fields and mineral oils. *Bioelectromagnetics* **22(3)**: 150-160.

Smerhovski, Z., Landa, K., Rossner, P., Brabec, M., Zudova, Z., Hola, N., Pokorna, Z., Mareckova, J. and Hurychova, D., 2001: Risk of cancer in occupationally exposed cohort with increased level of chromosome aberrations. *Environmental Health Perspectives* **109(1)**: 41-45.

Stodolnik-Baranska, W., 1967: Lymphoblastoid transformation of lymphocytes in vitro after microwaves irradiation. *Nature* 214, April 1: 102-103.

Svedenstgal, B-M., Johanson, K-L., Mattsson, M-O. and Paulson, L-E., 1999: DNA damage, cell kinetics and ODC activities studied in CBA mice exposed to electromagnetic fields generated by transmission lines. *In Vivo* **13**: 507-514.

Svedenstgal, B-M., Johanson, K-L., Mild, K.H. 1999: DNA damage, induced in brain cells of CBA mice exposed to magnetic fields. *In Vivo* **13**: 551-552.

Szmigielski, S.,1991: *International Scientific Meeting, Beograd* 8-11 April 1991, p34. Cited in Garaj-Vrhovac and Fucic (1993).

Szmigielski, S., 1996: Cancer morbidity in subjects occupationally exposed to high frequency (radiofrequency and microwave) electromagnetic radiation. *Sci Total Env* **180**: 9-17.

Szmigielski, S., Sobiczewska, E. and Kubacki, R., 2001: Carcinogenic potency of microwave radiation: overview of the problem and results of epidemiological studies on Polish military personnel. *Eur J Oncol* **6(2)**: 193-199.

Theriault, G., Goldberg, M., Miller, A.B., Armstrong, B., Guenel, P., Deadman, J., Imbernon, E., To, T., Chevalier, A., Cyr, 0., *et al.,* 1994: Cancer risks associated with occupational exposure to magnetic fields among electric utility workers in Ontario and Quebec, Canada, and France: 1970-1989. *Am J Epidemioogyl* **139(6)**: 550-572.

Tice R, Hook G, McRee Dl., 1999: Chromosome aberrations from exposure to cell phone radiation. *Microwave News*, Mar/Apr (1999) p7.

Tice, R.R., Hook, G.G., Donner , M., McRee, D.I., and Guy, A.W., 2002: Genotoxicity of radiofrequency signals. I. Investigation of DNA damage and micronuclei induction in cultured human blood cells. *Bioelectromagnetics* **23(2)**: 113-126.

Timchenko, O.T., and lanchevskaia, N.V., 1995: The cytogenetic action of electromagnetic fields in the shod-wave range. *Psychopharmacology Series*, Jul-Aug **(7-8)**: 37-39.

Tofani, S., Ferrara, A., Anglesio, L. and Gilli, G., 1995: Evidence of genotoxic effects of resonant ELF magnetic fields. *Bioelectrochemistry and Bioenergetics* **36**: 9-13.

Valjus, J., Norppa, H., Jarventaus, H., Sorsa, M., Nykyri, E., Salomaa, S., Jarvinen, P. and Kajander, J., 1993: Analysis of chromosome aberrations, sister chromatid exchanges and micronuclei among power linesmen with long-term exposure to 50 Hz electromagnetic fields. *Radiat & Environ·Biophysics* **32(4)**: 325-336.

Verschaeve, L., Slaets, D., Van Gorp, U., Maes, A. and Vanderkom, J., 1994: In vitro and in vivo genetic effects of microwaves from mobile phone frequencies in human and rat peripheral blood lymphocytes. Proceedings of *Cost Meetings on Mobile Communication and Extremely Low Frequency field: Instrumentation and measurements in Bioelectromagnetics Research*. Ed. D, Simunic, pp 74-83.

Vignati, M. and Giuliani, L., 1997: Radiofrequency exposure nerar high-voltage lines. *Environmental Health Perspectives*, **105** (Suppl 6): 1569-1573.

Vijayalaxmi BZ, Frei MR, Dusch SJ, Guel V Meltz ML, Jauchem JR. 1997: Frequency of micronuclei in the peripheral blood and bone marrow of cancer-prone mice chronically exposed to 2450 MHz radiofrequency radiation. *Radiat Res* **147**: 495-500.

Wertheimer, N, Leeper E., 1979: Electrical wiring configurations and childhood cancer. *Am J Epidemiol* **109(3)**: 272-284.

Wertheimer, N. and Leeper, E., 1982: Adult cancer related to electrical wires near the home. *Int J Epidemiology* **11(4)**: 345-355.

Wilson, B.W., Wright, C.W., Morris, J.E., Buschbom, R.L., Brown, D.P., Miller, D.L., Sommers-Flannigan, R. and Anderson, L.E., 1990: Evidence of an effect of ELF electromagnetic fields on human pineal gland function. *J Pineal Research* **9(4)**: 259-269.

Wood, A.W., Armstrong, S.M., Sait, M.L., Devine, L. and Martin, M.J., 1998: Changes in human plasma melatonin profiles in response to 50 Hz magnetic field exposure. *J Pineal Research* **25(2)**: 116-127.

Yao KT., 1978: Microwave radiation-induced chromosomal aberrations in corneal epithelium of Chinese hamsters. *J Hered* **69(6)**: 409-412.

Yao KT., 1982: Cytogenetic consequences of microwave irradiation on mammalian cells incubated in vitro. *J Hered* **73(2):** 133-138.

Zaret MM., 1977: Potential hazards of Hertzian radiation and tumors. *NY State J Med*: 146-147.

Zmyslony, M., Palus, J., Jajte, J., Dziubaltowska, E. and Rajkowska, E., 2000: DNA damage in rat lymphocytes treated in vitro with iron cations and exposed to 7mT magnetic fields (static or 50Hz). *Mutation Research* **453** (**1**): 89-96.

CHAPTER 19

Static Electricity in the Modern Human Environment

Jeremy Smallwood

19.1 INTRODUCTION

The presentation will review some documented electrostatic aspects of static electricity experienced by people in their modern home and working environment. Modern people experience wide fluctuations in their body voltage in their daily life, ranging over thousands of volts. This electrostatic voltage is the source of widely experienced electrostatic discharge shocks. Various factors influencing this phenomenon will be presented. Although static electricity effects on health are not widely studied or documented, some reported health effects, as well as areas where we might expect effects to occur, will be discussed.

In preparing this paper I have analysed over 70 enquiries I have received from ordinary people in many countries over the period 1999-2001, as well as contributions from the files of other consultants (Jones 2001, Hearn 2001). Many of the enquirers appeared to be actively looking for solutions to electrostatic effects that were causing them some discomfort or problem. Often they reported experiencing electrostatic effects in more than one situation. This paper has not considered enquiries from industries such as electronics manufacture, and petrochemicals (explosion hazards), where electrostatic discharges have for a long time been of concern.

Table 19.1 provides a simple analysis of the enquiries, and Table 19.2 gives some excerpts from their reports. What becomes clear from these is that the electrostatic effects are common and widespread. While for many people, these effects are mild and no more than an inconvenience, for others the effects become a source of distress and can affect the way people behave in day to day life. In the office or workplace, they can cause malfunction in IT equipment as well as discomfort. By far the most common reports involved the home (19) or office (17), followed by car (12) and other workplaces (7). I would suggest that the high frequency of the first three categories is in part because electrostatics complaints in these situations are often given low importance, as they do not normally have safety or production economics consequences.

In extreme cases, electrostatic discharges have been reported to be the source of burns to the body. In some cases these are the result of particular industrial circumstances, but occasionally they arise as a result of everyday circumstances. One particular case we investigated reported severe shocks, including burns, experienced by users of a UK car park when reaching from their cars to take tickets from a machine. The case had reached a certain notoriety in the local media.

Table 19.1 Enquiries received reporting electrostatic effects during 1999-2001

Report	No. of reports
Shocks in the outdoor environment	
Unspecified	1
Car	12
Public car park, ticket operated barrier	1
Shocks in the indoor environment	
Explosion risk	1
Home	19
Office	17
Retail sites	4
Workplace, severe shocks	2
Workplace, other	7
Using vacuum cleaner	1
From charged vehicle in workplace	1
Other electrostatic effects	
Brushing hair	1
Static in clothes	2
Contamination of product by particles	1
Affecting panel meter readings	1
Coffee grounds erupt from container!	1
Paper product manufacture & handling problems	3

Table 19.2 Excerpts from enquiries received reporting electrostatic effects during 1999-2001 - effects on health or wellbeing

I now notice the hair on my arms stand up on occasion while cleaning or working on the bumpers. The other day .... I received a static shock on the side of the booth with my other hand.
What happens is that I get electric shocks all the time, from practically everything in our house: Metal doorknobs, light switches, water taps and on one occasion, from the water itself!........ I hope you don't mind me writing, it's just that I'm at my wits end.
My son who is 11 years old is able to be charged with static electricity to the point that when you touch it, it is very discomforting to feel the current as the same happens when he plays basket *(sic)* and gets close to a working T.V
Co-workers and I have experienced some static discharges at work (and at home) in various office locations that not only stung but have locked up a company *(sic)* or personal PC or two.
.... all my life I have suffered from electric shocks but over the last eight to ten years they are becoming stronger...... Once I finished work and walked to my car I got belts from all metal no matter how small I also wore rubber soles. I too have nearly crashed a friends computer when I went near it, and one time I kissed my daughter on the lips goodbye and a blue spark and a loud crackle occurred.... It is now becoming more painful as well as embarrassing not just by my scream but the crackle sound before I scream.
Is it possible to ground my car in some way so that I won't be shocked every time I get out of it?
Specific query:: severe problems with static in a newly extended and refurbished supermarket.

please send information on static shock prevention. during the winter everything metal and steel shocks me that i touch . rubber gloves did not help at all. Please help

One of our divisions manufactures expanded polyurethane panels. Our personnel have started to experience problems associated with static electric shocks. We need to find an effective solution to this problem urgently.

I seem to have suddenly developed static sensitivity. I discovered this when I got zapped each time I opened my office door (Metal door knob) ... It was the chair I sit on. It must be something to do with the fabric. ... it even discharges when I touch somebody. It's quite unpleasant and I am wondering if it could be harmful to me in anyway. Others have tried my chair and environment, but do not build up static.

Hi, I was wondering if there is any way that I can get rid of the static electricity that I have, every time that I touch something metal or get out of my car, I get a shock. Last week I was rolling up tissue paper streamers and I got a shock so bad that I burnt my finger around my ring.

We have an office in Charleston SC. that has a terrible static problem. The office space is two stories tall, and is approximately 3500 sq.ft. in area.

I hope you may be able to shed some light on my wife's static problem. She seems to suffer from a high build up of static which is sometimes painful on discharge to earth. ... touching her hand or skin a charge was passed to myself or whoever was in contact with her at that time. When she kissed our grandchild on the lips a charge was passed and pain was felt. .... She also gets charges from shopping trolleys, car doors etc etc .

...the real reason I'm writing is a dog who suffers from shocks when she wakes in the morning. She sleeps on a pillow on the carpet in a very dry climate (Reno, Nevada) and crackles when she leaves her bed in the morning.

I'm a student nurse and am having huge problems with static shocks. We have to wear rubber soled shoes to minimise disturbance however when i make beds etc. or come into contact with metal objects - particularly the beds, I get a rather painful shock. This shock is now being conducted to the patients when i come close to them.

I have an office of about 2000 square feet that seems to be generating a lot of static electricity. Almost every time someone touches a computer, stamp machine, or counter top with metal trim a substantial shock is received. Even when people inter-act, shocks are exchanged.

I am a person who gets static electricity shocks all the time, and its making me CRAZY. .....People I work with are starting to drop papers on my desk instead of handing them to me, to avoid being the recipient of my "electric" personality.

Can you please tell me if it is possible to reduce/eliminate static electricity from personnel who drive vehicles? My staff often complains that they get shocks from door handles and body work of vehicles.

Specific query:: we are in software field, we are having about 50 computers, we are getting problem of shock some times what could be the reason and give the solution *(sic)*

HELP! Every time I get out of my car go into or come out of a shop with metal door handles touch anything metal when vacuuming go into a supermarket I get a static shock.....I even feel the shock through the plastic covering on the car door key (and through clothing) and most times there is a very bright spark along with the shock. I'm fed up with it. ....I would be grateful for any advice, especially as my daughter is now suffering the same experiences!

Please can you help - I have always had 'isolated' incidents of experiencing electric shocks but over the last few days virtually everything I touch I get a big shock off, now my husband and son are reluctant to touch me as I give them a shock as well!

What can one do to protect self from static electricity while shopping in a store? I need HELP!

I have a Recording Studio in Australia, lately band members have been getting static shocks off the microphones(via PA's), with an effect like touching a 9V battery against your tongue.

I am hoping that you may have some suggestions for me. I have installed a laminated wooden floor in my house that trys its best to kill me every time i touch anything earthed after walking across it. I have contacted the manufacturers who seem to think i am lying.

The plant just installed an epoxy coating over its concrete warehouse floor to protect the concrete from chemical attack. Since the coating was installed, most times when the forklift touches a drum, an electric spark/arc is generated.

How to avoid electrostatic charging of people walking over a glass floor?

I am getting very concerned and fed up with these nuisance shocks whenever I get out of my car. It is sometimes very painful.
I work in a recording studio and we have a huge problem with static. Any time we touch anything we get shocked.
I have recently bought a Suzuki Wagon R, since then I have continuously been affected by shocks when getting out of the car! At night they are bright enough to light up the car, there is an audible "snap" as the discharge hits whatever I touch. The shocks are very painful, my nine year old grand daughter was crying with the pain when I held her hand to cross the road!
I am very sensitive to.static electricity discharge. Compared against surrounding people, I am also much more subject to it. In the office, when I shake hands with other people, when I open doors or event faucets I would be hit with evil blue sparks with audible cracking sound. Sometimes it is so strong that people several yards away would hear it and the shock would always surprise me. This has made me very nervous with anything that can potentially trigger the shock.
I am getting static electricity shocks from my stereo receiver and CD player since I transferred to the 15th floor of a condominium a year ago.
I .... have been receiving severe static shocks as I get out of my car and close the door. [This has never happened to me in Madras in my native India.] One shock was severe enough to cause severe wrist pain for a day.
The 'leaving the car' shocks I have generated over the last few years make me fearful of getting out of the car.
I have been at my job for over a year and ever since I have been here I get or give shocks every time I get out of my chair.
.... the girls have to change over rolls and receive a severe 'belt'... of static electricity. .... they have to fold the material into job lots. ..... again the static gives them a 'belt'. ... these shocks can occur up to 10 times a day and by the end of their shift they feel physically sick. ...Two of the girls who do most of this work have at times experienced what would appear to be small burns over their chest area where the shock has left their bodies.
Please help me.... I have such a serious static electricity problem that I cannot even sleep at night. I get little stings up and down my body all night long. ....I work in front of a computer all day, and I do build up a charge there, as well. I touch my co-workers, and they literally jump across the room from the discharge. I feel like my skin is crawling a lot of the time, and I can see my hair move, all by itself. My computer at home is the worst. When I use it, it causes my head to itch from the static!
How can I cut down or eliminate the static electricity in my home office? I feel like I am being bitten by fleas, but there are no bites. It's driving me crazy!

19.2 HOW CHANGES TO THE HUMAN ENVIRONMENT HAVE CONTRIBUTED TO ELECTROSTATIC EFFECTS

Static electricity effects have been known for thousands of years. In early times Thales (640-548b.c.) discovered that an amber spindle attracted the silk which had rubbed it. Knowledge scarcely changed until Gilbert (1540-1603), physician to Queen Elizabeth, found that other substances such as glass, but not metals, could also be electrified. In 1734 du Fay showed that metals could be electrified if held using a glass handle. In 1729 Gray had shown that electric charges could be conducted by metals, water and the human body: glass and amber are examples of insulating materials.

Static electricity is the presence of an excess of positive or negative electrical charges on an object or body. All matter is made up of electrical charges: equal numbers of positive and negative charges exist in the universe, and when present in balance no static electricity effects are observed. Even a relatively small local imbalance, however, can give rise to high electrical potentials (high voltage). Such electrostatic charges imbalances tend to dissipate due to like-charge repulsion forces and attraction to opposite polarity charges.

Electrostatic charges are separated whenever two materials are in contact. One material becomes positively charged, and the other holds an equal negative charge. If the materials are separated in this state then they may remain charged after separation. This is a major source of electrostatic charge generation in practice, known as triboelectrification.

When one considers that in the real world contacts between different material surfaces are constantly being made and broken, one can understand that charge separation by triboelectrification is ubiquitous. So why do we not observe electrostatic effects all the time? Part of the answer lies in the balance between charge generation and dissipation rates. If charges are dissipated faster than they are generated, then no electrostatic charge accumulation occurs. If charge generation is even moderately faster than the dissipation rate, then high voltages of thousands of volts can rapidly build up. Many natural materials dissipate charge sufficiently that build-up is prevented.

A second part of the answer lies in our poor sensing of electrostatic charges. If a human body is charged up and then discharged to another object, typically no sensation is felt unless the body was charged to at least 2000 volts or more. Small discharges of static electricity from other objects cannot be felt until higher voltages. The threshold of hearing electrostatic sparks is also around this level, and unless in dark conditions, the threshold for seeing electrostatic discharge sparks is still higher. Very high electrostatic voltages may also be sensed by the prickling sensation created as the body hairs are moved by electrostatic forces.

Non-insulating materials allow electrical charges to move and therefore allow electrostatic charges to dissipate or recombine. Charges do not move easily in insulating materials, however. An electrostatic charge can remain on such a material for long periods. Until the 20th century, highly insulating materials were relatively uncommon. This started to change with the development of materials such as ebonite and bakelite. The massive use in the latter half of the 20th century of polymers such as polyethylene, polypropylene, PTFE, polyurethane and engineering plastics has made insulating materials extremely commonplace in the everyday human environment. This change has made static electricity charge build-up equally ubiquitous.

19.3 ELECTROSTATIC CHARGE GENERATION AND THE HUMAN BODY

An important way of dissipating excess electrostatic charge is by electrical conduction to the earth. Two centuries ago most people's bodies would have been in electrical contact with the earth most of the time. Bare feet or leather soled shoes dissipate charge easily to the floor, and many natural floor coverings allow dissipation to earth. Nowadays, most people stand on polymer shoe soles, often on a polymer floor covering. Any electrostatic charge built up on the body cannot easily flow through these insulating materials to earth. Moreover, people wear many other man-made materials that have similar insulating properties.

The making and breaking of contact between worn materials during everyday activities such as walking, continually generates static electricity which remains on the body for extended period. A high voltage can rapidly build up on the body, especially under dry atmosphere conditions. Body voltages commonly reach thousands of volts, and voltages in excess of 30,000 V have been recorded. Other actions such as rising from a seat can generate high voltages. These voltages commonly give rise to electrostatic discharges that are felt as shocks. A shock is not felt unless the body is charged to over 2000 V. A brief survey in any group of people will quickly demonstrate that few have not experienced these "nuisance" shocks in a variety of circumstances.

19.3.1 The Effect of Atmospheric Humidity

Water is a good conductor of electrostatic charge, and so the presence of moisture can have a dramatic effect on whether electrostatic charges build up. Many natural materials absorb moisture and have a conductivity which varies over orders of magnitude with atmospheric relative humidity. Similarly, many insulators dissipate charge through a surface moisture layer under humid atmospheric conditions. Under dry atmospheric conditions (relative humidity less than about 30%) electrostatic charge build-up is often considerably enhanced. There is therefore a tendency for electrostatic problems to appear and disappear with climatic, seasonal or artificial atmospheric humidity changes. Electrostatic effects often appear during the winter months, particularly when cold outside air is heated to give warm dry indoor conditions. Modern central heating systems can create very dry indoor ambient conditions which can enhance electrostatic charging.

19.3.2 The Relationship between Electrostatic Charge and Voltage, and Capacitance

Electrostatic charge does not in itself give rise to shocks, sparks and other effects. Many of these arise from high electrostatic fields and voltages. Electrostatic voltage V is related to charge by a simple equation

$$CV = Q \tag{19.1}$$

The variable C is known as capacitance, and is dependent on the body size and geometry, proximity to other objects, and electrostatic properties of materials.

As an example, a part of the human body capacitance is given by the capacitance C_f between the feet and the floor. When standing this is approximately

$$C_f = \frac{\varepsilon A}{d} \tag{19.2}$$

where A is the surface area of the soles of the feet, d is the distance between the foot lower surface and the floor, and ε is the effective permittivity of the material between the foot and floor. It is clear that this is a variable capacitance as d can vary during walking and other activity. Other contributors to body capacitance can vary similarly due to body proximity to furniture and other objects.

19.4 ELECTROSTATIC SHOCKS TO PEOPLE AND ASSOCIATED RISKS

19.4.1 Generation of Body Voltage During Walking

When walking, electrostatic charge is separated by repeated contact and lifting of the foot from the floor. Some excess charge remains on the floor, while an equal and opposite charge resides on the shoe sole. This charge in turn induces charge on the sole of the feet and gives an apparent charge and voltage on the body. The resulting body voltage varies during the walking movements, being greatest when the total body capacitance is least.

Charge separation, and hence body voltage build-up, is best effected when the materials of the floor and the shoe soles are highly insulating. This is the case with many common modern shoe and floor materials.

19.4.2 Generation of Body Voltage Rising from a Seat

When a person is seated, contact between the seat surface and body clothing results in electrostatic charge separation. While remaining seated, this charge does not give rise to significant body voltage as the separated charges remain in close proximity. The seat surface acts as one electrode of a capacitor, and the body a second electrode.

On rising from the seat, the effective capacitance between the seat surface and body is rapidly reduced as the distance between them increases. The stored charge on the body, if it cannot dissipate, rapidly results in a very high body voltage rise.

In the office, this can lead to a shock on rising to shake hands with, or take documents from, a visitor. On exiting a car, leaving the seat commonly creates a very high body voltage that is felt as a shock when the person touches the door to close it. The person commonly believes the car was charged and was the source of the shock - the reverse is generally true. Chubb (1998) has measured voltages over 10,000V on people after getting out of a car.

When refuelling cars, this high body voltage can lead to fuel ignition risk, especially in countries where a filler nozzle may be left in the tank aperture while the operator returns to the vehicle (UK Petroleum Industry Assoc. *et al.* 2001, Renkes 2000a&b).

Hearn (1999) has suggested that high voltage build-up on car occupants may have contributed to accidental airbag pyrotechnic ignitions.

19.4.3 Voltages Arising on Vehicles or Trolleys

A vehicle rolling on a surface generates electrostatic charges due to the repeated contact and separation of the tyres and road surface. This can lead to high potential on the vehicle if the road surface is sufficiently insulating. Fortunately, car tyres, and many road surfaces, are usually sufficiently conductive to dissipate electrostatic charges.

Some recent tyre formulations were, however, found to cause cars to accumulate charge and give electrostatic discharge ignition risks on refuelling (Von Pidoll *et al.*, 1996). A similar effect can arise if the road surface is insulating. In an investigation of electrostatic shocks experienced by drivers touching a ticket machine at the entrance to a car park, we have attributed the shocks to electrostatic charging of the vehicle when on a highly insulating epoxy floor surface. We have observed a similar effect on fork lift trucks in a warehouse.

In the retail environment, trolleys or carts rolling on a floor can rise to high voltages. If the shopper is holding the trolley by a metal frame part, then both may discharge when the shopper touches an earthed conductive object. If the shopper has been holding the trolley via an insulating handle, then the shopper may get a shock when the trolley is touched. We have investigated a retail outlet where a particularly sensitive employee habitually wore rubber gloves to avoid shocks! Trolleys were shown to reach over 8000V on the approach to a lift - typically a strong shock was felt when the lift button was pressed. Site engineers had, of course, failed to find any electrical fault despite regular staff protestations that "the lift is giving us shocks"!

19.5 OTHER ELECTROSTATIC EFFECTS

19.5.1 Effects Due to Clothing

Modern man-made fibre clothes often charge highly during wearing. This can give rise to textile "cling". Removal of a charged garment can cause the body voltage to rise by thousands of volts.

In the electronics industry the electrostatic field from charged clothing is considered a risk of electrostatic discharge damage to sensitive components, and in the petrochemical industry loose clothing has been suspected of causing fuel vapour ignition. Hearn (2001) has investigated a domestic incident in which electrostatic sparks from clothing were suspected of causing ignition of the flammable butane propellant from a deodorant aerosol.

19.5.2 Electrostatic Effects on the Skin

Several reports we received have mentioned tickling or itching sensations to the skin. We have not investigated these reports, but it is easily shown that high electrostatic fields can cause body hair to move perceptibly.

Pinto (1989) has reported that high levels of static electricity can cause carpet fibres and other particles to jump to the skin on arms and legs. He attributes pest bite-like sores and rashes to these impacted particles. Skin irritation and static electricity can both be made worse by dry atmospheric conditions.

VDU operators have reported problems such as rashes, itching and peeling, spots and a glowing sensation like sunburn. In older types of VDU, the screen often carried a high positive charge. An electric field was created between the worker's face and the screen, which can attract airborne particles to the face. Many modern VDUs are relatively free from electrostatic fields.

19.5.3 Air Ion Effects

The effect of air ions on the human organism has been the subject of much debate (Hawkins 1985, Smallwood 1998). Commercial units are available which claim beneficial effects through the generation of negative air ions into the atmosphere. Conversely, it has been claimed that an excess of positive ions can cause discomfort and symptoms such as headaches and nausea in some people. Natural air ions are produced by natural radioactivity and cosmic rays, waterfalls, or wind action.

In particular warm dry winds such as the Sirrocco (Italy), Sharkije (Egypt), Santa Ana (California), Hamsin or Sharav (Middle East) or the Foehn ((Central Europe), about 30% of the population were reported to suffer from migraine, depression, moodiness, lethargy or respiratory symptoms, with an increase in accidents and psychological illness. These health changes have been attributed to atmospheric air ion concentrations, and the ion polarity ratio (in fair weather, circa 1000 ions/cc for negative, and 1200 ions/cc for positive ions). Serotonin has been suggested to be involved in a mechanism of interaction between ions and biological systems. The main action seems to be through entry via the respiratory system.

Vincent (1986), studying aerosols such as asbestos fibres in industrial hygiene, found that in the workplace particle charge levels were relatively high. Effective charge levels were higher for fibrous particles and the effect of charge was to enhance lung deposition. Charge on aerosol particles has also been demonstrated to influence their deposition in the lung tissues in drug delivery experiments (Bailey 1997, *Bailey et al.* 1998).

Natural ion balances could be altered in man made environments, especially with the modern use of polymer materials which are prone to electrostatic charging to high voltages. Hawkins (1982) reported depletion of ions in an air-conditioned office, with a 3:1 excess of positive ions. He has found decreased incidence of headaches, nausea, dizziness and complaint rates in office workers exposed to a corona negative ion source in double-blind trials.

Given the known repulsion or attraction forces of like or opposite polarity charges, it seems likely that the electrostatically charged human body may have a modified ability to take in charged ions or aerosol particles. This may be a factor in the variability of air ion experimental results.

If an electric field exists between a VDU operator's face and the screen, then ions can be expected to migrate in response to this field. If the operator's face is at relative negative potential, then positive ion concentrations at their face can be expected to be enhanced, and negative ions depleted. The reverse is true for fields of the opposite polarity. Such fields could therefore influence the VDU operator's response to air ions.

19.6 CONCLUSIONS

In the past century the electrostatic conditions of the modern environment have changed radically with the common usage of highly insulating polymer materials.

One major aspect of this change is that use of highly insulating floor and footwear sole materials has, for many people, caused insulation of their body from earth. Electrostatic charges are now able to build up on the body and high body voltages are common. Use of insulating man-made materials in clothing and furniture, and drying of the atmosphere by heating systems, exacerbate this. Similar problems occur in vehicles.

The frequency with which we receive enquiries about prevention of electrostatic shocks is testament to the ubiquitous nature of static electricity today. While for many people this remains a minor nuisance, it appears that for some the effects may become scarcely tolerable, reduces their quality of life and may even cause minor injury and other risks.

Some research has linked air ion concentrations with health effects, which could form part of the complex "sick building" effects. High human body voltage and ambient electrostatic fields could influence the local air ion concentrations and contribute to these effects. Skin irritation conditions, including "pest bite"-like sores, have also been attributed to electrostatic fields through the attraction and impact of dust particles.

Static electricity effects in the human environment are worthy of greater attention. Designers and manufacturers of living and working environments and furniture could alleviate many of the electrostatic effects through suitable specification of materials. More subtle effects of electrostatic fields and air ions on human health may be worthy of further investigation.

19.7 REFERENCES

Bailey, A.G. 1997, *The inhilation and deposition of charged particles within the human lung.* Journal of electrostatics, **42**, 25-32

Bailey, A. G., Hashish, A. H., Williams, T. J. 1998, *Drug delivery by inhilation of charged particles,* Journal of Electrostatics, **44**, 3-10

Chubb, J., 1998, *The control of body voltage getting out of a car,* http://www.jci.co.uk/Carseats2.html John Chubb Instrumentation, Unit 30 Lansdown Industrial Estate, Gloucester Road, Cheltenham, GL51 8PL, UK

Hawkins, L.H., 1982, *Air ions and office health.* Occupational Health March

Hawkins, L.H., 1985 *Biological significance of air ions,* Proceedings of the IEE Colloquium on Ions in the Atmosphere, Natural and Man-Made, London, UK, 1985, BLL Conf. Ind. 3315.470 No 88

Hearn, G. L., 1999, 2001, Private communications, Wolfson Electrostatics, University of Southampton, Highfield Southampton, SO17 1BJ, UK

Jones, T. B., 2001, Private communications, University of Rochester, Rochester, NY 14627, USA

Jonassen, N., 1998, *Human body capacitance. Static or dynamic concept?* Proc. EOS/ESD Symp., 111-117, ESD Association. Rome, NY, ISBN 1-878303-91-0

Pidoll, U, von; Kramer, H., Bothe H., 1997 *Avoidance of electrostatic hazards during refuelling of motorcars.* Journal of Electrostatics, **40 & 41** 523-528

Pidoll, U. von Kraemer, H., Bothe, H., 1996, *Avoidance of ignition of gasoline/air-mixtures during refuelling of motor cars at filling stations,* Deutsche Wissenschaftliche Gesellschaft fur Erdoel, Erdgas und Kohle e.V., Hamburg (Germany) ISSN: 0937-9762. Report Number: DGMK-508. ISBN 3-928164-99-6

Pinto, L., 1989, *Paper mites, cable mites and other mystery bugs,* Pest control, 16 - 24

Renkes, R., 2000a, Special Report: Fires at Refuelling Site that Appear to be Static Related., http://www.pei.org/FRD/fire_summary.htm (viewed May 2000). Petroleum Equipment Institute, P. O. Box 2380 Tulsa, OK 74101, USA

Renkes, R., 2000b *Summary of Fires at Refuelling Site that Appear to be Static Related,* http://www.pei.org/FRD/fire_reports.htm (viewed May 2000). Petroleum Equipment Institute, P. O. Box 2380 Tulsa, OK 74101, USA

Rubber & Plastics News, 1995, *Opel Astra fires lead to study of Michelin's silica-based tyres,* April 10, p. 6, ISSN: 0300-6123

Smallwood, J. M., 1998, *The effect of air ions on human performance,* ESL Report No. 9809-01, Electrostatic Solutions Ltd., 13 Redhill Crescent, Bassett, Southampton, Hampshire, SO16 7BQ, UK

Smallwood, J. M., 2001, unpublished report. Electrostatic Solutions Ltd. 13 Redhill Crescent, Bassett, Southampton, Hampshire, SO16 7BQ, UK

UK Petroleum Industry Assoc. Society of Motor Manufacturers and Traders Ltd, Institute of Petroleum, May 2001, *Report on the risk of static ignition during vehicle refuelling. A study of the relevant research.* ISBN 0 85293 306 1

Vincent, J. H., 1986, *Industrial hygiene implications of the static electrification of workplace aerosols,* Journal of Electrostatics, **18**, 113-145

Screen Dermatitis and Electrosensitivity: Preliminary Observations in the Human Skin

Olle Johansson

The theme of this chapter as well as of the whole conference is elegantly summarized in the following quotation: "In this highly technical world, employee working conditions are of increasing importance to companies and the pressure placed on them to provide acceptable working environments means that potential employee health risks cannot be overlooked. Electromagnetic radiation fields and their effects on humans are of great interest at present because of the increasingly large number of people using visual display terminals and mobile phones, both in the workplace and at home. This is a potentially emotive area but everyone agrees that there needs to be further research and dissemination of knowledge. This has particular relevance to the health of occupants in buildings (From the conference program organized by Abacus Communications in conjunction with the University of Reading and the University of Bristol).

An ever increasing number of studies have clearly shown various biological effects at the cellular level of electromagnetic fields, including power frequency and radio frequeny ones as well as microwaves. Such electromagnetic fields are present in your everyday life, at the workplace, in your home and at places of leisure. At a recent *in vivo* experiment, using in *vitro* immunohistochemistry, we could demonstrate that cutaneous mast cells, from normal healthy volunteers, may change their appearance, location and number after exposure to ordinary PC- and television screens (Johansson *et al.*, 2001). These results are, naturally, of paramount importance. To understand the observations to be made by anything else than electric and/or magnetic fields, maybe in synergy with certain environmental chemicals, such as flame retardants, is very hard. Since no one of the volunteers reported any subjective sensations or experiences what so ever, it ought to be quite impossible to claim a "psychological explanatory model". Previous investigations of the phenomenon "screen dermatitis/electrosensitivity" have also pointed out the mast cells as targets for change. We would like to pursue these early demonstrations and concentrate on the impact of visual display terminals (VDTs) and mobile telephones, respectively. In summary, we are aiming at studies of such equipments on the release of mast cell-derived histamine as well as other biologically active substances (employing *in vivo* microdialysis and immunohistochemistry with histamine, tryptase, chymase and IgE surface marker antibodies, respectively). Also the viability of lymphocytes (using methylene blue- or trypane blue-based release techniques), as well as alterations in the serotonin content of the newly described 5-HT-containing melanocytes (or a 5-HT-like molecule; employing *in vivo* microdialysis and immunohistochemistry with polyclonal 5-HT antisera and monoclonal melanoma-associated antigen antibodies, respectively) will be in focus. The latter experiments are of particular importance because of the recent claims of electromagnetic field and light sensitivities that has been discussed, and their dependence of, for instance, melatonin. Furthermore, we also would like to continue the investigation of the above-mentioned markers (as well as markers for cell traffic, proliferation and inflammation) in healthy volunteers versus different clinical entities (including electrosensitive persons with subjective symptoms as well as with subjective and objective ones, respectively). Finally, double-blind provocation tests of these groups are needed, utilizing visual display terminals, mobile telephones and other modern utilities.

Persons claiming adverse skin reactions after having been exposed to computer screens very well could be reacting in a highly specific way and with a completely correct avoidance reaction, especially if the provocative agent was radiation and/or chemical emissions -- just as you would do if you had been exposed to e.g. sun rays, X-rays, radioactivity or chemical odours. The working hypothesis, thus, early became that they react in a cellularly correct way to the electromagnetic radiation, maybe in concert with chemical emissions such as plastic components, flame retardants, etc., something later focussed upon by Professor Denis L. Henshaw and his collaborators at the Bristol University (Fews *et al.*, 1999a,b) [This is also covered in great depth in Gunni Nordström's book "Mörklaggning - Elektronikens rättslösa offer" (2000).]

Very soon, however, from different clinical colleagues, and in parallel to the above, a large number of other 'explanations' became fashionable, e.g. that the persons claiming screen dermatitis only were imagining this, or they were suffering from post-menopausal psychological aberrations, or they were old, or having a short school education, or were the victims of classical Pavlovian conditioning. Strangely enough, most of the, often self-made, 'experts' who proposed these explanations had themselves never met anyone claiming screen dermatitis/electrosensitivity and these 'experts' had never done any investigations of the proposed explanatory models. The explanations were soon revealed to be excuses of a scientifically fraudulent nature! It is interesting to see that science at that time was mere witchcraft. It remains for skilful journalists to inquire how this came about.

Myself, I introduced the term "screen dermatitis", a clinical term to explain the cutaneous damages that developed in the late 1970's when office workers, first mostly women, began to be placed in front of computer monitors. Many of them became ill and developed cutaneous and neurological problems. Several clinical dermatologists, headed by the late professor Sture Lidén, instead talked about union-driven fears, mass media-based psychoses, imagination phenomena, Pavlovian conditioning and so forth. Personally, I refused to reduce people to an ill-defined psychologic homemade diagnosis, without any support even among experts in psychology and psychiatry. Instead, I called for action along lines of occupational medicine, biophysics and biochemistry, as well as neuroscience and experimental dermatology.

I support the democratic principle that citizens are allowed to be ill even in a disease, i.e. a new diagnosis, that is not yet acknowledged by the medical establishment. All diseases were once a "new diagnosis", and it should be remembered that the medical profession strongly has doubted asbestosis, cold urticaria, AIDS, the mad cow disease, skin lice, etc. I usually end my lectures with a quotation from Albert Einstein "The important thing is to not stop questioning". I have never stopped asking questions and I am using the answers to put into place the ever-growing number of pieces of a very, very complicated and enigmatic puzzle.

Neuropeptides in the central and peripheral nervous system may also be involved in the reactions in the skin of the electrically sensitive, and this is something we have wanted to study for many years, but so far we have not been able to pursue these lines of interest due to lack of funding. Since the persons claiming electrosensitivity/screen dermatitis report cutaneous sensations, such as itch, pricking pain, redness, etc., of course the peripheral as well as the central nervous system must be involved. And, by understanding alterations in the chemical neurotransmitter or neuromodulator levels, synthesis, break-down, release and re-uptake, much could be learnt and understood about the basis for these avoidance reactions based upon signals transmitted via the classical sensory and autonomic pathways.

The reaction pattern definitely points to a true biophysical effect, and not to anything else. And, finally, if you take into consideration the large number of publications showing severe changes or damages from low- or high-frequency irradiation of cells, tissues and non-human experimental animals, such alterations cannot ever be understood as "post-menopausal stress reactions", "imagination" or "techno-stress alterations"!

For instance, already in 1953, the Swedish Nobel Laurate, professor Ulf von Euler had shown that peripheral nerves biochemically could contain histamine. It was argued at that time that it only was due to a contamination of histamine from mast cells present around the peripheral nerves. However, further physiological experiments indicated that maybe there could be both central neurons containing histamine (recently proved) as well as peripheral nerves in various target organs.

Using a histamine-based immunohistochemistry we could then, in 1995, show images revealing the presence of histamine-immunoreactive nerves in the skin (Johansson *et al.*, 1995). Naturally, such a finding is of paramount importance, since all studies on histamine effects in the skin have been based on the assumption that the histamine only is released from local mast cells. So, for instance regarding itch, now we have had to reconsider the function of nerve terminal-derived histamine, something also of the greatest impact for areas such as electrosensitivity.

We are right now in the process of examining a larger number of facial skin samples, and from them the most common finding is a profound increase of mast cells. Nowadays we do not only use histamine, but also other mast cell markers such as chymase and tryptase, but the pattern is still the same as reported previously for other electrosensitive persons (Johansson and Liu, 1995). Furthermore, increases of similar nature have now been demonstrated in an experimental situation employing normal healthy volunteers in front of visual display units, including ordinary household television sets (Johansson *et al.*, 2001).

Among earlier studies, one paper (Johansson *et al.*, 1996) ought to be mentioned. In it, facial skin from so-called screen dermatitis patients were compared with corresponding material from normal healthy volunteers. The aim of the study was to evaluate possible markers to be used for future double-blind or blind provocation investigations. Differences were found for the biological markers calcitonin gene-related peptide (CGRP), somatostatin (SOM), vasoactive intestinal polypeptide (VIP), peptide histidine isoleucine amide (PHI), neuropeptide tyrosine (NPY), protein S-100 (S-100), neuron-specific enolase (NSE), protein gene product (PGP) 9.5 and phenylethanolamine N-methyltransferase (PNMT). The overall impression in the blind-coded material was such that it turned out easy to blindly separate the two groups from each other. However, no single marker was 100% able to pin-point the difference, although some were quite powerful in doing so (CGRP, SOM, S-100). However, it has to be pointed out that we cannot, based upon those results, draw any definitive conclusions about the cause of the changes observed. Whether this is due to electric or magnetic fields, a surrounding airborne chemical, humidity, heating, stress factors, or something else, still remains an open question. Blind or double-blind provocations in a controlled environment (Johansson *et al.*, 2001) are necessary to elucidate the underlying causes for the changes reported in this particular investigation.

In one of the early papers (Johansson *et al.*, 1994) we made a sensational finding when we exposed two electrically sensitive individuals to a TV monitor. When we looked at their skin under a microscope, we found something that surprised us. In this article, we used an open-field provocation, in front of an ordinary TV set, of patients regarding themselves as suffering from skin problems due to work at video display terminals. Employing immunohistochemistry, in combination with a wide range of antisera directed towards cellular and neurochemical markers, we were able to show a high-to-very high number of somatostatin-immunoreactive dendritic cells as well as histamine-positive mast cells in skin biopsies from the anterior neck taken before the start of the provocation. At the end of the provocation the number of mast cells was unchanged, however, the somatostatin-positive cells had seemingly disappeared. The reason for this latter finding is discussed in terms of loss of immunoreactivity, increase of breakdown, etc. The high number of mast cells present may explain the clinical symptoms of itch, pain, oedema and erythema. Naturally, in view of the present public debate, the observed results are highly provocative and, I believe, have to be taken much more seriously.

Such studies mentioned above fits very well with other observations, such as the ones reported by Dr. John Holt, in Australia, who in a recent letter to Leif Södergren in Göteborg, Sweden, wrote that when working with microwaves (to irradiate cancer cells) he had observed that the microwaves from cell phones cause a doubling of histamine (which are released from mast cells) and that such electrosmog from mobile phones could be the cause of the ever increasing asthma and other allergies. I had put forward this hypothesis several years earlier, in public here in Sweden, and I am now happy to finally see more and more data gathering to support this idea.

I and my collaborator, Dr. Shabnam Gangi, have also addressed this idea in two recently published papers of theoretical nature (Gangi and Johansson, 1997, 2000), using a model for how mast cells and substances secreted from them (e.g. histamine, heparin and serotonin) could explain sensitivity to electromagnetic fields. The model bounces off from known facts in the fields of UV- and ionizing irradiation-related damages, and use all the new papers dealing with alterations seen after e.g. power-frequency or microwave electromagnetic fields to propose a simple summarizing model for how we can understand the phenomenon of electrosensitivity. I strongly recommend the readers of this present short summary to familiarize themselves with these publications, since I fully believe they have a lot to offer as food for further thoughts.

In the first paper, in the journal Experimental Dermatology (Gangi and Johansson, 1997), we describe the fact that an increasing number of persons say that they get cutaneous problems as well as symptoms from certain internal organs, such as the central nervous system and the heart, when being close to electric equipment. A major group of these patients are the users of video display terminals, who claim to have subjective and objective skin- and mucosa-related symptoms, such as pain, itch, heat sensation, erythema, papules, and pustules. The central nervous system-derived symptoms are, e.g. dizziness, tiredness, and headache. Erythema, itch, heat sensation, oedema and pain are also common symptoms of sunburn (UV dermatitis). Alterations have been observed in cell populations of the skin of patients suffering from so-called screen dermatitis similar to those observed in the skin damaged due to ultraviolet light or ionizing radiation. In screen dermatitis patients a much higher number of mast cells have been observed.

It is known that UVB irradiation induces mast cell degranulation and release of TNF-alpha. The high number of mast cells present in the screen dermatitis patients and the possible release of specific substances, such as histamine, may explain their clinical symptoms of itch, pain, oedema and erythema. The most remarkable change among cutaneous cells, after exposure with the above-mentioned irradiation sources, is the disappearance of the Langerhans' cells. This change has also been observed in screen dermatitis patients, again pointing to a common cellular and molecular basis. The results of this literature study demonstrate that highly similar changes exist in the skin of screen dermatitis patients, as regards the clinical manifestations as well as alterations in the cell populations, and in skin damaged by ultraviolet light or ionizing radiation.

In the second publication (Gangi and Johansson, 2000), from the journal Medical Hypotheses, the relationship between exposure to electromagnetic fields and human health is even more in focus. This is mainly because of the rapidly increasing use of such electromagnetic fields within our modern society. Exposure to electromagnetic fields has been linked to different cancer forms, e.g. leukaemia, brain tumours, neurological diseases, such as Alzheimer's disease, asthma and allergy, and to the phenomenon of electrosensitivity/screen dermatitis. There is an increasing number of reports about cutaneous problems as well as symptoms from internal organs, such as the heart, in people exposed to video display terminals.

These people suffer from subjective and objective skin and mucosa-related symptoms, such as itch, heat sensation, pain, erythema, papules and pustules (cf. above). In severe cases, people cannot, for instance, use video display terminals or artificial light at all, or be close to mobile telephones. Mast cells, when activated, release a spectrum of mediators, among them histamine, which is involved in a variety of biological effects with clinical relevance, e.g. allergic hypersensitivity, itch, oedema, local erythema and many types of dermatoses. From the results of recent studies, it is clear that electromagnetic fields affect the mast cell, and also the dendritic cell, population and may degranulate these cells. The release of inflammatory substances, such as histamine, from mast cells in the skin results in a local erythema, oedema and sensation of itch and pain, and the release of somatostatin from the dendritic cells may give rise to subjective sensations of on-going inflammation and sensitivity to ordinary light. These are, as mentioned, the common symptoms reported from patients suffering from electrosensitivity/screen dermatitis. Mast cells are also present in the heart tissue and their localisation is of particular relevance to their function. Data from studies made on interactions of electromagnetic fields with the cardiac function have demonstrated that highly interesting changes are present in the heart after exposure to electromagnetic fields.

Some electrically sensitive have symptoms similar to heart attacks after exposure to electromagnetic fields. One could speculate that the cardiac mast cells are responsible for these changes due to degranulation after exposure to electromagnetic fields. However, it is still not known how, and through which mechanisms, all these different cells are affected by electromagnetic fields. In this article (Gangi and Johansson, 2000), we present a theoretical model, based upon the above observations of electromagnetic fields and their cellular effects, to explain the proclaimed sensitivity to electric and/or magnetic fields in humans.

The assistant professor and histopathologist Björn Lagerholm, at the Karolinska Hospital in Stockholm, already in the middle of the 1980's took biopsies of electrically sensitive individuals. He also found an increase in the mast cell number, but, unfortunately, he could never publish it.

As a matter of fact, his interest very much started with a female bank employee who had received a work injury compensation for skin changes after sitting in front of a visual display monitor. Björn Lagerholm described in great detail her skin changes, which turned out to be very similar to the kind of cutaneous alterations you may encounter in connection with ultraviolet light or X-ray damage. It is to be noted that Björn Lagerholm's reputation as a histopathologist was, and is, undisputed. He had examined at least 10,000 biopsies from other skin diseases before this particular case. In addition to her, he also examined nearly 100 further screen dermatitis cases, all having the same skin changes.

Björn Lagerholm wrote an article in the Swedish Medical Journal ("Läkartidningen") to describe his observations. Apart from this he was never able to pursue his ground-breaking and very elegant studies. They would be buried for several years, until I and my collaborators re-initiated them in the early 1990's.

One argument some doctors put forward was that the radiation from a computer monitor could not possibly affect the human skin. They claimed the nerves not to be that superficial, but this is completely wrong! The idea of the deeply buried nerve fibers were put forward by the late professors David Ingvar and Bernard Frankenheuser. However, we shortly after published the first true demonstration of the epidermal nerves in human skin (Wang *et al.*, 1990), followed by an ultrastructural identification (Hilliges *et al.*, 1995) as well as a detailed description and quantification of these very superficial nerves (Johansson *et al.*, 1999b). Exactly how superficial are these nerves in the skin? The nerves come as close as 10-40 micrometers from the stratum corneum, which could be, in e.g. the face, in itself very thin, thus, these nerves are very superficially located.

The skin is the largest organ of the body. It is also our foremost protector against the outside world. The skin is a very sensitive 'antenna system' containing, in addition, special sensory organs, such as the eyes, the nose and the ears. The function of the skin is, among many, to always guide us in an ever-changing environment, thus, enabling us to avoid tissue-damaging threats, such as heat, cold, UV-light, X-rays, radioactivity, etc. In the centre of this avoidance system is, of course, our nervous system which will help us to go in the right direction, away from some situations, maybe including e.g. electromagnetic fields from computer screens and cellular telephones? The future will tell us if I was right or wrong.

It is alarming that so many people have reactions in their skin that point to the skin having defensive reactions from say computer work. The risk of developing skin problems, including cutaneous cancers, is obvious if the skin is always in a defensive mode.

The whole concept of skin reactions is frightening, especially since the skin cancer forms, such as malignant melanoma and basalioma, are so quickly increasing their incidence. I have asked, over and over again, many colleagues if they really can rule out the surrounding electromagnetic fields as an important background factor for such cancers, but I have never received any clear answers...

Many electrically sensitive have experienced and reported a severe and lengthy light sensitivity after working with computers. Light sensitivity is increasing as a general problem in the population, and reports have been published in several countries about this. The reason behind it is not known, but from our work one could just speculate around the heat-, light- and UV-adsorbing cellular layer in the epidermis, the so-called melanocytes and their production of the pigment melanin. In one of our case-reports (Johansson *et al.*, 1999a), it was evident that this layer, for some unknown reason, was, more or less, completely gone. We used protein S-100 and HLA-DR (human histocompatibility complex class II (subregion DR)) as markers, and it was found that the immunoreactive dendritic cells were dramatically decreased in number, especially in the epidermis.

One could imagine that e.g. increased levels of light or UV-light, or increased levels of other frequencies of electromagnetic fields, such as microwaves, have led to a wear-down of the protective cellular shield in the skin after a long-term continuous irradiation period. If such a damage takes place, maybe the first sign would be light sensitivity in parallel to a modest electrosensitivity. However, if the damage proceeds naturally the situation could be very difficult for the patient, finally leading to a life in basically complete darkness. Several such cases have been reported, but too few studies have been done, again due to lack of funding. In our own case report (Johansson *et al.*, 1999a), we could also demonstrate that vitamin A was effective as a treatment for the patient. During the vitamin A treatment, the patient was to a large extent rehabilitated regarding her general light sensitivity, however, she was still sensitive to the presence of electric equipment, although not as much as before. The metabolism of vitamin A should be considered, since, in the human visual system, vitamin A is converted to alpha-cis-retinol, which is an essential chromophore component of rhodopsin, the photoreceptor protein of the retinal rods and is therefore essential for human vision. Maybe vitamin A influences cutaneous (as well as other) cellular systems similar to the retina. One explanation is that the patient for a time lost her melanocytes (or melanocytic content) as seen with the S-100 immunofluorescence, in response to an external or internal provocation. As a reaction to this, also her HLA-DR positive dendritic cells were affected. The vitamin A may have been capable of restoring this balance, as least partially.

Finally, I, and my collaborators, have done a series of blind tests to see if electrically sensitive persons reacted to microwaves from mobile phones, some done here in Stockholm, some in Göteborg and also some in Linköping. The experiments in Stockholm and Göteborg failed, maybe due to the fact that the surrounding environment could not be controlled from the point of low- and high-frequency signals, which may have interacted with the tests subjects. However, the study in Linköping (Johansson, 1995) was done in the countryside, more than 1 kilometer from the nearest live electric power source. One person was actually able to respond correctly to a mobile phone-based double-blind provocation experiment, 9 times out of 9 tests (p<2/1000), both in the 'acute' phase as well as in the 'chronic' phase (p<1/1000). This would mean that there may very well be negative health effects from such mobile telephones, most likely depending on their high-frequency fields. So, it does worry me that children use mobile telephones. Especially, since the British government, in December 2000, has taken firm action in respect to this question.

In addition, it should be noted that in a recent and on-going study (Sromová *et al.*, 2001, 2002), electrically sanitized premises (called the "ELRUM") in Skellefteå (in the northern part of Sweden) were used to study possible ways of rehabilitating electrosensitive persons. In summary, the results of our present study of persons hypersensitive to electricity clearly points to that intensive electrical environments cause their ailments, and that a reduction of electromagnetic fields in the living and workplace environment seems to be highly positive as a mean for rehabilitation. Other factors, such as exhaust fumes, certain chemicals, moulds, etc., probably also contribute to the symptoms. Our study, thus, can therefore lead to future recommendations regarding necessary adaptations, in the form of electrical (and chemical) sanitation, of a person's work and home environment.

As we have pointed out in earlier documentation, a majority of the guests at the ELRUM premises had been sick-listed for long periods of time (several years). The rehabilitation process can therefore, in its turn, be expected to take a long time and be demanding. This process can be initiated at such a rehabilitation centre as the ELRUM, but must be continued at home, assuming that the person receives full support from all parties involved (social insurance office, employer, doctor, etc.). Finding constructive solutions with regard to the person's work situation requires creativity, patience and perseverance on the part of all. When following up our recommendations, we see that it is very difficult to get economic help for carrying out an electrical sanitation of the home environment. Only on exceptional occasions municipalities provide funds for making necessary adaptations at home, and it is almost as difficult to get economic resources for adapting work environments. There are, however, companies that make an effort to carry out an electrical sanitation of their workplaces and adapt work assignments in order to retain staff.

The majority of the persons who took part in this project at the ELRUM rehabilitation centre, at the one-year follow-up, say that is has led to a positive change in their situation. The reason why some have answered "no" seems more to be a reflection of their personal disappointment that the recommendations we gave them have not been carried out back home.

How would one depict the world the electrically sensitive live in? In essence, it must be a very tough daily life, having to always (very much as an allergic or asthmatic person) look out for situations of provocative nature. And, where today would you find an electric environment equal to e.g. the 1950s? Or, even more mind-boggling, where would you find a high-frequency milieu the same as last year? Nowhere, I guess, since the growth of all such systems is so rapid and quickly covers us all. Therefore, to enable the basic freedom of choosing where to live, where to work, etc., is impossible in relation to the electrosensitive persons' requirements. And, thus, the question of electrosensitivity becomes a question about democracy!

Why do some people become electrically sensitive and others do not? This is also a most important and interesting question. As you know, in any kind of disease, not everyone is ill, and not at the same time. Everyone will not get cancer, everyone will not break a leg, and everyone will not have malaria. This is governed by the biological statistical rules of the natural variation. But, maybe you should turn the issue around somewhat. Perhaps all healthy persons, i.e. in the sense not being electrosensitive, ought to be extra happy for the electrosensitive ones, since they have acted as a warning for all of us? It could be, that we would owe them a lot since they reacted in time to something which the main bulk of mankind did not. Furthermore, the possibility is also that the electrosensitive persons will turn out to be tomorrow's great winners, given the fact that this Summer, twentyone world-leading scientists during a gathering in the French city Lyon, within IARC's (IARC = WHO's International Agency for Research on Cancer) expert panel, have concluded low-frequency magnetic fields as a possible cancer risk (=group 2B, containing in addition i.a. diesel and petrol fumes, chloroform, welding fume, lead and DDT). For children exposed to such low-frequency magnetic fields above 0.4 microTesla the cancer risk is doubled. (Therefore, I ask myself: How will people feel after having spent their everyday working hours at or around Vasagatan in the centre of Stockholm where the low-frequency magnetic field 1.2 meters above ground is between 0.3 and 2.2 microTesla, or in the commuter trains having levels between 1 and 100 microTesla in the traveller's compartment!?)

But, could all of this be pure speculation? Well, to people who suggest that electric sensitivity is purely imagined or psychological, I want to ask them to explain all the peer-review-published results around effects of, often very weak, electromagnetic fields on molecules, cells, tissues, organs and various non-human experimental animals, i.e. situations which cannot at all be understood in terms of imagination or psychology. If failing this task, I would then ask them to return to the first statement regarding humans, and to scrutinize and re-evaluate it. Generally, people at that moment suddenly lack scientifically sound arguments, and most of them also confess this.

I am, and have always been, very surprised to see how sloppy some of my colleagues address important issues such as the above. Very often one has to realize that all 'experts' are not true scientists and scholars. Furthermore, it is also very annoying to see that 'experts' claiming, for instance, "that the best way to treat electrosensitive persons is to completely ignore them through silence", did not have to face any personal consequence...!? Nothing happened to them, their position was not questioned, their competence as physicians was not questioned, their suitability as representatives for the medical profession was not questioned. Nothing! What kind of society is that?

I am also very disturbed by the fact that even if the electrosensitive persons were victims of an illusion, where in the health and law system does it say that you can treat them so badly as several have done? When I attended the medical school I was taught the very opposite: You should always address patients with kindness, a will to learn and help, support them, meet them and their concerns in a most respectful way, and so on. Where did that disappear? It seems as our world-famous health insurance policy contains very big gaps through which electrosensitive people, as well as other new diagnoses, fell, and still fall, head down!

The author and journalist Gunni Nordström has in an interview expressed herself in the following way: "The government seems to listen to those who have the right message, the message they wish to hear. Sometimes one wonders if the authorities have these reports custom-made or if someone in the background is masterminding all important positions and is handing out investigations to those with the correct beliefs or to the untalented. The independent thinkers get their heads chopped off as soon as possible at any rate." - From a philosophical point of view this can prove to be sad and badly wrong, since, without the latter, companies, authorities as well as governments can be fooled and tricked into the completely wrong corner. Personally, if I were the prime minister, I would be very afraid never to listen to the 'whistle-blowers' since it could be a big, big mistake, from a public health point of view, not to!

Within this context, one very interesting, and on-going, movement is the EU-based "The European framework for protection from exposure to electromagnetic fields". About it, to begin with, there are some general comments to be made. The EU does not seem to be interested in yet another 'BSE scandal', therefore they are carefully keeping an eye on the issues around health effects and health risks (N.B. Note the difference between health effects, identified and calculated health risks and unknown health risks) from electromagnetic fields, especially from high-frequency telecommunication, such as mobile telephony. Furthermore, the EU does not regard the above-mentioned irradiation systems to be proven safe. On the contrary, I believe they strongly understand it could be a major mistake to whole-body irradiate the whole European (as well as the world's) population, 24 hours around.

You often hear about "safe levels" of exposure and that there is "no proof of health effects", but my personal response to these seemingly reassuring statements is that it is very important to realize, from a consumer's point of view, that "no accepted proof for health effects" is not the same as "no risk". Too many times, 'experts' have claimed to be experts in fields where actually the only expert comment should have been: "I/we just do not know". Such fields were e.g. the DDT, X-ray, radioactivity, smoking, asbestos, BSE, heavy metal exposure, depleted uranium, etc., etc., etc., where the "no risk"-flag was raised before true knowledge came around. Later on, the same flag had to be quickly lowered, many times after enormous economic costs and suffering of many human beings. Along those lines, it is now (regarding "the protection from exposure to electromagnetic fields" issue) very important to clearly identify the background and employment (especially if they sit, at the same time, on the industry's chairs) of every 'expert' in different scientific committees, and likewise. It is, of course, very important (maybe even more important?) to also let 'whistleblowers' speak at conferences, to support them with equal amounts (or even more?) of economical funding as those scientists and other 'experts' who, already from the very beginning, have declared a certain source or type of irradiation, or a specified product, to be 100% safe.

In the case of "protection from exposure to electromagnetic fields", it is of paramount importance to act from a prudence avoidance/precautionary principle point of view. Anything else would be highly hazardous! Total transparency of information is the key sentence here, I believe consumers are very tired of always having the complete truth years after a certain catastrophe already has taken place. For instance, it shall be noted, that today's recommendation values for mobile telephony, the SAR-value, are just recommendations, and not safety levels. Since scientists observe biological effects at as low as 20 microWatts/kg, is it then really safe to irradiate humans with 2 W/kg (i.e., with 100,000 times stronger radiation!), which is the recommendation level for us? And, furthermore, it is very strange to see, over and over again, that highly relevant scientific information is suppressed or even left out in various official documents, as high up as at the governmental level of society. This is not something that the consumers will gain anything good from, and, still, the official declaration or explanation (from experts and politicians) very often is: "If we (=the experts) would let everything out in the open, people would be very scared and they would panic." Personally, I have never seen this happen, but instead I have frequently seen great disappointment from citizens who afterwards have realised they have been fooled by their own experts and their own politicians...

Another misunderstanding is the use of scientific publications (as the tobacco industry did for many years) as 'weights' to balance each other. But you can NEVER balance a report showing a negative health effect with one showing nothing! This is a misunderstanding which, unfortunately, is very often used both by the industrial representatives as well as official authorities. The general audience, naturally, easily is fooled by such an argumentation, but if you are bitten by a deadly poisonous snake, what good does it make for you that there are 100 million harmless snakes around?

In ten years, how will we look on in-door and out door electromagnetic fields from high-tension cables, visual display terminals, mobile telephones, and so forth? Hopefully without any remaining questions, scepticism or fear. I look forward to see the question marks around this technology resolved, and a well-documented and 100% responsible, human-friendly technology being presented. And, hopefully, from a patriotic point of view, tomorrow's human-friendly technology will be made by Swedish companies, in that way creating a 'healthy wealth' for my own country.

Thus, the future is not dark, not at all, but bright for all kind of "human-friendly technologies", including low-irradiation and low-emitting products. For, after all, who could sell a computer screen today with the slogan: "THIS IS A HIGH-LEVEL IRRADIATION SCREEN"!?

[For readers interested in more general details around these issues, I warmly recommend the books by Gunni Nordström and Carl von Scheele (Nordström, 2000, 2002, Nordström and von Schéele, 1989, 1995). They are of great value for persons wanting to acquaintance themselves with the political implications and impact of the phenomena of electromagnetic fields and of new diagnoses in our society.]

This study was supported by grants from the Cancer and Allergy Foundation and the Karolinska Institute. For expert technical and secretarial assistance, Ms Marianne Ekman and Ms Eva-Karin Johansson are gratefully acknowledged.

REFERENCES

Fews, A.P., Henshaw, D.L., Keitch, P.A., Close, J.J. and Wilding, R.J., 1999a, Increased exposure to pollutant aerosols under high voltage power lines. *Int J Radiat Biol*, **75**, pp. 1505-1521.

Fews, A.P., Henshaw, D.L., Wilding, R.J. and Keitch, P.A., 1999b, Corona ions from powerlines and increased exposure to pollutant aerosols. *Int J Radiat Biol*, **75**, pp. 1523-1531.

Gangi, S. and Johansson, O., 1997, Skin changes in "screen dermatitis" versus classical UV- and ionizing irradiation-related damage--similarities and differences. Two neuroscientists' speculative review. *Exp Dermatol*, **6**, pp. 283-291.

Gangi, S. and Johansson, O., 2000, A theoretical model based upon mast cells and histamine to explain the recently proclaimed sensitivity to electric and/or magnetic fields in humans. *Med Hypotheses*, **54**, pp. 663-671.

Hilliges, M., Wang, L. and Johansson, O., 1995, Ultrastructural evidence for nerve fibers within all vital layers of the human epidermis. *J Invest Dermatol*, **104**, pp. 134-137.

Johansson, O., 1995, *Elöverkänslighet samt överkänslighet mot mobiltelefoner: Resultat från en dubbel-blind provokationsstudie av metodstudiekaraktär*, (Stockholm: Enheten för Experimentell Dermatologi, Karolinska Institutet, Rapport nr. 2, ISSN 1400-6111 [in Swedish]).

Johansson, O. and Liu, P.-Y., 1995, "Electrosensitivity", "electrosupersensitivity" and "screen dermatitis": preliminary observations from on-going studies in the human skin. In *Proceedings of the COST 244: Biomedical Effects of Electromagnetic Fields - Workshop on Electromagnetic Hypersensitivity*, edited by Simunic, D., (Brussels/Graz: EU/EC (DG XIII), pp. 52-57.

Johansson, O., Hilliges, M., Björnhagen, V. and Hall, K., 1994, Skin changes in patients claiming to suffer from "screen dermatitis": a two-case open-field provocation study. *Exp Dermatol*, **3**, pp. 234-238.

Johansson, O., Virtanen, M. and Hilliges, M., 1995, Histaminergic nerves demonstrated in the skin. A new direct mode of neurogenic inflammation? *Exp Dermatol*, **4**, pp. 93-96.

Johansson, O., Hilliges, M. and Han, S.W., 1996, A screening of skin changes, with special emphasis on neurochemical marker antibody evaluation, in patients claiming to suffer from screen dermatitis as compared to normal healthy controls. *Exp Dermatol*, **5**, pp. 279-285.

Johansson, O., Liu, P.-Y., Enhamre, A. and Wetterberg, L., 1999a, A case of extreme and general cutaneous light sensitivity in combination with so-called 'screen dermatitis' and 'electrosensitivity' - a successful rehabilitation after vitamin A treatment - a case report. *J Aust Coll Nutr & Env Med*, **18**, pp. 13-16.

Johansson, O., Wang, L., Hilliges, M. and Liang, Y., 1999b, Intraepidermal nerves in human skin: PGP 9.5 immunohistochemistry with special reference to the nerve density in skin from different body regions. *J Peripher Nerv Syst*, **4**, pp. 43-52.

Johansson, O., Gangi, S., Liang, Y., Yoshimura, K., Jing, C. and Liu, P.-Y., 2001, Cutaneous mast cells are altered in normal healthy volunteers sitting in front of ordinary TVs/PCs - results from open-field provocation experiments. *J Cutan Pathol*, **28**, pp. 513-519.

Nordström, G., 2000, *Mörklaggning - Elektronikens rättslösa offer*, (Stockholm: Hjalmarson & Högberg Bokförlag, ISBN 91-89080-41-6 [in Swedish]).

Nordström, G., 2002, *Cover-Up* (accepted for publication).

Nordström, G. and von Schéele, C., 1989, *Sjuk av bildskärm* (Stockholm: Tidens Förlag, ISBN 91-550-3484-5 [in Swedish]).

Nordström, G. and von Schéele, C., 1995, *Fältslaget om de elöverkänsliga* (Stockholm: Tidens Förlag, ISBN 91-550-4083-7 [in Swedish]).

Sromová, L., Larsson, M. and Johansson, O., 2001, *Verksamheten vid ELRUM 1998-2000*, (Umeå: Arbetslivstjänster Västerbotten, 12 pp [in Swedish]).

Sromová, L., Larsson, M. and Johansson, O., 2002, *ELRUM 1998-2000 - Results and conclusions*, (Umeå: Arbetslivstjänster Västerbotten, in press).

Wang, L., Hilliges, M., Jernberg, T., Wiegleb-Edstrom, D. and Johansson, O., 1990, Protein gene product 9.5-immunoreactive nerve fibres and cells in human skin. *Cell Tissue Res*, **261**, pp. 25-33.

CHAPTER 21

Theoretical and Experimental Evidences where Present Safety Standards Conflict with Reality

V.N. Binhi and M. Fillion-Robin

21.1 INTRODUCTION

Why do some people feel unwell during a magnetic storm? Why is there a correlation between the level of electromagnetic background and the incidence of cancer [Portier *et al.*, 1998]? Why do so many medical centres use electromagnetic exposures to treat a wide variety of disorders in humans? People are continually immersed in electromagnetic fields both of natural and technological origin, such as those from power lines, domestic appliances, mobile phones. Human sense these fields, just as do other living beings. This fact is supported by an immense amount of scientific evidence [Bersani, 1999]. The World Health Organization now considers that enhanced electromagnetic contamination in occupational and residential areas is a stress factor for people.

At the same time, however, many authors note that the physical origins of the phenomenon are, as yet, unclear — the phenomena themselves often seeming paradoxical. This allows people to speculate about the safety aspect of EM radiation in ways that are not always compatible with science. In particular, manufacturers of widely used electrical appliances, such as cell phones, computers, and TVs continue to insist that their devices are safe, based solely on the fact that the radiation from such equipment is generally not intense enough to cause an adverse level of heating of biological tissues. At the same time, however, there are many experiments showing that weak and even hyperweak (*see below*) EM fields can influence living tissues, and even whole organisms. Generally, particularly in the case of time-dependent fields, such influences are characterized by "windows" in biological responses. Such findings undermine the philosophy on which existing EM safety standards are based — namely that EM fields can lead to biological effects only if they cause heating of biological tissues.

Both experimental findings and theoretical approaches indicate that EM fields — even when they are too weak to heat tissues — may result in a variety of different biological effects, some useful and some probably noxious. In this article, we show that hyperweak fields — even of intensities very much less than those emitted by cell phones — affect biology; thus it cannot longer be maintained that cell phone radiation cannot cause harmful biological effects. Accordingly, existent Western EM safety standards afford a totally inadequate level of protection, unlike those in Eastern Europe (and in the Russian Federation, in particular) which are much more stringent, having been developed with the benefit of experience with *non-thermal* biological effects, rather than solely heating. Here we will discuss in more detail the inadequacy of thermally-based guidelines, and study a new theoretical approach *via* which electromagnetic biocompatability might be realized.

21.2 DO BIOEFFECTS OF HYPERWEAK EM FIELDS EXIST?

Normally, biological systems dwell in the natural terrestrial magnetic field, the geomagnetic field, whose value varies with latitude, from approximately 40 µT at the Equator to 70 µT at the Poles. This provides a natural biological benchmark against which DC fields of technological origin can be compared. A magnetic induction of value $B_0 = 50$ µT is referred to in the scientific literature as "weak". Therefore, magnetic fields of strength $B \ll B_0$ — say below 1 µT — may be classed as hyperweak. There are good reasons to connect many biological effects of EM fields or magnetobiological effects[1] with the Zeeman and Stark effects at the atomic and molecular levels [Binhi *et al.*, 2000]. Since electric fields of the order of 100 V/m result in similar changes in the atomic structure as do magnetic fields of intensity B_0, we may take $E_0 = 100$ V/m as a natural benchmark, and consider electric fields below 1 V/m as hyperweak.

To characterize AC fields one may use their heating effect on biological tissues. In the ELF range this leads to approximately the same as given above for DC fields. Because thermal effects are directly proportional to EM frequency at frequencies of order 1 GHz, EM flux densities between 1–100 mW/cm^2 may be regarded as weak (causing the lowest discernible degree of tissue heating), and powers below 100 µW/cm^2 as hyperweak.

It is now necessary to consider whether such hyperweak EM fields can cause biological effects.

21.2.1 Experiments

Below, we list some of the results that confirm hyperweak EM fields can cause biological effects.

Early experimental evidence for biological detection of hyperweak signals, both magnetic (up to 1 nT) and electric (up to 0.1 mV/m), was given by Presman [Presman, 1970]. Bastian [Bastian, 1994] discussed the sensitivity of sharks and skates to electric fields as low as 0.5 µV/m. Peroxidase activity was changed from 9% to 72% after 3 h exposure to 8 Hz magnetic fields of magnitudes 0.02, 0.2, 1, and 2 nT. This was observed [Vladimirsky *et al.*, 1971] in measurements of

---

[1] The magnetobiological effect (MBE) is a generic term to designate any weak magnetic or electric field–dependence of any biological object.

biochemical parameters of neutrophils. Keeton, Larkin, and Windsor [Keeton *et al.*, 1974] reported that natural 100 nT geomagnetic fluctuations affect the orientation of pigeons. Delgado, Leal, *et al.* [Delgado *et al.* 1982] studied effects of 48 h exposure to pulsed magnetic fields on the growth of chicken embryos, statistically significant effects being found for ELF magnetic field of the order of 0.12 μT. Agadjanyan and Vlasova [Agadjanyan *et al.*, 1992] observed an effect of sinusoidal magnetic fields between 0.05–5 Hz, and approximately 100 nT on neuron spike activity in the cerebellum of mice; their experiments were made in a magnetically shielded room.

Jacobson, 1994, used hyperweak magnetic fields between 5 pT and 25 pT to treat epilepsy and Parkinson's disease. Application of a sinusoidal field to the human head was found to affect the pineal gland. Magnetic field stimulation induced changes in the production of melatonin, which is known to maintain anti-tumor activity. Work by Novikov and others, see for example [Fesenko, 1997], was devoted to investigation of policondensation of some amino-acids in aqueous solutions subjected to AC magnetic fields of about 20 nT in a geomagnetic-like static magnetic field. Effective frequencies were found to be of the order of a few Hz. Blank and Soo, 1996 researched Na, K-ATPase enzymatic activity on microsomes, and found that it appeared to depend on the 50 Hz magnetic field, the limit of susceptibility being 200–300 nT. Cell cultures of mouse epidermis grew twice as fast as the control when exposed to a 60 Hz, 1 μT magnetic field, superimposed on the background static laboratory field, [West *et al.*, 1996]. Akerstedt, Arnets, *et al.*, 1997, studied the effect of overnight exposure of human subjects to a 50 Hz, 1 μT magnetic field, it being found that such field significantly shortened the duration of the so-called "slow sleep".

Various scientific groups [Hyland *et al.*, 1999] showed that special microcrystalline solutions placed into small metal boxes appreciably reduced the risk of diseases arising from proximity to computer and TV monitors. It is likely that this protection is afforded by the magnetic field arising from precessing proton spins in water, the emitted field being in the pT range at half-meter distance.

There is much experimental data published by many investigators and reviewed by Binhi [Binhi *et al.*, 2000], in which electric fields were applied either externally — by capacitive or inductive coupling — or internally by direct coupling to electrodes. In all the cases, biological effects were observed for ELF electric fields ranging from 5 to 500 mV/m in magnitude.

Biological effects of hyperweak microwaves were also observed in several studies: 10^{-2} μW/cm^2 [Aarholt *et al.*, 1988], $5 \cdot 10^{-6}$ μW/cm^2 [Grundler *et al.*, 1992], 10^{-12} μW/cm^2 [Belyaev *et al.*, 1996], 10^{-12} μW/cm^2 [Kuznetsov *et al.*, 1997]. It is interesting to note that flux densities of the order of 10^{-10}–10^{-11} μW/cm^2 correspond to the thresholds of vision and hearing.

Some of these data are schematically depicted in Figure 1, which shows the domains of both the experimentally observed effects and various theoretical limits, and indicates the non-thermal physical nature of the effects.

21.2.2 Theory

To date, the physical nature of the biological sensitivity to weak and hyperweak EM fields remains unclear, although significant insight has recently been developed [Binhi, 2002]. A unified foundation is proposed which is claimed to account for the biological effects of EM fields. The interference of quantum states of ions and molecular groups explains many of the paradoxes surrounding the non-thermal action of EM fields. This theory is based on "primary" physical principles and agrees with experiments.

In many cases, biological effects display "windows" in biologically effective parameters of the fields. Most dramatic is the fact that relatively intense fields sometimes do not cause appreciable effect while smaller fields do. Linear resonant physical processes, as well as any kind of heating, cannot, of course, explain the existence of frequency windows.

It has been suggested [Binhi, 1997] that a nonlinear effect, involving the interference of quantum states of ions and molecules bound within some proteins — in particular, calcium and calmodulin — is a general molecular target for the external EM fields. The ion interference mechanism predicts multipeak biological effects in many cases: magnitude modulated magnetic fields, magnetic vacuum, pulsed magnetic fields [Binhi, 1998], weak AC electric fields, shift and splitting of MBE spectral peaks under the rotation of biological samples, combined action of different magnetic fields and magnetic noise, bioeffects of modulated microwaves. The consistency between theory and experiments indicates that what underlies the MBE is most likely the interference of ions.

In accordance to the ion-interference mechanism, the threshold field for biological responses to the ELF electric field falls into the range of hyperweak electric fields [Binhi *et al.*, 2000]. Ion interference mechanism applied to rotating biophysical structures, such as DNA-RNA fragments, provides a basis for understanding how weak EM fields affect biology [Binhi, 2000]. Of special interest is the existence of molecular gyroscopic degrees of freedom, because these degrees of freedom are not thermalised on biologically relevant time scales. Therefore, mechanisms that involve molecular gyroscopes can account for the biological effects of hyperweak EM fields [Binhi, 2002]. It is important to keep in mind that the possibility of the interference mechanism depends on the value of the local static magnetic field. Since this field varies in a complicated manner throughout the interior of modern buildings, the interference effect at the molecular level and consequently the biological endpoint, may not be reproducible in different places, even when all other EM fields are the same.

All living matter is built from the same molecular bricks — amino-acids and proteins. Despite their inherent differences they have very similar biophysical structures. Therefore, it is clear that the interference mechanism (a molecular physical theory) is equally applicable to biological systems having different levels of complexity. If an effect exists for one biological system at given EM fields, one may expect one for another biological system exposed to the same EM fields. The only condition for this is the presence in both systems the *same* molecular EM target.

Molecular mechanisms of biological sensitivity to EM fields are consistent with the fundamental quantum limit of sensitivity to electromagnetic radiation.

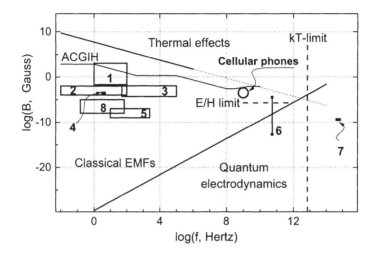

Figure 21.1 This figure illustrates various limits and areas of EMF biological effects as functions of two variables, EMF frequency f, in Hz, and amplitude B, in G = 100 μT. The large variation in both variables is chosen to show qualitatively different cases of the effects and theoretical approaches. A quantum of EMF with energy less than kT lies to the left of the dashed vertical line. This line defines a paradoxical area, wherein the biological effects are not possible from an orthodox viewpoint. The upper, diagonally downward directed line separates (very approximately) the domains of thermal and non-thermal EMF bioeffects. The lower, diagonally upward directed line is the QED limit. EMF must be described as a quantum system below this line. The step-inclined line is the ACGIH (see refs) – limit of safe levels of the EMF exposure. Areas marked by digits denote ranges of the parameters characterising: 1 — ELF EMFs used in most of magnetobiological experiments; 2 — EMFs produced by magnetic storms, which are known to correlate in time with a peaking of cardiac-vascular diseases; 3 — background EMFs produced by the variety of home appliances, video display terminals such as TV and computer monitors; 4 — MFs that affect some amino-acid solutions [Fesenko *et al.*, 1997]; 5 — MFs supposedly induced at 0.5 m from the TecnoAO units by VDT and cell phones EM radiation; 6 — EMFs below the quantum electrodynamics limit, which significantly affect cell cultures [Belyaev *et al.*, 1996]; 7 — limit of sensitivity of the human eye; 8 — magnetic fields used therapeutically to treat certain diseases. The open circle shows MF fields emitted by typical cellular phones.

Figure 21.1 shows several theoretical limits defining different mechanisms and descriptions of EMF bioeffects. The *kT* and thermal limits are well known, being used repeatedly in many scientific works on EMF standardization. The quantum electrodynamics limit, on the other hand, needs some comment.

The nature of the interaction between EM fields and a substance depends on whether the EMF can be treated classically, or whether its quantisation in photons has to be taken into account. Quantum electrodynamics gives the conditions for the validity of a classical description — namely, that the field must have a certain minimum intensity (*i.e.* amplitude), the value of which depends on frequency (as indicated by the lower, upwardly directed diagonal line in Figure 21.1). As can be seen, all "low frequency" effects except the hyperweak can be described within classical EMF approach.

However, this does not define a minimum EM field intensity that is detectable by a biological system. The natural constraint on the electromagnetic susceptibility of a biological reception, as well as any receiver of a physical nature, is given by the general laws of the quantum mechanics, which relates the minimal energy change, ε, and the time, t, required for its registration, according to $\varepsilon t > h$, where h is the Planck constant. For example, in order that an ELF EMF photon of frequency $f \sim \varepsilon/h$ to be registered by any system, including biological one, requires a time interval, t, of at least $1/f$.

At the same time, this relation does not set a lower limit of the susceptibility to EM fields. Apparently, there is no general theoretical constraint defining a lower limit for the intensity of EM fields capable of affecting a biological system. All physical constraints that have been suggested to date are based on specific primary mechanisms of the EM field signal transduction, and not on fundamental physical principles, which do *not* forbid biological hypersensitivity to EM fields. Only the microscopic design of a biological receptor and the time of its coherent interaction with EMF define the level of hypersensitivity in each specific case.

The indicated E/H limit shows the level of EM fields, in a plane wave approximation, below which atomic magnetic effects dominate electric ones. As is seen, such EM fields should be referred to as hyperweak fields. There are corresponding experimental observations [Belyaev *et al.*, 1996].

21.3 TOWARDS EM SAFETY STANDARDS

For a quantity of heat, Q, per unit mass of a biological tissue, the rate equation may be written as follows

$$dQ/dt = R + P - Q/\tau ,$$

where R is the heat input due to metabolic processes, P is the contribution from EM field absorption, and Q/τ is the heat loss on a time scale τ. In stationary regime, $dQ/dt = 0$, and we obtain a relation defining the heat increment corresponding to the growth of P: $\Delta Q = Q - Q_0 = P\tau$, where $Q_0 \equiv R\tau$ is a quantity of heat in the absence of the EM field. A heat increment ΔQ causes the temperature rise ΔT according to: $\Delta Q = c \Delta T$, where c is the specific heat capacity. It is then easy to find value of heat input, P_0, required to achieve a given temperature rise:

$$P_0 = c \Delta T/\tau .$$

For a biologically significant temperature rise of 0.1°C, a 1-minute relaxation time for biological tissue, and a specific heat capacity 1 J/g °C, we calculate P_0 of order 1 W/kg. This threshold quantity may be further linked to the so-called specific absorption rate (SAR)

$$SAR = \sigma E^2/\rho,$$

which shows the quantity of heat that is produced by alternating electric field, E, in a dielectric medium of electric conductivity σ and density ρ (we do not take into account the dielectric heating at the character frequency 2 GHz because it is two times less than Joule effect). If the SAR exceeds P_0, biological tissue will undergo a potentially dangerous degree of heating. It is impossible to analytically derive a relation between output power of a cell phone and SAR distribution in a user's head because of the very complicated field configuration in the near zone of an antenna due to non-uniform electric and dielectric properties of the head tissues. Numerous numerical results show that each Watt of the output power produces a SAR[2] value of the order of 1 W/kg. This value can, however, vary significantly depending on the type, orientation, and closeness of a cell phone to the human head. At the same time, the ICNIRP general public safety SAR threshold is 2 W/kg, at 935 MHz for localised exposure. Therefore, for example, a GSM 900–1800 MHz cell phone with an average output of 0.25 W is just within the recommended safety level, or just a little below it — see Figure 21.1.

There are similar data for the ELF range indicating the presence of ELF magnetic fields about 1–10 μT inside the head. This raises the hotly debated issue as to whether the present safety standards afford adequate protection against the emissions of cellular phones. In this regard, we would like to emphasize that both the existing EM safety standards (based solely on considerations of heating), and the current fixation with "fine-tuning" them, diverts public attention from the real problem of the bio-significance of non-thermal effects of this kind of radiation, which should, instead, be the centre of focus.

The real problem is that biological effects may be caused by hyperweak EM fields, far below those that can heat tissues. Most bioeffects of domestic appliances (which are confirmed in many scientific works) are not caused by EM heating, but rather by other, *resonance-like* physical processes that are still possible for much less intense EM fields. This well-known fact is a very dramatic one for manufacturers of devices that utilize EM fields. Therefore, in the interest of market growth, there are compelling reasons to suppress this knowledge, so that it does not become widely known.

Effecting simply a possible reduction of the output power of cell phones is not really a solution of the problem. It is easy to see that such attempt would be based, implicitly, on the direct proportionality between the EM power and subsequent biological effects. Whilst such proportionality indeed holds in the case EM heating

---

[2] In order to ease connection with experimental data where EM fields were like a plane wave, we indicate that EM flux density at the surface of the head may constitute 1–100 mW/cm² per Watt of output power.

of biological tissues, it does not necessarily hold in the case on non-thermal effects. There are both theoretical and experimental grounds for *non-linear* power-dependences of MBEs, in which a reduction of the EM power may actually lead to the growth of a biological effect! This is why a totally different approach must be adopted if people are to be protected against cell phone radiation and other forms of sub-thermal EM environmental pollution. It follows that physical concepts other than those underlying thermal effects must be urgently studied at a fundamental level, so that they can ultimately be applied to the development of more comprehensive safety standards that incorporate the principle of electromagnetic biocompatibility.

21.3.1 Power Reduction

Researchers observed biological effects of very different EM fields that are distinct in frequency, amplitude, polarization, configuration, and so on. In addition, very different kinds of biological systems may be involved in the response to EM fields. Furthermore, many different endpoint parameters — physical, chemical, biological, and even social — are measured in such experiments. Is it possible to indicate some scales of the physical observables that could characterize EM bioeffects as a natural phenomenon? Because of the great variety of different cases, we may speak about such scales only in the sense of averaged values. Taking into account a lot of existing empirical data obtained under varying conditions, we may conclude "on average" that appreciable biological responses, say 10%, appear in a time period of 10 to 100 min as a consequence of exposure to ELF magnetic fields between 10 and 100 μT, or microwave EM fields at power densities between 10 to 100 μW/cm^2. Bioeffects of ELF electric fields, on the other hand, are generally less clear and decipherable.

Whilst exposure to these fields of these strengths does not necessarily cause bioeffects, bioeffects are, nevertheless, found at these intensities. It follows, in principle, that the common procedure of averaging over the whole ensemble of both positive and negative findings is unacceptable. From the viewpoint of a person who "caught" the EM field and got some associated disease, such averaging understandably seems amoral.

Can we use the above considerations to develop a new EM safety standard? Present EM safety standards make use of the notion of a "dose" of exposure to EM fields. The dose is the product of the field amplitude (or power) and the time of exposure. Indeed, many different safety levels can in this way be defined, according to the particular duration of exposure adopted — a minute, hour, and day[3]. Moreover, the greater the time interval, the lower can be the level. It is possible to consider the possibility of defining a "safe" dose as an upper limit (or threshold value) on the power-time product, which should not be exceeded during normal use of a particular device.[4]

---

[3] There are a few or no EM safety standards related to month and year duration of exposure. By this, they renounce possibility of the long-term biological effects of the EM fields that is inconsistent with many experimental data on chronic EM exposure.

[4] Of course, the question is much more complicated, for example the EM frequency is an essential parameter of the EM safety standards. For clarity, we hold only main features of the question.

Despite the non-linearity of the physical processes underlying MBE, the notion of the dose is still a useful one, although it should be applied to or based on the above-mentioned biological effects rather than on heating effects. From the theoretical viewpoint, nonlinear molecular interference effects may be characterized by the dose, but only when all possible local static magnetic fields are averaged over, so that subject motion is taken into account.

Taking the above mean values for microwave power of 30 $\mu W/cm^2$ and time scale 30 min, we obtain a threshold dose about 10^3 $\mu W \cdot min/cm^2$. This value is significantly less, by between one and three orders of magnitude, than the thresholds suggested by the different specialist committees, and is likely to be better for human health. However, it would be much better if the averaging was done over a particular class of subjects, such as cell phone users, for example. The decrease of dose does not necessarily mean, however, that a particular individual is necessarily immune, particularly if the above time and space averaging is not realizable in that particular case. It thus becomes apparent that such "direct" methods cannot, in general, be considered to guarantee safety at an individual level.

21.3.2 Biological Protection

It was recently demonstrated that TecnoAO technology, based on hyperweak magnetic emission of a microcrystalline solution electromagnetically treated (International Patent, see references), offers an efficient protection against biological effects of the EM fields. Chicken embryos, which were exposed to computer/TV screens or the EM fields of cellular phones, showed an enhanced mortality rate, but in the presence of TecnoAO protective units (metal tubes or other forms filled with a special solution) the mortality rate or the status of the immune system of young adult chicken and of mice remained the *same* as in the control group [Youbicier-Simo *et al.*, 1996, 2000].

For humans exposed to VDU emissions in their workplace, epidemiological studies reveal a statistically significant improvement in the condition of 119 individuals one month after the VDUs had been protected by TecnoAO technology: specifically there was a 15% increase of stress resistance, and a 23% increase of concentration [Fillion-Robin *et al.*, 1996]. The portfolio of evidence obtained from controlled studies of the efficacy of TecnoAO technology makes it possible to conclude that that solution behaves as a source of signals that mitigated the adverse effect of device radiation on biological processes. The solution emits propagating physical field that affects biophysical targets. The signal is, in turn, transmitted to the cellular level, to organs and to the whole body. This implies that the saline solution supports a number of metastable states of differing biological activity [Binhi, 1991; Binhi, 1998b]. The solution acts as a re-emitter of radiation, which, however, modified in the low-frequency range. The solution within the TecnoAO is "charged" by a preliminary EM treatment, which transforms the solution into another metastable state, the re-emissions from which are characterised by the modified low-frequency EM spectrum that contains biologically significant information. In addition the EM fields, the solution may also radiate other physical agents of biological significance. Ultimately, the total emission targets particular biological structures preserving their integrity under irradiation from the device (cell phone, VDU), which, in turn, ensures that homeostasis is maintained.

Recently, the effect of exposure to cellular phone radiation on nitric oxide production by humans was observed [Fillion-Robin *et al.*, 2001]. An increase of nitric oxide concentration in exhaled air was found to be between 7 and 40% after two weeks intermittent exposure to GSM cell phone radiation. Endogenous nitric oxide plays an important role in a large number of biochemical processes in human body. The NO molecules are involved in the transmission of nervous impulses, regulation of vascular tension and the development of inflammation. Gaseous nitric oxide present in exhaled air is a bio-marker of inflammation processes in patients with respiratory diseases. Application of TecnoAO protective units during the following two-week exposure was found to return the NO concentration to the control level.

It is to be emphasized that TecnoAO technology protects at the biological level. It makes molecular targets of EM, such as ion-protein bonds, insensitive to EM fields, which could otherwise cause interference effects, resulting in abnormalities. Thus, this technology, which *compensates* harmful biological effects, rather than the EM field itself, implements the realization of electromagnetic biocompatibility.

21.4 CONCLUSION

Both published experiments results and recent theoretical advances show that EM fields may result in different biological effects, even if they are too weak cause any deleterious degree of heating. Present EM safety standards do not account this possibility. Therefore, in occupational and residential activities, people exposed to EM fields from industrial and domestic appliances, conform to existing safety standards, are *still vulnerable* to non-thermal biological effects having possible adverse consequences for human health.

The thermal effect of electromagnetic fields is the only factor that is currently used for the development of electromagnetic safety standards. There are, however, other influences — such as the molecular quantum interference, that can cause *non-thermal* biological effects that display resonance-like behavior. Levels of magnetic, electric, and microwave fields have been identified at which such non-thermal effects might be expected. These indicate that existing safety standards fail to provide adequate protection to the users of these devices.

New approaches to EM safety standards — based on the principle of electromagnetic biocompatibility — have been here proposed, from consideration of the *non-thermal* biological effects of EM fields of sub-thermal intensity.

21.5 REFERENCES

Aarholt, E., Jaberansari, M., Jafary-Asl, A.H., Marsh, P.N. and Smith, C.W., 1998, NMR conditions and biological systems. In: A. Marino (ed.), *Modern Bioelectricity*, Marcel Dekker, New York, pp. 75–105.

American Conference of Governmental Industrial Hygienists (ACGIH), 1994, Static magnetic fields, sub-radiofrequency (30 kHz and below) magnetic fields. In: 1994-1995 Threshold Limit Values for Chemical and Physical Agents and Biological Exposure Indices. Cincinnati, ACGIH, pp. 110-111.

Agadjanyan, N.A. and Vlasova, I.G., Infra-low-frequency magnetic field effects on the rhythm of nerve cells and their immunity to a hypoxia. *Biophysics*, **37**(4), pp. 681–689, 1992.

Akerstedt, T., Arnets, B., Ficca, G. and Paulsson, L-E., 1997, Low frequency electromagnetic fields suppress SWS. *J. Sleep Res.*, **26**, pp. 260–266.

Bastian, J., 1994, Electrosensory organisms. *Physics Today*, **47**(2), pp.30–37, February.

Belyaev, I.Ya., Shcheglov, V.S., Alipov, Ye.D. and Polunin, V.A., 1996, Resonance effect of millimeter waves in the power range from 10^{-19} to $3 \cdot 10^{-3}$ W/cm^2 on Escherichia coli cells at different concentrations. *Bioelectromagnetics*, **17**, pp. 312–321.

Bersani, F. (ed.), 1999, *Electricity and Magnetism in Biology and Medicine*, Kluwer Academic / Plenum Publishing Corporation, London.

Binhi, V.N., 1991, Induction of metastable states of water. *Preprint N3, CISE VENT*, Moscow, pp.35. [In Russian].

Binhi, V.N., 1997, Interference of ion quantum states within a protein explains weak magnetic field's effect on biosystems. *Electro and Magnetobiology*, **16**(3), pp. 203–214.

Binhi, V.N., 1998, Interference mechanism for some biological effects of pulsed magnetic fields. *Bioelectrochemistry and Bioenergetics*, **45**, pp. 73–81.

Binhi, V.N., 1998b, Structural defects of liquid water in magnetic and electric fields. Biomedical Radioelectronics, **2**, pp. 15–28. [In Russian]

Binhi, V.N., 2000, Amplitude and frequency dissociation spectra of ion-protein complexes rotating in magnetic fields, *Bioelectromagnetics*, **21**(1), pp. 34–45.

Binhi, V.N. and Goldman, R., 2000, Ion-protein dissociation predicts "windows" in electric field-induced wound-cell proliferation. *Biochimica et Biophysica Acta*, **1474**, pp. 147–156.

Binhi, V.N. 2002, *Magnetobiology: Underlying Physical Problems* (Academic Press, San Diego).

Delgado, J.M.R., Leal, J., Monteagudo, J.L. and Garcia, M.J., 1982, Embryological changes induced by weak, extremely low frequency electromagnetic fields. *Journal of Anatomy*, **134**, pp. 553–561.

Fesenko, E.E., Novikov, V.V. and Shvetsov, Yu.P., 1997, Molecular mechanisms of the biological effects of weak magnetic fields. *Biophysics*, **42**(3):742–745.

Fillion-Robin, M., Binhi, V.N. and Stepanov, E.V., 2001, Influence of the cellular phone on nitric oxide production by humans with and without TecnoAO protection. *Abstracts of the 23rd Annual meeting of the BEMS*, St.Paul, Minnesota, USA, June 10–14.

Fillion-Robin, M., Marande, J.-L. and Limoni, C.M., 1996, Protective effect of TecnoAO antenna against VDU electromagnetic field as a stress factor. *Abstract Book of the 3rd Int. Congress of the EBEA*, Nancy, France.

Grundler, W. and Kaiser, F., 1992, Experimental evidence for coherent excitations correlated with cell growth, *Nanobiology*, 1, pp. 163–176.

Hyland, G.J., M. Bastide, J.B. Youbicier-Simo, L. Faivre-Bonhomme, R. Coghill, M. Miyata, J. Catier, A.G.M. Canavan, M. Fillion-Robin, J. Marande, D.J. Clements-Croome. 1999, Electromagnetic biocompatibility in the workplace: Protection principles, assessment and tests. Results of an EMF protective compensation technology in humans and in animals. In *Nichtionisierene Strahlung: mit ihr leben in Arbeit und Umwelt*, N. Krause, M. Fischer, H.-P. Steimel (eds), IRPA, TÜV-Verlag GmbH, Köln, Germany, pp. 213–240.

TecnoAO technology (A.O.Autonomous Oscillator) international patent, reg. No.PCTWO93/25270. Tecnolab, France.

Jacobson, J.I., 1994, Pineal-hypothalamic tract mediation of picotesla magnetic fields in the treatment of neurological disorders, *FASEB Journal*, 8(5), p. A656.

Keeton, W.T., Larkin, T.S. and Windsor, D.M., 1974, Normal fluctuations in the earth's magnetic field influence pigeon orientation, *J. Comp. Physiol.*, 95, pp. 95–103.

Kouznetsov, A.P., Golant, M.B. and Bozhanova, T.P., 1997, Receipt by cell culture of the electromagnetic EHF radiation with the intensity below the noise. In: *Millimeter Waves in Medicine and Biology*, Institute of Radio-engineering and Radio-electronics RAS, Moscow, pp.145–147.

Portier, C.J. and Wolfe, M.S. (eds), 1998, *Assessment of Health Effects from Exposure to Power-Line Frequency Electric and Magnetic Fields*. Working Group Report, NIEHS/NIH, No. 98-3981.

Presman, A.S. 1970, *Electromagnetic Fields and Life*. Plenum Press, New York.

Vladimirsky, B.M., 1971, On possible factors of the sun activity affecting processes in biosphere. In: *Influence of the sun activity on the atmosphere and biosphere of the Earth*. Nauka, Moscow, pp. 126–141.

West, R.W., Hinson, W.G. and Swicord, M.L., 1996, Anchorage-independent growth with JB6 cells exposed to 60 Hz magnetic fields at several flux densities. *Bioelectrochemistry and Bioenergetics*, 39, pp. 175–179.

Youbicier-Simo, B.J., Lebecq, J.C. and Bastide, M., 1998, Damage of chicken embryos by EMFs from mobile phones: protection by a compensation antenna (TecnoAO). *Abstract Book of 20th BEMS Ann. Meeting*, St.-Petersburg, Florida.

Youbicier-Simo, B.J., Boudard, F., Cabaner, C. and Bastide, M, 1996, Bioeffects of continuous exposure of embryos and young chickens to ELF emitted by desk computer: Protective effect of Tecno AO antenna. *3rd EBEA International Congress*, Nancy, France.

Youbicier-Simo, B.J., Boudard, F., Cabaner, C. and Bastide, M., 1997, Biological effects of continuous exposure of embryos and young chickens to ELF emitted byvideo display units. *Bioelectromagnetics* 18, pp. 514–523.

Youbicier-Simo, B.J., Lebecq, J.C., Giaimis, J. and Bastide, M., 2000, Interference from GSM cellular phones with the production of stress hormones in Lewis Lung carcinoma-bearing mice : effectiveness of a protective device, *Proceedings of International Conference on Cell Tower Siting*, Land Salzburg, Austria.

Youbicier-Simo, B.J., 2000, Sensitivity of chicken embryos to portable computer radiation (LCD) and protective effectiveness validation of a compensation magnetic oscillator. *Abstract Book of VIIth Portughese Meeting of Protection against Radiation SPPCR*, IRPA, Lisbon, Portugal.

CHAPTER 22

Mobile Phones and Cognitive Function

Alan Preece

In the 100 years since Marconi first communicated by radio between countries and continents, the use of communications equipment has increased dramatically and so has the level of radio-frequency radiation in the environment (Uddmar 1999). In the last ten years the development of hand-held mobile phones has introduced a new element, namely a much higher level of exposure of the head and particularly the brain. Standards to control exposure of the body were introduced by the IRPA/International Non-ionising Radiation Committee for the protection of the occupationally exposed before mobile phones existed. The guidelines only considered thermal elevation and were set to be 10% of that shown to produce behavioural changes associated with elevated temperature, taking into account thermal distribution by blood flow. A number of countries developed their own guidelines, and levels of specific absorption rate (SAR) in the head were set to be 10 Wkg^{-1} (NRPB, UK), 2 Wkg^{-1} (ICNIRP, Europe) or 1.6 Wkg^{-1} (ANSI-IEEE, USA).

Mobile phones are now part of an extensive network based on several different providers. In the UK about 72% of us have one, but over 80% of children 11-15 years also use a mobile phone, and 10% of these spend more than 45 mins a day using one.

There are 4 main public systems in the UK:

- Vodaphone – 900 MHz GSM (and 1800MHz)
- Cellnet – 900 MHz GSM   ditto
- Orange – 1800 MHz DCS
- One-to-One – 1800 MHz DCS
- (plus DECT cordless phones)
- and one commercial/emergency services system called TETRA

These are in similar arrangements of base stations so spaced as to form "cells" which can be large (e.g. in rural low population areas), or small as in congested areas, or even as a low power micro-cell which may serve a single building such as a railway station.

It is largely the appearance of this network, and the "experiment" of millions of people holding a small powerful transmitter emitting microwaves next to their head that has caught public imagination, and led to anxiety, because this is:

- A new phenomenon that all can observe.
- New technology – pulsed microwave radiation

- Lots of base stations – appearing in neighbours back garden and being paid for the privilege
- Very little basic science – need to research this?

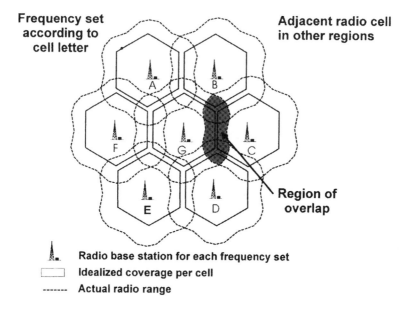

Frequency set according to cell letter

Adjacent radio cell in other regions

Region of overlap

🗼 **Radio base station for each frequency set**
☐ **Idealized coverage per cell**
------ **Actual radio range**

The ills that the public are concerned about as harmful outcomes are
•Cancer
•Ageing
•Neuro-degeneration
•Stress
•Epilepsy
•High temperature

As non-harmful effects:
•Cognition*
•Behaviour*
•Blood flow*
•Blood pressure*
•Evoked potentials
•Mild temperature rise
(*within normal limits)

In Scandinavia, which was the first to introduce extensive mobile phone use and the current leaders in phone use, a study of self-reported subjective symptoms in adults (Oftedal *et al* 2000) has suggested an exposure-related reporting of symptoms of headache, stress and warmth. The first provocation experiment in humans (Preece *et al* 1999) confirmed quite clearly that cognitive response to phone exposure at 915 MHz was occurring within the existing standards. Of a battery of responses, choice reaction time was significantly affected by one watt analogue signal but grouped reaction time tests were also affected by analogue or a weaker GSM simulation. This general observation was refined and improved in a study sponsored by a large manufacturer (Nokia) of mobile phones as a replication (Koivisto *et al* 2000). Further studies have indicated that the effect speeding up of response) is increased for increased cognitive load. A more recent small study of teenage users and non-users of phones (Lee *et al* 2001) now suggest that this enhancement of reaction response persists after exposure, as do the sleep studies (Borbely *et al* 1999). The changes recorded have been extremely controversial since they are such a small percentage of normal physiological rates. In the first study it was a 15msec change, and the replication study recorded 19msec. This actually equates to the level of change induced in the opposite direction by ageing - amounting to about 20 years. This is not insignificant and was detected by appropriate experimental design which used double blind repeated measures to remove inter-subject variation, balanced administration to remove training effect, a split study to reduce type 1 error, and finally comparison of single variable analysis with grouped variables to examine for a generalised effect. Such an enhancement is of no value to driving or other multi-tasks, because these studies took no account of the distracting effect of conversation and divided attention.

Other evidence of effects on brain function is also derived from sleep studies. Electroencephalograms (EEG), both for waking and sleep hours and at the time when the people were subjected to psychological tests were recorded, but the results have not been consistent, for example, the disturbance of slow wave brain potentials (Freude *et al* 2000) which has not been replicated by other research. Some results point to disturbed REM sleep phase (Mann and Roschke 1996), changed alpha, beta and delta waveforms (Reiser *et al* 1995, Kiltzing 1995), whereas other authors did not confirm those observations (Roschke and Mann 1997, Hietanen *et al* 1997). More extensive investigations (Huber *et al* 2000) of mobile phone exposure before sleep (Borbely *et al* 1999) or during sleep showed very consistent increases in EEG power, suggesting a mechanism similar to disturbance, and a reduction in the amount of REM sleep. This is in good agreement with the current studies (Krause *et al* 2000) that show augmentation of waking EEG, particularly at low frequency bands (alpha region) by exposure to normal communication levels of GSM 900 MHz. (See: http://www.unizh.ch/phar/sleep/handy/)

None of these experiments or studies suggests any evidence of adverse health effect for phone exposure. However, the mechanism for the consistently reported effects on reaction time and EEG is unknown. Temperature elevation is an unlikely mechanism as several studies show clearly that temperature elevation does not exceed 0.1° and has a time constant of 6 minutes. Other putative mechanisms are stress protein production such as heat-shock protein (HSP) series (de Pomerai, *et al* 2000) or a possible direct electrophysiological interaction as suggested by one group (Tattersall *et al* 2001) in studies of the rat hippocampus.

In the light of these reported effects, the UK Parliamentary Select Committee reviewed the evidence and set up the Independent Expert Group on Mobile Phones chaired by Professor Sir William Stewart which concurred not only that there was evidence of a biological effect, but also a lack of information for the public, and unregulated expansion of base stations against the public perception of risk (See http://www.iegmp.org.uk/IEGMP.txt.htm). In the absence of a known mechanism, the possibility that there may be long-term effects needs to be considered. Children are a special case because of anatomy and yet to be developed tissues. Nevertheless a small study in the UK showed that 85% of children aged 11-15 years use a mobile phone and 10% of these use their phone in excess of 45 minutes per day (without identifying intermittency or call length).

In addition to the reported cognitive changes, mobile phones, by acting on various brain structures, may disturb other neurologically controlled physiological functions, including heart rate and blood pressure. The radio frequency radiation can reach the cardiovascular centres in the medulla of the brain stem that control heart and circulation via the sympathetic and parasympathetic system and/or the receptors in the carotid body. The subjective reporting of stress with phone use suggests there may be further effects on circulatory stress hormones that could be affected by RF exposure. There are already some experimental data to suggest effects on the cardiovascular system. One study (Braune *et al* 1998) noted a small increase in arterial blood pressure of 5-10 mmHg and reduction of heart rate from 35 minutes exposure to 900 MHz RF.

A group of researchers (Thuroczy *et al* 1997) recorded a significant decrease in the diastolic blood pressure after termination of the exposure. Earlier research showed heart rhythm disturbances and circulatory neurovegetative regulation disorders (Bortkiewicz *et al* 1996, 1995) and disturbed circadian blood pressure variations (Szmigielski *et al* 1998) in people occupationally exposed to radio frequency EMF. The circadian rhythm of many physiological functions, including blood pressure, is controlled by melatonin. This is a possible mechanism for physiological disturbance by RF. The synthesis and secretion of melatonin from the pineal gland may be disturbed by radio-frequency electromagnetic fields through a direct influence on either the suprachiasmatic nucleus or the pineal gland itself. Very few reports are available on the effect of mobile phone EMF on human melatonin level although some authorities feel the effects of RF could mimic that of 50/60Hz which can alter the *in vitro* effects of melatonin (Harland and Liburdy 1997). Another study (De Seze *et al* 1999) did not record significant disturbances in the *in vivo* circadian rhythm of melatonin secretion after GSM or DCS exposure (2 h/day, 5 days/week).

The reported symptoms and the changes observed in the nervous and circulatory systems may be attributable to the disturbances at the cellular level. *In vitro* studies show that even SAR values below the guidance levels may be associated with changes in cellular membrane permeability (disturbed active transport of Na^+, K^+ ions and increased release of Ca^{++} ions). These processes affect tissue excitability, as they influence, among other things, the mechanism of neurotransmitter release (Dutta *et al* 1989 and Liu *et al* 1990). Ion pump disorders were shown to occur for a very wide range of SAR (0.2-200 W/kg) and frequency

2 MHz -10 GHz (Cleary 1995). Recent studies in Sweden have revealed that even with very low exposure levels (15 MHz, SAR 0.0016 W/kg), increased permeability of the blood/brain barrier could be observed(Salford *et al* 1994). Mechanisms suggested for these effects are heating, stress proteins and direct electrical stimulation. There is evidence for some of these ideas. The first is highly unlikely, unless the gradient is important ($1/e^2$ is about 13mm at 1800 MHz and 25mm at 900 MHz), since the change that plateaus at 6 mins is only $0.1°C$ or less. Heat shock proteins are slow to be expressed and not likely to be activated in time to account for the effects. That leaves direct electrical effects, that would require a non-linear element in the tissue at these high frequencies, and this phenomenon is disputed by many physicists.

In summary, effects on neurological function have been identified within the statutory exposure limits, without in reality any mechanism being identified. As far as we know these only apply to handset exposure and not base station levels. The comparison between the exposure situations can be assessed by examining the voltage levels generated at the head by a handset. This can be measured, or calculated by a simple approximation to:

$$E(V/m) = \frac{7\sqrt{P(watts)}}{d\,(metres)}$$

The end effect is that a phone will generate 25 V/m in air, whereas at 1 to 5m from a cell phone mast the field will be less than 3 V/m. Since the power deposition is proportional to the square of the field the relative difference in head exposure is enormous. However at the present state of knowledge we do not understand what sets the thresholds for cognitive effects and these need to be researched. Even the humble cordless DECT phone base emits a surprising level of RF even when not actively linked to a handset and may, if there is a large number in an office, contribute to the RF environment.

We can restrict exposure by sensible choice of handset, which vary by a factor of 20 as to the brain cortex exposure, or by use of hands free devices to separate head and phone, or even restrict the length of calls!

REFERENCES

Borbely AA, Huber R, Graf T, Fuchs B, Gallmann E, Achermann P. Pulsed high-frequency electromagnetic field affects human sleep and sleep electroencephalogram. *Neurosci Lett.* 1999 Nov 19; **275(3)**:207-10.

Bortkiewicz A., Zmyslony M., Palczynski C., Gadzicka E., Szmigielski St.: Dysregulation of Autonomic Control of Cardiac Function in Workers at AM Broadcast Stations (0.738-1.503 MHz), *Electro-and Magnetobiology*, 1995, **3**, 177-191

Bortkiewicz A., Gadzicka E., Zmyslony M.: Heart rate variability in workers exposed to medium-frequency electromagnetic fields. *J. Autonomic Nervous System* **59**, 1996, 91-97

Braune S., Wracklage C, Raczek J. and Lucking C.H. Resting blood pressure increase during exposure to a radio-frequency electromagnetic field, *Lancet*, 1998, **351**, 1857-1858.

Cleary S.F. Effects of radio-frequency radiation on mammalian cells and biomolecules in vitro. W: Blank M. (ed) *Electromagnetic fields: biological interaction and mechanisms*. Washington, American Chemical Society, 467-477, 1995.

de Pomerai D, Daniells C, David H, Allan J, Duce I, Mutwakil M, Thomas D, Sewell P, Tattersall J, Jones D, Candido P. Non-thermal heat-shock response to microwaves., *Nature*, **405(6785)**, 417-8, 2000

De Seze R., Ayoub J., Peray P., Miro L., Toutou Y. Evaluation in humans of the effects of radiocelluar telephones on circadian patterns of melatonin secretion, a chronobiological rhythm marker. *J. Pineal Res.* **27**, 237-242, 1999.

Dutta S.K., Ghosh B., Blackman C.F.: Radiofrequency radiation induced calcium ion effux enhancement from human and other neuroblastoma cells in culture, *Bioelectromagnetics*, **10**, 197-202, 1989.

Freude G., Ullsperger P., Eggert S., Ruppe I. Microwaves emitted by cellular telephones affect human slow brain potentials. *Eur.J.Appl.Physiol.*, **81(1-2)**, 18-27, 2000.

Harland J.D. and Liburdy R.P.: Environmental magnetic fields inhibit the antiproliferative action of tamoxifen and melatonin in a human breast cancer cell line. *Bioelectromagnetics*, 18, **555-562**, 1997

Hietanen M., Kovala T., Hamalainen A.M. Human brain activity during exposure to radiofrequency fields emitted by cellular phones. Scandinavian Journal of Work, *Environment & Health*, 26(2), 87-**92**, 1997

Huber R, Graf T, Cote KA, Wittmann L, Gallmann E, Matter D, Schuderer J, Kuster N, Borbely AA, Achermann P. Exposure to pulsed high-frequency electromagnetic field during waking affects human sleep EEG. *Neuroreport.* 2000 Oct 20;**11(15)**:3321-5.

International Radiation Protection Association/International Non-ionizing Radiation Committee. Guidelines on limits of exposure to Radiofrequency electromagnetic fields in the frequency range from 100kHz to 300GHz. *Health Physics* **54**: 115-123; 1988

Kiltzing I. Low-frequency pulsed electromagnetic fields influence EEG in man. *Phys. Med.* **11**: 77-80, 1995.

Koivisto M., Revonsuo A., Krause Ch., Haarala Ch., Sillanmäki L., Laine M., Hämäläinen Effects of 902 MHz electromagnetic field emitted by cellular telephones on response times in humans. Cognitive Neuroscience and *Neuropsychology* 2000, **11(2)**, 413-415.

Krause C.M., Sillanmaki L., Koivisto M., Haggqvist A., Saarela C., Revonsuo A., Laine M., Hamalainen H. Effects of electromagnetic field emitted by cellular phones on the EEG during a memory task. *Neuroreport*, **11(4)**, 761-764, 2000.

Lee T.M.C., Ho S.M.Y., Tsang L.YH., Yang S.Y.C., Li L.S.W., Chan Ch.C.H. Effect on human attention of exposure to the electromagnetic field emitted by mobile phones. *Cognitive Neuroscience and Neuropsychology* 2001, **12(4)**, 729-731

Liu D.S., Astumian R.D., Tsong T.Y. Activation of Na and K pumping modes of Na, K-ATPase by an oscillating electric field. *J. Biol. Chem.*,**265**, 7260-7267, 1990.

Mann K., Roschke J. Effects of pulsed high frequency electromagnetic fields on human sleep. *Neuropsychobiology*, **33**, 41-47, 1996.

Oftedal G., Wilen J., Sandstrom M., Mild.K.H. Symptoms experienced in connection with mobile phone use. *Occupational Medicine* (Oxford), **50(4)**, 237-245, 2000.

Preece A.W., Iwi G., Davies-Smith A., Wesnes K., Butler S., Lim E., Varey A. Effects of a 915-MHz simulated mobile phone signal on cognitive function in man. *Int. J. Radiat. Bio.* 1999, **75(4)**, 447-456.

Reiser H.P., Dimpfel W., Schober, F.J. The influence of electromagnetic fields on human brain activity, *Eur.J.Med.Res.* **1**, 27-32, 1995.

Roschke J., Mann K. No short-therm effects of digital mobile radio telephone on the awake human electroencephalogram. *Bioelectromegnetics* **18**, 172-176, 1997.

Salford L.G., Brun A., Sturesson K., Eberhard J.L., Person B.R. Permeability of the blood-brain barrier induced by 915 MHz electromagnetic radiation, continous wave and modulated at 8,16,50,200 Hz. *Micros Res Tech.*, **27**, 535-542, 1994.

Stuart W. *Mobile Phones and Health.* Independent Expert Group on Mobile Phones. http://www.iegmp.org.uk/IEGMP.txt.htm

Szmigielski St., Bortkiewicz A., Gadzicka E., Zmyslony M., Kubacki R. Alteration of diurnal rhythms of blood pressure and heart rate in workers exposed to radiofrequency electromagnetic fields. *Blood pressure monitoring* 1998, **3**, 323-330.

Tattersall J.E., Scott I.R., Wood S.J., Nettell J.J., Bevir M.K., Wang Z., Somasiri N.P., Chen X. Effects of low intensity radiofrequency electromagnetic fields on electrical activity in rat hippocampal slices. *Brain Res*, **904(1)**, 43-53, 2001

Thuroczy G., Kubinyi G., Sinay H., Bakos J., Sipos K., Lenart A., Szabo L.D. Human electrophysiological studies on the influence of RF exposure emitted by GSM cellular phones. In: *Electricity and Magnetism in Biology and Medicine*, ed. F. Bersani, Kluwer Academic/Plenum Publisher, New York, 1997.

Uddmar T. RF *Exposure from Wireless Communication.* MSc Thesis. Chalmers University of Technology, Gothenburg. 1999

Questions & Discussion

Malcolm Charsley (ETS Lindgren, Herts., UK)

This question is addressed to the last speaker Anne Silk. Given your thoughts on the triggering of electro-sensitivity via electric shocks, what do you understand the Government's announcement to issue tasers to police forces as a means of passive resistance instead of plastic bullets. These give a high voltage shock and knocks people unconscious. Based on what you say this sounds entirely the wrong thing to do.

Anne Silk (Vice Chairman of Electromagnetic Biocompatibility Association)

It is not a good thing to do. It may stop crime at the time but from the individuals point of view, especially given biogenetic background and their precursors to sensitivity, who knows what are the long-term effects of this action on people.

Dr Elizabeth Cullen (Irish Doctor Environmental Association, Dublin)

Are there any children suffering from electrical hypersensitivity? Their nervous system is not fully developed and it may be that they are more likely to be more susceptible to the effects of EMF than adults are.

Dr Jean Monro (Breakspear Hospital, Herts., UK)

I would suspect so. There are a lot of points Dr. Silk made which actually indicate the fact that membrane channels are disturbed. We know that nearly 70% of the body's energy is used in maintaining membrane integrity. That entails sodium pumps pushing electrons out of the cells and sodium and potassium exchange. It is my belief that in electro sensitivity states, people actually have a channelopathy being disturbed and therefore these people can be extremely vulnerable to a very broad range of responses. If the channelopathy occurs at the blood brain barrier, I think that these people are very much more vulnerable.

You mention children. I think children are now exposed to so many more chemicals as well as noxious food containing chemical contaminants.
In my talk I said that people with electro sensitivity states had food sensitivity states also and develop chemical sensitivity. In the literature with regard to migraine, it has been known for half a century that people can react to zigzag lights and therefore they can react to physical effects as well. There is a concordance between all of these and the effects on the brain. I feel sure that you are right to be worried.

Dr Cyril Smith (Formerly University of Salford, UK)

The problems of health effects due to the electromagnetic environment will remain until we have a World Health Organisation International Classification of Diseases Code (ICD Number). Diseases do not exist in modern medicine until they have an ICD number against which they can be reported. Germany has now got as far as adding " Chemical- Sensitivity-Syndrome Multiple" to its ICD coding framework under the code T78.4. Do you agree that an ICD code is urgently needed for health effects related to the electromagnetic environment and if so, how this can be brought about internationally?

Dr Jean Monro

There has been a denial for a long time that there is such a thing as multiple chemical sensitivity but it is now finally recognised. I believe that we have been looking at the wrong system to try to describe chemical sensitivity. For example, if one looks at the blood for allergens producing antibodies, often the test is irrelevant for food sensitivities unless people have a very acute sensitivity state. Many people with food sensitivity have antibodies in their stools, just as they do to helicobacter which is an organism which lives in the stomach. They will get food sensitivity antigens and antibodies, which you can identify in the stool. So if you look at that medium you can identify food sensitivity. Hence in chemical sensitivity one might need to look at a different system to evaluate abnormalities, just as in food sensitivity one might identify antibodies from the stool rather than the blood.

In a similar way people have been negating the possibility that people actually have chemical sensitivities and it is very foolish to keep denying this. A large group of people have complained that every time they smell scent or perfume or if they go into a room full of cigarette smoke, they would get a headache or whatever symptom they complained of when they were exposed to the chemicals. It would be foolish for doctors to dismiss the symptoms and say that they are going to mask them with a steroid or other drug.

There is a condition of multiple chemical sensitivity. It is now recorded in the German International Classification of Diseases that chemical sensitivity is a real entity and it has associations with allergy and also with toxicology. It is not just a psychiatric label.

Now we have got to establish that there is a real entity called electrical sensitivity and to have an ICD identification code for that. Many of you would know there was an international conference on electrical sensitivity run by International Congress on Occupational Health, in September 2000 and the President, Dr. Bengt Knave, gave a speech on electrical sensitivity.

Dr Martin Bootman (The Babraham Institute, Cambridge)

I asked earlier about the mechanism by which EMF affects biological tissues. In my lab in Cambridge we actually have a grant to study the effects of radio frequency and low frequency electromagnetic fields on cells and tissues. We are trying to do replication studies and we are also trying to do novel studies. It is absolutely right that if you apply a strong enough electric field to cells then you can electroporate the cells. But in your lifetime, unless you touch a power cable, you will not experience an electric field that is strong enough to actually electroporate your cells. The cell biology Anne Silk showed was really very poor. Proteins do go across the cell membrane, but your cells are also full of proteins. Some of the proteins, which are also very important buffers for the potentially dangerous things like superoxide, we normally produce as natural messengers.

A lot of people have probably come here today to discover a balanced argument about what really is going on. The last two talks were extremely negative. There are scientists around the world who have done very good studies with absolutely negative findings. On the other hand, many of the positive results that have been reported cannot be replicated in other laboratories. I am not denying that people might be highly sensitive to their environment and there is clearly a condition that needs to be looked into. The mechanisms described here today are not valid.

Dr Jean Monro

When a patient presents with electro-magnetic exposure, as a clinician I have to look into that individual's symptoms and address things as well as I might. If there is something that one cannot do anything about, one can just annotate and symptoms, and leave the matter for another scientist to resolve. However, we have reversed electrical sensitivity in some 500 patients, when they came to us with a set of symptoms which we graded on a symptom scoring chart, and we were able to reverse the severity and frequency of symptoms. We can use neutralising treatment to address problems.

Dr Martin Bootman

I believe an epidemiologist would have great difficulty with the data that Anne Silk showed us. Her few case studies seem to me to be very anecdotal and highly selective. Even in the village where she was considering 19 people, who are clearly very poorly, she made no account of other factors in their lives. The question and interview method must be carried out very carefully. We cannot just say that EMFs are bad and are the sole cause of these problems. People who are hypersensitive may well be hypersensitive to EMFs, but is that the cause or the consequence of something that they are suffering from? The cell biology is extremely weak because there are no known universally accepted mechanisms to explain the interaction of EMF with cells.

Anne Silk

I disagree. There are many many factors. I or the GP concerned have not yet gone through the procedure of interviewing patients on what is in the water they are drinking, what is in the milk they are drinking, and so on. I am just illustrating something that has accrued in the environment within the last six months and its worrying a lot of people besides costing the health service a lot of money. These people have been referred to various clinics. One does need to do a thorough analysis of all the confounding factors but I was not in a position to do that. I came today to show you something is happening and worrying people. The Japanese have similar case experiences and they believe they are related to diet. As scientists you need to carry out research to explain these case experiences.

Eva Marsalek (PMI, Austria)

It is not proved that EMF is the cause of poor health reported after mobile-phone-antennae installations have been completed. People are contacting and informing us that they are suffering after the installation of a mobile-phone-antennae system and that they have not changed anything in their homes, having lived there for many years without any problem. The problem is that nobody is willing to look at the causes of the reported low well-being. These people are going from doctor to doctor, causing a lot of costs to the health-system. We have many people contacting us and we are working without any payment. We hear these complaints nearly everyday but unfortunately we cannot find a scientist carrying out any research on this topic which will help these people.

Vic Clements (Radio Frequency Investigation, UK)

Dr Monro said that her patients showed sensitivity to phials, which they held, in vaccines, and you then described some electromagnetic influences. It was not clear how you arrived at that conclusion. Did you actually try to measure the electromagnetic fields in the vaccines?

Dr Jean Monro

No, we did not measure the electro-magnetic fields in the vaccines themselves. I was explaining that a proportion of patients, whom we saw, experienced symptoms when they held phials containing neutralising vaccines. In my account of this there is a chronological story of my means of discerning these problems. These patients did not react to some contaminant on the phial, they did not react to the coldness, they did not react to any of the ordinary things one might suspect in a medication. But they did react with symptoms when they held phials.

They were therefore reacting to the contents of the phials. The symptoms occurred through some interactive effect to whatever was inside that phial and which was emitted through the glass of the phial. What I did was to send the phial to the National Physical Laboratory for an opinion on what could get through the phial. They told me the range of frequencies which could penetrate the glass. There is an interactive effect between whatever is in the phial in that diluted antigen and the patient. A lot of scientific evaluations have been done since then as to what could be in water which is acting as a solvent for different things. A considerable amount of work from Dr. Emilio Del Giudice in Milan has illustrated this. He has reported that there are coherent groups of molecules which can have an effect. There was also one paper quite recently in which it was noted that the more dilute the coherent groups are, the larger the effect form them. There is a physical effect of chemicals, or whatever one is diluting, on the water molecules and their behaviour. I believe this is what we need to look at from the physics point of view.

Dr Gerd Oberfeld (Government of Salzburg, Austria)

I am an epidemiologist. What do we do with many exposures from different sources? When you do a study, you should take all possible exposures into account. Therefore you have to measure the places where people sleep, live and work. If you put the readings from low frequency electromagnetic fields and radiofrequency waves, in one model, you can sort out the different risk factors. It is quite usual to do this by a multivariate statistical analysis.

Dr Jane Cuthbert (Private Individual, UK)

I am a research scientist and I believe there are symptoms, which are caused by equipment generating electromagnetic fields. As scientists perhaps we are not yet sufficiently sophisticated to be able to determine what mechanisms are at work. Perhaps, we are not yet at that point where we can work out good working models. The fact remains that these symptoms exist and they are linked specifically to electromagnetic field emitting equipment.

Alan Meyer (Bircham Dyson Bell, London)

I would like to know when NRPB will take Dr Cuthbert's anecdotal story seriously.

Dr Zenon Sienkiewicz (National Radiological Protection Board, Oxon)

Are you an anecdotal story?

Dr Jane Cuthbert

I am not. We are all human beings. We have to be more serious about research issues. Anecdotal evidence is dismissed by scientists but it is only like listening to people reporting to their doctor when they attend a surgery. Anecdotal or case study evidence is important.

Dr Zenon Sienkiewicz

If Mr Meyer is asking if NRPB will focus on the issue of establishing hypersensitivity as a recognised disease, then I do not believe that we are the right organisation to do this. I think these questions should be directed towards the various elements of the National Health Service with more clinical expertise in identifying diseases and formulating treatments. I do not believe that such a role is within the remit of NRPB.

Dr Jane Cuthbert

The government has given the NRPB public money over the years to find out whether there are effects from electromagnetic fields on human beings. I believe it is the responsibility of NRPB to carry out such research.

Dr Zenon Sienkiewicz

I think recognising and classifying diseases are dealt within the NHS, not by NRPB.

Dr Jane Cuthbert

I repeat, anecdotal evidence should be studied not dismissed.

Alan Meyer

Sir William Stewart in May 2000 made it very clear that the well-being of people is part of the problem but NRPB will not listen to anecdotal evidence. It is a gap in what they are doing. It does not seem that much has changed. I would have thought that this problem was something, which you could take back to "experts" at the NRPB. NRPB is meant to be there, investigating these things in order to help people like Dr. Cuthbert.

Elizabeth Kelly (Council on Wireless Technology Impacts, California)

If you do not ask the questions you are not going get the answers regarding symptoms. Many people are suffering which are termed "electrical sensitivity" or "microwave sickness." These questions were asked recently in two California studies. The California State Heath Department EMF Programme issued a draft report in 2000 as part of a major EMF report that included a survey on electrosensitivity symptoms. They reported that 3.5% of those surveyed had these symptoms for which they were seeking medical care. Based on the sample size, 3.5% translates into 770,000 Californians! The EMF Programme states that based on these findings, electrosensitivity needs to be addressed as a public health issue. The Marin County Health Department in California, replicated the questions used by the EMF Program in a health survey it did last year and found that 7% of respondents surveyed reported they had electrosensitivity symptoms and were seeking medical care. Seven percent of survey respondents translate in 20,000 Marin County residents. These surveys yield a subjective response but indicate the large numbers of persons who have these symptoms. We need to keep asking questions about electrosensitivity.

Dr Danuta Sotnik (Applied Immunology, Poznan, Poland)

Would it be helpful to check retrospectively whether, and if so, how many psychiatric patients have developed an electrical sensitivity as a result of electro-convulsive therapy?

I wonder whether the clinical picture, presented by Dr Monro, can be attributed to an anaphylactic reaction, e.g. it is known that tiny amounts of nuts can trigger anaphylaxis in sensitive individuals, leading in some cases even to death. Besides, bearing in mind the extremely low magnetic fields of our organs (e.g. heart & brain 10^{-9} & 10^{-12} Tesla) other factors should also be taken into consideration, including the heavy metals in modern diet and the air quality we breathe.

Dr Jean Monro

Dr Sotnik asked whether I am concerned about anaphylactic reactions. I am the Medical Director of Breakspear Hospital in the UK and we also have a sister hospital in Germany where patients are tested using skin-testing techniques for various factors. They can present with anaphylactic reactions. We have had patients with peanut sensitivity or nut sensitivity coming into the Hospital who had collapsed with anaphylaxis on previous occasions. I would prefer to treat them with a neutralising vaccine rather than with adrenalin or the other things available to treat anaphylaxis because it reverses the reaction much more quickly. One can ameliorate an anaphylactic reaction almost instantly with a correct neutralising vaccine. So in my view it is a further proof that this is a different mechanism from what is acknowledged which is an antibody reaction. It is anaphylaxis and total body responses are often not simply an antibody antigen reaction.

Eva Marsalek

The WHO definition of health is not only the absence of disease but also well-being. Even if EMFs are not the reason there is an increasing number of people feeling a loss of well- being.

Alistair Philips (UK Powerwatch, Sutton)

My one concern with your study is the metal magnetic screening of the containers. There are quite a few EMF effects. Possible mechanisms include the interaction of the static background magnetic field and AC magnetic field like Larmor and Cyclotron resonance effects. How do you allow for these?

Professor Barry Michael (Gray Cancer Institute, Middlesex)

We are trying to achieve conditions where the steady field from the Earth is virtually eliminated inside the double skinned screened mu-metal container and the only field that we are exposing our samples to is the 50 Hertz applied field.

Alistair Philips

There is a lot of evidence in the scientific literature that the combination of the levels of the AC and DC fields affects the biological outcome.

Dr Gerard Hyland (Formerly University of Warwick, UK)

This point is extremely important in connection with replication experiments. There are a lot of so-called replications where some essential aspect of the experimental protocol is different from that of the original experiment – a difference that undermines the fidelity of the intended replication. Did you use exactly the same protocol as obtained in the original work? If not, then you have performed a different experiment!

Professor Barry Michael

Yes. We have consulted with the original researchers (Walleczek *et al.*) and repeated our experiments a number of times. We are using cells from exactly the same source. The protocol we have used so far reproduces the one originally used.

Perhaps these are two things to remember; Despite having followed the original protocols as closely as possible, we actually find the mutation frequencies are substantially lower than those reported by Walleczek. The second point is that as a consequence of the low mutation frequencies, the counting statistics are poor due to the low numbers of mutants. We are therefore planning to scale everything up. We will increase the numbers of exposed cells substantially, growing them to somewhat higher densities, and we will also increase the dose of radiation. In the end, the objective is to obtain results, which are statistically significant. We consulted with the agency who are funding the study (EMF Biological Trust) and their view is that if the effect reported by Walleczek is reproducible, then if we scale up your experiment to achieve statistical significance we should still be able to observe the same effect.

Professor Olle Johansson (Karolinska Institutet, Stockholm)

It is also important to remember that it is extremely difficult to get funding for any kind of replication of previous data. Also I think most scientists find when they are trying to publish replications that many scientific journals also turn them down saying that the work has already been published. The problem of replication should be addressed by the scientific community. Should there be a special journal for replication studies?

Professor Dennis Henshaw (University of Bristol)

I know you tried to replicate Walleczek's experiment, but in vivo you would have melatonin present at around 10^{-9} to 10^{-11} molar. As I have said before, there are many experiments showing (i) the suppression of melatonin production in the human pineal gland at magnetic fields as low as 0.2 µT and (ii) a reduction in the oncostatic action of melatonin in vitro in the presence of 1.2 µT magnetic fields at physiologically relevant melatonin concentrations (10^{-9} to 10^{-11} molar). I understand that magnetic fields actually slow down the action of melatonin on the DNA itself. I wonder whether you see any scope in introducing melatonin into your experiment. Of course this would then be a different experiment to Walleczek's.

Professor Barry Michael

I think that based on what you described and put forward about melatonin there could well be an effect. However, our experiments, which use one cell type fairly sparsely seeded, are far removed from the in vivo situation where intercellular signalling besides the involvement of tissue architecture and the presence of different cell types in close contact are likely to play an important role, as we heard yesterday in Prof. Adey's introductory talk about the influence of cell-to-cell communication, (See Part I discussion).

Dr Enda D'Alton (Irish Doctors Environmental Association, Dublin)

Has anyone investigated or assessed the effects a cocktail of frequencies on mammalian cells. I mean a mixture of frequencies right across the EMF spectrum. If not, has anyone thought of doing it?

Professor Barry Michael

Could you be a little bit more specific on what you mean by cocktail?

Dr Enda D'Alton (Irish Doctors Environmental Association, Dublin)

Most work refers to one particular frequency but what of the other frequencies right across the spectrum?

Professor Ross Adey (Loma Linda University, California)

You mean a true life situation like we are sitting here experiencing all sorts of frequencies.

Professor Barry Michael

I am sure there are other people who can probably answer that. You are quite right and our immediate priority is to concentrate on the effect of a single frequency.

Professor Ross Adey

There may be interaction between the emitters. Does anybody look at that? There may be interaction for example between microwaves and mobile telephones or power lines that emit 50 Hz.

Professor Barry Michael

I think we have got to concentrate first of all on obtaining meaningful results with 50 Hertz. Once we have achieved this, we can see if there are any such interactive effects, for example, we can superimpose a DC field on the 50 Hertz field.

Dr Enda D'Alton

My second question is concerning transmission lines which carry 50 Hertz or 70 Hertz currents. Is it not true to say that they also carry microwave control signals?

In Ireland I have measured considerable levels of microwave radiation from these lines several miles from the actual lines away from the powerlines which generate the fields.

Professor Olle Johansson

In Sweden there are full scale experiments in this field for instance on the island of Gotland in the Baltic See, where they do not only have control microwave signals but they send whole broadband spectra of Internet communications. There are already conferences in Sweden on the possible health effects from such systems.

Dr Gerard Hyland

Professor Johansson mentioned that Sweden is trying to introduce a human bio-compatibility standard in Sweden. What is this standard, and on what is it based?

Professor Olle Johansson

So far, there have just been discussions in different newspaper. Hopefully standards will continually be reviewed. As we heard from Dr. Vladimir Binhi the standards we have nowadays are probably not sufficient.[2]

[2] **Editor's note:** Russian EMF Standards have much lower thresholds than European ones.

Eva Marsalek

Dr Binhi says that the standards are not low enough, and I agree with him, but the problem is that the biocompatibility products on the market cannot solve the problem. The democratic rights of people are being violated by the erection of antennae where they are not wanted. This situation could also have health-implications due to the toxico principle where no biocompatibility-product-solution can help.

Dr Vladimir Binhi (Russian Academy of Sciences, Moscow)

I think we have really good devices now to protect human beings against electromagnetic pollution. I would like to say that I know of at least two devices, which can be effective. One device is made by Professor Litovitz and another device is made by Dr Fillion-Robin. Both of these devices use electromagnetic radiation which interferes with the electromagnetic fields that cause the biological effects. The principle of the device of Theodor Litovitz is in use of torsion magnetic fields, and according to my theory, it really reduces the biological effects. For the other device, we have no proven explanation as yet, but there is a lot of evidence confirming its effectiveness.

Ana de Oliveira Rodrigues (PhD student at Imperial College of Science, Technology and Medicine)

Would Professor Johansson please comment on the phrase: "Our society can not survive without electricity". You said that "..it can, just look at the cows and the flowers...". Nowadays to have the flowers and cows we use lots of electricity in fact. I am not denying the possible effects of electromagnetic fields. Criticising electricity will not help; it will only scare the general public and increase the selling of "miraculous" products. These reports will say to the general public: "We are in real danger". We must realise that advances in medicine based on magnetic fields such as x-rays and magnetic resonance scanning have increased the quality of life for many people. It is necessary to carry out more research to provide better standards for electromagnetic devices.

Professor Olle Johansson

The increase in life span has very little to do with electricity. Hygiene and diet has increased that. The lifetime of Swedes has increased but the consumption of drugs has also. I would argue that all species could survive without man-made devices including electricity. That is a philosophical point of view. If we talk about the best life for the Earth than it would be better if humans disappeared!

Dr Jane Cuthbert

I was surprised to hear that Dr Preece was looking for mechanisms in animals rather than studying people who are particularly sensitive to using phones. I am a research scientist but at the moment I cannot use a single type of phone. Studying with sensitive people are more likely to show multitudinal effects because if you look at the range of reactions it is large. Is there anyone here who can suggest solutions for people like myself who can no longer use a phone of any type whether landline or mobile?

Dr Alan Preece (University of Bristol, UK)

The ethics of carrying out research on human beings mean it is very difficult to investigate a mechanism without being invasive. A provocation study is possible but even that is quite difficult. The Stewart Committee for example said that children should be discouraged from using phones; therefore, it becomes unethical to expose a child, who has never used a mobile phone, to mobile phone radiation because you may be possibly doing some harm. Experiments on children therefore defy the rules of ethics. The other thing is I do not know any way to use a human to find out what is actually happening at a cellular level. We need to carry out much research on animals and simple systems otherwise the whole thing will be confused. Experiments on electrosensitive people have been done in a number of cases (there has been quite a good study in Finland and that was absolutely negative). Electrosensitive people are not able to detect the presence of the mobile phone radiation. I keep an open mind and if you could suggest any experiment then I would certainly be interested to design it.

Dr Jane Cuthbert

Dr Gro Harlem Brundtland, medical doctor, current director of the World health Organisation and former Prime minister of Norway, made it public in March 2002 that she is electrosensitive and that she reacts adversely to mobile phones, cordless phones and computers. Electrical Hypersensitivity (EHS) has been formally recognised in Sweden for over fifteen years and it is thought that there are 3% of the population who are highly electrosensitive. People effected with EHS in Sweden receive medical, technological and benefits' assistance.

Unlike Sweden where medical and research establishments have entered into dialogue with EHS people and sought to help them by carrying out research in co-operation with them, the UK has so far refused to open this dialogue, concentrating instead on animal research as a consequence. EHS remains officially unrecognised, leaving EHS affected people in a total medical, technological and benefits vacuum. General Practitioners generally dismiss the condition as imaginary or assume it is psychiatric. There is not a single NHS consultant dealing with EHS. Incapacity benefit is generally refused to sufferers. The recording of symptoms and causes as reported by EHS sufferers is not when being carried out by general practitioners.

Symptoms of EHS include pain in various parts of the brain, the sensation of brain cells completely irritated, the sensation of burning skin and burning brain tissue, brain numbness, depressed mental functioning, depression, nausea... In order to try and escape the electromagnetic causes of such symptoms, EHS people may become prisoners on their own homes (if they have a detached house), or struggle to survive in shacks, tents, caravans; some commit suicide. Because electromagnetic radiation from the equipment of the neighbours (fridges, freezers, washing machines, fish tank pumps...) can go through floors, ceilings, walls... Cheaper housing options such as flats, studios... are not an option. The earning powers of EHS people can be very poor or nil (They may be unable to use telephones, computers, fax machines). They may be unable to tolerate fluorescent lights, car, bus and train journeys. Social isolation sets in with the inability to use a phone, to enter friends' houses, to leave "home" in some cases. Some EHS people have to leave their families. All the suffering made worse by the lack of recognition and therefore lack of understanding, sympathy and assistance. Some EHS people do not report their symptoms, as they fear misdiagnosis as psychiatric cases and the possibility of sectioning; the UK government has not fully recognised the condition. Some EHS people have been wrongfully sectioned because of ignorance on electrosensitivity among medical staff. Hospitals are particularly damaging and stressful places for EHS people (given the fluorescent lights, many sensors...). Many EHS people could avoid a worsening of their condition if they received help and there was more education on the issue.

There is an urgent need for government to enter into dialogue with EHS people and specialists who have been studying the condition: with a view to recording symptoms and causes, recognition, assisting affected people, initiating research, trying to prevent increasing numbers of people being effected. Government has a duty to do this.

John Steed (Department of Trade and Industry, UK)

The UK Government is taking Professor Henshaw's theories seriously – a sub-group of the Advisory Group on Non-Ionising Radiation (which reports to the main Board of the National Radiological Protection Board) are independently assessing both the production of corona ions by power lines and the likely retention of aerosols (together with any pollutants) by the human body.

Two questions for Professor Henshaw:

Firstly is Professor Henshaw aware of any similar work on corona ions from powerlines taking place anywhere else in the world? And secondly, Professor Henshaw's last slide referred to most powerlines being in constant state of low-level corona. There is an inevitable association with high voltage engineering. From all his experimental work, would Professor Henshaw be able to quantify typical values for "low level corona" and then postulate what he might regard as a level for corona ion emission above which action should be considered?

Professor Denis Henshaw (University of Bristol, UK)

The phenomenon of corona ion emission from high voltage powerlines has been known since the 1950s and has been discussed in a number of power industry publications. To our knowledge we are the first to have suggested that corona ion emission results in increased risk of illnesses associated with air pollution, due to increased lung deposition of inhaled pollutants with added electric charge. We have had discussions with a number of groups regarding collaborative research and in conjunction with Dr Giles Harrison in the Department of Meteorology at the University of Reading. We are planning to organise a special one-day conference in Spring 2003 on the subject of electrified aerosols and their health implications.

Corona ion emission from powerlines is typically of a magnitude to cancel out the Earth's natural DC field of 100 V m^{-1}. Direct ion measurements based on a small number of sites suggest that this corresponds to unipolar ion densities in the range 400 - 500 per cm^3. However, our main concern is with a number of powerlines emitting high levels of corona, such as the 400 kV line at Lower Godney, Somerset. Here the average unipolar ion density is estimated at ~6000 per cm^3, which is sometimes effective up to more than one kilometre from the line. Our current estimate of the number of excess cases of ill health attributed to corona ion emission in the UK is based on a 30% increase in lung deposition of inhaled aerosols as a result of a 17% increase in aerosol charging by these corona ions. Details of our risk assessment have recently been published. (See D L Henshaw, *"Does our electricity distribution system pose a serious risk to public health?"*, Medical Hypotheses 2002, 59: 39-51)

Part V
Awareness

CHAPTER 23

Electromagnetism and the Insurance Industry

Alastair Speare-Cole

A review of how the industry views potential problems with electromagnetism, the precautions it is taking and what coverage it may offer.

In 1996 Swiss Re, a company that ranks amongst the largest risk carriers in the world and a company that regularly publishes discussion papers on academic aspects of insurance, wrote in its discussion document.

"Electrosmog – a phantom risk?"

"a prospective hazard, the magnitude of which cannot be gauged and which perhaps does not even exist but which is nonetheless real – only in that it causes anxiety and provokes legal actions".

The insurance industry today gazes upon a scene which is as uncertain as it was in 1996.

Leaving aside the diminution of property values near EMF sources in the USA, there has been no real red letter day, created by legal case or generally accepted scientific research that has removed the state of the art defence and made its producers liable for damages caused by EMF.

Since no one can predict the likely cost, if any, to the industry no one can "price for it". However the insurance industry is a competitive market place and EMF is just another factor in the daily brawl of short term v long term goals, individual underwriter's career aspirations, the interface of insurers and reinsurers and the ghosts of other problems such as pollution, asbestos, toxic mould and terrorism

Asbestos stands as a convenient model for what could happen with EMF.

This is the story in brief. In the 1950's there was some concern in academic circles about the relationship between blue asbestos and cancer. Asbestos usage had been growing steadily from the early part of the century and suggestions existed that mesothelioma, a range of cancers, scarring of lung tissue (asbestosis) and other pleural damage might result.

The history of the subsequent asbestos litigation on both sides of the Atlantic is complex and fascinating. I have only time to comment here that much of the cost to the manufacturers and their insurers is as much related to the costs of the legal process as it is to the true cost of compensating the victims. Several manufacturers lie bankrupt and many insurers would have also failed had they not been able to spread the losses over many years. In short, asbestos has made a lot of US lawyers very rich.

In the early 1970s the first case in the US (The Borel case) found an asbestos manufacturer liable for the bodily injury caused by their product. Usage of asbestos was rapidly curtailed. But the period between exposure and manifestation ranged from 15 to 40 years and so, whilst the cause of the problem was understood and dealt with, the cost to insurers has not even begun. Thirty years after that first case the industry has not yet reached the peak point of paying compensation.

As an example, the Manville Trust (formed after the bankruptcy of one manufacturer) received 58,600 new claims in 2000, 81% more than the 32,000 it received in the previous year.

The jargon for this vast temporal displacement between cause, manifestation and financial compensation is called "latency". It is something anyone managing anything but the youngest of insurers dreads. Imagine, as a senior manager, you have invested a career in an insurance company and suddenly a problem begins to emerge emanating from a time thirty years before when you were at school! Not only that but the problem is of a size which threatens the existence of the company, the existence of the insured, and of human rather than perhaps commercial concern, the compensation of the victims. Without a solvent insurer and insured there is little or no compensation for the victim.

The cost of asbestos to the insurance industry? No one knows exactly but A.M. Bests the rating agency calculated the US insurance industry alone would end up paying US$65 billion.

The worst case scenario for EMF would be far in excess of this figure and is life threatening to many companies and their insurers. In fact it would be so profound as to require a radical rethink of the use of energy in Western society. The scenario more often examined by insurers is the possibility that specific types of EMF may prove to be harmful. Currently these are the ones which receive the most press speculation such as mobile telephone use or power transmission.

So what is the industry doing to protect itself? I could say flippantly that, in many situations, very little.

If EMF takes as long from initial exposure to payment of compensation as Asbestos, then claims against policies written today will occur long after the current underwriters have moved on to managerial roles or have retired. The temptation for insurers is first and foremost to follow their competitors in the market and if this means taking an optimistic approach to EMF which is, after all, still perceived as a "phantom risk", so be it. For saving the company from a theoretical but highly unlikely oblivion in thirty years time contemporary shareholders are unlikely to be thankful. For missing the next twelve months income and profit projections there are immediate and tangible consequences. If governments have difficulty tackling the problems of an ageing population and inadequate pension funding when it is a known problem it is possible to sympathise with insurers when the problem of EMF may not even materialise.

But the insurance industry will tackle and is already tackling EMF in the following ways.

Firstly the industry can exclude claims arising from EMF. The diminution of property values is already a common exclusion. This approach is already to be seen on some original policy forms. One problem the industry has is that by putting on an exclusion on today's policies an insurer lends credence to the assertion that cover was given on policies that have been issued for decade after decade. There are ways to avoid the damned if you do and damned if you don't situation, but it is difficult problem to solve.

A second approach is to try to take a selective approach to risk as far as EMF is concerned. Mobile telephone manufacturers and power transmitters may attract a more precautionary approach when compared to small retailers or householders. Since the science is hot available for a qualitative approach let alone a quantitative approach to EMF risk there is no other practical solution. Especially since a competitive insurance market will not allow a blanket across all classes solution.

Another approach which is already common is to change the nature of the annual insurance contract to only pay claims that are made against the insurer during the actual year in question. The jargon is "a claims made policy". Those medical practitioners will be familiar with this approach. It matters not when the error was made. It is when the claim is made that counts. If you cease to buy the insurance and then a claim from many years before arises you have no cover. It can be contrasted to the "losses occurring policies" typically issued to almost all commercial insurers to this day and is exactly the policy form issued to asbestos manufacturers.

In summary, claims made policies deal with the problem of latency for the insurers but leave the insured with the possibility of no ongoing insurance when a problem begins to emerge.

The losses occurring policy is great for the insured (provided they buy enough sum insured to cater for a few decades of prospective inflation) but leaves the insurer carrying the burden of latency.

The claims made approach is already common on power generators and power transmission companies and increasingly any heavy users of electricity. Similarly if cover is given to the cell phone industry it is also likely to be on this basis.

But these last comments are all concerned with prospective coverage. The real fear of the industry is that, as the film makers used to say, the problem is already "in the can".

For instance the UK's electricity usage has risen from 50 GWh in 1950 to 300,000 GWh in the early 1990's. During this time the insurance industry here, in Europe and in the US has been issuing occurrence based liability policies (the one where the insurer carries the latency) and it has been doing so without any reference to EMF.

For the industry the concern is that a health risk or more particularly a liability risk emerges relating to long term exposure to EMF. (Remember that a lot of the cost of asbestos to the industry was legal costs rather than compensation to the victims.)

The defences that the insurance industry has probably fall into the following simplistic statements.

- There is no causal link between the victim's injury and any EMF received through the actions of the insured.

- Whilst there is a causal link between the victim's injury and EMF there is no proof that it was this particular insured's actions that are to blame
- Whilst the insured's EMF may have caused the victim's injuries it was completely unforeseeable and therefore the insured cannot be held responsible at law.
- Whether the insured was liable or not the policy gives no coverage for this peril.
- The exposure was spread out over many years and the insurer only insured the insured for one of those years.

Some of these defences come through science, others through the law. Was the insured liable under the law of that country for the consequences of the product? The answer will be different in different countries. In the UK the bulk of any exposure to the industry will have to come through the law of tort. There is no time to get into details here but phrases such as negligence and forseeablity will be at the core of any legal fight. I referred to "red letter days" earlier and one can imagine a plaintiff's barrister saying something like " It was common knowledge from the Spring of that year that your product was then causing harm to your clients and you did nothing either to curtail the sale of that product or to try to prevent the use of that which had been already sold."

The defence implicit in this is that until Spring of that year some sort of defence exists although the strength of that defence may be very dependent upon the nature of the product and the circumstances surrounding its sale.

In summary there is a long road ahead before Swiss Re's phantom risk of Electrosmog turns into a serious problem for the insurance industry if problem there is at all. Insurers monitor the science carefully. Insurers have fought successfully so far the attempts to convince courtrooms of a link between specific victim's injuries and their exposure to EMF. If there is going to be a problem it remains to be seen if it is going to be on a broad front or a specific area. Different national jurisdictions, and the interplay of the various types of defence outlined above make speculation easy but prediction impossible. The industry, having been badly mauled by asbestos and pollution and with the prospects of tobacco in the wings, wants to take some basic precautions for the future tempered always by short term competitive pressure. What may already been have incurred but the industry does not know about is to some extent the subject of this book.

CHAPTER 24

PTSD or PTSD^2
Assessing and Responding to Pre and Post Telecoms Stress Disorder

Ray Kemp

24.1 SYNOPSIS

Every day in the UK, newspaper reports around the country raise "human interest" stories about local communities opposed to the siting of radio base stations ("mobile phone masts") in their areas. Objections arise for many reasons, but on the whole, these are not mere "Not In My Back Yard" objections.

In some communities, people hold deeply-held fears and concerns that mobile phone masts may give off high levels of harmful radiation - Electro Magnetic Fields (EMF). This issue was partly addressed in the Report of the Independent Expert Group on Mobile Phones (the "Stewart Report" after its chairman, Sir William Stewart) which was published in May 2000. However, government responses, revised public exposure guidelines, research by the National Radiological Protection Board (NRPB), and new planning procedures all fail properly to address the central issue. Namely, that certain individuals and communities undergo an experience that is not dissimilar to Post Traumatic Stress Disorder (PTSD) in many respects. This Disorder manifests itself once telecoms mast development proposals are discovered - "Pre-Telecoms Stress Disorder" - and may continue once the mast has been switched on - "Post Telecoms Stress Disorder" hence PTSD^2.

Fortunately, recent initiatives by the UK mobile phone operators have led to altered attitudes and approaches to network development by the mobile telecoms industry itself. This has gone a long way towards reducing stress in the community, but there is much still to achieve and further research is needed into the longer-term effects of post telecoms stress on communities.

This presentation describes the underlying issues, outlines recent changes in mobile phone network planning and consultation procedures that have taken place as a response to those issues, and suggests some key areas for future action.

24.2 REFERENCES

DETR 1998, *Code of Best Practice. Telecommunications prior approval procedures as applied to mast/tower development.* (London and Cardiff:

Department of the Environment, Transport and the Regions and The National Assembly for Wales).

Dolan M, Nuttall K, Flanagan P, and Melik G 1999 The application of prudent avoidance in EMF risk management. In Repacholi M H and Mue AM (editors) *1999 EMF Risk Perception and Communication*, (Geneva: WHO).

Federation of Electronics Industry, 2001, *Guide to Using Traffic Light Rating Model for Public Consultation.* (London: Federation of Electronics Industry).

Independent Expert Group on Mobile Phones. 2000, *Mobile Phones and Health.* (The Stewart Report). (Didcot: National Radiological Protection Board).

Kemp R 1997, Modern Strategies of Risk Communication : Reflections on Recent Experience. In Matthes R, Bernhardt, J H, and Repacholi M H, (editors) 1998 *Risk Perception, Risk Communication and its Application to EMF Exposure*, (Geneva: WHO).

Mann, S M, Cooper, T G , Allen, S G , Blackwell R P , and Lowe A J 2000, *Exposure to Radio Waves near Mobile Phone Base Stations*, (Didcot: National Radiological Protection Board, NRPB-R321)

Science and Technology Committee (1999). Third Report. Scientific advisory system: mobile phones and health. Volume 1, Report and Proceedings of the Committee. (London, HMSO).

CHAPTER 25

Electromagnetic Environments and Health in Buildings

Nicole Hughes and Michael Dolan

25.1 RESPONDING TO COMMUNITY CONCERNS ABOUT MOBILE PHONE BASE STATIONS AND HEALTH

Today in the UK, there are 45 million users of mobile phones. It has been an incredible growth industry – just over two years ago there were 23 million users.

It is a surprising fact that under normal circumstances, a mobile phone base station or mast can only support an average of between 80-120 calls at any one time. Therefore more radio base stations are needed where there are more users, such as in cities or on major transport arteries. With 45 million users, significant network infrastructure is required. Without the infrastructure the services will not work and customers either lose their connection or are not able to get through in the first place.

It is a significant irony therefore, that the overwhelming growth in numbers and economic impact of this technological icon has been matched by the growth of community concern regarding mobile phone radio base station development and increased sensitivity over public health issues.

It is an issue which has brought increasing constituency concern for politicians working both locally and nationally. In fact public concerns led the UK Government to launch an independent inquiry three years ago under the chairmanship of former chief scientific adviser Sir William Stewart to review the science and make recommendations.

In taking evidence from members of the public, the perception that masts were a health risk and the belief that the planning system was not taking sufficient account of people's concerns was made plain to the expert group. So the Government decided to significantly strengthen the planning system in this respect by adopting a precautionary approach and harmonising community consultation procedures in line with those stipulated under full planning.

The UK mobile operators themselves also recognised the need for change and the need for an industry code of best siting practice. As a result, in 2001 the network operators, BT Cellnet, Hutchison 3G, One2One, Orange and Vodafone launched their Ten Commitments on best mast siting practice under the auspices of the Federation of the Electronics Industry (FEI).

The Ten Commitments have three main aims: to address concerns in the community about mobile phone masts, to balance this with the need for further mast development and to provide more support to a planning system not designed or intended to address possible health risks.

25.2 A PROBLEM OF PERCEPTION

There is a widely held perception of a health risk from base stations, even though the radio waves emitted from them are measured at hundreds or thousands of times below international exposure (ICNIRP) guidelines. This is also despite the Stewart Report's conclusion that:

"...the balance of evidence indicates that there is no general risk to the health of people living near base stations on the basis that exposures are expected to be small fractions of the guidelines." *(par 1.33)*

Public perception of risk has understandably led people to assume that the planning system should take these concerns into account in planning decisions. But as we all know the planning system is not designed to handle this kind of public debate, and full planning procedures for all types of masts – which many have called for – would not give the public any greater control over such development than already exists.

In England the Government's PPG8 planning guidelines therefore say:

"It remains central Government's responsibility to decide what measures are necessary to protect public health. In the Government's view, if a proposed mobile phone base station meets the ICNIRP guidelines for public exposure it should not be necessary for a local planning authority, in processing an application for planning permission or prior approval, to consider further the health aspects and concerns about them." (par 30)

However, as long as people have concerns about base stations, elected members will be under pressure to refuse applications for them, putting them at odds with internationally accepted science based exposure guidelines and standards, and the principles of the planning system, thus leading to unnecessary and costly appeals.

25.3 WHAT IS THE ANSWER?

Is there a way out of this cycle? The mobile phone operators believe the answer lies in more up-front and early contact with local people – undertaken by the operators themselves – in order to address directly people's continuing concerns and to respond where the planning system cannot.

In essence it means occupying the middle ground! This means that where there is some uncertainty about a possible health risk, you should: be open-minded about science and research; take on board and address public concerns; engage stakeholders in dialogue; pursue solution-driven policies; and, bear in mind the public benefits of mobile.

The operators' Ten Commitments were introduced to address these issues directly and were developed in consultation with a variety of local government and local community stakeholders to ensure they are workable on the ground. The key to this approach is that the public is involved at an earlier stage in the siting process than ever before.

The Ten Commitments are:

- develop, with other stakeholders, clear standards and procedures to deliver significantly improved consultation with local communities
- participate in obligatory pre-rollout and pre-application consultation with local planning authorities
- publish clear, transparent and accountable criteria and cross-industry agreement on site sharing, against which progress will be published regularly
- establish professional development workshops on technological developments within telecommunications for local authority officers and elected members
- deliver, with the Government, a database of information available to the public on radio base stations
- assess all radio base stations for international (ICNIRP) compliance for public exposure, and produce a programme for ICNIRP compliance for all radio base stations as recommended by the Independent Expert Group on Mobile Phones
- provide, as part of planning applications for radio base stations, a certification of compliance with ICNIRP public exposure guidelines
- provide specific staff resources to respond to complaints and enquiries about base stations, within ten working days
- begin financially supporting the Government's independent scientific research programme on mobile communications health issues
- develop standard supporting documentation for all planning submissions whether full planning or prior approval

There has been significant work carried out so far. By the end of October 2001, all the operators had shared their strategic rollout plans with UK local planning authorities, and offered follow up meetings. The operators now seek local planning authority views on potential site locations and the operators have committed to engaging in significant pre-application consultation. In addition, operators will issue declarations of compliance with the ICNIRP public exposure guidelines and provide information on site need, alternatives considered and other technical details with planning applications. In order to respond to community concerns the operators now have specific staff resources to deal with complaints and enquiries within ten working days.

The operators have also entered into an agreement to publish clear, transparent and accountable criteria and cross-industry agreement on site sharing, against which progress will be published. Transparency is one of the vital elements of the Ten Commitments and, as recommended by the Stewart Report, the operators have assisted Government in setting up the national site finder database of all mobile base stations. This can be found at the Radiocommunications Agency (RA) website, www.radio.gov.uk. Additionally, the operators have ensured that all sites were ICNIRP public exposure compliant by 31 December 2001.

As well as continual internal auditing through industry workshops and MORI research, the industry has actively encouraged independent random auditing of its sites by the RA. The RA has audited over 100 base stations located at schools and at all sites the emissions have been found to be small fractions of the ICNIRP public exposure guidelines. These results are consistent with earlier work carried out by the National Radiological Protection Board and the Stewart Report conclusions.

The industry also acknowledges that further scientific research into the health issue must be pursued and therefore has committed to fund half of a joint £7.4 million Government/industry research programme. The commencement of the programme was announced by its project management committee chairman Sir William Stewart on 25 January 2002.

Overall, the operators believe this new approach to base station siting, by providing for more and earlier community consultation in network development, will ensure local people are better informed and have a better opportunity to voice their concerns and have them considered. Local councils at all levels will certainly have an important role to play in this process, ensuring the right balance is struck between the UK's access to world class mobile services and environmental responsibility.

Reaction to the initiative so far from central and local government, planning officers, the Local Government Association and activist groups such as Mast Action UK and Powerwatch has been positive and the industry ensures it is accountable through regular meetings with key stakeholders including Ministers. However, there is a great deal yet to be achieved and the operators are continually reviewing their progress with a view to ensuring the 10 Commitments remain effective.

Questions & Discussion

Peter Sudworth (Private individual, UK)

I am representing more than 1000 people in my home town who have expressed their concern about the siting of mobile phone base stations near their homes. (If Ray Kemp needs to do a case study on Pre and Post-Telecoms Stress Disorder, he needs look no further!)

Over the past two years we had a series of planning applications for phone masts in our town and there has been great public concern. Last September, our Local Planning Authority hosted a seminar at which Mike Dolan of the FEI (Federation of the Electronics Industry) gave a talk very similar to the one that Nicole Hughes has given today.

In November last year, NTL wrote asking for our views on a proposal to extend their masts in the town and add Orange and Airwave TETRA to the three phone networks on that mast. We immediately replied asking for information about the individual and collective outputs of the various antennae existing and those proposed for the mast. We wrote to Orange asking for similar information. Orange referred us back to NTL, who, they said was handling pre-application consultation on their behalf. We heard no more from NTL and we were surprised when, in February 2002, the planning application was put in without any pre-application consultation having taken place. Since then, there has been a chain of correspondence, initiated by us, between ourselves, Orange and NTL in which we have reiterated our requests for information but they have not given it. We complained to FEI and they, despite the fact we had given them documentary evidence of Orange and NTL's failures, merely told us that Orange has assured them that they were committed to the "Ten Commitments".

It is our experience that, contrary to what Nicole Hughes has stated, the code operators are now giving less information with planning applications then they were two years ago and are even less willing to enter into any form of meaningful dialogue or debate with our community.

Nicole Hughes (Federation of Electronics Industry, London)

One of the key issues is the point, which I raised in my presentation. While we are saying this any particular action is happening, there are instances where it is not occurring, as we would like. Perhaps yours is one of these examples. I do not know the circumstances surrounding your case. We hope that through the auditing process that we are putting in place, we will be able to look at the procedures being adopted by the companies and come forward with some advice on how perhaps things could be improved. The Ten Commitments were introduced in September last year, but, it does take a period of time for processes to actually work right across the board. I would be happy to communicate with you and try to get some interaction between you and ORANGE.

Eva Marsalek (PMI, Austria)

The problem is the same everywhere. Just a question to Nicole Hughes (Federation of Electronics Industry.) Your Ten Commitments involve the planning authorities but not residents. Why?

Nicole Hughes

Certainly part of the Ten Commitments is the traffic light model which identifies what consultation needs to be carried out as part of that site. That may mean, that the community liaison officers within the companies will then go out and talk to those particular communities and find out who are the acquisition agents or the operators to identify who the people should be talking to. There are community liaison people here in the audience who go out and have drop in sessions with local communities.

Professor Ray Kemp (Galson Sciences Ltd, Rutland)

This issue is European wide. If you look at the actual planning procedures that are used across Europe, there is a great variation in the different levels of involvement that local people have. Operators in Switzerland, for example, can theoretically do what they want to do with very little involvement by local communities. There are always going to be instances where things fail and this generates mistrust. In the UK, the "Ten Commitments" is a voluntary code of practice. So how do you actually enforce and regulate it? In some parts of Europe operators are being forced down a particular route while others are not. Local communities do need to be involved and consulted on how the code will be implemented.

Dr Gerd Oberfeld (Government of Salzburg)

In 1998, twelve GSM masts sites had been erected in the city of Salzburg in an agreement between the local people and the network provider. The planning value for the sum total of all GSM base stations was 1 mW/m² which is equal to 0.6 V/m on the outside of the buildings in the vicinity. This was the beginning of the Salzburg Model. The community of Salzburg continued to negotiate with the network providers in planning new sites. Because the Salzburg Model began to spread to other countries the network providers decided to stop the process in October 2001. As a response the city of Salzburg decided to allow no more antennae.. This decision is based on landscape and aesthetic issues. I would like to do epidemiological studies on health effects from base stations and other EMF sources, but the federal government of Austria is not interested in giving any financial support.

Nicole Hughes

Do you mean emissions as in emissions from the radio base stations? The operators themselves are actually looking at this issue in terms of the emission levels coming from the radio base stations. It is an issue that is becoming a little bit more prevalent with pressure on operators to provide levels. It has however been felt that the people who are in the best position to carry out those emission studies are independent bodies such as the Radio Communications Agency (RCA) in the UK who have just completed an audit of base stations effects on schools. They have found the highest reading was $1/279^{th}$ of the ICNIRP public exposure guidelines. The RCA is continuing that audit for the next 12 months. As far as the transparency aspect goes, the industry believes that the RCA can provide because it is an independent body verifying the emissions that come from radio base stations. We are looking into the problem to see if there is any other ways that can help to address the issue.

Professor Olle Johansson (Karolinska Institute, Stockholm)

In Sweden there is a large insurance company called Scandia. Some years ago they actually ceased to cover damages from electromagnetic fields. They have also instructed other insurance companies in Sweden to follow their example. In Sweden last year was published a government report similar to the Stewart Report in the UK. Two out of the three experts of that report were at the same time employed by the industry. There was even a Swedish representative who was employed by Ericsson, which is one of the largest manufacturers in the world, and one employed and several by Tele-mobile the largest telecom operator in Sweden!

Eva Marsalek

I just would add that National Limit Value Setting Commission of Austria has employees from the mobile phone companies and they can vote. Institutions that decide the threshold values for EMF's should be free of commercial influences but they are not.

Part VI
The Future

CHAPTER 26

Exposure Guidelines for Electromagnetic Fields and Radiation: Past, Present and Future

Zenon Sienkiewicz

26.1 INTRODUCTION

The National Radiological Protection Board (NRPB) has responsibility in the UK for providing advice on restrictions on the exposure of people to electromagnetic fields and radiation. In 1993, NRPB published advice on restricting human exposure to static and time-varying electromagnetic fields and radiation (NRPB, 1993). These restrictions were based on a rigorous assessment of the possible effects on human health derived from biological information, from dosimetric data, and from epidemiological studies of people exposed at work and at home.

There has been increased activity and interest in electromagnetic fields in the intervening years since this advice was first published. NRPB is now engaged in a comprehensive re-assessment of the advice in order to provide improved guidance on limiting exposure to electromagnetic fields.

26.2 PERSPECTIVE

NRPB published its advice on restrictions on human exposure to electromagnetic fields in 1993 following a thorough and detailed assessment of the scientific data relating to the effects of electromagnetic fields (NRPB, 1993). In this advice, it was considered that the epidemiological data available at that time did not provide a basis for restricting human exposure, and therefore guidance was based on experimental data describing thresholds for well-established direct and indirect biological effects of acute exposure.

Specifically, these recommendations were designed to prevent acute, direct effects such as vertigo and nausea caused by exposure to static magnetic fields, the detrimental effects of induced electric currents on the functions of the brain and nervous system arising as consequence of exposure to low frequency fields, and to prevent adverse responses to increased heat load and elevated tissue temperature resulting from exposure to radiofrequency (RF) fields (NRPB, 1993). The possibility of other field-induced biological responses were noted, but none were considered sufficiently well established to be used as a basis for setting standards. Guidance was also given for the avoidance of the annoying effects of direct perception of surface electric charge, and indirect effects such as repeated micro-shocks, electric shock and RF burns.

These recommendations applied to both workers and members of the public. Patients undergoing clinical investigations or medical diagnosis were excluded.

The International Commission on Non-Ionizing Radiation Protection (ICNIRP) is an independent scientific organisation responsible for providing guidance and advice on the health hazards of non-ionizing radiation exposure. In 1998, ICNIRP published its advice on limiting exposure to electromagnetic fields (ICNIRP, 1998). The biological bases and scientific rationale for this were very similar to those of NRPB. The major difference, however, was the application of an additional reduction factor of five by ICNIRP in deriving the basic restrictions for members of the public from the restrictions for those of workers.

A thorough comparison between NRPB and ICNIRP guidelines has been addressed by NRPB (1999a). In summary, NRPB concluded that there was no convincing scientific support for the introduction and choice of these additional reduction factors. Further, it was believed that the existing UK advice for the general public already provided adequate protection, and ICNIRP had failed to demonstrate that additional health benefits had been gained from further reductions in exposure. Therefore the Board of NRPB saw no scientific justification for altering its previous advice on exposure guidelines for members of the public, but it was accepted that other factors may need to be taken into account by government in establishing generally accepted exposure guidelines for the public (NRPB, 1999b).

More recently, the Independent Expert Group on Mobile Phones (IEGMP, 2000) supported the analysis by NRPB of the scientific data (relating to possible biological and health effects from RF fields) but considered that the two tier approach to restrictions adopted by ICNIRP was preferable to that of NRPB. This was put forward as a precautionary approach to reflect some uncertainties in the knowledge about possible biological effects of exposure to RF fields at exposures below guidelines.

The proposal to adopt ICNIRP guidelines for public exposure to RF fields had also been recommended by the Select Committee on Science and Technology (1999) and the Scottish Parliament Transport and the Environment Committee (2000). Similar views have been expressed by various institutions within the UK in connection with planning issues concerning the development of telecommunications masts.

Furthermore, the ICNIRP guidelines have provided the basis for the EU Council Recommendation (1999) on limiting public exposure to electromagnetic fields (from 0 Hz to 300 GHz). In May 2000, the Board of NRPB accepted that the ICNIRP Guidelines for restricting exposures of the public should be adopted for mobile phone frequencies.

Notwithstanding these developments, well over a thousand studies on biology and epidemiology, and many others on dosimetry, physics and engineering, have been published in the peer-reviewed literature since the NRPB guidelines were first produced in 1993. Many reviews of this literature have been published. Summaries of the major findings of these reviews are presented here. Unfortunately, the experimental database still contains many largely phenomenological reports that use a wide range of exposure conditions and biological models, and attempts to replicate positive results remain uncommon. Many of the reported effects also tend to be small in magnitude, making their significance and implications for health unclear.

In an attempt to help clarify these uncertainties, the International EMF Project was established by WHO in 1996 and is concerned with all frequencies from 0 Hz to 300 GHz. Through this project, WHO is pooling resources and knowledge concerning effects of exposure to electromagnetic fields in order to work towards an international consensus and resolution on the health concerns.

The Project aims to identify gaps in knowledge, recommends research programmes that allow better health risk assessments, and conducts critical reviews of the scientific literature. Individual countries also have their own research agendas, and substantial programmes of work investigating biological and health effects of RF fields are underway in France, Finland and other countries. The UK government recently launched the LINK Mobile Telecommunications and Health Research programme to help resolve and clarify some of the particular uncertainties surrounding RF fields associated with mobile telephony.

Overall, these projects and programmes should provide valuable information to allow better risk estimates to be made and thus facilitate the provision of improved human exposure standards.

26.3 RECENT EXPERIMENTAL DATA

26.3.1 Low Frequency Electric and Magnetic Fields

Several expert groups and committees have reviewed in great detail the possible effects of low frequency electric and magnetic fields with emphasis on their carcinogenic potential. These include NRC (1997) NIEHS (1998) IARC (2001) and the NRPB Advisory Group on Non-ionising Radiation (AGNIR) (NRPB 1993, 1994a, 1994b, 2001a). Pooled analyses of the epidemiological studies investigating magnetic fields and childhood leukaemia have also been published (Ahlbom *et al.*, 2000; Greenland *et al.*, 2000).

Overall, prolonged exposure to high magnetic fields in the home has been associated with an increase in the risk for childhood leukaemia. The risk appears to double with exposure to fields above 0.4 µT. No plausible explanation for this increase in risk has been offered. It was felt that this result was unlikely to be due to chance, but may have been affected by some unintentional bias in the original selection of subjects. The evidence for other health effects, such as childhood cancers other than leukaemia, adult cancers, adverse effects on reproductive function, and neurobehavioural dysfunction are generally considered too inconsistent and contradictory to support possible associations with magnetic fields although the data are frequently very limited.

Biological studies with animals and isolated cells in culture do not provide any convincing evidence of an increase risk of cancer from magnetic fields of less than 100 μT, although the possibility exists that some modest changes in cell biochemistry may occur in higher fields. Consistent behavioural changes in animals exposed to electric fields in excess of about 5 kV m^{-1} are attributed to surface charge effects and subtle and reversible neurobehavioral changes may occur in fields above 100 μT or so. Other parameters have been assessed but there is little evidence for any consistent detrimental responses. Electric or magnetic fields do not appear to impair reproduction or development in animals.

26.3.2 Radiofrequency Fields

Several expert groups have also reviewed the effects of RF fields including those associated with mobile telecommunications. These include McKinlay *et al* (1996, updated in 1999), ICNIRP (1996), the Royal Society of Canada (1999), IEGMP (2000) and Direction Général de la Santé (2001). Elwood (1999) Moulder *et al.* (1999) and Krewski *et al.* (2001) have also reviewed these data, while AGNIR (NRPB 2001b) examined the possible health effects posed by terrestrial trunked radio (TETRA) signals.

Apart from the increased incidence of motor vehicle accidents when drivers use mobile phones, most likely caused by the distracting effects of holding a conversation (see IEGMP, 2000) there is general agreement by these expert groups that exposure to RF fields at values found in the home and at work does not cause disease in people. In particular, the inconsistency of the reported increases in risk for cancers makes it unlikely that substantial causative effects occur.

However this research has many limitations and it is not possible to give complete reassurance that there are no hazards. Older studies tended to suffer from poor exposure assessment or other methodological deficiencies. There are also relatively few studies and these have examined a wide variety of exposure conditions. The lack of data is particularly true for investigations into mobile phone frequencies, which are only now beginning to be appear in the literature. For example, fatigue, headaches and other subjective symptoms have often been reported during or shortly after mobile phone use, and although experimental studies do not support these possibilities, the available data are too limited to dismiss them completely.

It seems unlikely that exposure to RF fields at levels too low to induce heating can substantially increase the risk of cancer in animals, although some subtle effects on cell biochemistry cannot be disregarded. However, it is possible that low level RF fields may influence brain function. Specific effects have been reported in volunteers and animals including changes in cognitive function, measures of brain activity, and on sleep. The significance of these effects and any relevance to health, however, remain unclear. Other subtle biological effects have been reported with exposure to RF fields, although the overall pattern of response is too diffuse to clearly suggest any obvious risks to health.

26.4 DISCUSSION

NRPB issued guidance on human exposure to time-varying electromagnetic fields in 1989 (NRPB GS-11). This advice was revised and extended in 1993 following detailed analysis of the scientific data available at that time and covered all frequencies from 0 Hz to 300 GHz (NRPB, 1993).

Following recent scientific and other advances in electromagnetic fields, NRPB is now engaged in a comprehensive re-assessment of the science covering the areas of biology, epidemiology and dosimetry in order to provide improved guidance on limiting exposure to electromagnetic fields. This revision will also consider any need to invoke a precautionary approach and what that might be. If appropriate, new advice will be issued in due course after consultation.

It is not possible to prejudge the outcome of this assessment. However, from the foregoing summaries of the biological and epidemiological evidence, there would appear to be no obvious alternatives to the current biological bases for setting guidelines on human exposure to electromagnetic fields. In the view of the various expert groups which have examined these data, epidemiological studies have not indicated any consistent and appreciable increases in risk for any disease, nor have robust and reproducible biological effects suggestive of any adverse effects been identified using fields at levels commonly encountered in the environment.

The possibility still exists that intense and prolonged exposures to high levels of magnetic fields can increase the risk of leukaemia in children. Different groups have placed slightly differing interpretations on this finding, however it is the view of AGNIR, that in the absence of a clear carcinogenic effect in adults or of a plausible explanation from experiments on animals or isolated cells, this evidence is not sufficiently strong in itself to justify the conclusion that such field cause leukaemia in children (NRPB, 2001a). Unless further research indicates that this finding is due to chance or some currently unrecognised factor, this association between magnetic fields and childhood leukaemia will remain.

Therefore surface charge effects, induced current effects and responses to heat and elevated temperatures appear to remain the most appropriate bases for formulating restrictions for human exposures.

It must be emphasised, however, that not all issues and possibilities regarding the effects of electromagnetic fields have been resolved, and it is important to consider where information may be lacking, and where more research needs to be performed. This applies not only to identifying possible novel biological effects and exploring the circumstances under which they may occur, but also to investigating uncertainties in the biological bases adopted for guidance. This is particularly true in regard to variation in sensitivity between subgroups in the population. Particular issues include increased thermal sensitivity to RF fields and heightened sensitivity to low frequency currents induced within the nervous system. Other uncertainties exist, including the cumulative effects of long-term exposure, and the effects of static fields.

These, and other considerations in epidemiology and dosimetry, will form a significant part of the NRPB review process. Expert opinion, including that of

AGNIR, as well as scientists and others with interests in these matters, will be sought as part of the consultative process.

26.5 CONCLUSION

NRPB is conducting an extensive review of the standards it recommends for occupational and public exposure to electromagnetic fields and radiation. It is inappropriate to attempt to prejudge the outcome of this assessment. Nevertheless, simply from a biological perspective, well established, acute effects still appear to remain the most appropriate and valid scientific bases for restricting human exposure.

26.6 ACKNOWLEDGEMENTS

I am indebted to my colleagues at NRPB, and especially to Richard Saunders and Roger Cox, for providing invaluable discussion and advice.

26.7 REFERENCES

Ahlbom, A., Day, N., Feychting, M., Roman, E., Skinner, J., Dockerty, J., Linet, M., McBride, M., Michaelis, J., Olsen, J.H., Tynes, T. and Verkasalo, P.K.., 2000, A pooled analysis of magnetic fields and childhood leukaemia. *British Journal of Cancer*, **83**, pp. 692-698.

Council of the European Union, 1999, Council Recommendation of 12 July 1999 on the Limitation of Exposure of the General Public to Electromagnetic Fields (0 Hz to 300 GHz). *Official Journal of the European Community*, **L199**, pp 59 (1999/519/EC).

Direction Général de la Santé, 2001, Les telephones mobiles leurs stations de base et la Sante, Rapport au Directeur General de la Santé, Denis Zmirou, président du groupe d'experts, available from www.sante.gouv.fr.

Greenland, S., Sheppard, A.R,, Kaune, W.T., Poole, C. and Kelsh, M.A., 2000, A pooled analysis of magnetic fields, wire codes, and childhood leukaemia. *Epidemiology*, 149, pp. 624-634.

Elwood, J.M., 1999, A critical review of epidemiological studies of radiofrequency exposure and human cancer. *Environmental Health Perspectives*, **107** (Suppl. 1), pp. 155-168.

IARC, 2001, Non-ionizing Radiation, part I: Static and Extremely Low Frequency Electric and Magnetic Fields, (19–26 June 2001), IARC Monographs on the Evaluation of Carcinogenic Risks to Humans, Volume 80 (in preparation).

ICNIRP, 1996, Health issues related to the use of hand-held radiotelephones and base transmitters. *Health Physics*, **70**, pp. 587-593.

ICNIRP, 1998, Guidelines for limiting exposure to time-varying electric, magnetic and electromagnetic fields (up to 300 GHz). *Health Physics*, **74**, pp. 494–522.

IEGMP, 2000, *Mobile Phones and Health*, Independent Expert Group on Mobile Phones, Chairman Sir William Stewart. ISBN 0-85951-450-1.

Krewski, D., Byus, C.V., Glickman, B.W., Lotz, W.G., Mandeville, R., McBride, M.L., Prato, F.S. and Weaver, D.F., 2001, Recent advances in research on radiofrequency fields and health. *Journal of Toxicology and Environmental Health, Part B*, **4**, pp. 145-159.

McKinlay, A.F. Andersen, J.B., Bernhardt, J.H., Grandolfo, M., Hossman, K-A., van Leeuwen, F.E., Mild, K.H., Swerdlow, A.J., Verschaeve, L. and Veyret, B., 1996, *Possible Health Effects Related to the Use of Radiotelephones*, Proposal for a research programme by a European Commission Expert. (Brussels, European Commission). With update, 1999.

Moulder, J.E., Erdreich, L.S., Malyapa, R.S., Merritt, J., Pickard, W.F. and Vijayalaxmi, 1999, Cell phones and cancer: what is the evidence for a connection? *Radiation Research*, **151**, pp. 513-531.

NIEHS, 1999, Health *Effects from Exposure to Power-line Frequency Electric and Magnetic Fields*, NIH Publication No. 99-4493, available from www.niehs.nih.gov.

NRC, 1997, *Possible Health Effects of Exposure to Residential Electric and Magnetic Fields*, (Washington DC: National Academy Press).

NRPB, 1989, Guidance as to restrictions on exposures to time varying electromagnetic fields and the 1988 recommendations of the International Non-Ionizing Radiation Committee, NRPB-GS-11 (Chilton: NRPB).

NRPB, 1993, Board Statement on Restrictions on Human Exposure to Static and Time Varying Electromagnetic fields and Radiation. *Documents of the NRPB*, **4**(5), pp. 7-68.

NRPB, 1994a, Electromagnetic fields and the risk of cancer, Supplementary report of an Advisory Group on Non-ionising Radiation (12 April 1994). *Documents of the NRPB*, **5**(2), pp. 77–81.

NRPB, 1994b, Health effects related to the use of visual display units, Report of an Advisory Group on Non-ionising Radiation. *Documents of the NRPB*, **5**(2), pp. 1–75.

NRPB, 1999a, 1998 ICNIRP guidelines for limiting exposure to time-varying electric, magnetic and electromagnetic fields (up to 300 GHz): advice on aspects of implementation in the UK. *Documents of the NRPB*, **10**(2), pp. 5–59.

NRPB, 1999b, Board Statement: Advice on the 1998 ICNIRP guidelines for limiting exposure to time-varying electric, magnetic and electromagnetic fields (up to 300 GHz). *Documents of the NRPB*, **10**(2), pp. 1–3.

NRPB, 2001a, ELF electromagnetic fields and the risk of cancer, Report of an Advisory Group on Non-ionising Radiation. *Documents of the NRPB*, **12**(1), pp. 5–179.

NRPB, 2001b, Possible health effects from Terrestrial Trunked Radio (TETRA), Report of an Advisory Group on Non-ionising Radiation. *Documents of the NRPB*, **12**(2), available from www.nrpb.org.uk.

Royal Society of Canada, 1999, *A Review of the Potential Health Risks of Radiofrequency Fields from Wireless Telecommunication Devices*, An Expert

Panel Report prepared at the request of the Royal Society of Canada for Health
    Canada, RSC.RPR 99-1, (Ottawa, Ontario: Royal Society of Canada).
Science and Technology Committee, 1999, Third Report.   Scientific advisory
    system: mobile phones and health. Volume 1, Report and Proceedings of the
    Committee.
Scottish Parliament Transport and the Environment Committee, 2000, Third
    Report. Report on inquiry into the proposals to introduce new planning
    procedures.

CHAPTER 27

Research on Mobile Phones and Health

Sakari Lang

27.1 INTRODUCTION

It has been estimated that by the year 2010, the number of mobile phone users will be around 2.2 billion (Telecompetition Inc. 2001). The increasing use of mobile phones and the increasing amount of mobile phone users has lead to concerns that exposure to electromagnetic (EM) waves emitted by the phones may lead to adverse health effects. This concern is further magnified by the fact that the most common human exposure is to the head of the handset user. In addition, even the low-level (whole-body) exposure from mobile phone base stations emissions has raised public concern about a possible health risk.

It is not widely known that biological and health effects of RF emissions have been studied for about 50 years. More than a thousand studies, from biophysical theoretical analyses to human epidemiological studies, have been carried out and published in peer-review scientific literature. This is often overlooked in media debates and – unfortunately – even in scientific discussions. These studies have, however, enabled the establishment of current safety guidelines, such as established by the International Commission on Non-ionizing Radiation Protection (ICNIRP 1998) and IEEE (ANSI 1992). Large safety margins are incorporated in these guidelines to protect both the general public and workers. These guidelines are based on scientifically established health effects determined by reviewing the extensive database. In case of RF exposure, the only scientifically established interaction mechanism is thermal. No adverse health effects have been shown to occur below current safety levels.

Why investigate potential health effects of mobile phone exposure if the phones and their base stations already meet the current safety standards? The main reason is that a more complete scientific database leads to more definitive assessments by public health authorities, which in turn increases public confidence in safety standards and the use of mobile telephony products. In addition the number of people using phones is large and growing, thus even the possibility of a minor potential health risks must be clarified.

At present, there is a large number of biological and health studies completed or ongoing related to mobile telephony frequencies. This paper presents the current status of this research and discusses the applicability (portability) of the current bioelectromagnetics research findings to future wireless technologies such as UMTS.

27.2 CURRENT STATUS OF MOBILE TELEPHONY HEALTH RESEARCH

About 300 studies have been initiated using cell culture, animal, and human models to investigate whether exposure to analog, CDMA, or TDMA-modulated mobile telephony signals can cause adverse health effects. A Summary of these studies is shown in Table 27.1. At present, almost 200 studies have been completed and about 100 are ongoing. About 50% of the studies are directly or indirectly related

Table 27.1 Current mobile telephony related studies

Cancer relevant or related	Ongoing	Completed	Total
Epidemiological studies	20	9	29
Standard bioassays	8	6	14
Sensitized *in vivo* studies	6	17	23
Acute *in vivo* studies	5	20	25
In vitro studies	27	52	79
Total cancer studies	66	104	170
Non-cancer studies			
Epidemiological studies	1	6	7
Acute *in vivo* studies	10	35	45
In vitro studies	9	15	24
Human studies	20	47	67
Total non-cancer studies	40	103	143
Grand totals	106	207	313

All studies are listed on the WHO web site: http://www-nt.who.int/peh-emf/database.htm

Table 27.2 Examples of mobile telephony signals

Signal type	Frequency	Modulation
TETRA	380 – 470 MHz	17 Hz (TDMA)
MIRS	806-821 MHz	11.1 Hz (TDMA)
NADC	824-849 MHz	50 Hz (TDMA)
CDMA	824-849 MHz	Multichannel (800 Hz)
ANALOG	824-849 MHz	FM
GSM	890-915 MHz	217 Hz (TDMA)
PDC	929.2 MHz/1.5 GHz	50 Hz (TDMA)
IRIDIUM	1616-1626 MHz	11 Hz (TDMA)
CDMA	1765 MHz	Multichannel (800 Hz)
PCS	1805-1880 MHz	217 Hz (TDMA)
GSM	1800 MHz	217 Hz (TDMA)
GSM	1900 MHz	217 Hz (TDMA)

to RF exposure and cancer. Since the database is already relatively extensive we aim not to review all studies in detail but to focus on certain key issues. All the listed studies can be found on the WHO web site (http://www-nt.who.int/peh-emf/database.htm). These studies cover a wide range of frequencies and modulations used by current mobile telephony instrumentation as shown in the Table 27.2. All of these studies utilise non-thermal levels of exposure and are testing the hypothesis of whether a frequency dependent or modulation dependent non-thermal effect exists.

27.3 CANCER STUDIES

A summary of mobile telephony related studies on cancer are shown in Table 27.1. The studies presented here as well as other relevant studies will be used by the International Agency for Research on Cancer (IARC) in approximately 2005 to evaluate radiofrequency (RF) emissions as a potential human carcinogen. In such evaluations, epidemiological studies carry the most weight. Animal studies will play an important role when epidemiological studies are weak or not definitive. *In vitro* studies generally have a supporting or clarifying role in these evaluations. Three fairly recent epidemiological studies, published in qualified peer-review journals (Muscat *et al*, 2000; Inskip *et al*, 2001; Johansen *et al* 2001.), do not support the hypothesis that long-term exposure of humans to mobile phone emissions causes cancer of any type which has been suggested by Hardell *et al* (1999). A significant problem for all epidemiological studies, however, is exposure assessment (Valberg 1997; Moulder *et al* 1999). This difficulty often leads to inclusion of confounding factors and false positive results. More recently designed studies have attempted to address this problem. This includes a multi-center (about 16 institutions) case control study coordinated by IARC. This study will produce an extensive amount of data in a few years, and will play a significant role in the evaluation of RF as a carcinogen.

The recent negative epidemiological observations are supported by long-term animal studies (for example, see Imaida *et al* 1998; Chagnaud, *et al* 1999; Adey *et al* 2000; Zook and Simmens 2001; Heikkinen *et al* 2002) that do not either show association between RF exposure and cancer. The only published study included in the list above showing an increased cancer incidence has been carried out using transgenic lymphoma-sensitive PIM-u mice (Repacholi *et al* 1997). This study is currently being replicated in two independent laboratories.

An important aspect in cancer-related research has been whether RF emissions are capable of inducing DNA damage which has been shown to be associated with cancer induction e.g. by certain geotaxis chemicals and ionizing radiation. A large number of *in vitro* studies, including studies at mobile phone frequencies, have shown that RF emissions are not genotoxic (for review, see Verschaeve and Maes 1998; Brusick *et al* 1998, Moulder *et al* 1999). Widely debated exceptions are the experiments by Lai and Singh (1995, 1996) proposing that RF emissions able to induce DNA single and double strand breaks in rat brain (*in vivo*). The authors used the so-called Comet-assay in their experiments.

However, this observation has failed to be replicated in 4 independent studies (Malyapa *et al* 1997a, 1997b, 1998; Li Li *et al* 2001).

27.4 NON-CANCER STUDIES

A summary of mobile telephony related studies on non-cancer studies is shown in Table 27.3. Even this database is relatively extensive: about 100 studies have been completed and 38 are ongoing. A key issue in this area has been whether RF emissions from mobile phones are capable of affecting human brain function. Studies on brain electrical activity (EEG) have been inconsistent (Eulitz *et al* 1998; Freude *et al* 2000; Wagner *et al* 1998; Hietanen *et al* 2000). It has also been proposed that mobile phone exposure may affect human cognitive functions, such as reaction time measured by psychological tests (Preece *et al* 1999, Koivisto *et al* 2000; Jet *et al* 2001). A third key area under investigation has been whether mobile phone exposure is able to disturb sleep (Wagner *et al* 1998; Wagner *et al* 2000; Borbely *et al* 1999; Hubér *et al* 2000). These studies have not either produced consistent data. In summary, none of these proposed effects have been scientifically established; i.e., the observations have not been replicated successfully in several independent laboratories. A review of Hermann and Hossmann (1997) has concluded that there is no evidence that RF emissions at environmental levels cause adverse effects in human central nervous system.

Table 27.3 Non-cancer studies

	Completed	Ongoing
Epidemiology		
Eye pathology	1	1
Headache & fatigue	5	0
Human studies		
Sleep and EEG	22	7
Cognitive function & memory	7	3
Headache and fatigue	2	2
Hypersensitivity, blood pressure, heart rate	5	5
Other (including eye pathology)	4	1
Human and animal studies		
Hearing and auditory pathology	3	3
Hormonal changes	12	1
Nerve conduction, muscle contraction & Ca changes	3	0
Animal studies		
Behavior, brain chemistry & neuropathology	9	4
Blood brain barrier	5	1
Teratogenicity, reproduction & development	5	1
Other animal studies	5	2
Tissue culture studies	15	9
Total non-cancer studies	103	40

Moreover, evaluation of these results is difficult due the fact that there is no plausible, established biophysical interaction mechanism explaining the proposed effects. Animal studies have not either produced consistent data. For example, studies on possible effects of RF emissions on blood brain barrier are contradictory (Salford *et al* 1994; Tsurita *et al* 2000), and the same is true for studies on behavior (Wang and Lai 2000; Zienkiewicz *et al* 2000). Again, these observations lack a plausible biophysical interaction mechanism.

27.5 REPLICATION AND REPRODUCIBILITY ARE NECESSARY FOR SCIENTIFIC PROGRESS

The above studies have failed to produce any consistent non-thermal findings. Studies that have reported findings have either failed to be replicated or the replication attempts have not been completed to date. Furthermore, these reported findings are by in large not supportive of one another and thus do not provide a basis for proposing low-level effects. This is particularly true of *in vitro* studies.

Why are there a number of low-level effects that have not been replicated? The fault may lay with the original investigator, the scientist performing the replication or both. In either case, particularly for small effects, the experimental procedures must incorporate specified elements and appropriate laboratory practices which can be followed in a second laboratory. What are these reasons that can lead to non-repeatability?

Biological experiments generally compare two distributions, an exposed group and a control group, which are statistically analyzed for any difference. Well-done studies are always subject to statistical variation. If 100 studies are conducted and statistically evaluated on a 95% confidence level then 5 will be expected to differ because of statistical variation (give false positive or false negative results). The occurrence of statistical variation creates the necessity for replication. Factors that can lead to additional incorrect findings and additional problems in replication include:

- Inadequate controls
  Investigators must determine the system variability by conducting sham/sham experiments. This should be followed by positive controls to test the system sensitivity (How small of a variation is possible to observe). Finally, investigators must include (in a blinded fashion) sham controls with every aspect identical to the exposed animal or cells. Other negative controls, such as cage controls for *in vivo* experiments, are desirable.
- Imprecise or overly complex hypothesis (not testable)
- Exposure range inadequate to define dose response
- Inadequate statistical power
  - Statistical tests not defined *a priori*
  - Number of repeat experimental runs
  - Multiple comparisons

- Temperature control
  - "Athermal" effects at high SAR *in vitro*
  - Thermal stress in studies with laboratory animals
  - Subtle thermal effects
- Microwave hearing as artifact in animal studies
- Poor or improper dosimetry
- Improper handling techniques
  - Time sequence of treatments
  - Stress due to improper handling of samples

27.6 PORTABILITY REQUIRES A MECHANISM

A major problem in the RF bioelectromagnetics research today is that other plausible biophysical interaction mechanism than thermal have not been established at mobile phone frequencies (Valberg 1997; Foster 2000; Pickard and Moros 2001; Repacholi 2001).

A popular claim has been, even in the scientific community, that modulated RF signals, used widely in current wireless technologies, may give rise to "special" biological effects different from those of continuous RF signals. Proponents of this claim provide as support the questionable statement that "special" biological effects are caused by "non-thermal" interactions" of RF energy with biological systems. As a consequence, single non-established biological effects are loosely linked to various human physiological functions suggesting that the non-established biological effects may lead to adverse health effects in humans. Furthermore, these suggestions of non-thermal mechanisms lead to suggestions of potential averse health outcome from new technologies.

Even though the approximately 300 studies addressing mobile telephony and health comprehensively cover current technologies they do not and cannot cover the infinite number of possible frequencies and modulations in use. For example, the majority of these studies utilize generic type GSM signals, but by no means cover all the possible GSM modulation characteristics. Although 3G will utilize a combination of current technologies the future may bring additional variations. Thus, is it necessary to test all possible combinations of frequency and modulation to determine whether or not each signal type will or will not cause adverse health effects, or is current research portable to new technologies? The answer to this question depends on the mechanism of interaction of the RF with the biological system.

A thermal mechanism depends only on the amount of energy absorbed and thus its frequency dependence is predictable. The amount of energy absorbed will depend on the electrical properties of the tissue and the geometrical interaction with the biological object, both of which will cause well-established frequency variations. There is no modulation dependence for a thermal mechanism. A non-thermal mechanism, on the other hand, would be expected to exhibit frequency dependent responses, modulation dependent responses or both. The current 300 mobile telephony studies using all technologies test the hypothesis of whether there is a frequency dependent or modulation dependent response. A workshop was held in May 2001 in Washington DC where a group of experts considered the plausibility of various proposed mechanisms (For a summary see the Research section at http://www.mmfai.org/). This group concluded in part, "For exposures to RF energy from sources in the general environment and from use of mobile telephone devices, the only clearly plausible mechanisms for RF interactions with biological systems involve heating." This workshop evaluated physical mechanisms of interaction from a theoretical basis only. Subjects considered included: temporal and spatial temperature gradients, alteration of membrane potential, membrane rectification, polarization of structures or molecules, RF pumping and chemical kinetics, magnetic dipole interactions, coherence and cooperative interactions. None, except for thermal gradients, were considered plausible at environmental exposure levels but most required further theoretical evaluation to determine their limitations.

Theoretical examination of proposed mechanisms is one approach that is necessary. A second approach to a determination of possible mechanisms is to study a repeatable biological effect and establish the biochemical and biophysical event that causes this response. An examination of the RF biological effects literature does not provide a consistent body of data that can be used as the basis for formulating theoretical postulates other than a thermal mechanism. There are a number of publications that report effects at non-thermal levels. These reported findings do not build a consistent or connected body of data and thus do not support one another. Support must come from either independent replication or from established biological or biochemical connections in which the occurrence of one finding would predict the second. Science builds upon the mechanistic knowledge gained in one experiment by using it to generate the hypotheses for another experiment. No reported "non-thermal" experimental RF biological effect has been able to be repeated in independent laboratories and connections have not been established between reported findings. In the absence of any plausible mechanism to explain reported non-thermal "positive" findings and in the absence of any consistent or repeatable biological results one must conclude that the only currently acceptable mechanism is thermal.

27.7 CONCLUSIONS

Current cancer study database is extensive (more than 150 studies ongoing or completed) and fully addresses the WHO agenda. The majority of the studies show no effect and those few studies showing an effect are being addressed through replication. The replication studies that have been completed have failed to confirm the original findings. There are a large number of non-cancer studies that have not produced an established effect. However, there are also several hypothesis-generating studies that require further study in order to confirm or reject the original findings. Studies on RF effects on the human central nervous system, such as sleep and cognitive functions (reaction time), are examples of reported effects that require follow up.

As new wireless technologies entering the market a number of health related questions have been raised. The most common the applicability of the current database to possible health risk related to new technologies. A fundamental question is whether there is/are other biophysical interaction mechanism(s) other than thermal. A careful analysis of theoretical biophysical studies to date and lack of replicable biological effects strongly suggest that the only plausible interaction mechanisms at mobile telephony frequencies and emission levels is thermal.

At present there is no established scientific evidence that mobile phone or base station emission cause adverse health effects in humans.

27.8 REFERENCES

Adey, W. R., Byus, C. V., Cain, C. D., Higgins, R. J., Jones, R. A., Keran, C., Kuster, N., MacMurray, A., Stagg. R., Zimmerman, G., Phillips, J. and Hagren, W., 2000, Spontaneous and nitrosourea-induced primary tumors of the central nervous system in Fischer 344 rats exposed to frequency-modulated microwave fields. *Cancer Research*, **60,** pp.1857-1863.

American National Standards Institute (ANSI), 1992. Institute of Electrical and Electronics Engineers, Inc. (IEEE). Safety levels with respect to human exposure to radiofrequency electromagnetic fields, 3 kHz to 300 GHz. Princeton, NJ: IEEE C95.1-1991 (revision of ANSI C95.-1982).

Borbély, A. A., Huber, R., Graf, T., Fuchs, B., Gallmann, E., Achermann, P, 1999, Pulsed high-frequency electromagnetic fields affects human sleep and sleep electroencephelogram. *Neuroscience Letters,* **275**, pp. 207-210.

Brusick, D., Albertini, R., McRee, D., Peterson, D., Williams, G., Hanawalt, P. and Preston, J., 1998, Genotoxicity of radiofrequency radiation. *Environmental and Molecular Mutagenesis* **32**, pp.1-16.

Chagnaud, J. L., Moreau, J. M. and Veyret, B, 1999, No effect of short-term exposure to GSM-modulated low-power microwaves on benzo(a)pyrene-induced tumours in rat. *International Journal of Radiation Biology,* **75**, pp. 1251-1256.

Eulitz, C., Ullsperger, P., Freude, G. and Elbert, T., 1998, Mobile phones modulate response patterns of human brain activity. NeuroReport, **9**, pp. 3229-3232.

Foster, K., 2000, Thermal and nonthermal mechanisms of interaction of radiofrequency energy with biological systems. IEEE Transactions on Plasma Science, **28**, pp. 15-23.

Freude, G., Ullsperger, P., Eggert, S. and Ruppe, I., 2000, Microwaves emitted by cellular telephones affect human slow brain potentials. *European Journal of Applied Physiology,* 81, pp. 18-27.

Hardell, L., Näsman, Å., Påhlson, A., Hallquist, A. and Mild, K. H., 1999, Use of cellular telephones and the risk of brain tumors: a case-control study. *International Journal of Oncology*, **15**, pp. 113-116.

Heikkinen, P., Kosma, V-M, Hongisto, T., Huuskonen, H., Hyysalo, P., Komulainen, H., Kumlin, T., Lahtinen, T., Lang, S., Puranen, L. and Juutilainen, J., 2002, Effects of mobile phone radiation on x-ray induced tumorigenesis in mice (In press, *Radiation Research*).

Herrman, D.M. and. Hossmann, A., 1997, Neurological effects of microwave exposure related to mobile communication. *Journal of Neurological Sciences*, **152**. pp.1-14.

Hietanen, M., Kovala, T. and Hämäläinen, A-M., 2000, Human brain activity during exposure to radiofrequency fields emitted by cellular phones. *Scandinavian Journal of Work & Environmental Health*, **26**, pp.87-92.

Huber, R., Graf, T., Cote, K. A., Wittmann, L., Gallmann, E., 2000, Exposure to pulsed high-frequency electromagnetic field during waking affects human sleep EEG. *Neuroreport* **111**, pp. 3321-3325.

ICNIRP, 1998, International Commission on Non-Ionizing Radiation Protection. Guidelines for limiting exposure to time varying electric, magnetic, and electromagnetic fields (up to 300 GHz). *Health Physics* **74**, pp. 494-522.

Imaida, K., Taki, M., Yamaguchi, T., Ito, T., Watanabe, S., Wake, K., Aimoto, A., Kamimura, Y., Ito, N. and Shirai, T., 1998, Lack of promoting effects of the electromagnetic near-field used for cellular phones (929.2 MHz) on rat liver carcinogenesis in a medium-term liver bioassay. *Carcinogenesis* **19**, pp. 311-314.

Inskip, P. D., Tarone, R. E., Hatch, E. E., Wilcosky, T. C., Shapiro, W. R. and Selker, R. G, 2001, Cellular-telephone use and brain tumors. *New England Journal o fMedicine*, **344**, pp. 79-86.

Jech, R., Onka, K., Rika, E., Nebuelský, A., Böhm, J., Juklíková, M. and Nevímalová, S., 2001, Electromagnetic field of mobile phones affects visual event related potential in patients with narcolepsy. *Bioelectromagnetics,* **22**, pp. 519-528.

Johansen, C., Boice, J., McLaughlin, J. and Olsen, J., 2001, Cellular telephones and cancer - a nationwide cohort study in Denmark. *Journal of National Cancer Institute*, **93**, pp. 203-207

Koivisto, M., Revonsuo, A., Krause, C., Haarala, C., Sillanmäki, L., Laine, M.,and Hämäläinen, H., 2000, Effects of 902 MHz electromagnetic field emitted by cellular telephones on response times in humans. *Neuroreport* **11**, pp. 413-415.

Lai, H. and Singh, N., 1995, Acute low-intensity microwave exposure increases DNA single-strand breaks in rat brain cells. *Bioelectromagnetics* **16** pp. 207-210.

Lai, H. and Singh, N., 1996, Single- and double-strand DNA breaks in rat brain cells after acute exposure to radiofrequency electromagnetic radiation. *International Journal of Radiation Biology*, **69**, pp.513-521.

Li Li, Kheem, Bisht, K., LaGroye, I., Zhang, P., Straube, W., Moros, E. and Roti Roti, J., 2001, Measurement of dNA damage in mammalian cells exposed In vitro to radiofrequency fields at SARs of 3-5 W/kg. *Radiation Research* **156**, pp. 328-332.

Malyapa, R. S., Ahern, E. W., Straube, W. L., Moros, E. G., Pickard, W. F. and Roti Roti, J. L., 1997, Measurement of DNA damage following exposure to 2450 MHz electromagnetic radiation. *Radiation Research*, **148**, pp. 608-617.

Malyapa, R. S., Ahern, E. W., Straube, W. L., Moros, E. G., Pickard, W. F., Roti Roti, J. L. Measurement of DNA damage following exposure to electromagnetic radiation in the cellular communications frequency band (835.62 and 847.74 MHz). *Radiation Research*, **148**, pp. 618-627.

Malyapa, R. S., Ahern, E. W., Straube, W. L., LaRegina, M., Pickard, W. F. and
    Roti Roti, J. L., 1998, DNA damage in rat brain cells after in vivo exposure to
    2450 MHz electromagnetic radiation and various methods of euthanasia.
    Radiation Research, **149**, pp. 637-645.

Moulder, J. E., Erdreich, L. S., Malyapa, R. S., Merritt, J. H., Pickard, W. F. And
    Vijayalaxmi, 1999, Cell phones and cancer: What is the evidence for a
    connection? *Radiation Research*, **151**, pp. 513-531.

Muscat, J. E., Malkin, M. G., Thompson, S., Shore, R. E., Stellman, S. D, Mcree,
    D.,Neugut, A.,and Wynder, E., 2000, Handheld cellular telephone use and risk of
    brain cancer. *Journal of American Medical Association*, **284**, pp. 3001-3007.

Pickard, W. and Moros, E. 2001, Energy deposition processes in biological tissue:
    Nonthermal biohazards seem unlikely in the ultra-high frequency range.
    Bioelectromagnetics, **22**, pp. 97-105.

Preece, A. W., Iwi, G., Davies-Smith, A., Butler, S., Lim, E. and Varey A, 1999,
    Effect of a 915-MHz simulated mobile phone signal on cognitive function in
    man. *International Journal of Radiation Biology*, 75, pp. 447-456.

Repacholi, M. H., Basten, A., Gebski, V., Noonan, D., Finnie, J. and Harris, A.,
    1997, Lymphomas in Eμ-Pim1 Transgenic Mice Exposed to Pulsed 900 MHz
    Electromagnetic Fields. *Radiation Research*, **147** pp. 631-640.

Repacholi, M., 2001, Health risks from the use of mobile phones. *Toxicology
    Letters*, **120**, pp. 323-331.

Salford, L. G., Brun, A., Sturesson, K., Eberhardt, J. L., Persson, B. R., 1994,
    Permeability of the blood-brain barrier induced by 915 MHz electromagnetic
    radiation, continuous wave and modulated at 8, 16, 50 and 200 Hz. *Microwave
    Research Technology*, **27**, pp. 535-542.

Sienkiewicz, Z. J., Blackwell, R. P., Haylock, R. G., Saunders, R. D., Cobb, B. L.,
    2000, Low-level exposure to pulsed 900 MHz microwave radiation does not
    cause deficits in the performance of a spatial learning task in mice.
    Bioelectromagnetics, **21**, pp. 151-158.

Telecompetition Inc., 2001, Declining ARPU- Worldwide subscriber insights
    2000-2010
    (website:http://www.wowcom.com/market_research/market_research_category.c
    fm?research_typeID=41)

Tsurita, G., Nagawa, H., Ueno, S., Watanabe, S. and Taki, M., 2000, Biological
    and morphological effects on the brain after exposure of rats to a 1439 MHz
    TDMA field. *Bioelectromagnetics*, **21**, pp. 364-371.

Wang, B., and Lai, H., 2000, Acute exposure to pulsed 2450-MHz microwaves
    affects water-maze performance of rats. *Bioelectromagnetics,* **21**, pp. 52-56.

Wagner, P., Röschke, J., Mann, K., Hiller, W. and Frank, C., 1998, Human sleep
    under the influence of pulsed radiofrequency electromagnetic fields: a
    polysomnographic study using standardized conditions. *Bioelectromagnetics*, **19**,
    pp. 199-202.

Wagner, P., Röschke, J. *et al.* 2000, Human sleep EEG under the influence of pulsed
    radio frequency electromagnetic fields. Neuropsychobiology, 42, pp. 207-212.

Valberg, P., 1997, Radio frequency radiation (RFR): the nature of exposure and
    carcinogenic potential. *Cancer Causes and Control,* **8**, pp. 323-332.

Verschaeve, L. and Maes, A., 1998, Genetic, carcinogenic and teratogenic effects of radiofrequency fields. *Mutation Research* **410**, pp. 141-165.

Zook, B and Simmens, S., 2001, The effects of 860 MHz radiofrequency radiation on the induction or promotion of brain tumors and other neoplasms in rats. *Radiation Research*, **155**, pp. 572-583.

CHAPTER 28

New Site Sharing Technology for 3G and the City of Tomorrow

Mike Smith

28.1 INTRODUCTION

The 3GSM World Congress in Cannes, February 2001 saw the launch of Quintel S4 Ltd, following an agreement signed between QinetiQ (formally the Defence Evaluation and Research Agency), and the Rotch Property Group. This exciting new joint venture brings together the ground breaking new mobile telecommunications technologies from QinetiQ and the property and financial skills of the Rotch Property Group.

28.1.1 Background

It is estimated that to date €116 billion has been spent on 3G licences in Western Europe. The mobile telecommunications operators are now facing the demanding challenge of rolling out networks as quickly as possible, whilst credit ratings have declined and have an adverse effect on cost efficient procurement.

The major issues are the ability to deliver high quality, high data rate services. This is very closely linked to the availability of suitable sites, especially in urban and business areas. Other key factors are the general level of public concern over health and safety of mobile phone masts and the lack of confidence in current scientific evidence and the concerns of local authorities. These public concerns include the proliferation of mobile base stations and masts and specific health, safety, and environmental issues. However, there is the real and growing demand from the general public and business sector for good quality, affordable services from their mobile phone. These services need to be available at all times and locations. It is also of significant interest to the local authorities to enable the latest mobile phone technology, especially within the business districts of our major towns and cities. The obvious financial needs of the 3G licence holders are clear to all. 3G will provide the next generation of digital, mobile personal communications services with provision of a full range of multimedia services to the mobile user. It is a new air interface delivered through a new or updated base station infrastructure and provides a world-wide standard, removing most technical obstacles to unconstrained roaming by users.

As the demand for multi-media service grows, delivery, via enhanced 2G (2.5G) and 3G technology, will be needed. There is projected to be a huge increase in subscriber demand for mobile communications. Currently there are around 500 million subscribers world-wide and the forecast is for ~1500 million by 2010. Understanding the concerns of the general public on health, safety and environmental issues is key. These concerns may be considered by some as just perceptions but the consequences of these perceptions are very real. The result may be that base stations are not built or installed and the roll out of the 3G infrastructure is put at risk.

Quintel aims to provide a satisfactory solution to the issues of 3G roll out, minimising the health and environmental impact and demonstrating control of non ionising radiation from our 3G base station sites. Through these carefully selected, professionally managed, shared sites, Quintel technology will enable all national operators to provide 3G mobile high speed data, video and voice services within major cities and towns. Quintel brings together an extensive property portfolio, comprehensive RF health and safety expertise and technology that will deliver an environmental friendly, fully managed safe shared site.

All reports and information suggest that the mobile operators are finding it difficult to meet the challenge. The construction of new base stations is proving to be one of the limiting factors in the rollout of new services by the operators. Construction of sites is further constrained by non-technical factors such as planning consent, lack of sites, a growing backlash from the general public, lobby groups etc. and UMTS cells' being considerable smaller than GSM cells just exasperates the situation. Potentially UMTS network operators will require 50% more sites than for current GSM networks to achieve the same coverage and less than 60% of existing GSM sites are suitable for UMTS conversion. In the UK alone, this would indicate a requirement for 36,000 new base station sites, a figure, which may be much higher once 'in building' (micro & pico level) requirements are met.

The availability of suitable sites for base stations is key to the success of the 3G mobile business. These sites are chosen to give the best possible coverage in a given geographic area to support the planned numbers of users and the value added services to be offered. An extensive portfolio of potential sites is therefore essential to give the network planners the maximum flexibility.

However, local communities are becoming increasingly concerned by the potential health and environmental impact of mobile base stations and do not automatically accept the construction of new mobile phone masts by operators and specialist site providers.

Optimum site sharing will therefore prove to be one of the crucial elements for a successful future roll out of 3G mobile networks. At present, 2G radio masts are occasionally shared but have separate antennas sets, with the necessary 5 metre separation requirement and subsequent increase in the height of the tower. Other forms of sharing are possible but require network operators to share many more business critical components of their network and in some cases fall outside national licence regulations.

28.1.2 Antenna Sharing Solution

In response to this growing dilemma, Quintel set itself the target of achieving a paradigm shift in the sharing of mobile base station infrastructure, to the benefit of both the operators and the community. Quintel has designed an Antenna Combiner Unit (ACU) which allows multiple UMTS base station 'Node B' transceivers to share a single antenna system. A schematic of the system is detailed in Figure 28.1. The unit combines both transmit and receive channels and the combining and distribution system will allow a number of UMTS Node B transceivers to share common antennas and feeder cables within a W-CDMA FDD base-station. The transmit path multiplexing is provided by very high quality passive filters, whilst the receive chain has active devices.

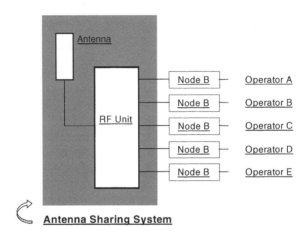

Figure 28.1 Antenna Combiner Unit

The performance of this type of system is very sensitive to intermodulation products, and this problem is particularly acute for the UMTS licensed frequencies due to the receive bands falling within the spectrum of the seventh and ninth harmonics of the transmit frequencies. Other characteristics that the operators look for are minimal power loss and preservation of antenna functionality, including polarisation/spatial diversity.

Using the innovative technology developed for Quintel by QinetiQ, each antenna will be capable of supporting all of the licence holders or a single rooftop or mast antenna. This is an ideal site sharing solution with 12 RX and TX channels combined into one antenna. The ACU can be configured to support both spatial and polarisation diversity.

28.1.3 Specifications

End to End system testing with Node B equipment and antennas has taking place at QinetiQ's Defford Satellite Communications Ground Station Site and has proved the RF interface and detailed performance characteristics. The antenna-combining unit was fully stressed using full power, WCDMA, 5 MHz channels tested simultaneously replicating a combination of operators' bands in use together. The testing assessed the performance of individual carriers in the FDD WCDMA mode sharing a common 3G antenna and Quintel combining unit. The trial fully exercised all elements of the Quintel solution with an active Node B replicating the operator's networks.

Transmit RF Characteristics	
Transmit Frequency Range	2110 – 2170 MHz
Transmit Insertion Loss	1 dB
Isolation Between Transmit Ports	> 35 dB
Return Loss at Transmit Port	> 15 dB
Power Handling	120 W (average) per input port
Frequency Response	
100 kHz – 2100 MHz	-35 dB
2180 MHz – 12.75 GHz	-35 dB
Intermodulation products (two 20 Watt carriers)	
Transmit band	-80 dBc
Receive band	-160 dBc
Harmonics	-90 dBc

Receive RF Characteristics	
Receive Frequency Range	1920 – 1980 MHz
Noise Figure	2.8 dB
Input 3rd-order Intercept Point	+10 dBm
Gain	6 dB
Isolation Between Receive Ports	>30 dB
Isolation Between Transmit and Receive Ports	>90 dB
Frequency Response	
100 kHz – 1880 MHz	-60 dB
2010 MHz – 2110 MHz	-20 dB
2110 MHz – 12.75 GHz	-60 dB

General	
Operating Temperature Range	-20°C to +50°C
Operating Humidity	0% to 90% (non-condensing)
Dimensions of 3-sector RF Unit	900 mm H x 500 mm W x 500 mm D
Transmit port connector	7/16 DIN (F)
Antenna port connector	7/16 DIN (F)
Receive port connector	N-type (F)

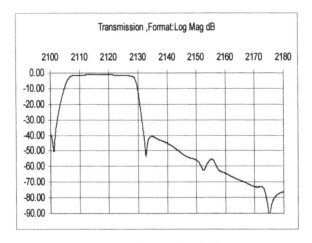

Figure 28.2 Wideband insertion loss

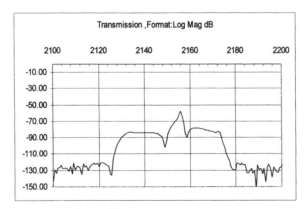

Figure 28.3 Narrow band isolation

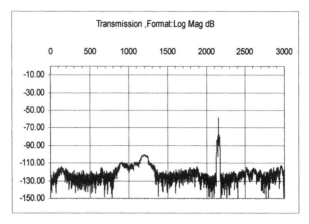

Figure 28.4 Wideband isolation

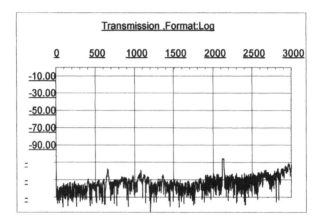

Figure 28.5 Transmit receive isolation

28.1.4 Government and EU Acceptance

The Department of Trade and Industry, Radiocommunications Agency and OFTEL have been briefed on the Quintel technology solution. A positive response has been received from all quarters especially with regards to the anti-competition issues, which have also been discussed with the EU with tacit agreement that the Quintel solution does not appear to pose a problem.

28.1.5 Pilot Network

A pilot 3G network is being planned for late 2002 with the participation of a number of operators, OEMs, industrial partners and the local authorities. This will demonstrate a fully installed system within a public area, ensuring all Quintel's procedures meet the specific needs of the operators and of course the local authorities and the general public. These sites will be available for demonstration and evaluation for up to a period of 6 months.

Roll out plans are heavily dependent on operator timescales but current discussion's with the operators would suggest full scale roll out commencing mid 2003.

28.2 INDIVIDUAL ELECTRONIC BEAM CONTROL

The successful development of the Quintel 3G antenna combiner unit technology has overcome a major problem facing the operators. The concept of reducing the size and visual impact of a 3G base station can only be successful if the operators agree to use a common antenna. However the antenna combiner unit does not allow operators to individually adjust the elevation (tilt) of the transmit and receive beam and this may limit its application to the smaller cells where tilt is not required.

To overcome this concern Quintel, in partnership with QinetiQ, have developed technology to give individual beam control. The results of the research are that an antenna with individual operator electronic beam control is entirely feasible and a prototype is now under development. The development program has passed the preliminary design stage and laboratory tests have been successfully undertaken. The development of the individual electronic beam control capability has also meant that antennas themselves will not be tilted and thus they will thus be less obtrusive, fitting closer to the building or mast.

28.3 UNOBTRUSIVE ANTENNAS

Several 3G base station antenna manufacturers are already supplying unobtrusive/disguised antennas. The nature of the disguise and the specification of the antenna are very specific to each installation scenario. This may be in the form of church windows, flag poles, Scots pine trees, roadside advertising etc. Normally the means of disguising the antenna is to construct a special enclosure around an existing antenna design. This will change the antenna profile to that of a common object or the background around the mounting structure. The colour of the enclosure may also be altered to match the background. This option is normally achieved either by painting the enclosure to match the surroundings by imprinting a photograph of the surroundings on the enclosure at the manufacturing stage.

28.3.1 Technology

However, this approach to the problem is not the final solution. In keeping with QinetiQ's research and development background and using cutting edge technology other exciting solutions are available.

Quintel has instigated research in the use of Polymer Composites in the manufacture of telecom masts and control cabins. QinetiQ has a 30-year track record in evaluating and designing with commercially available GRP (Glass Reinforced Polymer) and CFRP (Carbon Fibre Reinforced Polymer) composite systems suitable for these applications. Advancements have recently been made in the use of natural fibre composite systems, and are in use within the automotive industry and for MOD systems. The advantages come when disposal of the product is required since the polymer resin/natural fibre combination can be treated as a polymer blend, eliminating additional cost and hazards with the disposal of glass and carbon fibres (restrictions on land-fill, problems with fibre filtering and shorting of electric's in incinerators).

28.4 STRUCTURES

Composites have been proven to offer 40-75% weight saving over equivalent metallic structures of this type (mono-pole or lattice).

Benefits from less material needed, compensating for higher relative composite material cost.

Easy ability for disassembly in the future if modifications to the structure are required due to advancement in telecoms technology (i.e. ability to reduce height without major re-work if technology advancement no longer required such high structures).

Novel jointing options for faster assembly – click-together on-site. Variable stiffness along the mast to compensate for sway – reduce weight of structure even further and still meet stability needs of antennae combined with vibration damping of structures for further improved stability.

In the construction of cabins and control boxes similar benefits can be achieved by making use of sandwich construction where thin composite skins are combined with lightweight foam or honeycomb cores. This will also minimise the passive intermodulation effects found with metal structures which limit overall communication performance.

28.5 MINIMISING THE VISUAL INTRUSION OF MASTS

Quintel in partnership with QinetiQ are conducting research in the area of technologies for minimising the visual intrusion of masts in the urban and rural environment, and assessing the electromagnetic properties of materials that may be used in mast construction or concealment. The main offerings are:

28.5.1 Camouflage Modelling

QinetiQ has modelling codes used for many years in the development of camouflage schemes for military vehicles. Masts can be modelled in their environment for all seasons of the year and in day/night scenarios to develop optimised paint schemes for each particular environment.

28.5.2 Adaptive Colour Control

In addition to the passive camouflage schemes outlined above, it is also possible to use paints with active pigments that adapt to stimuli such as light, heat or electric current to undergo a colour change. These can be used to obtain a better match with background over a wider range of conditions. Simple versions of this technology are well established, for example coatings on coffee mugs that clear when hot water is poured in. QinetiQ is working on a broad range of materials of this type.

28.5.3 Diffractive Coatings

Profiling of the surface of the mast to give diffraction effects is another way of controlling the visual appearance of a mast, and can be used to improve background matching. QinetiQ is actively involved in work on this type of material, a well-publicised example being the study of diffractive structures, which give colour to butterfly wings.

28.5.4 Integrated Circuitry

QinetiQ has patented technology by which circuitry can be screen-printed onto structural fabric reinforcements. These can then be incorporated into standard composite structures by conventional manufacturing techniques. The integrated circuitry can be used for sensing, actuation or lighting. For mast camouflage integrated lighting may be attractive because the lighting can be adapted to match the light level of the background. The technology is being considered for applications such as vehicle control panels, interior lighting, and robust deck landing lights.

28.5.5 Unobtrusive Techniques

Before discussing what must be done to hide a radio mast in a given environment, it is worth considering what draws the eye to it in the first place. Human vision is an excellent tool for distinguishing subtle differences between images, and there are a variety of visual cues that will reveal the presence of supposedly camouflaged items.

At the simplest level, the light that is travels from an object to the viewer must match that of the neighbouring environment. This means that the *colour* and *intensity* of the light must be controlled in order to reduce the contrast. The most obvious method for achieving this is the use of appropriately coloured pigments, either in a surface coating (e.g. paint or appliqué sheet) or embedded within the material from which the object is comprised. It is possible to produce materials that simultaneously perform other functions, such as radio wave absorption and damage monitoring, adding to the functionality of the treatment.

It is extremely unlikely that a perfect match can be found with single-tone paint, however, especially when the background contains visual texture (e.g. patterns from leaves and branches, windows and paperwork, etc). This is a particular issue for a tall and thin object such as a mast, which may present a unique shape in the environment. In such cases it is necessary to use two or more carefully chosen colours in patterns that break up the outline of the object such that it is harder to distinguish, a process known as *disruptive camouflage patterning*.

The task becomes more complicated if the background varies in its appearance, however. The above approaches may be sufficient if the mast is always viewed against a particular man-made backdrop (clamped to the side of a tower, perhaps), but in many cases the background will exhibit seasonal variations. When set against deciduous vegetation, for example, it may be necessary to decide whether to adopt an all-season compromise in the camouflage, or to apply different schemes at set intervals. Even worse, if the mast is set against the sky then the lighting conditions could fluctuate on an hourly basis, and the mast may be silhouetted when the sun moves to a certain point. The only solution to such problems is the use of an active dynamic camouflage technology that adapts to its surroundings. For example, QinetiQ has recently developed pigments that exhibit negative photochromism. Photochromic materials change colour when exposed to light, the most common example being sunglasses that darken when the sun is bright.

Our new pigments react in the opposite manner, tending to paler shades as the ambient light levels increase in intensity, and vice-versa. This is useful since it means that a mast painted with such materials will attempt to match the skies behind them: diffusely reflective in the day, but dark at night, reducing their visual contrast. These pigments still require further development, but come in a range of colours and would provide low-maintenance adaptive camouflage paint.

A more direct way to match fluctuating light-levels involves the use of emissive camouflage. It is possible to produce appliqué sheets that contain light-emitting phosphors and polymers that act as lamps, matching their background by emitting the correct colour and intensity of light. Light-emitting camouflage has been considered for many years, and was experimented with in a variety of ways during WWII, but the technology has been restrictive until recent years. This method has the advantage of being able to negate silhouettes, and could be used to eliminate shadows cast against neighbouring objects if used with care (e.g. for masts held near to the sides of buildings)[1]. Emissive camouflage has the disadvantage of requiring a power supply, although new developments in solar cell technology may be of assistance in this area. Light sensors will also be required in order to monitor the ambient light levels that are to be matched, but these can be relatively cheap and unobtrusive.

Even when all of the above factors have been taken account of, there are problems such as *glint*. When light strikes an object at near-grazing incidence it is likely to be strongly reflected, producing a flash of light that can be observed over

---

[1] It should be noted that shadow effects could also be dealt with via counter shading – the use of lighter tones of paint in shadowed regions – although this is obviously a static treatment.

great distances. This can be alleviated via the use of diffraction gratings and surface texturing of the object, effects that may also be of use at other angles.

In summary, visual camouflage is concerned with matching the appearance of an object to that of its background, but it is not expected that perfect agreement will be attained. It is usually the case that an approximation must be made, and this is often sufficient. QinetiQ has vision model codes that can predict the time taken for an observer to detect a hidden object, and these can determine the level of approximation that can be accepted (e.g. masts next to a road will only be in the observer's view for a set time-span). If practical and effective camouflage is to be achieved then it is necessary to take careful stock of the conditions under which the camouflage scheme is to operate.

28.6 CONTROL OF ELECTROMAGNETIC RADIATION IN BUILDINGS USING MATERIALS

With the increasing use of wireless technology for communications and data transfer, there is a growing need to adapt the environment to optimise the performance of electronic systems working within it. Different environments have different requirements, and it may be necessary to adapt building materials to perform shielding, absorption, transmission or filtering functions. The mechanisms by which these functions can be achieved are different, but all are feasible with minimal changes to existing building materials.

Figure 28.6 A representative church indicating the three paths under investigation. Interior paths are represented by a dotted line.

The following sections describe situations where each type of functionality may be required, and some of the practical ways in which these can be realised.

Propagation within the indoor environment has been observed to be strongly influenced by the building layout, the construction materials, as in the following church example (Figure 28.6) item such as church bells and the building layout etc.

In this example, three different scenarios have been investigated: Path 1 – an unobstructed line of sight path, Path 2 – a path vertically downwards passing through a single wooden floor, and Path 3 – a path passing through a brick wall and a wooden floor.

28.6.1 Propagation Models

The simplest to model is path 1, where only the free space path loss, L_{FS}, needs to be considered. For a frequency of 2.4 GHz, this simplifies to

$$L_{FS} = 40 + 20 \log_{10} d \tag{28.1}$$

In order to model the interior paths within the Church, use is made of two simple indoor propagation models (Keenan-Motley, and UMTS). These models predict the average path loss likely to occur within a building. The Keenan-Motley path loss, L, is given by,

$$L_{KM} = 40 + 20 \log_{10} d + n_w L_w + n_F L_F \tag{28.2}$$

where d is distance in metres, and n_w and n_F are the number of walls and floors, respectively, and L_w and L_F are the attenuation of the walls and floors, respectively. The measured attenuation, at 2.4 GHz, for a concrete and single brick wall is 2.0 dB and 2.5 dB, respectively, whilst a concrete floor is 14 dB.

The UMTS model, derived from the COST 231 indoor model, predicts the path loss to be,

$$L_{UMTS} = 40 + 30 \log_{10} d + L_{FAC} n^{((n+2)/(n+1)-0.46)} \tag{28.3}$$

where L_{FAC} is the attenuation factor for different material types. For internal walls this is 3.4 dB, whilst concrete and brick walls the value is 6.9 dB. Floors are 18.3 dB.

28.6.2 Line of Sight Scenario (Path 1)

This path considers line of sight propagation to the Church surroundings (i.e. churchyard and beyond the Church boundary). The predicted field strength for this scenario is shown in Figure 28.7. The levels of field strength for the four categories defined by the Quintel Design Standard (see *Annex A*) are indicated by the horizontal dotted lines. For paths greater than 7 m, the predicted field strength is below the level defined by category 3 (6.14 V/m). This would imply for a 10 m high antenna installation, field strengths encountered at ground level would be within the safety limits of this category.

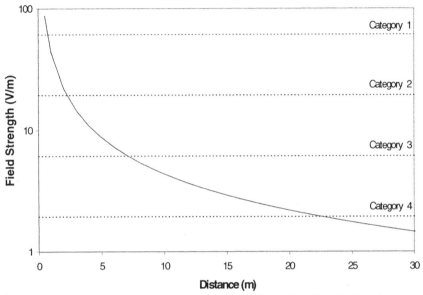

Figure 28.7 Predicted field strength as a function of distance for a line of sight path.

28.6.3 Single Wooden Floor (Path 2)

This scenario considers the additional propagation losses through a single wooden floor. The predicted field strength for the two models is shown in Figure 28.8. At distances of greater than 2 m the field strength is less than level defined by Category 2 (limited access areas). At distances of greater than 5 m the field strength has dropped to below the level defined for Category 3 (unlimited access). For a 10 m high antenna installation these levels would be reached only half way down the Church tower.

28.6.4 Wooden Floor and Brick Wall (Path 3)

This scenario considers propagation through a wooden floor and brick wall. The propagation losses for this scenario will be greater than previously considered, and hence predicted field strengths will be less. Predicted levels, for both models, are below the level defined by Category 4 (Special care areas) within 10 m of the antenna installation, and within just 3 m for Category 3.

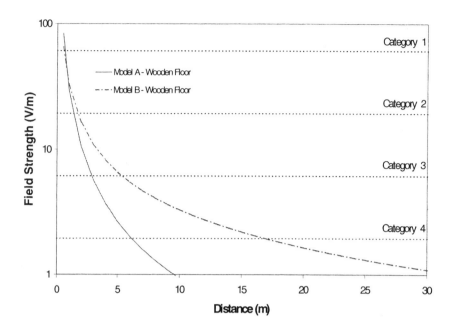

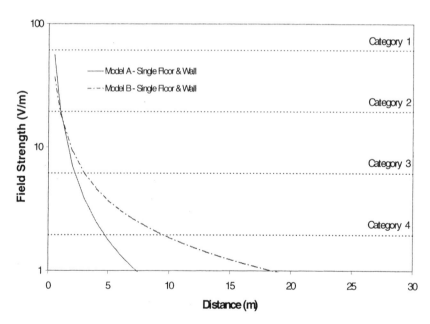

Figure 28.8 Predicted field strengths as a function of distance for the Keenan-Motley model (Model A) and the UMTS model (Model B).

28.6.5 Conclusions

Three different scenarios within and around a Church building have been modelled. The two indoor propagation models (Motley-Keenan and the UMTS) tended to predict similar path losses at close distances (<2m) but these rapidly deviated from one another as the distance increased. The MK model predicted lower path losses (resulting in higher field strengths) than the UMTS model.

The attenuation losses occurring for wood were not known at this time and an estimate was made of the expected attenuation likely to occur. The majority of Churches tend not to be built with brick and therefore more representative field strength values could be obtained if the attenuation losses through stone were known. An estimate of the sidelobe level of an UMTS antenna was made. This is likely to be highly dependent upon location relative to the antenna boresight.

The predictions given here should be used with care as they represent only illustrative levels of expected field strength. The received power (dBW) is given by

$$P_r = P_t + G_t - L + G_r \tag{28.4}$$

where P_t and G_t are the power (in dBW) and gain of the transmitting antenna, respectively, L is the total path loss (in dB) and G_r is the gain of the receiving antenna. The power flux density S in decibels relative to 1Wm^{-2} is related to the received power by

$$P_r = S a_e \tag{28.5}$$

where a_e is the effective aperture of the antenna which is given by $\lambda^2/4\pi$ for an isotropic antenna. The field strength (V/m) is related to S by

$$e = \sqrt{120\pi \ S} \tag{28.6}$$

Predicted field strengths were calculated assuming a transmit power of 40W and a sidelobe gain of 2 dBi for the transmitting antenna, and 0 dBi for the receiving antenna.

28.7 MATERIALS WITH RF PROPERTIES

In addition to materials and technologies for reducing the visual impact of masts, QinetiQ have extensive expertise in the design of microwave absorbers. It is considered that where there are concerns about the level of emissions from antennas and base stations, materials can be developed to absorb, shield or filter the electromagnetic radiation. Discussions are being held with building material manufacturers to incorporate these technologies into standard building products.

QinetiQ has extensive expertise in the field of electromagnetic design codes for the modelling of multi-layer microwave absorbing structures and frequency selective surfaces used in selective filtering of microwave radiation. These codes have been validated over several years of design, manufacture and measurement of prototype panels for the Ministry of Defence and can be readily applied to the design of absorbers operating at mobile telecommunications frequencies.

QinetiQ also has facilities for the measurement of transmission through and reflection from absorber panels. These facilities can be used in the verification of absorber designs.

28.7.1 RF Shielding

Shielding materials operate by providing a continuous conductive surface that reflects electromagnetic radiation and does not allow the transmission of any energy to occur. Electromagnetic shielding is required in situations where there is a need to prevent electromagnetic radiation entering (or leaving) an enclosure, usually to prevent interference between operating systems in close proximity. Examples of where this may become an issue are offices near airports, where the airport radar may operate at a similar frequency to the speed of the computers, and in hospitals where there are potential interference problems between mobile phones and some hospital equipment. In recent years, electromagnetic shielding has been applied to equipment housings following European directives on electromagnetic compatibility. For instance, many equipment housings are extruded from plastics. Shielding is either introduced by metallising the internal surface of the housing or by using conductive filler within the plastic. For the latter approach, the plastic must be conductive and hence the filler concentration needs to be above the percolation threshold (the limit at which continuous conducting pathways are formed in the material). Manufacturing, mechanical and aesthetic requirements tend to impose a need for minimising the percolation threshold to ensure as low a concentration of conductive filler as possible is used.

Similar approaches can be adopted within the built environment. If a shielded enclosure is required, the walls, floor, ceiling and any apertures within them either need to be covered with a continuous conductive layer, or conductive fillers need to be blended into the construction materials themselves. To achieve very high levels of shielding and produce a Faraday cage, there needs to be continuous electrical contact between all surfaces. In practice this is difficult to achieve, but shielding levels of between −20 and −30 dB can be obtained by relatively simple measures. Materials that can be used for such purposes include aluminium foil, which can be easily applied to many construction materials, flexible conductive fabrics which can be used as curtains or underlay, and thin, transparent metallised coatings that can be applied to windows (such as those already used as solar reflective treatments). The level of shielding achieved in practice is very sensitive to the electrical contact and gaps between different treatments, and if high levels of shielding are required it will be necessary to exercise tight control over joints between different types of material and gaps around features such as doorframes.

28.7.2 RF Absorption

The second mechanism that can be employed to control electromagnetic radiation is absorption. This can be used in similar situations to shielding materials, preventing unwanted radiation from reaching equipment by absorption, rather than by reflecting it away. In addition, absorbers can be used where there is a specific need to suppress reflected radiation. An example of where this type of material has been used is in buildings situated close to, or within airport perimeters. The issue with these buildings is that they may cause unwanted reflections of the airport surveillance radar system, which may in turn give rise to ghost images on tracking systems, Figure 28.9.

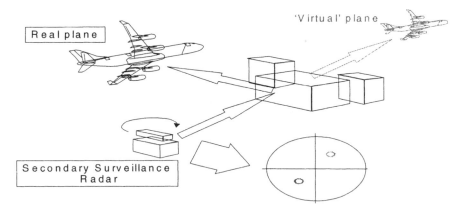

Figure 28.9 A schematic diagram showing how reflections from buildings can give rise to false radar images

In these situations, these reflections can be suppressed by incorporating absorbers in cladding panels, with minimal change to the existing construction. Absorbing treatments of a similar type are used on the BA World Cargo Centre at Heathrow Airport for this reason.

Several approaches can be used to create absorbing materials. A commonly used technique is to blend lossy additives into an electrically insulating host material. Unlike the case for shielding, where such fillers need to be as conductive as possible (e.g. copper, silver or nickel powder), for absorption resistive fillers, such as carbon black, are preferred. This is because the presence of the filler allows dissipation of the energy by resistive heating, and the concentration used is below the percolation threshold preventing the formation of continuous conducting pathways. Alternatively, fillers with magnetic loss mechanisms (e.g. carbonyl iron) can be used instead, and such fillers can be blended into building materials such as insulation foams, plaster or mortar to give an inherently absorbing building material. Another option is to incorporate resistive layers into constructions of walls, ceilings or floors at positions calculated to give destructive interference between radiation reflected from the resistive layer and that transmitted through it and reflected from a conductive back face, Figure 28.10.

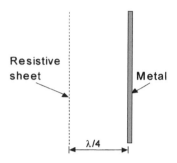

Figure 28.10 Schematic diagram of a Salisbury Screen absorber

This type of absorber is known as a Salisbury Screen, and is essentially a narrow band absorber, operating at a wavelength determined by the spacing between the lossy layer and the metal reflector. Broader bandwidth performance can be obtained by using more lossy layers, or by using inherently lossy materials (such as carbon black loaded foams) as spacers between the lossy layer and the reflector. Absorbing materials can be designed to give levels of absorption ranging from −10 to −20 dB over broad frequency ranges. They may be preferable to shielding materials in applications where it is unlikely that full electrical contact can be achieved around the whole area to be shielded, in which case radiation will leak into the chamber. Absorbers can be used in strategic positions to absorb energy passing through incomplete shielding treatments, for example as movable divider panels placed in front of computers.

28.7.3 Transmission

Transmission is a requirement in cases where buildings are used to house antennas, and the antenna cannot be mounted on a roof or externally for reasons of aesthetics or security. In this case the material used as a façade in front of the antenna needs to have as low a dielectric constant as possible, to give good matching to free space, and low electrical and magnetic loss to allow as much signal to be transmitted as possible. In practice, many structural polymers and glass fibre reinforced composites meet these criteria, and these are the preferred options in such scenarios, although it may be necessary to apply surface finishes so that the transmissive panel looks similar to adjacent brickwork.

28.7.4 Filtering

In addition to shielding, absorption and transmission, it is also feasible to filter electromagnetic radiation. Future wireless requirements may be more complex, and could necessitate materials being more electromagnetically intelligent. For instance, it may be necessary to allow mobile phone frequencies to enter a building but to prevent emission of signals from internal networks to provide data protection and to prevent eavesdropping. Fortunately, technology exists to achieve this, based on the use of periodic arrays of conducting elements or periodic arrays of slots in a conducting sheet, known as Frequency Selective Surfaces (FSS). Filters can be designed with high, low or band pass (or band stop) characteristics, Figure 28.11.

This technology is maturing in forms that are compatible for inclusion within fibre-reinforced polymer composites, as membranes within building structures, or for use as appliqués (wallpaper). Conventional processes such as screen-printing can deposit such patterns, and thus any cost increase associated with modified products can be minimized. Examples of FSS applied to a flexible fabric and printed on paper substrates are illustrated in Figure 28.12.

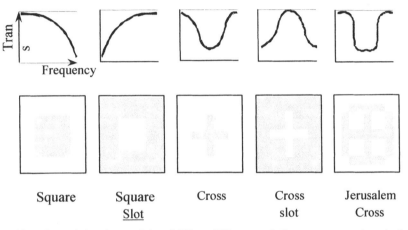

| Square | Square Slot | Cross | Cross slot | Jerusalem Cross |

Figure 28.11 Transmission characteristics of different FSS patterns (yellow areas represent conductive material, light areas insulating material)

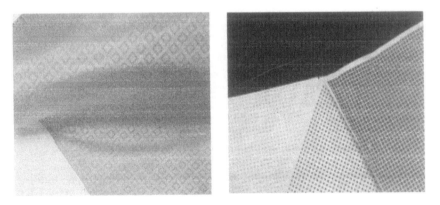

Figure 28.12 Examples of FSS designs applied to different substrate materials

28.7.5 Materials Summary

It can therefore be demonstrated that there are an increasing number of scenarios where control of electromagnetic radiation in the building environment is becoming important, and this will only grow as an increasing number of systems utilise wireless technologies. In some cases current building materials and practices are not adequate to meet requirements. However, technologies do exist for providing the desired characteristics (shielding, absorption, transmission or filtering) and by modelling of the environment at an early stage in building design, building materials can be adapted to meet performance needs. It is also possible to use modified building materials as a 'retrofit', to bring older buildings within a specification, but this may be more difficult.

28.8 ANNEX

The Quintel Design Standard Zoned areas around a base station by category of access and usage. The categories are:

- Category 1 Restricted access: These areas are normally only accessible by Quintel authorised members of staff and contractors involved in installation and maintenance tasks. This category would also cover a landlord or building users who are required by their day to day duties to have access to all parts of the building i.e. key holders. Access is strictly controlled (the area would normally be locked) and limited to authorised persons who have been briefed on the safety implications of the area. In a building these areas would be the locations where Quintel equipment is installed for example plant rooms. Some areas within close proximity to the antenna will have clearly defined physical exclusion zones, which will delineate areas within which exposure guidelines may be exceeded. These areas will be clearly marked and personnel will be protected by safe work procedures and safety interlocks.

- Category 2 Limited access: Areas that are in general use but where vulnerable groups (such as children) will not be present for significant times (typically offices, kitchens, belfries etc).

- Category 3 Unlimited access: All areas within a building that may be used by vulnerable groups for significant times, all areas accessible to the general public and everywhere accessible beyond the building boundary.

- Category 4 Special Care Areas: Areas where there is a high concentration of vulnerable groups for extended exposure times (for example primary schools, hospitals etc).

The Quintel Design standard categorises zones around a base station according to who has access and their expected exposure times. Electromagnetic radiation limits are then set to ensure exposure levels are well below the precautionary principle requirements as detailed in the ICNIPR Guidelines.

CHAPTER 29

Mobile Communications and Health

Peter Grainger

29.1 INTRODUCTION

Mobile telephones are the fastest growing consumer product with the telecommunications industry experiencing rapid growth on a global scale. Mobile telephones have become an essential part of life for many millions of people. This is a direct consequence of rapid technological development which in turn has facilitated the application of new technologies. There are a number of other advantages to be derived from the application of this technology. Mobile or wireless telecommunications play an increasingly important role in general commercial activity and thereby make an indirect contribution to any national economy. Currently Finland is the world leader in the percentage of the population owning mobile telephones (in excess of 80%). Throughout the world it can be seen that people are anxious to keep in touch whilst on the move. As we will see later in this chapter it is not just people who wish to keep in touch with each other but appliances/machines to appliances/machines and people to appliances/machines. It is projected that within the next five years we will have our refrigerators, central heating systems, security systems all communicating with each other and us via the central home/office computer and our mobile communication device.

Concurrent with this rapid uptake of wireless communication devices has been an increasing number of reports of adverse health effects associated with the use of these devices. Research and published peer-reviewed papers to date offer no satisfactory explanation for these reported adverse health effects.

Obviously, the assessment of any health risk resulting from exposure to radio-frequency (RF) fields and waves depends on the results of well conducted and reproducible scientific studies and the need for these is all the greater because any effect of exposure to RF fields and waves for the levels encountered from mobile wireless telecommunications devices is likely to be subtle.

This chapter considers the effects on humans within the present exposure standards, and the importance of understanding sources of exposure and their interactions with tissue and the possible effects of new technology.

29.2 REPORTED ADVERSE HEALTH EFFECTS TO RADIO FREQUENCY EM FIELDS

Most countries around the World have set radio-frequency exposure guidelines (see section 29.4) to protect against thermal effects in the form of tissue heating (heating can lead to permanent de-maturing of proteins and tissue burning) and electric shock. Other biological effects are deemed not to exist due to the low quantum energy at these frequencies. Nonetheless, there is growing public concern that other adverse biological effects do occur below these set limits, i.e. athermal and non-thermal effects.

There tends to be considerable confusion about the terms thermal, athermal and non-thermal when considering electromagnetic energy and biological tissue interactions. This confusion and uncertainty over athermal/non-thermal effects and definitions was first raised in the 1930s (Schwan 1992). The generally accepted definition of the terms are given below:

- **athermal** – physiological changes whereby the core temperature of the body is not elevated but sufficient energy is absorbed to trigger thermoregulatory receptors (Elder 1987).
- **non-thermal** – this where the energy absorbed by the tissue is at a level insufficient to trigger the thermoregulatory receptors.
- **thermal** – the tissue absorbs sufficient energy for a temperature rise to be detected –usually taken as $0.1°C$. At around $1°C$, a person will begin to sweat more freely, reversible protein denaturing commences. At approximately $4°C$ rise an individual will begin to feel general distress, nauseous and headaches develop. Protein denaturing is on the verge of being permanent resulting in cell damage and death.

29.2.1 Reported Adverse Health Effects

Several health effects and physiological phenomena have been reported to be caused through exposure to RF energy at levels well below the known thermal effects level. Some of these effects have been scientifically investigated. These concerns essentially stem from the use of mobile telephones.

In May 2000, the UK Government set up a special committee, the "Independent Expert Group on Mobile Phones" to examine these concerns. A comprehensive report was issued on mobile phone safety issues (The Stewart Report 2000).

On the general issue of radio-frequency radiation safety, the Independent Expert Group concluded that:

"The balance of evidence to date suggests that exposures to RF radiation below NRPB and ICNIRP guidelines do not cause adverse health effects to the general population."

"There is now scientific evidence, however, which suggests that there may be biological effects occurring at exposures below these guidelines. This does not necessarily mean that these effects lead to disease or injury, but it is potentially important information..."

Lai and Singh (1995), reported rat brain DNA damage when exposed to both continuous wave and pulsed 2.45 GHz signals of between 0.6 and 1.2 W Kg^{-1}. Using the COMET assay technique they observed single strand DNA breaks. This observation was and still is very controversial as conventional wisdom states that there is insufficient quantum energy at 2.45 GHz to effect DNA strand breakage.

This effect however, was not observed by Malyapa *et al* (1998) who put the Lai and Singh results down to an experimental artefact.

Cancer is probably the main concern expressed by the general public in connection with the use of wireless communications. The head and hand are the nearest parts of the body to the phone aerial. Therefore, it is not surprising that people are associating phone use with brain, ocular and auditory tumours. Cellular phone use is little more than a decade old and therefore there has been limited opportunity to examine long-term health effects. However, several recently published large cohort studies have compared cell phone use among brain cancer patients and individuals free of brain cancer.

The 1996 paper by Rothman *et al* followed 250,000 phone users for one year. No evidence was found to suggest that the death rate between users of hand-held phones was different compared with users of "suit-case" type mobile phone.

In three case-control studies, Hardell *et al* (1999) compared 233 brain cancer patients diagnosed between 1994 and 1996 in the Stockholm and Uppsala regions of Sweden, and 466 controls. Muscat *et al* (2000) compared 469 brain cancer patients diagnosed between 1994 and 1998 in New York, Providence, and Boston, with 422 controls. Inskip *et al* (2001) compared 782 brain cancer patients diagnosed in Phoenix, Boston, and Pittsburgh between 1994 and 1998 with 799 controls.

All three case-control studies found similar results. Brain cancer patients did not report more cellular phone use overall than the controls. In fact, most of the studies showed a tendency toward lower risk of brain cancer among cellular phone users, for unclear reasons.

When individual types of brain cancer were considered, none was consistently associated with cell phone use. However, Muscat *et al* reported a non-significantly increased risk of neuroepitheliomatous tumours (odds ratio of 2.1 with a 95% confidence interval of 0.9 to 4.7), Inskip *et al* found a non-significant decrease (odds ratio of 0.5 with a 95% confidence interval of 0.1 to 2.0).

When specific locations of tumours within the brain were considered separately, no associations with cell phone use were observed.
None of the studies showed a "dose-response relationship" between cell phone use and brain tumour risk.

In a Danish cohort study, (Johansen *et al* 1999) linked data on all of the 420,095 cellular telephone users in Denmark between 1982 and 1995 to the Danish

Cancer Registry. The results are in agreement with the findings of the above three case-control studies. Cellular phone use was not associated with an increased risk of developing brain tumours overall; nor was there an association with any brain tumour sub-types or with tumours in any anatomical location within the brain.

There is now considerable epidemiologic evidence that shows no consistent association between cellular phone use and brain cancer. However, the big problem with cancer is the long lead time between exposure to a carcinogenic assault and the physical manifestation of the disease. This time lag can be as long as 30 years.

A number of perception and cognitive effects have been reported. Long before mobile telephones were in use some people were reporting they could "hear" microwave signals. This auditory phenomena has been investigated and established by Frey (1961) and Sommer & von Gierke (1964). The frequency range that appears to stimulate the human auditory system is 200 to 8,000 MHz. More importantly for today's technology, it is not the carrier wave that causes the problem but pulses. The threshold for human audibility at 2,450 MHz is approximately 400 mJ m^{-2} with a pulse of less than 30 μs (Guy *et al* 1975, Lin 1989).

The sounds that are heard, range from audible clicks, buzzing, chirping, hissing to popping noises. Elder and Hill (1984) consider that the mechanism for this phenomena is thermoelastic expansion in the head. This produces a pressure wave which is detected by the hairs in the cochlea via bone conduction. Sufferers find this effect irritating but it has not been found to be damaging. However, Lin (1989) advises caution until it is known whether radiofrequency pulses at power levels above threshold levels pose a health risk.

Memory effects have also been reported by many users of mobile phones. The Stewart report stated:

"...the widespread use of mobile phones by children for non-essential calls should be discouraged and... that the mobile phone industry should refrain from promoting the use of mobile phones by children."

The recommendation that children be discouraged from using phones was based largely on the cognitive effect studies of Preece *et al* (1999), Koivisto *et al* (2000) and the European Union "Precautionary Principle" (Foster *et al* 2000). The studies by both Preece *et al* and Koivisto *et al* used a comprehensive range of tests on human volunteers to study cognitive performance. Both groups exhibited a speeded up reaction time.

When an organism is subjected to an abrupt temperature rise, it produces heat shock proteins (HSP) to enable it to adapt to this challenge. In an interesting experiment de Pomerai *et al* (2000), found HSPs in nematodes even though they had been exposed to non-thermal levels of RF. If true, this observation could have profound implications for human health.

29.3 TISSUE INTERACTIONS

The interaction between RF waves and biological tissue is complex due to the composition of biological tissue and involves reflections at tissue interfaces, energy absorption and electromagnetic energy propagation. Tissue composition comprises membranes around cells, ions in solution and macro-molecules ranging from lipoproteins and nucleic acids (large) to amino acids (small). The mechanisms of interaction are a function of the electrical and magnetic properties of matter. This interaction is known as permittivity (ε). The electric field acts upon the charges found in biological material and the term complex relative permittivity (ε_r) has been developed. Electromagnetic theory gives the equation $\varepsilon_r = \varepsilon' + j\varepsilon'' -$ a complex equation. Where ε' is the real term representing the ability of the material to store energy in the charge distribution. The second term, ε'', represents the energy dissipated in the tissue. Consequently, the dielectric properties of biological tissue vary with frequency in a complex form as shown in Figure 29.1 below.

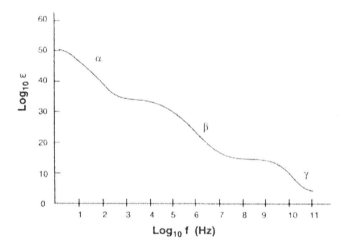

Figure 29.1 Schematic demonstrating the frequency dependence of the complex permittivity and the three major dispersion zones in typical biological tissue.

As can be seen from Figure 29.1 there are three well defined dispersion zones, alpha, beta and gamma.

Alpha is the least well understood section. The relaxation of counter ions surrounding charged cell membranes dominates and probably accounts for the migration of ions through cell membrane holes.

Some texts show the beta section in two parts (β and $\beta 1$), the demarcation being very indistinct. The first 75% of the β section is due to the inhomogeneity of the tissue (Maxwell-Wagner effect). Charge builds up at boundaries separating regions of different ε' and σ (conductivity). A finite time is required for the charging process and the resulting ε' and σ of the medium depends on the temporal behaviour relative to the imposed field. The information given by this portion of the dispersion curve indicates the structure and width of cell membranes. Whereas, the end 25% section is due to the rotation of biological molecules other than water, proteins for example.

The gamma region is due to the relaxation of free water molecules, characterised by parameters close to pure water. Amino acid rotational motion is also important. In the 2 GHz region, any large proteins will also increase the permittivity.

29.4 RADIO-FREQUENCY EXPOSURE GUIDELINES

29.4.1 Introduction

As a direct result of a number of "health scares" (electric power lines and sub-stations, electrical appliances, mobile telephones and their base stations), the general public have become more aware of sources of electromagnetic energy and exposure standards.

There are National and International safety guidelines covering both general public and occupational exposure to electromagnetic energy. In the United Kingdom, the body charged with this task is the National Radiological Protection Board (NRPB). The International Commission on Non-ionizing Radiation Protection (ICNIRP) examines exposure safety guidelines internationally.

29.4.2 Guideline Philosophy

Organisations around the world responsible for advising on exposure guidelines for electromagnetic fields and radiation are agreed that data should be based upon comprehensive assessments of the science. Any exposure limits should be set to prevent only well established adverse human health effects. Therefore, anecdotal evidence of adverse human health effects cannot be taken into consideration. However, anecdotal evidence can point the way to future research.

This is a topic where the science is moving very rapidly and it is therefore essential that guidelines are kept under constant review and update.

29.4.3 NRPB Guidelines

The NRPB was originally set up by the UK Government to advise on ionizing radiation exposure. Their remit was later extended to additionally advise on non-ionizing radiation. NRPB guidelines (NRPB 1993) on exposure do not distinguish between occupational and general public exposure. The UK has no national legislation on exposure but organizations are expected to adhere to NRPB guidelines (investigation levels) through such legislation as the Health & Safety at Work Act.

29.4.4 ICNIRP Guidelines

ICNIRP is the successor organisation to The International Non-Ionizing Radiation Committee of the International Radiation Protection Association (IRPA/INIRC). ICNIRP guidelines (reference levels) were published in 1998 (ICNIRP 1998) and differentiate between occupational and general public exposure.

29.4.5 Technical Standards Organisations

To comply with exposure guidelines, dosimetry needs to be undertaken. These organisations produce protocols by which calculations and/or measurements may be made to enable guideline compliance. These protocols and equipment emissions standards are undertaken by for example, British Standards Institution (BSI) for the UK, the European Committee for Electrotechnical Standardisation (CENELEC) and the International Electrotechnical Commission (IEC).

29.5 ELECTROMAGNETIC WAVE SOURCES AFFECTING THE INDOOR ENVIRONMENT – CURRENT AND FUTURE TECHNOLOGIES

29.5.1 Background

In 1864, a paper "A Dynamical Theory of the Electromagnetic Field" was presented to the Royal Society by James Clerk Maxwell. Building upon the work of OERSTEAD, Faraday and Henry, Maxwell set out his four fundamental equations: the famous *Maxwell's Equations*. These equations in their differential form are given below:

$$\nabla \times E = -\frac{\partial B}{\partial t} \tag{29.1}$$

$$\nabla \times B = \mu_o \left(J + \varepsilon_o \frac{\partial E}{\partial t} \right) \tag{29.2}$$

$$\nabla \cdot E = \frac{\rho}{\varepsilon_o} \tag{29.3}$$

$$\nabla \cdot B = 0 \tag{29.4}$$

These equations postulated the probability of electromagnetic waves and how they may be generated. Three years later in 1887, Heinrich Hertz realised the predictions of Maxwell by producing and detecting radio waves with a frequency of approximately 1 GHz. Radio (wireless) communications over large distances was pioneered by Marconi between 1895 and 1901.

The term wireless communications has been in use since the first radio (wireless) sets came onto the consumer and industrial markets but no longer solely applies to radio receivers. Wireless communications require a transmitter and receiver, neither of which need to be in the building. Thus the indoor electromagnetic environment is affected by devices within the building and outside the building.

Since the end of the Second World War, electronics has developed very rapidly. During the 1940s, the concepts of information theory, cellular telephones and digital communications were first developed. This was followed by the rapid development of the transistor, microelectronics (integrated circuits etc.) and optical fibre technology. "Mobile" telephones were first used by the police and emergency services and were large, bulky and heavy appliances. The public saw these and wanted them. The public service started in the 1980s.

There are many indoor wireless devices we are all very familiar with, for example, mobile (cellular) telephones, cordless telephones, cordless headphones and head-sets, cordless computer mice and keyboards. Other devices will not be so familiar or obvious, for example, wireless networks, security devices for personnel and goods. External devices affecting the indoor environment include radio and television transmitters, cellular telephone base stations, land-line microwave links, pagers plus many other services both commercial and Military.

We can now fit more computing power on to the palm of our hand than was used for the first manned flight to the moon – and this power is still rapidly increasing.

29.5.2 Bluetooth

Currently, the main system for bringing about this indoor wireless communications world revolution is known as Bluetooth and over the next few years is expected to be the major wireless communication system linking equipment within the home and office, though there is a rival system being developed.

The name originates from Harald Bluetooth, a 10th century Danish King. Harald unified Demark and Norway and was renowned for his ability to get people to talk to each other. Bluetooth is still a relatively rare system despite the fact that work first started in 1994. Critics consider the design to be overly complicated. Equipment supporting the standard requires a radio, baseband controller, logical link control and adaptation protocol specification, link manager, host controller interface and an application program interface library. However, integrated circuit manufacturing technology has now advanced to a stage where all these requirements can now be incorporated onto a single integrated circuit.

Bluetooth devices communicate at a frequency of 2.4 GHz using a packet system for data transfer. This frequency is an unlicensed globally available frequency in the ISM (Industrial, Scientific and Medical) band. Frequency hopping is also incorporated into the standard so as to mitigate data corruption in noisy environments. Communication distance between devices in different rooms is approximately 10 metres. There is however, provision within the protocol for a higher power system giving a range of up to 100 metres so allowing the user to wander over a whole home or office. The base station typically having an RF output of 1 – 10 mW. Microsoft is expected to have Bluetooth support in its forthcoming Talisker operating system.

29.5.3 Wireless Domestic Devices

Headphones for listening to home entertainment systems are now quite common. Cordless telephones are common in homes today. A new variation on this devices is the Bluetooth headset. This plugs into a standard telephone socket and can communicate not only with the telephone network but with other devices which recognise speech. More and more homes now have more than one computer which are being networked using cables. Wireless networking is set to be the preferred method. Computer keyboards and pointing devices are already becoming more common in wireless form – this may be RF of infra-red. See section 29.5.5. covering the "smart" home.

29.5.4 Wireless Devices in Commerce and Industry

Do you have trouble finding clothes that fit properly? Soon you will be stepping into an infra-red scanning booth. Here you will strip down to your underclothes and in approximately 12 seconds your body will be scanned using a low level infrared beam and typically 300,000 data measurements made. A three dimensional hologram of the body is then created. These data are then stored onto a smart card. This personal three dimensional digitised hologram on the smart card will enable the holder to down-load the data over the internet. Thus enabling on-line clothes stores to "clothe" the hologram and so allowing the shopper to ascertain their suitability. The shopper will also be confident that the clothes will fit. This system will also enable clothes to be custom made. With the advances in computing power the shoppers head can be made realistic with appropriate musculature and hair. This will enable things like spectacles, hats, scarves and so on to be tried online.

Our cars are also likely to become Bluetooth devices and part of the wireless communications network. Data about the cars performance would be collected and analysed back at the service centre. This is expected to have major advantages for the solving of intermittent faults reported by the driver. The service station will also use Bluetooth for stores control. Items being tagged much in the sane way as items used in the smart home.

29.5.5 The Smart Home

"Smart" homes in different parts of the world are now being trialled. What does this mean? The homes generally look no different from a conventional home. The inside though is based on a complete integrated digital communications system centred around a central computer server. Through the use of a wireless network (Bluetooth) all household appliances are in constant communication with the central server. These appliances can be operated manually, through touch pads around the home, by other computers or via a mobile telephone from almost anywhere in the world. The home and its intelligent appliances are thus available for interrogation by the owner 24-hours a day from wherever they may be. With improvements being made in speech recognition software, it is now possible to wear a wireless head-set and speak commands to the house and its "smart" contents. In addition to control of appliances, control of lighting, curtains, heating/cooling and security would be included.

When food is removed from store (fridge/freezer, cupboard etc) this is noted by the server and by the cooking appliances. The smart tags contain cooking instructions and suggested menus. Cooking instructions are relayed to the microwave oven etc and menu suggestions displayed on a visual display screen or wireless tablet. We would be warned automatically when food had passed it's use-by date.

For those who do not like, or do not have the time, the house will "do" the shopping. Smart tags on the food packaging is read by the rubbish bin and transmitted to the shopping list located on the server. The server automatically places an order with the appropriate shop who then delivers the goods. For those who enjoy the visit to supermarkets can expect to use a Bluetooth enabled trolley. The shopper swipes their loyalty card through the trolley and it steers them to their favourite areas of the shop, pointing out special offers on the way.

Security becomes biometric based. That is the use of finger prints and iris analysis is made. This will be used for home security and security outside of the home. Cameras inside and outside of the home would monitor visitors/intruders and relay this to the central server for recording and passing onto home owner. This monitoring by the home owner could be carried out from almost anywhere in the world. Should delivery be made and no one is home, the identity of the caller could be verified and the house door opened remotely and the safe delivery of the goods into the home monitored. Conversely, the owner can be notified of intruders and appropriate action taken.

The technology used in the smart home is also applicable to shops, offices and industry.

29.5.6 What can We Expect in the Future?

There is a rival to Bluetooth called 802.11 (IEE 802.11) and IEEE 802.11b(WiFi). This system is reputed to have a greater data transmission rate but weaker security features compared with Bluetooth. Another development is tri-band devices. The frequencies used would be 2.4 GHz, 5.2 GHz and 5.8 GHz. The standards covering these devices would be IEEE 802.11a,b and HiperLan. This will provide a single system for simultaneous data, media and voice operations.

The general public are still very concerned over the safety of mobile phone base stations and their proliferation. At least one company is working on using an airship base station. This would cover a large area and be kept in place by solar powered electric motors with a small internal combustion engine for emergency use. The signal power density at ground level would be expected to be very much less than that from terrestrial base stations.

Satellite base stations are expected to play a greater role, covering a larger area than airship base stations.

Current texting is very restrictive. An electronic "pen" that can send hand written notes over the internet using Blue-tooth is being developed. This will also allow sketches to be sent in addition to text.

There are many other potential uses for wireless communication systems but in these troubled times security has become a major issue. For example, airports and railway stations could provide Internet access terminals and ticket reservation and purchase terminals using personal smart cards being read remotely and biometric interrogation.

Looking a little further into the future we could have "smart" clothing in the form of body-WLAN (body wireless area network). The system would monitor our health, exchange personal information with other people as we passed them by (like exchange of business card without physical contact being needed), how far we have walked and where and who we met). We would pick up and exchange information automatically as we walked around buildings and the streets. In the office it would be business information, around the streets this would include entertainment information.

As can be seen, wireless devices are becoming more and more complex, not only in the functions they offer the user but also in their communications protocols. This is going to make dosimetry more complex and it will probably need to take the following into account:

- carrier frequency
- mean output power
- distance to sensitive tissue
- exposure time
- carrier wave modulation

This will require more complex mathematical and physical models.

29.6 CONCLUSION

There is little indication that there will be a slow down in the demand from both the domestic and commercial/industrial markets for wireless communication devices. The technology governing public wireless networks is fast moving with no indication yet which system will become the dominant one. Physiological effects other than heating are being reported with greater frequency in the scientific and popular literature. However, no convincing and universally recognised mechanisms have been agreed upon. Additionally it is far from certain as to whether these reported physiological effects are detrimental human health.

What is certain is that further research into health effects is required and that many more uses of wireless and mobile communication devices will be developed. We particularly need to know more about athermal and non-thermal physiological responses and cell signalling mechanisms.

29.7 REFERENCES

De Pomerai D, Daniells C & David H 2000 Non-thermal heat-shock response to microwaves *Nature* **405(6785)**:417

Elder J 1987 Radiofrequency radiation activities and tissues *Health Phys* **53**:606-611

Foster K P Vecchia P & Repacholi M 2000 Science and the precautionary principle Science **288**:979-981

Frey A 1961 Auditory system response to radiofrequency energy *Aerospace Med* **32**:1140-1142

Guy A W, Chou C K, Lin J C & Christensen D 1975 Microwave-induced acoustic effects in mammalian auditory systems and physical material *Ann N.Y. Acad Sci* **247**:194-218

Hardell L Nasman A Pahlson A, Hallquist A & Hansson Mild K 1999 Use of cellular telephones and the risk for brain tumors: A case-control study *Int J Oncol* **15(1)**:113-116

ICNIRP 1998 Guidelines for limiting exposure to time-varying electric, magnetic and electromagnetic fields (up to 300 GHz) *Health Physics* **74(4)**:494-522

Inskip P D Tarone R E Hatch E E Wilcosky TC, Shapiro WR, Selker RG, Fine HA, Black PM, Loeffler JS, Linet MS.2001 Cellular telephone use and brain tumours *N Engl J Med* **344(2)**:79-86

Johansen C, Boice J D Jr., McLaughlin J K, Olsen J H 2001 Cellular telephones and cancer - a nationwide cohort study in Denmark *J Natl Cancer Inst* **93**:203-207

Koivisto M, Revonsuo A, Krause C, Haarala C, Sillanmaki L, Laine M, Hamalainen H 2000 Effects of 902 MHz electromagnetic fields emitted by cellular phones on response times in humans *Neuroreport* **11(2)**:413-415

Lai H and Singh N P 1995 Acute low-intensity microwave exposure increases DNA single-strand breaks in rat brain cells *Bioelectromagnetics* **16**:207-210

Lin J C 1989 Pulsed radiofrequency field effects in biological systems In *Electromagnetic interaction with biological systems* J C Lin ed pp 165-177 New York:Plenum Press

Malyapa R S Ahern E W, Bi C Staube W L LaRegina M Pickard W R & Roti Roti J L 1998 DNA damage in rat brain cells after *in vivo* exposure to 2450 MHz electromagnetic radiation and various methods of euthanasia *Radiat Res* **149**:637-645

Muscat J E Malkin M G Thompson S *et al* 2000 Handheld cellular telephone use and risk of brain cancer. JAMA **284**:3001-3007

NRPB 1993 *Board statement on restrictions on human exposure to static and time-varying electromagnetic fields and radiation* vol.4 no.5 (Chilton:NRPB)

Preece A W, Iwi G, Davies-Smith A, Wesnes K, Butler S, Lim E, Varey A 1999 Effects of a 915 MHz simulated mobile phone signal on cognitive function in man *Int J Radiat Biol* **75**:447-456

Rothman K J, Loughlin J E, Funch D P & Dreyer N A 1996 Overall mortality of cellular telephone customers *Epidemiology* **7**:303-305

Schwan H P 1992 Early history of bioelectromagnetics *Bioelectromagnetics* **13**:453-467

Stewart Report 2000 Independent Expert Group on Mobile Phones: Report on Mobile Phones and Health NRPB:Chilton Online at: http://www.iegmp.org.uk/IEGMPtxt.htm

Sommer H & von Gierke H 1964 Hearing sensations in electric fields *Aerospace Med* **35**:834-839

CHAPTER 30

Moving Beyond EMF Public Policy Paralysis[1]

B. Blake Levitt

30.1 INTRODUCTION: NEW REVOLUTION IS NEEDED

In writing this presentation, I wondered what I could possibly offer to such an august group. Clearly you all know the scientific issues that bring us together at this conference. In fact, as a science journalist, I have often cited the work of many of the speakers in my own writing. In so doing, I have been astounded at the clarity, the vision, and the courage that many of you have shown over the years. So, since the science is well covered, I have the opportunity to address some of the broader societal issues — the kind that are often left out of academic gatherings.

Albert Einstein once said: "No problem can be solved by the consciousness that created it." If we are to move beyond the societal paralysis we experience today in which no sane EMF public health policy recommendations are made across nations, we need a new consciousness to take over this subject once and for all. We need a new international revolution — not like the EMF protest committees of the past in which a handful of people have doggedly thrown themselves at this problem with more or less success, but rather a full-blown revolution of scientists and lay people alike. Professionals in the biological sciences must seize control of this subject and firmly restructure the dialogue into a permanent biomedical and environmental perspective. That's what has been lacking so far. We cannot, and should not, continue to take this one technology at a time, with the same professions that develop the technologies also controlling the research, the recommendations and the standards. A much broader vision and course of action is needed.

Do not rely on the U.S. to take the lead in this, or to push for any serious reform or research. Even before September 11th, American EMF policy was hamstrung by military influence. It is even more so now. America is also too corrupted by industry dollars in our political process. There is virtually no unbiased research being conducted in America today. Five independent bioelectromagnetics research laboratories have closed in the U.S. in recent years due to a lack of funds. I would hazard to say that this is not an accident. Without the European sphere of influence to counterbalance this, we will all continue to be hapless participants in

---

[1] This speech was delivered to conference attendees after dinner and therefore has a conversational tone rather than a formal academic style.

what has been described as the greatest global experiment in the history of the human race.

30.2 PHYSICS V. BIOLOGY: HEART OF THE PROBLEM

I was heartened to see this conference held at the Royal College of Physicians. Many of the great physicians of the past whose portraits don these walls were alive during the great Vitalist v. Mechanist debates. We soldier on in a direct line from those debates but with many more twists and turns today. The stakes are far higher now.

The reason it is important that the conference is at the College of Physicians is because, at its core, this is a medical issue. It isn't a physics or an engineering problem. The general principles of energy propagation are well established and understood. As societies, we know what we need to know about those areas of science which have brought us the spectacular technologies of the last century. But the questions now on the table concern the consequences to the living organisms in the path of these technologies. The urgent questions now are all biological in nature.

It is increasingly obvious that the situation industrialised countries find themselves in — with burgeoning technologies unlike anything we have ever known being afforded an automatic clean bill of health — is due to the different methodologies and standards of proof between the physical and the biological communities. I suspect that if the biologists controlled the subject, safety recommendations would be very different and they would have been made years ago.

The physics and engineering disciplines come from the non-living sciences and they can never explain the astonishing complexity of a single living organism. The wonder of life will never be encompassed by a mathematical equation or a computer model. Yet, it is the physics and engineering communities that have controlled the dialogue on the biological aspects of this issue for decades; and it has become increasingly obvious that this is utterly inappropriate. Those from the non-living branches have influenced everything from the genesis of the technology, to its myriad applications, to how and what kind of safety standards are set. They have also determined how government and lay-people understand — or more precisely do not understand — the risks of these technologies. The physics and engineering communities have also controlled what research is funded, determining everything from the questions that are deemed legitimate to ask, and ultimately how all aspects of the dialogue are framed.

This is not the way it should be. As a science journalist for over 20 years, I can attest to the fact that nowhere else in any area of science does an unrelated discipline exert such significant influence over that of another. That this had evolved unquestioned for decades is downright odd. To use the reverse example, it would be as if medical doctors assumed all control over the funding, research, intellectual debate, publication and peer review for all aspects of the Superconductor Supercollider. Such professional preemption outside of one's expertise is absurd, yet that is precisely what has occurred with the biomedical and environmental regulation of EMFs. To society's detriment, the subject has historically — and continues to be — in the wrong professional hands. In saying this, please know I am not talking about individual physicists or engineers, many of whom have been of enormous help with understanding this complex subject and who have worked tirelessly for more responsible public health policies.

30.3 PROFESSIONAL BIASES

There are many reasons this situation came about, including, to some extent, an utter absence of interest by biologists. In order to minimally understand what is said about this subject, one needs a passing knowledge of physics and most biologists shy away from that. The subject continues to fall de facto into the physics camp.

But the simple truth is this: while physicists and engineers usually get the calculations right, they sometimes get the conclusions very wrong. A stunning example from the 1950's is that of Edward Teller, the father of the hydrogen bomb and more recently the author of America's missile defence shield, popularly called "Star Wars." In the 1950's, Teller offered to use nuclear devices to reconfigure the coastline of Alaska in order to create a better shipping harbour, and to level the tops of mountains to create new airport sites. I think we would all agree it was a good idea we didn't take him up on the offer. While using nuclear explosions as excavation tools would certainly have done the job, it would also have rendered much of the Northern Hemisphere permanently uninhabitable. There are numerous other examples in history of the same gleeful myopia in the physical sciences. The abiding lesson in this should be: Just because we can technologically accomplish something, doesn't mean that we should.

There is of course the very promising interdisciplinary field of bioelectromagnetics, which is the most fascinating area of science I have ever found. One could say that bioelectromagnetics is where "the rubber meets the road." Bioelectromagnetics is the "crash point" between the living and the non-living sciences, rather like positive and negative charges seeking to balance each other, and in fact there are no areas of science, public policy or the law that are untouched by it. But unfortunately, most members of the bioelectromagnetics community wander into it from the physics community, not from the biological sciences. By the time a physicist takes an interest in the "softer science" (as biology is sometimes called), their professional biases, perspectives, and methodologies are already formed — often without the individual's awareness. And it is also possible that those who are drawn to the physical sciences have a particular temperamental mindset beforehand. Add to this the fact that throughout the last century, physics has been at the very top of the scientific food chain, considered to be the very zenith of what science should be. Physicists have become accustomed to a great deal of societal respect and nearly unlimited research money. They have become accustomed to issuing dictums on the fundamental nature of just about everything.

It should come as no surprise that there would be little hesitation to cross professional boundaries. But it's time to rein that in now. This is a biological issue.

The fact that the physics and engineering disciplines have controlled the focus on EMFs likely accounts for the subject of environmental effects getting so bogged down in a discussion of "mechanisms" and the long-running competitive silliness of thermal v. non-thermal effects — they both exist and with the exception of a few old hardliners, everyone knows it. (The only question is whether non-thermal effects are adverse, and if so, are they reversible.) Physicists and engineers are "think & do" people who make things happen. Understanding mechanisms is critical for them. But it is important to understand that this is only one approach to problem solving. There are others that work as effectively, given the task assignment.

The living and the non-living branches have a completely different approach to problem solving. Biologists are far more comfortable with uncertainty, for example, than the average physicist or engineer. In fact, most uncertainty drives physicists and engineers bonkers, which for those professions and society in general is a very good thing. No computer engineer would design a system without knowing what every single circuit was for. And no one wants any uncertainty whatsoever when it comes to electrical wiring. We need to know with absolute precision what will, and will not, occur, each and every time we throw a light switch. Physics and engineering are, by definition, linear. They have to be. Mankind has no use for machines or fundamental theories that don't work.

Biologists, on the other hand, aim for reasonable accuracy rather than precision and have always been more comfortable with the mysteries and enigmas of life. Biology, by its very nature, is non-linear. Physicists would have us think that non-linearity is a kind of fuzzyheaded illusion; a temporary state of unreality until the underlying linearity is calculated. But just think of the weather. That's as real and as non-linear as things get and that's likely going to be a permanent state of affairs with weather.

Embracing uncertainty has never stopped good, thorough biological research, or hampered intelligent public policy formation. Biologists have mapped the entire human genome, but they still have little understanding of how the 40,000 or so genes work together to form a living human being. Nor do we actually know the real role of sleep, or why the human brain reverses polarities during sleep. But that doesn't stop physicians from recommending that people get a full night's rest. Nor did uncertainty about actual mechanisms stop health officials from warning patients against cigarette smoking. There was an association with smoking and lung disease and that was enough. A comparable analogy to the EMF situation would be if the tobacco industry had convinced everyone that no change in policy could be made — including the recommendations of private physicians — because adverse effects didn't fit the tobacco industry's research model for how lung tissue is supposed to behave. Talk about the fox guarding the henhouse.

In fact, we have far more data showing adverse effects from myriad EMF sources than we ever had for smoking and lung disease. Plus, some of those associations are for diseases like leukaemia for which no other environmental trigger, other than EMFs, has ever been found. Since no one has the vaguest idea what causes leukaemia, that observation alone is intriguing and should inspire far more curiosity in biologists and many more research dollars. But industry spin and a distortion of the dialogue away from a biological model (in which uncertainty is allowed) into a mechanistic physics model has created utter paralysis. We cannot make reasonable health recommendations for EMFs because the standard of proof that has been created around the subject is anomalous to our normal public health formation models. Societies would have no such trouble with policy changes or health recommendations with a toxin or chemical that had demonstrated multi-system effects, whether the underlying mechanisms were understood or not.

Part of the problem is that most medical doctors know very little about the subject. Bioelectromagnetics simply isn't taught in medical schools as part of clinical training. It's an adjunct of physics departments. I can speak with personal experience on this. I have 12 medical doctors in my immediate family. Some are quite distinguished in their specialities, but not a one of them knows anything about environmental bioelectromagnetics. Ironically, they all use any number of therapeutic EMF devices in their practices, including cardiac defibrillators, electrical bone stimulators, laser surgical tools, CT scans, MRI's and ultrasounds to name but a few — all of which, by the way, rely on non-thermal effects. But in some odd way, the MD's in my family just don't make the leap that this is bioelectromagnetics. They seem to think that the use of such devices is just standard practice for their areas of expertise. That's because the devices come from the university physics departments and are shepherded into clinical applications. There is an informational disconnect for the average medical practitioner that turns up in society at large.

The fact that bioelectricity is utterly fundamental to living organisms somehow has not got across. Living organisms are coherent energetic systems. That should be the bedrock understanding of physiology but the chemical-mechanistic model of living systems is still dominant in most people's thinking. Even those who know better constantly revert to it as their baseline understanding. Perhaps bioelectricity is so fundamental that we overlook it even in our generalisations, the same way we take for granted the ability of humans to move about.

But we leave bioelectricity out of our basic biological understandings now at our own peril. The advent of so many new technologies entering the environment unquestioned for safety is a serious mistake. The poet Dante Alighieri wrote, "The hottest places in hell are reserved for those who in time of great crisis maintain their neutrality." It is the ethical responsibility of the biologists and clinicians among us who do know better, to say once and for all that this must stop — and then do something about it.

30.4 INCREASING AMBIENT EXPOSURES AND ENVIRONMENTAL EFFECTS

Our background ambient exposures to countless EMFs are increasing at an alarming rate. This is especially true with the radio frequency bands. In its natural state, very little RF reaches the earth's surface. In the last 50 years, for the first time in our evolutionary history, we have utterly infused the earth's surface with a blanket of artificial energy exposures with no clear understanding of what the consequences may be. The exposures contain propagation characteristics such as modulation, complex pulse signalling, unusual wave forms such as square, spike and sawtooth shapes, and power intensities that simply do not exist in nature. These are all man-made artefacts.

Wireless applications these days are almost dizzying. Everything from remote control garage door openers and channel changers, to automatic car door locks, cordless phones, Palm Pilots, wireless computer networks in primary schools, wireless domestic stereo systems — you name it and an engineer has designed a way to unplug it. In America, unless an individual objects, both the gas and electric utility companies are removing the old mechanical meters on houses that measure how much gas or electricity is used within a set period of time. The old meters require a live human being to drive up to the house, get out of the vehicle, and check the meter. The new meters are small RF transmitters. A van with an antenna mounted on top can now drive through a neighbourhood and pick up the signal with the information from each house. No one has to leave the vehicle. Unfortunately, this is creating a whole new layer of low-level continuous RF exposure in residential areas. Ironically, these systems can only be installed in relatively "clean" environments for other kinds of RF pollution, or the signals would be obscured by interference.

Every new application, though within various guidelines for those particular technologies, just adds that much more to the aggregate. Rarely is the aggregate measured. But one European study a few years ago found that background RF in several cities had increased by 3000% over the previous decade. By anyone's reckoning, that constitutes a serious environmental alteration. The primary cause was thought to be mobile phone technology. Yet the scientific community is paralysed with a conversation about 'How much RF is too much?' 'Where's the data?' 'Is it reliable?' 'Has it been replicated?' 'Peer-reviewed?'

Researchers in England, France and Spain are reported to be having difficulties expressing their concerns about cell phone base stations. It is apparently easier for governments everywhere to protect the status quo than it is to open that subject up. No one knows what secrets will come flying out when the EMF Pandora's Box is fully breached. Whose national security and profits will be harmed? Who has liability? Responsibility? And the rationale protecting the status quo is: "What can be done anyway? People will never give up their conveniences." It is further reasoned that "Advances in medical treatments will cure whatever damage may be caused by such ubiquitous exposures." Wouldn't it be lovely if that were true? But I doubt it.

Now consider this... For the first time since medical records have been kept, brain cancer this year will become the #1 cancer death among people under the age of 20 in industrialised countries. Just 50 years ago, overall cancer incidence was 1-in-12. Now it is nearing 1-in-2. Dr. Robert O. Becker, one of the medical pioneers in this area, said in a recent radio interview that we may soon see a time when everyone experiences two or more unrelated cancer incidences in their lifetimes. Clearly something is turning our DNA off and on in abnormal ways, and it isn't likely that it is just environmental chemicals and hormone disrupters or increased longevity. Clearly there is an energy relationship here that is central to this disease process. All cell division — abnormal and normal — involves electrical activity. What is jamming the growth switch in the "on" position? What have we changed so ubiquitously in the environment within the last 50 years that might be responsible? One doesn't have to look very far to figure that out.

It is high time we introduced non-ionizing energy exposures into the environmental dialogue where it belongs. Non-ionizing radiation is an air pollutant like any other. Some states in America actually regulate it that way. But the energy modalities are not like our traditional toxins models. Energy modalities are non-linear. We may discover that some exposures are unsafe at any intensity. There is some evidence to suggest this is true. We already know that some living systems are exquisitely sensitive to the non-ionizing bands. We are dealing with a whole new model. It is our responsibility to keep making that point, and to educate traditional toxicologists and environmentalists who are only accustomed to dealing with the classic dose-response relationship models.

We've barely begun to investigate man-made EMF effects to other species, but consider this... Rates of extinction in plants and animals have accelerated to a point never seen before. While there are many complex factors involved, increases in artificial ambient energy levels are not even part of the dialogue — and should be. There are also unusual mutations in bacteria and viruses that are perplexing to researchers. HIV-like syndromes are now found in numerous species all over the globe, including domestic cats and sea mammals. Basic research projects, as well as military R&D, continue to infuse the world's oceans with massive amounts of ELFs in an effort to map the contours of ocean floors. We do this with no understanding of the effects to other species.

In the visible light bands, the nascent area of artificial-light ecology is finding that artificial light-at-night is disrupting nocturnal animals in devastating ways. As is true with so much of EMF research, all someone had to do was look to find effects. Observations include disorientation and disruption in breeding and migration cycles in turtles, flying insects, birds, butterflies and a host of other wildlife — mammals included. Even illumination reflected off of clouds — known as "sky glow" — can produce unnaturally bright conditions at night. It has been found that changing the colour of the light can help one species, yet harm another. Low-pressure sodium lights, which have more yellow in their spectrum, reduce moth deaths around the bulbs, but salamanders cannot navigate from one pond to the next under yellow or red light. Frogs have been observed to freeze for hours, even after lights have been turned off, and to suspend both feeding and reproductive behaviours.

And consider bird deaths and communications tower construction. One tower at 150 feet is thought to account for as many as 3000 songbird deaths a month in some flyways. It has been known for years that the songbird population of North America is plummeting. Only recently were towers considered a factor. But is the problem one of construction in migratory pathways? Or is something more interesting going on? What if RF is acting in a resonant capacity with the magnetite in avian eye areas, or by a more direct action with the avian brain? In a new about-to-be-published study, noted American ornithologist Robert Beason and Peter Semm of Germany found that a pulsed RF signal at 900 MHz, modulated at 217 Hz — similar to that of mobile phone technology — resulted in changes in neural activity in more than half of the avian brain cells being tested. Seventy-six percent increased their rates of firing by an average of 3.5 fold.

Also consider what we are doing to the ionosphere with basic research heating projects. The High Altitude Auroral Research Project — called HAARP — is the best known, but there are about 13 others in various parts of the world, including northern Europe. We know that what happens in the ionosphere directly affects our weather patterns. Consider that thunderstorms increased 25% over North America between 1930 and 1975, versus between 1900 and 1930. That period directly parallels our first introduction of serious environmental EMFs. Global warming may not be just about green house gases, and RF may not be the only energy culprit. As far back as 1975, a team of researchers at the Stamford University Radioscience Laboratories, then headed by Robert Helliwell, found evidence that power line emissions are amplified within the magnetosphere, causing a veritable rain of electron precipitation into the ionosphere, which could theoretically lead to global changes in weather patterns. The Helliwell Phenomenon was subsequently confirmed and reported in Physics Today. Now consider what we have added since 1975, and think what global weather patterns are doing. These may not just be interesting theories. There may already be real consequences. Yet energy is not part of the global warming dialogue. Why not?

As for the human population, aside from astronomical increases in cancers of all varieties, there are sharp increases in motor neuron diseases such as Amyotrophic Lateral Sclerosis (ALS) and Multiple Sclerosis (MS). Increases in immune system diseases such as HIV, Chronic Fatigue Syndrome and allergies of all kinds. Asthma is increasing at an alarming rate in metropolitan areas where some of the highest ambient concentrations of environmental RF are now measured. Studies have found that RF elevates serum histamine levels. Why isn't anyone connecting the dots for public health officials, or recommending further study of this possible relationship?

And consider electrical sensitivity... that most perplexing allergic-like reaction that some people experience when exposed to certain electromagnetic fields. Electrical allergy is not even recognised in America as a *bona fide* syndrome. People who report such symptoms are often referred to psychiatrists. At least it is taken more seriously in Europe as a physiologically based phenomenon. Hippocrates, the father of western medicine, once taught his students: "Listen to your patients. They are handing you the diagnosis." People usually know what makes them ill. The electrically sensitive may be our new canaries-in-a-coal mine. They may be our sentinel population, forewarning us of what is on the horizon for others.

Also consider the rise in violence and suicide and depression, especially among the young. Then consider the incredible increases in learning and behavioural disabilities in children. Perhaps if someone looked, they would find a relationship between the use of prenatal ultrasound during the time of fetal brain development and learning disabilities later in life. Some preliminary studies may have already found such an association.

Now let's consider the newest popular technology, the one that will likely tip the scales for adverse effects in the most amount of people. Cell phone radiation — both from the handsets and the base stations — is just like secondary cigarette smoke but worse because the radiation has the ability to alter DNA. The person using the cell phone has a voluntary exposure but no one around that person does. One recent measurement taken by Robert Kane, a former engineer at Motorola, found that a person standing 10 feet away from an in-use portable phone had an exposure of 0.001 mW per square centimetre. In the normal, non-radiated environment — if there is such a place today — that exposure would be 0.000000000000001 mW per square centimetre.

A recent study from Japan reported by BBC News Online found that radiation levels inside of trains can exceed international safety limits if even a small number of passengers are using their phones. The radiation just bounces around the interior of the metal structure. That should come as no surprise. All someone had to do was measure. Unfortunately, we keep approaching this one consumer product at a time, when we need talk about energy as a broad-based environmental pollutant. Then people would understand that when they use a cell phone, they are irradiating everyone else around them too.

And the final thing that should no longer be left out of the environmental dialogue is perhaps the most frightening of all. It's the next generation of military applications. Most people, myself included even though I am a journalist and supposed to be fearless about such things, recoil from getting too involved with this side of the subject. It isn't called "Black Research" in bioelectromagnetics circles for nothing. There is something fundamentally corrupt about manipulating people through their own biorhythms. But to leave military applications out is naïve, especially since many of the researchers in the bioelectromagnetics community have worn different hats in their careers, crossing back and forth between classified work and civilian research. These are the same people controlling the committees and the standards. But these decisions should rest exclusively in civilian hands. We are the ones who decide what levels of risk to assume. Military R&D and its applications must get closer responsible civilian scrutiny.

For the first time in history, in Afghanistan, America is conducting an almost exclusively remote-controlled war. After what happened on September 11th, terrorism must be stopped and I support what my country is doing toward that effort. But no country could conduct warfare in this way without complete autonomy regarding the liberal use of the non-ionizing bands. It is also a big factor in the developing weaponisation of space. No one has any idea what the environmental consequences of this will be. What is worse is that no environmentalist is even raising the question — because they do not know it is relevant.

There is an interesting situation right now brewing in the U.S. A lifelong military researcher, named Richard Albanese, who is also a civilian medical doctor, has broken ranks with the military over the safety of a particular kind of radar called "phased array." Phased array is being used to update an older over-the-horizon radar installation called Pave Paws, on the Northeast coast of America. The goal of the update is to link the facility to the controversial National Missile Defense System. Pave Paws, by the way, is aimed at Europe. The old system currently reaches to Spain. The proposed system will reach throughout the continent and beyond.

According to Dr. Albanese, phased array is a completely different technology. Unlike traditional radar, phased array contains millions of steep overlapping wave banks that hit the human anatomy at different times and from different directions. Health concerns have nothing to do with heating effects or signal strength. Signal shape is the only determinant, no matter how weak the intensity. There is nearly no research on phased array other than that done by Albanese, who has seen disturbing results in some therapeutic applications research he conducted at the U.S. Air Force's Electromagnetic Health and Safety Program, which he directs. Dr. Albanese has stated that the results of his team lead inescapably to the conclusion that phased array can injure a growing fetus, cause birth defects, and both initiate cancer as well as accelerate its growth. He also cautions against the development of hardening of the arteries, neurological disorders and autoimmune diseases like lupus erythematosis in the exposed population.

The Air Force's research program is classified, as is most of Dr. Albanese's work, which is why you will not find a long list of articles published after his name. But as a private citizen and medical doctor speaking out, he is acting bravely according to the Hippocratic oath to: "First, do no harm." As you can imagine, the powers-that-be are not happy with him. They have been successful thus far at keeping him off of a review panel at the National Academy of Sciences. Phased array is now being incorporated into several new civilian applications, particularly aboard commercial aircraft, with what little research that does exist, being kept classified. That cannot be justified.

30.5 WHAT A NEW EMF REVOLUTION WOULD ENTAIL

We have an environmental problem, and it may not be an insignificant one. The technologies we've created are supposed to be our partners, not our nemesis. For the most part, their benefits are obvious but the risks are hidden. Like an impressionist Monet painting, we have to pull far back to see the actual contours. The full range of environmental effects from EMFs may only come into focus when we keep our vision very broad. It is too easy to get bogged down in one place or another — like radiation in trains, or the power output of particular cell phone models, or building set-backs from tower masts, or specific absorption rates, or whether headsets do, or do not, reduce radiation. While these are all important aspects, they are nevertheless individual brush strokes, not the whole canvas.

Biologists and clinicians are the ones best able to see the large canvas and to keep it in focus. Physicists and engineers can help design the tools for measurement and mitigation, but the primary effort must come from the biological sciences if there is to be the kind of broad-based change in consciousness that we need. Fundamental control must be taken away from the physical and engineering communities. Those areas still have a critical role to play but when it comes to biological effects, they must hereafter be considered adjunct. But first, biologists must assume the responsibility to learn what the bioelectromagnetics community has been studying all of these years.

Other outlines of this revolution could entail a phalanx of both scientists and attorneys whose mandate is to protect the environment. The National Resources Defence Council in America is such an organisation but they do not address EMF issues. The Union of Concerned Scientists, as well as Physicians for Social Responsibility, are also good models, but they also do not address nonionizing radiation issues — partly due to an absence of knowledge. Immediate efforts should include educational outreach to medical schools, environmentalists, journalists, government officials, land planning associations, universities, and architectural design firms, among others. Keep in mind, most people really do not understand this subject and trust that someone more knowledgeable is protecting them.

Emphasis should be placed on continual funding from unbiased sources for appropriate research into the long-term, low-level effects of the nonionizing bands. Too much inappropriate research has been done, then used in an inappropriate manner. At the present time, there is increased funding for mitigation and therapeutic applications because profits can be made there. Unfortunately this is serving as a distraction from the hazards side, as if we can just skip over the risks and go directly to the benefits without understanding the tradeoffs. While seductive, that is an irresponsible approach.

There should be an international organisation of professionals ready to create informational papers to respond to every dumb thing that is said from now on about thermal v. non-thermal effects. All that leads to is paralysing, obsolete hogwash, to the benefit of RF industries. Biologists must insist on being seated at every standards and research review committee. Make a scene, if need be. Societies must also insist through legal channels that manufacturers be held accountable for the safety of their products and the processes by which they function. EMF industries must be encouraged in preventive ALARA principles — As Low As Reasonably Achievable. And the Precautionary Principle should be considered the reasonable person's approach.

Lastly, biologists and concerned others must get over the fear of professional ridicule. Keep in mind, you are on the cutting edge of this broader environmental understanding. Others will catch up. One day, you will be recognised for the visionaries that you truly are.

30.6 CONCLUSION

At the turn of the millennium, a Public Broadcast commentator in the U.S, summed up the recent centuries like this: "The 19th Century was the century of chemistry. The 20th Century was the century of physics." And he predicted that the 21st Century would be the "century of biology." He was referring to advances in genetic engineering and cloning and sophistication in organ transplantation and other such wonders of modern medicine.

I would make a slightly different prediction. I would say that the 21st Century will be the century of biology and physics. We are on the cusp of a radical shift. It is nothing short of the reintroduction of electromagnetism back into the biological paradigm where it had traditionally been. New wonders abound in everything from ion channel work to EMF therapeutics. But hopefully, by the time we understand the therapeutic side, we will not have done so much damage to ourselves and the other creatures of the earth that any therapeutic promise will be far too late.

30.7 REFERENCES

Adey, W.R., Bawin, S., and Lawrence, A.F. "Effects of Weak Amplitude-Modulated Fields on Calcim Efflux From Awake Cat Cerebral Cortex," *Bioelectromagnetics,* **3** (1982): 295-308.

Adey, W.R. Testimony, Ad Hoc Subcommittee on Consumer and Environmental Affairs, United States Senate Committee on Governmental Affairs, Hearing on Health Risks Posed by Radar Guns; The Extent of Federal Research and Regulatory Development of Microwave Emissions From Hand-Held Radar Guns, August 7, 1992.

Adey, W.R. "Cell and Molecular Biology Associated with Radiation Fields of Mobile Telephones," *Review of Radio Science, 1996-1999,* eds. W.R. Stone and S. Ueno, Oxford University Press, 1999, pp 845-872.

Adey, W.R. "Electromagnetic fields and the essence of living systems," *Modern Radio Science 1990,* ed. J. Bach Andersen, Oxford University Press, 1990.

Albanese, Richard. "Is PAVE PAWS Safe? Military needs phased-array radiation data to decide if radar facility is safe for Cape Cod," *Cape Cod Times,* January 27, 2002.

An Evaluation of the U.S. Navy's Extremely Low Frequency Communications System Ecological Monitoring Program, National Research Council, National Academy Press, Washington, D.C. 1997.

Andersson, Bengt *et al.,* "A Cognitive-Behavioral Treatment of Patients Suffering from 'Electric Hypersensitivity,' Subjective Effects and Reactions in a Double-Blind Provocation Study," *JOEM,* Vol. 38, No. 8, 1996.

"Atlas of Cancer Mortality in the United States 1950-94," *National Institutes of Health and National Cancer Institute,* NIH Publication No. 99-4564, September 1999.

Beal, Colleen, *et al.* "Brain Tumors among Electronics Industry Workers," *Epidemiology,* Vol. 7, No.2, March 1996.

Beason, Robert C. "The bird brain: magnetic cues, visual cues, and radio frequency (RF) effects."*Proceedings of Avian Morality at Communication Towers Conference, 117th Meeting of the American Ornithologists' Union,* Cornell University, Ithaca, N.Y., August 11, 1999.

Beason, Robert C., and Semm, Peter. "Neuronal responses to digital signals," accepted for publication, 2002.

Becker, Robert O. "The Hazards of Electroradiation, From the testimony of Robert O. Becker, MD, before the House Sub-Committee on Water and Power, 22 September 1987," *Earth Island Journal,* Winter 1988, pp. 27-29.

Becker, Robert O., and Selden, Gary. *The Body Electric,* William Morrow, N.Y. 1985.

Becker, Robert O. *Cross Currents, the Perils of Electropollution, the Promise of Electromedicine,* Jeremy Tarcher, Inc. Los Angeles 1990.

Beckman, Kristen. "New Metawave Products Address Capacity Issues," *RCR News,* December 14, 1998.

Bise, William. "Low Power Radio-Frequency and Microwave Effects on Human Electroencephalogram and behavior," *Physiol. Chem & Physics,* 10, 1978.

Blackman, Carl. "Is Caution Warranted in Cell Tower Siting? Linking Science and Public Health," *Cell Towers, Wireless Convenience? or Environmental Hazard? Proceedings of the Cell Towers Forum, State of the Science/State of the Law, December 2, 2000,* ed. B. Blake Levitt. pp. 50-64.

Borenstein, Seth. "Birds fall victim to new predator: Communication Towers." *Free Press,* August 5, 1999.

Borosodi, Bela. "The Fighting Next Time, why reformers believe that preparing the military for next-generation warfare is radical and crucial and one more casualty of 9/11," *The New York Times Magazine,* March 10, 2002.

Bowmen, Lee. "Telecom towers kill up to 5 million birds a year." *Denver Mountain News,* September 25, 1999.

Braile, Robert. "Proliferation of towers poses threat to birds." *The Boston Globe,* May 23, 1999.

Broad, William J. "Test of Underwater Sound System Showing No Harm to Animals." *The New York Times,* January 21, 1996, National, p.20.

Brooks, Clark. "Fatal Flaws; How the military misled Vietnam veternans and their families about the health risks of Agent Orange," *San Diego Union-Tribune,* November 1, 1998.

Burns, Robert. "Pentagon Unveils Non-Lethal Weapon," *Associated Press,* March 8, 2001.

Candedy, Dana. "Mysterious Virus at Source Drives Up the Price of Pearls." *The New York Times,* May 24 1998.

Cox, Jim. "Blinking lights mark scenes of death for birds." *Tallahassee Democrat Online,* 1997.

Daniells, C. *et al.* "Transgenic nematodes as biomonitors of microwave-induced stress," *Mutation Research,* 399:55-64, 1998.

"Does USAF Have Secret Health Studies on Phased Arrary Radiation? Tensions Surface at NAS-NRC Meeting on Pave Paws Radar," *Microwave News,* Vol. XXII, No. 2, March/April 2002.

"EMF Good for Trees?" *Science,* Vol 267, January 27, 1995.

Feagin, J.E. Wurscher, M.A., Ramon C., Lai, H.C., "Magnetic Fields and Malaria," *Biologic Effects of Light: Proceedings of the Biologic Effects of Light Symposium,* ed. M.F. Holick and E.G. Jung, Kluwer Academic Publishers, Hingham, MA. 1999. pp. 343-349.

Fisk, Stewart. "Comet Assays and RF cell damage..." *Electronics Australia,* April 1999.

Fitch, Charles S. "Tower Lighting, Much to Learn." *Radio World,* July 21, 1999.

French, P.W. *et al.* "Mobile Phones, heat shock proteins and cancer," *Differentiation,* **67**:93-97, 2000.

Glanz, James. "Cast of Star Wars Makes Comeback in Bush Plan," *The New York Times,* International, July 22, 2001, p. 10.

Glinz, Franz. "The Forest is dying from 'Electrosmog.' *auto-illlustrierte,* February, 1992.

Hademenos, George. "The Biophysics of Stroke," *American Scientist,* Vol. 85, No. 3, May-June 1997.

Hanley, Robert. "A Mystery in a Refuge: 600 Dead Geese." *The New York Times,* Metro, December 10, 2000, p. 60.

Harder, Ben. "Deprived of Darkness, the unnatural ecology of artifical light at night," *Science News,* Vol 161, April 20, 2002.

Hertel, H. "The Forest Dies as Politicians Look On," *Raum&Zeit,* May/June, 1991.

Hitt, Jack. "Battlefield Space," *The New York Times Magazine,* August 8, 2001.

Hyland, G.J. "Non-thermal bioeffects induced by low-intensity microwave irradiation of living systems," *Engineering Science and Education Journal,* December, 1998.

Hyland, G.J. "Physics and biology of mobile telephony," *The Lancet,* Vol. 356, November 25, 2000.

Hyland, G.J. Appendix 15: Memorandum submitted by Dr. G.J Hyland to the Select Committee on Science and Technology, House of Commons, UK, 2001.

Jamison, Kay Redfield. *Night Falls Fast, Understanding Suicide,* Alfred Knopf, New York 1999.

Kane, Robert. *Cellular Telephone Russian Roulette,* Vantage Press, New York, 2001.

Kane, Robert. *On Second-Hand RF Radiation,* author-published, 2002.

Kasevich, Raymond S. "Brief Overview of the Effects of Electromagnetic Fields on the Environment," *Cell Towers - Wireless Convenince? or Environmental Hazard? Proceedings of the Cell Towers Forum, State of the Science/State of the Law, December 2, 2000,* ed. B. Blake Levitt, New Century Publishing 2001, pp. 170-175.

Kellert, Stephen R. *The Value of Life, Biological Diversity and Human Society,* Island Press/Shearwater Books, 1996.

Kerber, Ross. "Making (radar) waves," *The Boston Globe,* July 2, 2001.

Knox, Richard A. "Science and secrecy: A new rift. Proprietry interests found to intrude on research disclosure," *The Boston Globe,* March 20, 1999.

Lai, Henry. "Biological Effects of Radio Frequency Radiation from Wireless Transmission Towers," *Cell Towers, Wireless Convenience? or Environmental Hazard? Proceedings of the Cell Towers Forum, State of the Science/State of the Law,* ed. B. Blake Levitt, New Century Publishing 2001. pp. 65-74.

Lai, Henry. "Genetic Effects of Nonionizing Electromagnetic Fields," Paper presented at the International Workshop of Biological Effects of Ionizing Radiation, Electromagnetic Fields and Chemical Toxic Agents, Sinaia, Romania, October 2, 2001.

Lai, H., Singh, N.P. "Melatonin and a spin-trap compound block radiofrequency electromagnetic radiation-induced DNA strand breaks in rat brain cells," *Bioelectromagnetics,* **18 (6)**:446-454, 1997.

Lai, H. Carino, M.A., Singh, N.P., "Naltrexone blocks RFR-induced DNA double stand breaks in rat brain cells," *Wireless Networks,* 3:471-471, 1997.

Leggett, Doreen. "Researchers begin new round of PAVE PAWS study," *The Cape Codder,* January 11, 2002.

Leggett, Doreen. "Physician will study PAVE PAWS impact on health," *The Cape Codder,* January 2, 2002.

Levitt, B. Blake, "Telecommunications Technologies - An Overview: Wireless Convenience? Or Looming Environmental Problem?" *Cell Towers, Wireless Convenience? or Environmental Hazard? Proceedings of the Cell Towers Forum, State of the Science/State of the Law, December 2, 2000,* ed. B. Blake Levitt, New Century Publishing, 2001. pp.1-49.

Levitt, B. Blake. *Electromagnetic Fields, A Consumer's Guide to the Issues and How to Protect Ourselves,* Harcourt Brace, 1995.

Levitt, B. Blake. "Towers of Power: Wireless Technology at What Cost?" *Litchfield County Times,* April 4, 1997.

Levitt, B. Blake. "Electromagnetic Fields: Their Impact on the Environment," *Quinnipiac Magazine,* Spring Edition, 1996.

Levitt, B. Blake. "Cell-Phone Towers and Communities, The Struggle for Local Control," *Orion Afield,* Autumn Edition, 1998.

Levitt, B. Blake. "Telecom Towers Tsunami, There are medical and political ramifications to cell tower siting," *The New Milford Times,* March 3, 2000.

Liakouris, A.G. "Radiofrequency (RF) Sickness in the Lilienfeld Study: An Effect of Modulated Microwaves?" *Archives of Environmental Health,* Vol. 53, No. 3, May/June 1998.

Libby, Sam. "When Homing Pigeons Don't Go Home Again." *The New York Times,* December 6, 1999.

"Light Behind the Knee Shifts Circadian Rhythms," *Microwave News,* Vol XVII, No. 1, January/February 1998.

Lohmeyer, Michael. "Environment contaminated by Microrwaves, Directional Radio Beams cut Breaks into Forests," *Die Presse,* July 31, 191.

Manville, Albert M. III. "Avian Mortality at Communication Towers: Steps to Alleviate a Growing Problem." *Cell Towers, Wireless Convenience? or Environmental Hazard? Proceedings of the Cell Towers Forum, State of the Science/State of the Law,* New Century Publishing, 2001, ed. B. Blake Levitt, p. 75-86.

Manville, Albert M. III. "Migratory Bird Conservation and Communication Towers: Avoiding and Minimising Conflicts." *Summary of Interagency Meeting,* Panama City, FLA. November 17, 1998.

Manville, Albert M. III. "Avian mortality at communication towers: background and overview." *Proceedings of Avian Mortality at Communication Towers Conference, 117th Meeting of the American Ornithologists' Union,* Cornell University, Ithaca, N.Y., August 11, 1999.

"Many Reports of Deformities Among Frogs are Puzzling," *The New York Times*, October 13, 1996, National, p. 33.

Marino, Andrew A. "Assessing Health Risks of Cell Towers," *Cell Towers - Wireless Convenience? or Environmental Hazard? Proceedings of the Cell Towers Forum, State of the Science/State of the Law*, December 2, 2000, ed. B. Blake Levitt. pp. 87-103

Marino, Andrew, and Becker, Robert O. "High Voltage Powerlines, Hazards at a Distance,"*Environment*, Vol. 20, No. 9, November 1978.

Marino, Andrew, and Ray, Joel. *The Electric Wilderness*, San Francisco Press, 1986.

McVicker, Dee. "RF Emissions Can Detoxify Soil." *Radio World*, February 22, 1995.

Meadows, Donella H. "Where Have All the Frogs and Toads Gone?" *Lakeville Journal*, December 17, 1998. Syndicated.

Meinesz, Alexander. *Killer Algae*, trans. Simberloff, Daniel, University of Chicago Press, 2000.

Milius, Susan. "Things That Go Bump, Vibrational messages that we're just beginning to get." *Science News*, Vol 159, March 24, 2001.

Mishler, Wayne. "Who's Playing Hell with HAARP?" *Monitoring Times*, October 1996.

Monastersky, R. "A sound way to take the sea's temperature?" *Science News*, CVol. 154, August 29, 1998.

Monastersky, R. "The Case of the Global Jitters, Even in seemingly stable times, climate can take an abrupt turn," *Science News*, Vol 149, March 2, 1996.

Morgan, R.W. *et al.* "Radiofrequency Exposure and Mortality from Cancer of the Brain and Lymphatic/Hematopoietic Systems," *Eipdemiology*, Vol 11, No. 2, March 2000.

"Natural Climate Variability on Decade-to Century Time Scales," *National Research Council*, Washington, D.C. 1995.

"Navy Withdraws From Hearing on Controversial Sonar." *Nature's Voice, National Resources Defense Council*, March/April 2001.

Noble, Barbara Presley. "Staying Bright-Eyed in the Wee Hours," *The New York Times*, June 13, 1993.

"Norwegian Investigation of Birth Defect Cluster Focuses on RF/MW Radiation from Electronic Warfare Systems," *Microwave News*, Vol. XVII, No.5, September/October 1997.

O'Harra, Doug. "HAARP's mixed signals, solid research or menace to Alaskans?" *We Alaskans, The Anchorage Daily News Magazine, April 7, 1996.*

"Pave Paws Radar on Cape Cod At Center of New Controversy," *Microwave News*, Vol. XX, No.2, March/April 2000.

Payne, Katy. *Silent Thunder, In the Presence of Elephants*. Simon & Schuster, 2000.

Pear, Robert. "Pentagon Misspending Spree," *The New York Times*, Week in Review, July 25, 1999, p. 2.

"Physicists and Biologists Butt Heads at First NIEHS EMF Risk Workshop," *Microwave News, Vol XVII, No. 2, March/April 1997.*

Raloff, Janet. "Does Light Have a Dark Side?" *Science News,* Vol 154, October 17, 1998.

Samet, Jonathan M. *et al.* "Fine Particle Air Pollution and Mortality in 20 U.S. Cities, 1987-1994," *The New England Journal of Medicine,* Vol 343, Number 24, December 14, 2000.

Sandstrom, M. *et al.* "Neurophysiological Effects of Flickeing Light in Patients with Perceived Electrical Hypersensitivity," *JOEM,* Vol. 39, No. 1, January 1997.

Santini, Roger. "Symptoms Experienced by People Living Near Mobile Phone Base Stations," *Pathologie Biologie,* (accepted for publication 2002.)

Seachrist, Lisa. "Shocking Rhythms, How do jolts of electricity bring people back to life after heart fibrillations?" *Science News,* Vol. 149, January 27, 1996.

"Secrecy vs. Public Health: Industry Must Not Set the Standard," *Microwave News,* Vol. XVII, No. 3, May/June 1997.

Shabnam, G. and Johansson, O. "A Theoretical Model Based Upon Mast Cells and Histamine to Explain the Recently Proclaimed Sensitivity to Electric and/or Magnetic Fields in Humans, *Medical Hypotheses, 54,* April 2000, pp.663-671.

Smith, Cyril W., Simon Best. *Electromagnetic Man, Health & Hazard in the Electrical Environment,* St. Martin's Press, N.Y.1989.

Smith, Hilary. "E-tenna uses military technology to develop cheaper, more efficient handset antennas - GAO report asks for more data," *RCR News,* May 28, 2001.

"Sounding Out the Ocean's Secrets," *Beyond Discovery,* National Academy of Sciences, March 1999.

Stevens, R.G., Wilson, B.W., and Anderson, L.E. eds., *The Melatonin Hypothesis, Breast Cancer and the Use of Electric Power,* Battelle Press, Columbus, OH, 1997.

Stine, Randy J. "Radio Station Towers are for the Birds." *Radio World,* October 13, 1999.

"Strong Evidence for an Alzeheimer's-EMF Connection," *Microwave News,* Vol. XVII, No.1, January/February 1997, p. 1.

"Suicide Linked to EMF Exposures Among Electrical Workers," *Microwave News,* Vol. XX, No.2, March/April 2000.

"The Incidence of Brain and Central Nervous System Tumors in Residents in the Vicinity of the Lookout Mountain Antenna Farm in Golden, Colorado," Prepared by the *Colorado Department of Public Health and Environment,* February 1999.

Travis, John. "Atomic Resolution snapshots illuminate cellular spores that control ion flow." *Science News,* Vol. 161, March 9, 2002, p. 152-154.

Travis, John. "Timely Surprises, Biological clocks sense light in obscure ways," *Science News,* Vol 154, July 11, 1998.

Travis, John. "Microbe linked to Alzheimer's disease," *Science News,* Vol. 154, November 21, 1998.

"U.S. Air Force Looks to the Battlefields of the Future: Electromagnetic Fields That Might 'Boggle the Mind'," *Microwave News,* Vol. XVII., No.1, January/February 1997.

United States Environmental Protection Agency, *An Investigation of Energy Densities in the Vicinity of Vehicles with Mobile Communications Equipment and Near a Hand-held Walkie Talkie,* Office of Radiation Programs, ORP/EAD 79-2, March 1979.

United States Environmental Protection Agency, *Radiofrequency Radiation Levels And Population Exposure in Urban Areas of the Eastern United States,* Office of Radiation Programs, EPA-520/2-77-008, May 1978.

"Urban Electrosmog Increasing," *Microwave News,* Vol. XX, No. 4, July/August 2000.

Vinocur, Barry. "Is Ultrasound Safe?" *IMTS News,* 1.11.14.83.

Volkrodt, Wolfgang. "Electromagnetic pollution of the environment," *Environment and Health; A Holistic Approach,* Proceedings of the International Symposium, Man... Health... Environment, Luxembourg, Germany, March, 1988, Avebury Press, England, 1989.

Volkrodt, Wolfgang. "Are Microwaves faced with a Fiasco Similar to that experienced by Nuclear Energy?" *Wetter-Boden-Mensch,* April 1991.

Watson, Tracy. "Towers, antennas silencing songbirds." *USA Today,* September 2, 1999.

Wee, Eric L. "Homing pigeons Race Off to Oblivion." *The Washington Post,* October 8, 1998.

Windrem, Robert. "The high-tech hunt for terrorist lairs, U.S. could employ satellites and antenna arrays to map caves," *MSNBC News Online,* November 27, 2001.

Wojtas, Joe. "Looking for Answes After a Dolphin Dies," *The New York Times,* March 25, 2001.

Yoffe, Emily. "Silence of the Frogs." *New York Times Magazine,* December 13, 1992.

Yoon, Carol Kaesuk. "Newly Found Fungus is Tied to Vanishing Species of Frog." *The New York Times,* June 28, 1998.

Yoon, Carol Kaesuk. "Swift Disease Killing Oaks in California." *The New York Times,* August 13, 2002, p. 1.

Questions & Discussion[1]

Janet Newton (The EMR Network, Vermont, US)

Dr Sienkiewicz stated that NRPB is looking for a succession of effects that could lead to recognition of adverse non-thermal effects. Have you considered blood-brain barrier studies such as those by Persson and Salford? (Persson BRR, Salford LG, Brun A, Blood–brain barrier permeability in rats exposed to electromagnetic fields in wireless communication. Wireless Network 3:455-461, 1997) Exposures can open the brain to injury whether the sources are viral, bacterial, or chemical toxins. If the blood-brain barrier effect comes from chemical exposure it is not questioned that the exposure will lead to adverse health effects. Blood-brain barrier effects have been observed in radio frequency exposures.

RF exposures may also cause DNA strands to break. Jerry Phillips has replicated the Lai/Singh studies and has demonstrated double strand DNA breaks.

A large build up of Heat Shock Proteins (HSPs) occurs after exposure to both ELF and RF radiation. The DiCarlo-Litovitz study (DiCarlo A, White N, Fuling G, Garrett P, Litovitz T. 2002. Chronic electromagnetic field exposure decreases HSP70 levels and lowers cytoprotection. Journal of Cellular Biochemistry 84:447-454) shows that after the first exposure the Heat Shock Protein levels are increased, and after repeated or prolonged exposures of several different periods of time Heat Shock Protein levels are then decreased. HSPs are a cellular substance whose function is to repair DNA breakage—and has been found in other RF exposure studies. When HSPs are depleted they are not available to make the needed repairs to the DNA breaks. Tice studied micronuclei formation and three times obtained approximately the same results. (Tice RR, Hook GG, Donner M, McRee DI, Guy AW. Genotoxicity of radiofrequency signals. I. Investigation of DNA damage and micronuclei induction in cultured human blood cells. Bioelectromagnetics **23**: 113-126, 2002.) When blood-brain barrier and Heat Shock Protein effects occur after a chemical toxic exposure they are seen as an adverse health effect. This same succession of effects after exposure to EMFs leads one to believe there are non-thermal EMF effects that adversely affect human health.

Dr Zenon Sienkiewicz (National Radiological Protection Board, Oxon)

That sounds a very persuasive argument. I do not have particular expertise in cellular biochemistry but various scientists who have looked critically at the

---

[1] Editor's note: Dr Sakari Lang (Nokia) was asked several questions but requested that this part of the discussion not to be publicly recorded.

studies reporting very low level effects on the blood-brain barrier concluded that these results are not particularly robust. Previous studies suggested you needed exposure at about 10 W/kg to elicit a response. Therefore any changes in permeability are believed to be thermal in origin. So these studies are inconsistent with these data. The same inconsistency applies to the studies investigating DNA strand breaks. There are studies that have failed to reproduce the effect on DNA. Overall I do not consider biochemical changes in isolated tissues can be taken as an indication of adverse health effect. It may be different for testing chemical toxicity: I do not know that field.

I think this also illustrates a possible difference in outlook between ourselves regarding the suitability of studies for setting standards for human exposures. I probably would like validation and corroboration before I would feel confident in using a particular result. This would include asking such questions as has the study been successfully replicated; are the results consistent with the results of similar studies, and are the results scientifically justifiable? Of course every scientist or expert group asks similar questions. They will all probably have slightly different opinions and so arrive at slightly different conclusions. It is an extremely difficult task assessing possible health risks from electromagnetic fields and there are no easy answers.

Dr Gerard Hyland (Formerly University of Warwick, UK)

If we did not understand the mechanism of microwave heating, would thermal effects be excluded from consideration in the formulation of safety guidelines?

Dr Zenon Sienkiewicz

I wish I knew the answer. The trouble is, when you can actually see some consistent effects, which can be understood, you need not look for any other mechanisms. When living tissues are exposed to a microwave field, you can get heating and everyone accepts that. End of story. Scientists sometimes seem unwilling to stretch their imaginations to consider other possibilities.

Arwell Barrett (Health and Safety Executive, UK)

Do we believe in or need a futuristic wish list? Computer software does not cope with the way that I want to do things. I certainly would be very interested to know how my smart kitchen will know when I go and take a couple of eggs, some cocoa chocolate and some butter out of the fridge, a packet of flour out of the cupboard and a lot of sugar and syrup, and that I am going to cook chocolate brandy.

Dr Peter Grainger (University of Bristol, UK)

You will be pleased to hear that at the moment Microsoft have not been able to bring out any Blue-tooth drivers for Windows, but they are working on it and they are promised for the next generation of Windows. The intention is for everything to have an identification tag so enabling the smart equipment to keep track on where items are moved from and to.

Professor Olle Johansson (Karolinska Institute, Stockholm)

In Sweden there is growing concern regarding teenagers. People are looking closely at their behaviour. Nowadays they mostly sit still in front of computers in school, after school and in the evening. Also from your presentation I thought it was a pretty dull world. When we aim to cook the roast beef to look exactly the same every time. We have the "Naked Chef" from Great Britain and he would be put out of business because we would not need him. Is this something we want? Some do not want to get in touch with their oven or their toaster. They want to lead a normal human life. As a representative of the medical profession, I do not want elderly people who are immobile. I do not want them to pay their invoices via Internet because I want them to walk several kilometres to the post office to exercise and to meet people and socialise. In that way they avoid osteoporosis.

Dr Peter Grainger

I also share your concerns. There are certainly 3 or 4 home of these smart homes to my knowledge that are being developed. People living in them reported how much they enjoy the experience. Your are concerned about the lack of exercise that children get but with the new generation of phones they will go out jogging carrying the phone with them and play computer games or watch their latest "soap" as they run along. But is this the way we want life?

With regard to your comments on food, I am sure there will be inherent inbuilt variability (un-documented features) within the software to ensure your meal will have interesting variability.

Professor Olle Johansson

In Regents Park just outside here young people are lying down on the grassland, they look more and more like Swedes who are over weight.

Dr Peter Grainger

Blue-tooth and its successors are looking at communication frequencies up to 8.5 GHz. We have to say whether this is what we want and let the manufacturers and developers know in no uncertain terms which way we want society to go.

Professor Olle Johansson

During the last year I have encountered on the Internet discussion groups who claim that the ICNIRP, which has been referred to many times as a private organisation. It is not connected in any way to the WHO nor the UN, or any other public organisation .Its made up of representatives either directly or indirectly from industry. The kind of data that has been put on the Internet is very persuasive. I talked with Kjell Hansson Mild for example. He said that "yes this is not a governmental agency". The Swedish representative on ICNIRP, died last year, He was member of the Telia Mobile and of Stockholm Energy. All this makes me worried. It would be better if people sitting on these bodies were not from the industry. And also we need a government protection system for the individual.

Dr Zenon Sienkiewicz

Each members of ICNIRP has acknowledged scientific expertise. That is why they have been appointed to the Commission. It is independent of industry but many of the members are still highly active researchers in their field. They are not retired. When scientists need funding they may turn to industry to help fund their research. Maybe that situation should not occur, and maybe that is a problem. I am not going to try and decide, or to suggest alternative funding strategies.

Professor Olle Johansson

That is very hard to say what is the most appropriate. Sweden and England are extremely rich countries. We should be able to persuade government to support this kind of research

Dr Zenon Sienkiewicz

Another problem is should a responsible industry keep their profits and leave the government or others to pick up the cost of doing the research into health effects?

David Gillett (Environmental Harmony, Suffolk)

We are studying social development and looking at the way we respond to the planet. These two days have been marvellous. There are many people working very hard looking at all these different problems. At the other end of the scale there are many ordinary people needing help who are living in high force fields and are suffering without being aware of the problems. There are a few representatives here that are suffering from a cocktail of environmental bombardments. We are never going to have all the answers. I wonder out of conferences like this, whether a worldwide organisation can actually start looking at worldwide problems. There is enough concern nowI would have thought, for a movement around the world where biologists, physicists and medics get together to draw up better standards and provide answers to questions discussed in this book.

Dr Zenon Sienkiewicz

There is the WHO EMF programme.

David Gillett

How is it organized? Because we are actually seeing the results of the EMF build up now. Is it too early to see the results of their work?

Dr Zenon Sienkiewicz

WHO will publish the results of their latest evaluation into the health effects of ELF fields in the next couple of years, and then they will publish their review of RF fields in about 2005-6. Within this framework they are holding various seminars, workshops and meetings. So yes, there is a lot of activity and interest into possible health effects. And WHO would seem to be the best suited organization to perform this function. Details of the entire WHO-EMF programme can be found on their website at http://www.who.int/peh-emf/en/

APPENDIX

Catania Resolution

September 2002

**The Scientists at the International Conference
"State of the Research on Electromagnetic Fields –
Scientific and Legal Issues"**

organized by ISPESL*, the University of Vienna, and the City of Catania, held in Catania (Italy) on September 13th – 14th, 2002, agree to the following:

1. Epidemiological and in vivo and in vitro experimental evidence demonstrates the existence for electromagnetic field (EMF) induced effects, some of which can be adverse to health.
2. We take exception to arguments suggesting that weak (low intensity) EMF cannot interact with tissue.
3. There are plausible mechanistic explanations for EMF-induced effects which occur below present ICNIRP and IEEE guidelines and exposure recommendations by the EU.
4. The weight of evidence calls for preventive strategies based on the precautionary principle. At times the precautionary principle may involve prudent avoidance and prudent use.
5. We are aware that there are gaps in knowledge on biological and physical effects, and health risks related to EMF, which require additional independent research.
6. The undersigned scientists agree to establish an international scientific commission to promote research for the protection of public health from EMF and to develop the scientific basis and strategies for assessment, prevention, management and communication of risk, based on the precautionary principle.

Fiorella Belpoggi, Fondazione Ramazzini, Italy
Carl F. Blackman, President of the Bioelectromagnetic Society (1990-1991), Raleigh, USA
Martin Blank, Department of Physiology, Columbia University, New York, USA
Emilio Del Giudice, INFN Milano, Italy
Livio Giuliani, University Camerino, Italy
Settimio Grimaldi, CNR-INMM, Roma, Italy
Lennart Hardell, Department of Oncology, University Hospital, Oerebro, Sweden
Michael Kundi, Institute of Environmental Health, University of Vienna, Austria
Henry Lai, Department of Bioengineering, University of Washington, USA
Abraham R. Liboff, Department of Physics, Oakland University, USA

Wolfgang Löscher, Department of Pharmacology, Toxicology and Pharmacy, School of Veterinary Medicine, Hannover, Germany
Kjell Hansson Mild, National Institute of Working Life, Umea, Sweden
Wilhelm Mosgoeller, Institute for Cancer Research, University of Vienna, Austria
Elihu D. Richter, Unit of Occupational and Environmental Medicine, Hebrew-University-Hadassah, Jerusalem, Israel
Umberto Scapagnini, Neuropharmacology, University of Catania, Member of the European Parliament, Italy, Stanislaw Szmigielski, Military Institute of Hygiene and Epidemiology, Warsaw, Poland

* = Istituto Superiore per la Prevenzione e la Sicurezza del Lavoro, Italy
    (National Institute for Prevention and Work Safety, Italy)

Index

Aalto, A.   5, 12
Abdou, O.A.   16, 18, 19
acupuncture   53–4
  background   87
  coherent frequencies   72
    other meridians   75
    Ting meridians   72–4
  Five Elements   88–91
  frequencies   90–1
    entrainment   79–80
  organs/meridians   88
  Yang and Yin   87, 88, 89
Adey, R.   35–51, 241, 245, 331
Ahuja, Y.R.   336, 337
air quality   3
Alzheimer's disease   219, 243
American National Standards Institute (ANSI)   37
amyotrophic lateral sclerosis   219
amytropic lateral syndrome   243
Aniolczyk, H.   324
Antenna Combiner Unit (ACU)   469
Antipenko, E.N.   342
antitheft systems   191–2
Arnetz, B.B.   350
asbestos   431–2

Bachler, K.   92
Balcer-Kubiczek, E.K.   344
Baldry, C.   7
Balode, Z.   342
Baranski, S.   338
Baron, R.A.   22
base stations   see masts/base stations
Bawin, S.M.   42
Bedford, T.   18
Bennett, W.R. Jr.   213
Berg, M.   350
Berglund, B.   7
Best, S.   93
Bialek, W.   45
Binhi, B.N.   391–403
biological effects   39, 41–2
  electromagnetic fields   195–204
  EMF   391–2
    safety standards   396–400
  existence of hyperweak EMFs   392
    experiments   392–3
    theory   394–6

low frequency EMF   207–8
  background information   208–10
  complications   214–15
  exposure   210–15
  induced current effects   196–8
  response   215–22
  surface electric charge   196
report bias/consistency   222–6
RF fields:
  localised responses   200–1
  surface heating   201
    pulsed radiation   201–2
  whole-body responses   199–200
role of free radicals in electromagnetic
    field   43–4
biological system sensors   254
Bluetooth   233, 495
body-WLAN (body wireless area network)
    497–8
Boyd, S.   325
brain:
  biochemistry/exogenous electric signals   288
  effect of mobile phones   407–8
  hotspots   288
  problems   287–8
  refraction of RF energy   287
British Standards Institution (BSI)   493
buildings:
  coherence perception   91
  control of EMR   477–8, 480–1
    line of sight scenario   478–9
    propagation models   478
    single wooden floor   479
    wooden floor/brick wall   479
  electromagnetic exposure:
    EAS systems   322–3
    hospitals   324–6
    mobile phone base stations   323–4
  electromagnetic health status   115–16
  endogenous frequencies   92–6
  spatial frequencies   96–108
Burch, J.B.   350
Burge, S.A.   8, 15, 20
Burnet, M.   333
Burt, T.S.   7

California EMF Program   208, 223–4
Callahan, P.   250

cancer   215, 243, 385, 391
  adults   216
  cell-to-cell communications   350
  cellular ion concentration   350
  children   215–16
  effects of mobile phones   457–8, 461
  and electrification   36
  ELF genotoxicity:
    chromosome damage   335
    comet assay method   336
    DNA strand breakage   336–7, 457
    EMF/EMR relationship   331–2
    conclusions concerning   353–4
    epidemiological evidence   350–3
    recommendations   354
    reviews/studies   334–5
  EMR genotoxicity   338
    chromosome aberrations conclusions   343–4
    chromosome damage from RF/MW
      exposure   338–43
    conclusions concerning   349–50
    direct evidence of DNA-strand
      alteration/breakage   344–6
    Motorola funded counter research on
      DNA breakage   346–9
    neoplastic transformation   344
  historical trends   332–4
  incidence near radio/TV transmitters   113–15
  low frequency EMFs   449
  melatonin   222, 350
  radio frequency effects   450
Carnac (Brittany)   111
CASTLE approach   281
Cavallo, D.   346
Cellular Operations Ltd (Swindon)   25
Chadwick, Philip   187–93
chakras   72
chemical toxicology   226
Cherry, N.J.   331–64
chi   7
Chow, K.-C.   337
chromosomal aberrations   222
Chubb, J.   371
circulatory system   408–9
Clements-Croome, D.J.   3–33
cognitive function   405–9
Cohen, B.S.   300
Cohen, M.M.   335
coherence:
  in biological systems   56–9
  diamagnetism and order in water   60–1
  as order by diffusion process   61
  QED theory for water   59–60
coherent frequencies:
  acupuncture:
    other meridians   75
    Ting meridians   72–4
  buildings:
    endogenous   92–6
    perception   91
    spatial   96–108

chakras   72
energy medicine   75–6
entrainment of environmental frequencies
    76–7
living environment   61–2
  electromagnetic resonators   62–3
  nature's resonators   63
  technical oscillations   62
people:
  chemical signature in EM sensitive
      patients   85–6
  chemical/electrical hypersensitivities   81–4
  chemical/electrical sensitive patients   84–5
  entrainment by EM sensitive patients   78–9
  signature of chemicals   79–81
water   63
  erasing imprints   65–6
  imprinting techniques   64–5
  reading imprints   67–8
computers   220
Consumer Protection Act (1987)   167
Cook   335
Court-Brown, W.M.   333
Cyfracki, L.   17
Czerski, P.   338

Davis, G.   325
Davis, S.   350
Dember, W.N.   22
Department of Trade and Industry (DTI)   472
depression   221
Diener, E.   10
Ding, G.K.   22
DNA:
  damage/repair   331, 457
  strand breakage   336–7, 457
  synthesis   222
Dolk, H.   353
Dolan, Michael   437–40
Doll, R.   248, 333
Dorgan, C.E.   13, 20, 21

Elder, J.   490
electric blankets   221
electric power/electronic systems:
  and appearance of office wireless
      networks   36–7
  effect on tissue components   37
  far-field exposure   37
  growth of fields at electric power
      frequencies   36
  near-field exposure   37
  permissible exposure limits (PELs)   37
  and Specific Absorption Rate (SAR)   37
electroclinical syndromes   282–3
  biological responses   290
  brain chemistry/exogenous electrical signals   288
  counfounding factors   292
  electrosensitivity:
    signal transduction pathways   289–90
    symptoms   284–5

hotspots in brain   288
  coupling efficiency   289
individual response to hotspots   291
migraine variants   283
perceived problems   287–8
pointers to hypersensitivity   292–3
refraction of RF energy in the brain   287
RF hotspot   286–7
electromagnetic:
  biocompatibility   249
  environment   119–20
  hypersensitivity   283
  radiation, ionisation/optical effects   133
  sensitivity   220
  waves:
    general description   120–3
    quantitative description   124–8
electromagnetic compatibility (EMC)
  61–2, 150, 233
electromagnetic fields (EMFs)   3, 7, 39, 46–7
  bioeffects   195–204, 391–6
    comparability with (un)modulated fields   41–2
  buildings   115–16
    base stations for mobile telephony   323–4
    EAS systems   322–3
    hospitals   324–6
  effects of 50hz on mammalian cells in
    culture   308–11
  exposure to   307
  health risks   386–7
  hyperweak   392–6
  initial transduction imposed at nonthermal
    energy levels   42–4
  interactions with human body   313–22
  in living environments   53–116
  low frequency:
    biological effects   196–8, 207–26
    effect on health   449–50
  magnitude/subtlety of problem   160
  masts/base stations   137–48
  metrics   241–2, 253–4
    sensitivity   242, 250–3
    viewpoints   242, 247–9
    windows   242, 244–6
  mobile phones/VDUs   152–60
  modern office building   128–33
  natural/man-made comparison   35–6
  possible effects of exposure   202–4
  protection from   387
  public scepticism/concern   160–1, 162
  questions and discussion   179–83, 257–63
  reports on   387–8
  research   161–2, 308
  role of free radicals   43–4
  Safety Guidelines   149–52, 160
  safety standards   396–8
    biological protection   399–400
    power reduction   398–9
  sources   187–93
  stress   116
  technogenic/natural origin difference   149

electromagnetic radiation (EMR):
  cancer link   331–2, 338–49, 350–4
  control in buildings   477–81
electronic article surveillance (EAS) systems   322–3
electronic systems   *see* electric power/
  electronic systems
electrosensitivity (ES)   381, 385–8
  as imagined/psychological   385–6
  and neuropeptides   379
  questions and discussion   413–27
  signal transduction pathways   289–90
  symptoms   284–5
electrosmog, environmental impact   149–62
electrostatic effects   374
  air ion   373–4
  background   365–8
  charge generation/human body link   369–70
    atmospheric humidity   369–70
    charge, voltage, capacitance   370–1
  clothing   372
  effect of changing human environment   368–9
  shocks to people/associated risks:
    during walking   371
    rising from a seat   371
    vehicles/trolleys   372
  skin   372–3
Elia, V.   68
endogenous frequencies:
  effect of exogenous on frequency activity   94–6
  multiple effects involving ultraviolet/
    microwaves   93–4
  stimulation of geopathic stress-line   92
energy   119–20
energy medicine   75–6
environment:
  coherent frequencies   76–7
  distance related effects near radio/TV
    transmitters   113–15
  over-head power lines   111–13
  see also indoor environment
equity theory   9
erasing frequency imprints in water:
  de-gaussing   65–6
  heating   66
  hiding   66
European Committee for Electrotechnical
  Standardisation (CENELEC)   493
European Union (EU)   448
exposure guidelines:
  National Radiological Protection Board   447–9
  NRPB   451–2
  perspective   447–9
  recent experimental data   449–51
extremely low frequency (ELF)   149, 150
  interactions with human body   314–18
  and mobile telephony   152–3

Falling Water house (Pennsylvania)   12
Fanelli, C.   350
Fathy, Hassan   12
Feng Shui   7, 54, 87

Fews, A.P. 295–306
Feychting, M.B. 218
fields:
    definition 55
    four types of 55
    magnetic vector potential 56
    quantum state 56
    random/oscillatory motion 55–6
Fillion-Robin, N. 391–403
Fine, B.J. 18
Fisk, W.J. 22
Folkard, Melvyn 307–12
Foster, K.R. 331
Franck, K.A. 11
Fraumeni 333
free radicals 43–4
Frey, A. 490
Frey, A.H. 226
Froggatt 25
Frölich, Herbert 56–9
Fucic, A. 340

Gangi, S. 381
Garaj-Vrhovac, V. 339–40, 343
Garcia-Sagredo, J.M. 335
General Product Safety Directive (GPSD)
    (2001) 171
General Product Safety Regulations
    (GPS Regulations) (1994) 171–3
Gilchrist, Stuart 307–12
Giuliani, L. 331
Goldhaber, M.K. 217
Goldsmith, J.R. 339
Gorlaev, P. 68
Graham, C. 350
Grainger, Peter 487–99
Green, L.M. 333
GSM telephony 151, 152, 156, 460
Gupta, V.K. 18
Guy, A.W. 331

Haider, T. 342
Hand, J.W. 313–29
Hardell, L. 457, 489
Harrison, G.H. 344
Hatch, E.E. 333
Havas, M. 207–32
Hawkins 21
health:
    adverse effects of RF/EMF 488–91
    effects of low frequency EMFs 449–50
    effects of radio frequencies 450
    mobile phones 487–93
    preventative medicine 116
    questions and discussion 519–23
    research on mobile phones 455–61
    stress 12, 13, 116, 141
    thermal comfort 13–14
    tissue interactions 491–2
    well-being 10–12, 24
Health and Safety Executive (HSE) 9, 25

healthy buildings:
    and animism of architecture 11–12
    design principles 25
    futuristic offices 25
    and importance of designing thinking
        environments 24–5
    indoor environment 6–10, 13–24
    materials used in 11–12
    meaning of 3–4
    and physical/technological infrastructure 25
    and the senses 4–6
    well-being/productivity 10–13
Hearn, G.L. 371, 372
Hedge, A. 7–8
Heller, J.H. 332, 338
Henshaw, D.L. 295–306
Hepatitis C case 169–71
Hermann, D.M. 458
high voltage powerlines 295–6
    air pollution related illness 300–2
    corona ion emissions 296–8
        health implications 298–300, 303
    oscillation of aerosols in 50hz electric
        field 302–3
Hill, A.B. 248, 284, 490
Holcomb, L.C. 17
Holl, S. 4, 12
Holt, J. 380
homeopathy 53
    frequency entrainment 79
    Proving-Symptoms 67
hospitals 324–6
Hossmann, A. 458
Hughes, Nicole 437–40
Human Rights Act (1998) 138, 142–3
Hyland, G.J. 149–65
hypersensitivity:
    and chemical toxicity 54
    electrical 81–2
        diagnosis 82–3
        therapies 83–4
    electromagnetic (EM) 78–9
    frequency of acupuncture electromagnetic
        stress 53–4
    pointers 292–3
    speed of reaction 53

Ianchevskaia, N.V. 335, 342
Ilgen, D.R. 24
Impact of the Office Environment on Productivity
    and the Quality of Working Life report 18
imprinting frequencies into water 64
    by contact 64
    chemical potentising 65
    clinical aspects 64
    magnetic fields 65
    proximity and succussion 64
    toroids 64–5
    voltage pulses 65
Independent Expert Group on Mobile Phones
    (IEGMP) 137, 448

indoor environment:
electromagnetic wave sources 493–4
Bluetooth 495
future expectations 497–8
smart home 496–7
wireless devices in commerce/industry 495–6
wireless domestic devices 495
interior finishes 7
physical features 6
and productivity 13–24
quality of 6–7
related illnesses 7–10
Inskip, P.D. 489
Institute of Technology (Otaniemi) 12
insurance industry:
asbestos litigation 431–2
EMF exposure 431–4
Interagency Radio Frequency Working
Group 37
International Commission on Non-ionizing
Radiation Protection (ICNIRP) 149, 439,
448, 455, 486, 493
International Electrotechnical Commission
(IEC) 493
ion fluxes 222

Jaakkola, J.K.J. 7
Jacobson, C.B. 338
Jacobson, J.I. 393
Jamison, K.R. 10
Johansson, O. 377–89
Johnson, C.C. 331
Jones, P. 7, 8, 9, 20
Juutilainen, J. 350

Kajima Corporation (Tokyo) 22
Kamon, E. 18
Karasek, M. 350
Katsouyanni, K. 300
Kemp, Ray 435–6
Khalil, A.M. 335
Kiasma Museum of Contemporary Art
(Helsinki) 12
Kline 24
Knasko, S.A. 22
Knave, B. 289
Kobrick, J.L. 18
Koivisto, M. 490
Kolomytkin, O. 42
Konz, S. 18
Koutras, R.A. 218
Koveshnikova, I.V. 342
Kraut, A. 333
Krebs, J. 284

Lagerholm, B. 382
Lai, H. 156, 336, 345, 349, 457, 489
Laithwaite, E. 250
Lander, H.M. 44
Lang, Sakari 455–65
Langston, C.A. 22

Le Corbusier 5, 12
Lee, C. 267
Leeper, E. 209, 217, 331, 334, 335
Lepori, R.D. 11
Lester, J.R. 335
leukaemia 36, 243, 244–5, 331, 449
Levallois, P. 350
Levitt, B. Blake 501–18
Li, S.H. 337, 350
light 5, 11, 14–15, 22
Lin, J.C. 490
Line, P. 324
LINK Mobile Telecommunications and Health
Research programme 449
Lisbet, B. 7
Lisiewicz, J. 335
Local Government Responsibilities Memorandum:
addendum 143–5
second addendum 145–6
third addendum 146–8
local-area-networks (LANs) 36–7
Lorsch, H.G. 16, 18, 19
Louvre Pyramid (Paris) 12
Ludvigson, H.W. 22
Lyznicki, J.M. 326

McKenzie, D.R. 353
Mackworth, N.H. 13
McLauchlan, K.A. 43
Maes, A. 342, 345
magnetic storms 391
Mailhas, J.B. 342
Malyapa, R.S. 156, 346, 347, 349
Mann, S.M. 323
Maoris 111
Marvin, A.C. 119–34
mass psychogenic illness (MPI) 7
Mast Action UK 440
legal services 137–48
masts/base stations 435, 497
directional/isotropic transmission 152
electromagnetic exposure 323–4
exposure to 192–3
legislation:
Human Rights Act (1988) 138, 142–3
PP8 (Revised) 137–8, 139
local government responsibilities
memorandum 143–8
planning applications/notifications 139–42
problem of perception 438
questions and discussion 441–3
responding to community concerns 437
structures 474
Ten Commitments of operators 438–40
unobtrusive 473–4
visual intrusion:
adaptive colour control 475
camouflage modelling 474
diffractive coatings 475
integrated circuitry 475
unobtrusive techniques 475–7

materials:
  control of EMR 477–81
  RF properties 481–2, 486
    absorption 483–4
    filtering 484–5
    shielding 482
    transmission 484
measurement:
  pragmatists 248–9
  sensitivity 242, 250
    earth 251–2
    not everyone is affected 252–3
  viewpoints 242, 247
    business 247
    governments 247
    public 247
    scientists 247–8
  windows 242, 244–5, 254
    endogenous, entrainment, interface
      245
    natural EMF noise 245–6
    resonance 245
melatonin 220–2
  cancer link 222, 350
  mobile phones 408
Meltz, M.L. 345
Mendell, M.J. 8
Merleau-Ponty, M. 4
Michael, B.D 307–12
microwave ovens 119
microwave radiation 35–6, 37, 39–42, 198,
    234–5, 281, 384
  biological effects of hyperweak 393
  interactions with human body
    318–22
  measuring 235
    calcium silicate units 239
    climaplus 238
    details about 240
    reinforcement mesh from ISPO 238
    results 236–40
    shielding gypsum wallboard from
      KNAUF 239
    spruce-fir 238, 239
    Swiss Shield 238
    thermal insulation 239
    timber frame construction 239
    vertical perforated bricks 239
migraine variants 283, 289
  drop attacks 283
  electroclinical syndromes 283
  exploding head syndrome 283
  SUNCT syndrome 283
  thunderclap headaches 283
Mild, K.-H. 289
Milham, S. 36, 244, 248, 333, 334
Miller, J.B. 267, 333
Minder, C.E. 350
Mitchell, Stephen 307–12
mitosis 222
Miyakoshi, J. 336, 337

mobile phones 129–30, 132–3, 141, 149, 187,
    233, 384
  cognitive function 405–9
  electromagnetic exposure 323–4
  exposure to ELF fields 152–3
  health 487
    adverse effects 157–9, 488–90
    tissue interactions 491–2
  health research 455
    cancer studies 457–8, 461
    current status 456–7
    non-cancer studies 458–9
    portability requires a mechanism 460–1
    replication/reproducibility 459–60
  heart rate/blood pressure 408
  melatonin 408
  Meyer, Alan 137–48
  nervous/circulatory system 408–9
  non-thermal effects 155–6, 460
  radiation exposure 192–3
  research studies 160–2
  RF exposure guidelines 492–3
  Safety Guidelines 154
  thermal effects of exposure 153–4
  see also masts/base stations
Mobile Telephone Health Research (MTHR) 241
modulation frequency-dependence 39
Moldan, Dietrich 233–40
molecular resonance 222
Monro, A.J 267–80
Monteagudo, J.L. 335
Moore, D.F. 335
Morris 335
Moser, C.C. 40
moving/static elements 3
Muscat, J.E. 489
Myers, D.G. 10

National Electrical Manufacturers
    Association (Washington) 16
National Institute of Environmental Health
    Sciences 207, 223
National Radiological Protection Board
    (NRPB) 440, 447
  guidance on human exposure to EMFs
    451–2, 493
  publications/recommendations 447–9
nervous system 408–9, 458
neurotoxic disorders (NTD) 7
neutralising vaccines:
  antigen effects 279
  apple/orange 274
  background 267
  cheese 271
  dairy products 269
  egg 271
  fish 273
  grain 269
  maize/oats 274
  method 267–8, 275–7
  milk 270

results 278
rice 272
tea 273
tomato 272
wheat 270
Nordenson, I. 335
Nordstrom, S. 218
Nyberg, F. 300

Obe, G. 335
occupational exposure to EMF 212–13
OFTEL 472
Olive, P.L. 346
Ossiander, E.M. 36, 333, 334

Pallasmaa, J. 4, 5
Pavli, Peter 233–40
Pavel, A. 342
Pearl, Simon 167–77
Pedelty, J.F. 17
Pepler, R.D. 14
Perez-Gomez, A. 4
Pershagen, G. 300
Petersen, R.C. 323
Petrie, K. 285
Pfluger, D.M. 350
Philharmonie (Berlin) 12
Philips, Alasdair 241–56
Phillips, J.L. 345, 349
photographs 110–11
Pinto, L. 373
Pittard Sullivan building (Los Angeles) 25
policy initiatives 501–2
    increasing ambient exposures/
        environmental effects 505–10
    meaning of new EMF revolution 510–11
    physics vs biology 502–3
    predictions 511–12
    professional biases 503–5
pollution 3, 7, 19
    EMF exposure 220
Poole, C. 218
post-traumatic stress disorder (PTSD) 435
power-frequency fields 187–9
powerlines *see* high voltage powerlines
PP 8 planning circular (revised) 137–8, 139
precautionary principle 175–6
Preece, A.W. 405–11
Preparata, Giuliano 59
Prise, K.M. 307–12
Product Liability Directive 167–9
    role in product regulation 174
product safety:
    civil liability cases 173
    Development Risks Defence 177
    legislation 171–2
    producer obligations 172–3
    provision of information 173–4
productivity:
    and air quality 17, 22
    and air-conditioning 17, 18, 19

comfort factors 16
and control over environment 15–16, 19
definition of 20
economic benefits 20–1
and GDP per worker 21–2
and indoor environment 13–24
and lighting 14–15, 22
measurement 24
measures 16–17
and odour/scents 22
personal, social, organisational, environmental
    factors 23
relationship to work environment 20
and respiratory illness 22
and temperature 13–14, 17–18, 20
and ventilation 19, 21
and well-being 10–12

Qassem, W. 335
Quantum Electrodynamics (QED) theory
    59–60
Quintel 4 Ltd 467–86

Radio Communications Agency (RCA)
    439–40, 443
radio frequency (RF) 35–6, 37, 39–42, 149,
    234–5, 281, 447
    biological effects 198–202
    biological responses 290
    effect on health 450
    exposure guidelines 492
        ICNIRP 493
        NRPB 493
        philosophy 492
        Technical Standards Organisations 493
    hotspot 286
        eddy currents 286–7
    interactions with human body 318–22
    materials with RF properties 481–6
    measuring 235
    refraction in the brain 287
    research 455
radio/television transmitters 129, 187
    distance related effects 113
        cancer incidence 113–15
        population density 113
Radiocommunications Agency 472
Randolph, T. 267
Raw, G.J. 7, 20
reading water frequency imprint 67
    clinical 67
    dowsing 67
    electrodes 67
    laser 68
    microcalorimetry 68
Reiter, R.J. 350
Renaissance 4
reproduction 216
    maternal VDT use 217
    paternal exposure 218
    residential exposure 217

residential exposure to EMF 210
  appliances 211–12
  components 212
  human reproduction 217
  indoor distribution system 210–11
  outdoor distribution system 210
Ricci-Bitti, P.E. 7
Robertson 8
Ronchamps Chapel 5, 12
Rosenfeld, S. 17
Rosenthal, M. 335
Rothman, K.J. 489
Roti, J. 346
Rotton, J. 22
Rundnai, P. 7

Sagripanti, J. 341
Sarkar, S. 344
Säynätsako Town Hall 5
Scharoun, Hans 12
Schwan, H.P. 331
Schweisheimer, W. 18
Scottish Parliament Transport on Science and
  Technology (2000) 448
screen dermatitis:
  background 377
  claims for 378
  experiments 377–8
  explanations 378
  as illness 379
  reaction patterns/studies 379–81
  reasons 381–3
  treatment/solutions 384–5
  UVB irradiation 381
seasonal affective disorder (SAD) 221
Select Committee on Science and Technology
  (1999) 448
senses 4–6
sensitive patients:
  chemical frequency signatures 85–6
  chemical/electrical 84–5
sewing machines 220
sick building syndrome (SBS) 7–9, 13, 17,
  19, 20, 22, 25–6
Sienkiewicz, Z.J. 195–205, 447–54
Silk, C.A 281–94
Singh, N.P. 156, 336, 345, 346, 349, 457, 489
Skyberg, K. 335
Smallwood, J.M 365–74
smart home 496–7
Smerhovsky, Z. 338
Smith, A.H. 8
Smith, C. 53–118, 289
Smith, Mike 467–86
Södergren, L. 380
Sommer, H. 490
space 5–6
spatial frequencies 96
  effect of orientations relative to
    north-south axis 102
  Molecular Model patterns 106–8

parallel wires 102–3
random arrangement of objects 102
range of coherence interactions 96–7
real building materials - bricks 105–6
regular patterns of cuvettes 103–4
resonant structures 97–8
simple coherent structures 100–2
velocity of propagation 99–100
waveform patterns formed in wire 105
Speare-cole, Alastair 431–4
static electricity *see* electrostatic effects
Sterling, E.M. 7
Stewart, A. 248
Stewart Report (2000) 137, 439, 443
Stodolnik-Babanska, W. 344
Sundell, J. 8
Sundstrom, E. 6
Svedenstal, B.-M. 336, 337
Swicord, M.L. 341

T-lymphocytes 80–1
t3G technology 460
  annex 486
  antenna sharing solution 469
  background 467–8
  control of EMR in buildings using materials
    477–81
  demand for 467–8
  government/EU acceptance 472
  individual electronic beam control 473
  materials with RF properties 481–6
  minimising visual intrusion of masts 474–7
  pilot network 472–3
  specifications 469–72
  structures 474
  unobtrusive antennas 473
    technology 473–4
tape erasures 187–8
Tattersall, J.E. 42
Teixeira-Pinto, A.A. 332, 338
telecommunications systems 192–3, 233
Terra Incognita 281
Tertishny, G. 68
Thansandote, A. 324
Tice, R. 343
Timchenko, O.T. 335, 342
tissue components:
  cell membranes as site of initial field
    transductive coupling 43
  comparability between bioeffects 41–2
  detection of low frequency fields 39–42
  directional differences in signal paths 40
  domain functions as general biological
    property 45–6
  effect of electric power/selectronic
    systems on 36–7
  electron tunnelling in transmembrane
    conduction 40
  excitable cells 45–6
  multicellular conductance pathways 38
  non-excitable cells 46

nonlinearities related to electric charge
distribution   40
RF/microwave fields amplitude-
modulated at low frequencies   39–42
role of cellular ensembles in setting thresholds
for intrinsic/environmental stimuli   44–6
sensitivities to nonthermal stimuli   38–42
structural/fnctional organization of
extracellular space   38–9
Townsend, J.   10
trains   190–1
trams   190–1
transportation exposure to EMF   213–14
TWA Chiat Day Company (Los Angeles)   25

UMTS   468, 469
US National Research Council   207, 223

Vernon, H.M.   14, 19
Verschaeve, L.   345
Vignati, M.   331
Vijayalxmi, B.Z.   342
Vincent, J.H.   373
visual display terminals (VDTs)   217, 377,
381
visual display units (VDUs)   119, 129,
149, 187
adverse health effects   159–60
electromagnetic radiation   153
emissions   189
non-thermal impacts   157
research studies   160–2
Safety Guidelines   154
visual stimulation:
cameras and stolen souls   110–11
flashing light   108–9
response to basic patterns   109–10
Von Gierke, H.   490

Walleczek, J.   308
Warr, P.   11
water:
coherent frequencies   63
erasing   65–6
imprinting   64–5
reading   67–8
diamagentism and order in   60–1
memory:
effect of a frequency imprint   70–1
frequency effects on dilution/sucussion   71–2
frequency ratios in bulk water   70
possible mechanism   70
trace water in n-alkanes   68–70
QED theory for coherence   59–60
Weiss, M.L.   12
Wertheimer, N.   209, 217, 331, 334, 335
Westinghouse Furniture Systems Company
(Buffalo, New York)   18
Whitely, T.D.R.   9
Wilkins, A.J.   14
Wilkins, J.R. III   218
Wilson, B.W.   350
Wilson, S.   8
wireless devices:
commerce/industry   495–6
domestic   495
wireless networks   36–7
Woods, J.E.   18
World Health Organisation (WHO)   7, 161, 282,
391, 449, 461
Wright, Frank Lloyd   12
Wyon, D.P.   15

Yao, K.T.   339

Zaret, M.M.   332
Zmyslony, M.   336, 337